Production Technology of Stone Fruits

Mohammad Maqbool Mir •
Umar Iqbal • Shabir Ahmad Mir
Editors

Production Technology of Stone Fruits

Springer

Editors
Mohammad Maqbool Mir
Division of Fruit Science
Sher-e-Kashmir University of Agricultural
Sciences & Technology of Kashmir
Srinagar, Jammu and Kashmir, India

Umar Iqbal
Division of Fruit Science
Sher-e-Kashmir University of Agricultural
Sciences & Technology of Kashmir
Srinagar, Jammu and Kashmir, India

Shabir Ahmad Mir
Department of Food Science & Technology
Government College For Women
Srinagar, Jammu and Kashmir, India

ISBN 978-981-15-8919-5 ISBN 978-981-15-8920-1 (eBook)
https://doi.org/10.1007/978-981-15-8920-1

This Springer imprint is published by the registered company Springer Nature Singapore Pte Ltd.
The registered company address is: 152 Beach Road, #21-01/04 Gateway East, Singapore 189721, Singapore

Contents

About the Editors

Mohammad Maqbool Mir, Ph.D. obtained his Ph.D. in Horticulture (Fruit Science) from Sam Higginbottom University of Agriculture, Technology and Sciences, Allahabad. At present, he is an Associate Professor cum Senior Scientist at the Division of Fruit Science, Sher-e-Kashmir University of Agricultural Sciences and Technology of Kashmir, Shalimar, Srinagar. He is associated with many externally aided projects and also with research group working on canopy architectural management, production technology, and standardization of production protocols for different fruits. He has supervised/co-supervised several M.Sc. and Ph.D. students besides postgraduate teaching. He is associated with many academic and professional societies and has more than 60 scientific publications in different reputed journals at national and international level and other 20 popular articles, book chapters, extension bulletins, and has edited 1 book.

Umar Iqbal, Ph.D. is an Assistant Professor cum Junior Scientist at the Division of Fruit Science, Sher-e-Kashmir University of Agricultural Sciences and Technology of Kashmir, India. He has done his Ph.D. Agriculture (Fruit Science) from SKUAST-Jammu. He was the recipient of a merit scholarship during his Ph.D. program. He has concluded satisfactorily two externally funded projects by HTM and MIDH as a Principal Investigator and been a Co-PI on another externally funded project by Potash Research Institute and International Potash Institute. Dr. Umar has supervised/co-supervised several M.Sc. and Ph.D. students in Fruit Science and has published more than 30 papers in Indian and foreign journals. Dr. Umar has contributed chapters to more than four books and is the coauthor of one book on Fruit Science.

Shabir Ahmad Mir, Ph.D. obtained his Ph.D. in Food Technology from Pondicherry University, Puducherry, India. At present, he is an Assistant Professor at the Government College for Women, Srinagar, India. He has received the Best PhD Thesis Award 2016 for outstanding research work by the Whole Grain Research Foundation. He has organized several conferences and workshops in Food Science and Technology. Dr. Mir has published numerous international papers, book chapters, and edited five books.

Varietal Diversification of Stone Fruits

1

Ali Gharaghani and Sahar Solhjoo

Abstract

Stone fruits, including apricot, cherries, peach, nectarine, and plums, are species of the *Prunus* genus, Rosaceae family. During the last century, numerous cultivars have been introduced for major stone fruit crops throughout the world, most of them coming from crossbreeding, either via open pollination or via controlled crosses and only a few percent from bud sports. Mutation breeding and intraspecific hybridization are among the other possible and more advanced breeding techniques utilized for these crops. Selection for yield and basic fruit quality attributes which have been practiced by ancient fruit growers for centuries is still the goal of modern fruit breeding programs. In addition, some major trends including increased resistance to abiotic and biotic stresses, simplified orchard practices, extension of the adaptation zones and harvest window, new fruit types, enhanced nutritional value, and eating convenience are of paramount importance in developing new cultivars. Stone fruits typically have long breeding cycles; thus, developing a new cultivar through traditional breeding may require many breeding cycles and dozens of years. Although recent advances in genomics and biotechnologies are able to accelerate the stone fruit breeding, more studies and efforts need to sustain the involvement of new tools in future breeding programs.

Keywords

Prunus · Genetic resources · Breeding · Genomics · Biotechnology

A. Gharaghani (✉)
Department of Horticultural Sciences, School of Agriculture, Shiraz University, Shiraz, Iran
e-mail: agharghani@shirazu.ac.ir

S. Solhjoo
Department of Horticultural Sciences, College of Agriculture, University of Tehran, Karaj, Iran

M. M. Mir et al. (eds.), *Production Technology of Stone Fruits*,
https://doi.org/10.1007/978-981-15-8920-1_1

1.1 Introduction

Stone fruits, species of the widespread genus *Prunus*, Rosaceae family, including apricots, cherries, peaches, nectarines, and plums, are grown mainly for their eatable fleshy mesocarps. Almond, which is grown for their edible nut, is also a member of this genus. *Prunus* comprises approximately 200 species and includes, aside from the stone fruits, other taxa grown as ornamentals and numerous wild species of local economic value, as well as wild relatives that have been useful in major crop studies and breeding programs (Potter et al. 2007). The species of stone fruits belong to temperate areas in the Northern Hemisphere, but *Prunus* also includes around 35 species from the old-world tropics and 25 from the new-world tropics (Potter 2012). *Prunus* species are trees and shrubs with typically 5-merous flowers with a single carpel that matures into a fleshy mesocarp drupe and a tough endocarp containing a single seed. The main stone fruit species, all of which arose from Asia or Europe but were commonly spread around the globe, are now discovered by individuals and significant producers on all continents except Antarctica (Janick 2005). Nutritionally, stone fruits are rich in vitamins and minerals. There is growing concern in their prospective importance as nutraceuticals owing to the existence of phenolic compounds with antioxidant characteristics (Wargovich et al. 2012). Prunus has been the target of comprehensive basic and applied research owing to the vast number of cultivated species and their economic significance, as well as the diversity and wide distribution of wild species, and substantial resources available for germplasm enhancement and genomics to be used in breeding programs (Kole and Abbott 2012).

1.2 Taxonomy of Stone Fruits

The genus *Prunus* spp. includes more than 200 species of deciduous and evergreen trees and shrubs including fruits and nut plants of economic importance. Rosaceae's latest classification grouped *Prunus* into an extended Spiraeoideae subfamily as the only genus in the Amygdaleae tribe (Potter 2007; Shulaev et al. 2008). The most commonly known infrageneric grouping of *Prunus* by Rehder (1940) comprises of five subgenera: *Amygdalus*, *Prunus*, *Cerasus*, *Laurocerasus*, and *Padus*, in which the commercial stone fruit cultivars belong to three of these subgenera (Fig. 1.1) as follows: diploid peaches, nectarines, and almonds (*P. persica* Batsch and *P. dulcis* (Mill.) D.A. Webb.), respectively, belong to the subgenus *Amygdalus*; the subgenus *Prunus*, which comprises *Prunophora* section consisting of diploid Japanese plums (*P. salicina* L.) and hexaploid European plums (*P.* × *domestica* L.) and *Armeniaca* section consisting of diploid apricots (*P. armeniaca* L.); and the subgenus *Cerasus* comprising diploid sweet cherry (*P. avium* L.), tetraploid sour cherry (*P. cerasus* L.), and ground cherry (*P. fruticosa* Pall.). However, some scholars segmented *Prunus* L. into only three subgenera: *Prunus* (comprising of almond, peaches, apricots, and plums which were categorized into the different sections), *Cerasus* (including cherries), and *Padus* (Potter 2012).

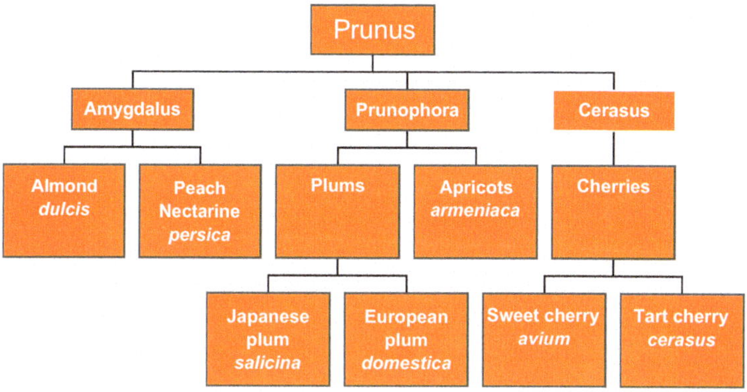

Fig. 1.1 Botanical classification of *Prunus* genera showing the relatedness of various stone fruit crops

While stone fruits have enough in common to be classified in the same genus, they are quite different in many tree, fruit, and flower characteristics. The flowers of the various stone fruits are quite characteristic for the respective groups. In peaches, nectarines, and apricots, they are borne singly, arising from one to three separate buds at each node. They are without stems in the peaches and nectarines and nearly so in the apricot. They are on long stems in the cherries and moderately long ones in the plums, but in both fruits the flowers are borne in clusters. The flowers of the edible plums are white or nearly so, while those of the apricot and the peach may be white, pink, or even reddish (Fig. 1.2).

When the fruits are ripe, the flesh of some varieties separates easily from the pit. Such fruits are called as freestones. Other varieties and species, in which the flesh adheres to the stone, are clingstones. The individual fruits may be hairy, as in the peach, or smooth, as in the nectarine, plum, apricot, and cherry. Fruits vary in size, shape, and color, with species and varieties. The flesh may be white, green, yellow, and red or show various combinations of these colors. The stones (pits) of the plum and cherry are relatively smooth, those of peach and nectarine are rough and grooved, and those of the apricot are somewhat intermediate (Fig. 1.3).

1.3 Origin and Domestication

Prunus species have been grown and extremely valued by individuals in Asia and Europe for thousands of years for their edible fruits, and the main cultivated species have spread extensively throughout the globe over many decades. Furthermore, wild species of these fruits have been locally important in Asia, Europe, and North America (Janick 2005). It can be extremely challenging to identify the accurate geographical origins of cultivated *Prunus* species, because interspecific hybridizations as well as their long history of cultivation and human-interceded dispersal have played a role in multitude varieties cultivated, so that each variety originated

Fig. 1.2 Flower bud types in stone fruits. (**a**, **b**) Simple buds—apricot and peach/nectarine, respectively. (**c**–**e**) Buds with multiple flowers—European plum, Japanese plum, sweet cherry, and tart cherry, respectively (Adapted from H.J. Larsen 2010)

Fig. 1.3 Diversity of fruit size, shape, and color in various stone fruits

from different ancestors. In most instances, it is possible to designate only wide regions of origin with certainty. Nonetheless, there was significant interest in monitoring the stone fruit species' origins and dissemination, and many insights were acquired (Potter 2012; Kole and Abbott 2012).

1.3.1 Peaches and Nectarines

Peaches (*P. persica* (L.) Batsch) and nectarines (glabrous-skinned varieties of *P. persica*) originated in China, with more than 4000 years of cultivation history, but has a delusive Latin name, indicating its early distribution from China to Persia via the Silk Road, from where it was later brought to Greece and Rome in the first or second century BC. Peaches then were transported to North and South America by European explorers and settlers during the sixteenth century (Scorza and Sherman 1996; Janick 2005). The origin center of Chinese wild peach (*P. consociiflora* Schneid.) and flat peaches (*P. persica* var. *platycarpa*) is also attributed to China (Potter 2012). As interspecific hybridization is common within *Prunus* species, it seems *P. persica* and some other species including *P. dulcis, P. kansuensis, P. ferganensis, P. scoparia, P. mira*, and *P. davidiana* have evolved from a common progenitor and all are closely related (Knight 1969; Gharaghani et al. 2017).

1.3.2 Plums

Various plum species originated and were independently domesticated on three continents. Europe is considered as the origin center of *P. domestica*, Western and Central Asia (the Caucasus and Crimea regions) for Myrobalan plum (*P. cerasifera*), China for the Japanese plum (*P. salicina*), and North America for species of the *Prunocerasus* section like *P. americana* Marshall, *P. hortulana* Bailey, *P. munsoniana* Wight & Hedr., *P. angustifolia* Marsh., and *P. maritima* Marsh. (Okie and Hancock 2008; Topp et al. 2012). Similar to other *Prunus* species, plums have a fundamental chromosomal number of eight, and the ploidy level is varied from diploid ($2n = 2x = 16$) to hexaploid ($2n = 6x = 48$).

P. cerasifera is indigenous to Middle East (refers to areas of Iran, Iraq, Caucasia, Anatolia), as well as the Balkan Peninsula and occasionally to Central Europe through Slovakia, Moravia, and Austria where it is just scattered and perhaps not native. It has been grown since 200 BC in the Mediterranean region and the Balkan Peninsula. People in West Asia from the Tien Shan and Pamir Mountains over to the Caucasus Mountains have used fresh and dried fruit of *Prunus cerasifera* for millennia (Okie and Hancock 2008; Gharaghani et al. 2017). Due to its native spectrum and wide range of graft- and cross-compatibility with many other species, *P. cerasifera* is proposed to be the progenitor of all plum species (Okie and Weinberger 1996). Asia Minor is supposed to be the place where natural hybrids originated first between *P. cerasifera* and *P. spinosa* (blackthorn or sloe, tetraploid, $2n = 4x = 32$), and the distribution of their seeds from Iran and Asia Minor might have been the ancestors of *P. domestica* in Europe (Crane and Lawrence 1956). Cytological studies show *P. spinosa* bears the genome of *P. cerasifera* as well as a second one from an unidentified parent (Reynders-Aloisi and Grellet 1994). Therefore, *P. domestica* may be derived from polyploid types of *P. cerasifera*, which has long history of selection and local application across the continent and also has a variety of fruit color (Okie and Hancock 2008).

Tetraploid ($2n = 4x = 32$) type of P. *spinosa*, which is called the blackthorn or sloe, is a wild species native to the Urals throughout Europe, north of Africa, north of Anatolia, Caucasus, north of Iran, and northwest of Turkmenistan. The Caucasian region is one of the centers of origin where $2n = 16$, 24, 32, 40, 48, 64, and 96 types of *P. spinosa* were found there (Erturk et al. 2009; Hartmann and Neumüller 2009; Topp et al. 2012; Gharaghani et al. 2017).

1.3.3 Apricot

According to the famous Russian botanist Vavilov (1951), there are three centers of origin for cultivated apricots that include the Chinese center (mountainous regions of Central and Western China), the Central Asiatic center (from Tien Shan to Kashmir), and the Near Eastern center (Iran-Caucasian). Due to cultivated varieties and the absence of wild forms of apricot, the Near Eastern center may be a secondary center for cultivated apricots (Kostina 1946). Most of the cultivated apricots refer to the species *Prunus armeniaca* L., common apricot, which emerged in Central Asia and China, where it has been grown for millennium and was subsequently spread to both East and West. More than 3000 years ago, *P. armeniaca* was cultivated in China and outspread across Central Asia. During the first century BC, apricot cultivation was introduced in the Mediterranean region from Iran or Armenia, albeit more recently, new Middle East introductions have been made especially in Southern Europe. In the seventeenth century, apricots were brought into England and the USA (Virginia) as a consequence of trading and commerce. Later, in the eighteenth decade, the Spaniards introduced apricot into California (Faust et al. 1998; Janick 2005; Zhebentyayeva et al. 2012).

1.3.4 Cherries

It is believed that *Prunus avium* (sweet cherry), *P. cerasus* (sour cherry), and *P. fruticosa* (ground cherry) originally emerged in the area beside the Caspian and Black Seas that includes Asia Minor, Anatolia, Southern Caucasus, and Northern Iran and were spread through Europe by animals, birds, and humans (Brown et al. 1996; Webster 1996; Gharaghani et al. 2017). The cultivation and domestication of diploid *P. avium* is proposed to have started in the region of Central Asia-Caucasia. The earliest records of its cultivation in Europe are from Greece, where *P. avium* was used not only as a fruit but also as a timber tree (Iezzoni et al. 1990). Cultivated forms of cherry were spread by Romans throughout the Mediterranean region, but in Central and Northern Europe sweet cherry arrived much later (Janick 2005). Cherries were transferred to the America by the nineteenth century (Brown et al. 1996).

The tetraploid sour cherry arose from the same area as sweet cherries or from Switzerland to the Adriatic Sea and from the Caspian Sea to the north of Europe as well (Iezzoni 2008). *P. cerasus* is thought to be the result of hybridization between *P. avium* diploid and tetraploid ground cherry, *P. fruticosa*, which spread across the main part of Central Europe, Siberia, and Northern Asia (Kappel et al. 2012). With a great probability, the sour cherry was brought to Eastern Europe with the Slavic tribes during the migration of the peoples, from their original settlement area in the north of the Persian Empire, probably triggered by the invasion of Mongolian tribes (the fourth to tenth century). As a result, Slavic tribes first migrated to north to the steppe between Don and Dnieper and later spread in three directions: North, Russians, Byelorussians, and Ukrainians; West, Poland, Sorbs, Slovaks, and Czechs; and South, Serbs, Slovenian, Croats, and other more (Faust and Suranyi 1997). It can be assumed that the origin of the cultivated sour cherry was derived from few initial genotypes forming the known local cultivar groups (Iezzoni et al. 1990).

1.4 Genetic Resources

1.4.1 Peaches and Nectarines

Peaches and nectarines are fruit species which are typically self-fertile and naturally self-pollinating. Although polyploidy is common in the *Prunus* genus, five species may be referred to as "peach": *P. persica*, *P. davidiana* (Carr.) Franch, *P. mira* Koehne, *P. kansuensis* Rehd., and *P. ferganensis* Kov. & Kost., all of which are diploid ($2n = 2x = 16$) (Knight 1969; Hancock et al. 2008). On the basis of fruit morphology, three varieties can be taxonomically recognized. The common peach (*P. persica* var. *vulgaris* Maxim.) has 5–7 cm in diameter with rounded and hairy fruits. Compared to peaches, the nectarines (*P. persica* var. *nectarina* Maxim.) have rounded glabrous fruits and more compact cells. The flat peach (*P. Persica.* var. *platycarpa* Bailey [syn: *P. persica* var. *compressa* Bean]) has flat fruit and a small pit (Byrne et al. 2012; Potter 2012).

Prunus persica is interfertile with *Amygdalus* subgenus (mentioned related species, *P. dulcis* and *P. scoparia*), and interspecific hybridization is common among them. These species especially are used as a source of plum pox virus (PPV), powdery mildew, peach aphid resistance, and adaptation characteristic in scion breeding (Gradziel 2003; Byrne et al. 2000b; Foulongne et al. 2003). Successful hybrids were also generated between peach and other *Prunus* species which are mainly sterile (Hancock et al. 2008). Peach is usually graft-compatible on its own; relative species such as *P. dulcis*, *P. davidiana*, *P. ferganensis*, *P. kansuensis*, and *P. mira* and interspecific hybrids of peach × almond and peach × *P. davidiana* are available (Byrne et al. 2012). Interspecific hybrids of peach × almond are highly fertile, vigorous, and tolerant to iron chlorosis, so they are useful rootstocks in calcareous, poor, and dry soils as well as in fruit tree replanting situations (Kester and Assay 1986; Felipe 2009). The hybrids of peach × *P. davidiana* are generally very productive for peach, and some of their selection is also resistant to root-knot nematodes (Edin and Garcin 1994). Some of the plum rootstocks are very useful for peach in waterlogging and/or replanting problems due to fungal disease tolerance including *Armillaria* root rot and *Phytophthora* crown rot (Byrne et al. 2012). Plum is also more resistant to root-knot nematodes (*Meloidogyne* genus) compared to other resistant sources of peach and almond. Rootstocks of different species of *Euprunus* have also been used as peach rootstocks (Dirlewanger et al. 2004a), including the hexaploid plums (*P. domestica* L.) or St. Julien plums (*P. insititia* L.), because they usually have excellent graft compatibility with peaches. Graft compatibility of peaches on fast-growing Myrobalan plum and interspecific hybrids with Myrobalan differs considerably based on the genotype being tested (Rubio-Cabetas et al. 1998; Zarrouk et al. 2006; Reighard and Loreti 2008).

In China, comprehensive collections of genetic resources have been collected for peach, so that about 1500 accessions are kept in three national repositories in Nanjing, Zhengzhou, and Beijing (Wang et al. 2002). Other major national collections would include more than 2000 accessions in Europe with the biggest collections in France, Spain, and Italy, 600 accessions in Japan, 300 accessions in South Korea, 280 accessions in the USA, 732 accession in Brazil, and about 1500 accessions in Ukraine (Byrne 2012).

1.4.2 Plums

Most plum cultivars belong to only two species: the hexaploid European plum (*P. domestica*) ($2n = 6x = 48$) and the diploid Japanese plum (*P. salicina*) ($2n = 2x = 16$). *P. domestica*, which categorized in European plums, is grown not only in Europe but also in other continents as the most common plum. Based on the fruit characteristics, this species can be divided into several groups including plums, prunes, gage plums, and mirabelles, as well as the wild plums like cherry plums, bullaces, damsons, and sloes (Okie and Hancock 2008). The differentiation between plum and prune group is difficult; the prunes are oval to elongate with mostly dark blue skin and smaller than plums. In contrary to plums, their shoots and leaves are never pubescent. During cooking, while the flesh of plums dissolves, prunes remain firm and they do not lose their shape. The most common prunes in Europe are "Prune d'Agen," "German prune," and "Italian prune" (Topp et al. 2012). *P. italic* Borkh. = *P. domestica* subsp. *italica* Gams ex Hegi, which are classified as gages or greengages, have aromatic round green fruit with a sweetish firm green flesh used for fresh consumption (Roach 1985). Botanically the mirabelles were often categorized as *P. insititia*. Nowadays, they are regarded as a subspecies of *P. domestica* which have round fruits, often yellow in color, mostly with red spots, but green and purple varieties are also available. Other attributes of mirabelle fruit include juicy, freestone, sweetish (Brix 18–20%), aromatic, and good quality which is used particularly for canning, brandy industry, and fresh consumption. The most popular cultivar in mirabelle group is 'Mirabelle de Nancy' (Jacob 2007; Topp et al. 2012).

The hexaploid *P. domestica* are generally incompatible with diploid species, though they can effectively be hybridized with *P. spinosa* (with tetraploid genome) and *P. cerasifera* (with diploid genome) (Minev and Balev 2002; Neumüller et al. 2009). The diploid *P. cerasifera* (Myrobalan or 'Cherry plum') is not only a progenitor of European plums but also a cross-fertile species with Asian and American diploid plum species. These species were not widely included in modern breeding probably because of small fruit size, except for breeding of 'Methley' (a chance hybrid) in South Africa and 'Wilson' in Australia, but are a valuable source for cold hardiness, earliness, and self-fertility (Okie and Hancock 2008). The European wild plums such as *P. cerasifera*, *P. spinosa* (sloe), and *P. insititia* (St. Julien A, damson, bullace, and mirabelle) are the most interesting species for hybridization in plum scion and rootstock breeding programs as environmental adaptability donor and resistance as described in Table 1.1.

The Asian plums including Japanese plum (*P. salicina*) and apricot plum (*P. simonii*) involved in improvement of new plum cultivars such as 'Blood plum of Satsuma', 'Santa Rosa' (complex hybrid of *P. salicina*, *P. simonii*, and *P. americana* with predominant character of *P. salicina*), 'Formosa', 'Beuty', 'Shiro', and 'Wickson', which mainly originated from Luther Burbank breeding works in the USA (Okie and Weinberger 1996).

The species of North American are considered as a precious resource for diploid plum and rootstock breeding. They have many essential horticultural characteristics and are suited to a wide variety of conditions (Table 1.1). *P. salicina* and American species display a strong degree of cross-compatibility which allows introgression of useful traits including crown gall resistance from black sloe (*P. umbellata* Ell.) and Allegheny plum (*P. alleghaniensis* Porter) and late bloom from *P. lanata* Mack. & Bush., *P. umbellata* Ell., *P. maritima*, and *P. Americana*. Also, other resources such as *P. americana* and *P. nigra* are valuable resources for confined root suckering and frost resistance; *P. hortulana* for late ripening; *P. subcordata* and *P. reverchonii Sarg* for their tolerance to drought; and *P. angustifolia* and *P. hortulana* for resistance to bacterial leaf spot (Okie and Hancock 2008; Topp et al. 2012).

There is generally a high degree of interspecific cross-compatibility within the subgenus *Prunophora* between the diploid plum and non-plum species. This involves *P. cerasifera*, *P. salicina*, and *P. simonii* in *Euprunus*, American plum species in *Prunocerasus*, and apricots and mumes in *Armeniaca* (Okie and Weinberger 1996). The diploid plum species are also able to be crossed with species from the subgenera *Amygdalus* (almond and peach) and *Cerasus* (cherry), but with less fertility, and are remarkably important for breeding of rootstock stone fruits (Lespinasse et al. 2003).

Many of the plum species and interspecific hybrids have been and are used as rootstocks. Myrobalan (*P. cerasifera*) (such as 'H29C', 'GF31', and 'B'), Marianna (*P. cerasifera* × *P. munsoniana* (such as '2624', 'GF8-1', and 'Buck'), *P. instititia* (such as 'Pixy', 'St. Julien A', and 'St. Julien GF655-2'), and *P. domestica* (such as 'Black Damas', 'Brompton', 'Common Mussel', 'Prune GF43', and 'Wangenheims')

Table 1.1 Characteristics of the most important genetic resources of plum

Group	Species	Common name	Fruit properties	Useful properties and utility
European species	*P.* × *domestica* L.	Garden plum, Europeanplum	Drupe, oval, or almost spherical in shape, up to 8 cm in length, green, yellow, red to purple and dark blue, with green or yellow pulp, sweet, easily or not easily detaching from the endocarp	High flavor and fruit quality
	P. insititia (P. domestica ssp. insititia) It can be divided into three groups (St. Julien, damson, mirabelles, and bullace)	St. Julien, damson, mirabelles, bullace	**Mirabelles:** small round fruit with 22–28 mm in diameter, mostly yellow colored with red spot but also green and more purple, very sweet with high aroma and quality **St. Julien**: small fruit with green-yellow color **Damson**: elliptic to spherical fruits, with a dark blue to the green surface and a bitter, spicy and sweet taste; aromatic and astringent **Bullace**: small, round fruits, with dark blue color and sweet taste	Used specially for canning and brandy industry; as a rootstock (GF655/2, Damas 1869, St. Julien GF 655/2) but have tendency to root sucker
	P. cerasifera Ehrh.	Myrobalan or cherry plum	Small round fruits, yellow to red and dark violet colored, with a diameter of 15–20 mm, soft, juicy, and sweet to subacid flesh; poor quality	Earliness; nematode resistant; good productive; resistant to diseases, drought, and heat; winter hardiness; as a rootstock (trees grafted on Myrobalan show vigorous growth but no root suckers) Problems: very sensitive to spring frost; the lower chilling requirements cause problems in regions with fluctuating winter temperatures
	P. spinosa L.	Blackthorn or sloe	Small fruit with black surface, green sour or bitter flesh	Very drought resistant; disease resistance; dwarfism and robustness in rootstock breeding; they're not consumed raw or fresh but are used to make sloe gin and folk medicine

Asian species	*P. salicina* Lindl	Japanese plum	Big and round or heart shaped, much lower sugar and acid content than European plum, good appearance and well appropriate for transportation	Good in size, appearance, and firmness; preserving quality at high temperatures; very hardy in winter; the most commercially oriental species
	P. simonii Carr	Apricot plum, Simon plum	Small flat fruit, 25–30 mm in diameter, dark purple-red color, firm aromatic flesh, and clingstone	Firmness and high volatiles; upright tree; cold hardiness
American species	*P. maritima* Marsh	Beach plum		Late bloom; high heat threshold
	P. hortulana	Hortulana plum	Small fruit with a diameter of 25 mm, red- to yellow-colored fruits, acidic flesh and clingstone	Late ripening; resistance to bacterial leaf spot; dwarf rootstock without suckering and compatible to plum and peach
	P. Americana	Common wild plum	Red fruit with a cling or free stone and an astringent skin	Tough skin; wide climatic adaptation including winter hardiness, late bloom Problem: suckering; hybridizations with *P. domestica* are rarely successful
	P. angustifolia	Chickasaw plum	Small and cherry-like fruit, bright to red colored, often even yellow	Bacterial leaf spot resistance
	P. munsoniana	Wild goose plum	Ovoid and shiny, red and yellow colored, with juicy flesh	Fruit quality; late blooming; cold hardiness
	P. nigra Ait.	Canadian wild plum	Red-orange to yellowish fruits with an astringent skin	Limited root suckering and winter hardiness
	P. subcordata	Sierra plum (Western or Pacific plum)	Globular to oblong in shape, with dark-red to the purplish surface, with semiacidic flesh	Drought tolerance

are common rootstocks for European plums (Ashton 2008). Non-plum *Prunus* species, including apricot, mum, peach, and almond, are also used as rootstocks for European plums; however, compatibility varies. GF677 is an almond-peach hybrids that used in alkaline soils. There is a graft compatibility between peach seedlings and *P. domestica* like 'Stanley', although they are not compatible with the German prune (Okie 1987). Japanese plums can be successfully grafted on Myrobalan and Marianna rootstocks, especially in soils with poorly drained surfaces. Peach seedlings like 'Elberta', 'Lovell', 'Nemaguard', and 'Flordaguard' are also frequently exploited as rootstock for Japanese plum, without any graft incompatibility issues, which is very common with European plum (La Rue and Johnson 1989).

The Chinese National Germplasm Repository for Plums and Apricots is located at the Institute of pomology, Chinese Academy of Agricultural Sciences, Xingcheng, Liaoning, China. This collection contains 717 accessions from nine plum species including *P. salicina* and *P. simonii* (Topp et al. 2012). The USDA-ARS National Clonal Germplasm Repository for fruit and nut crops at Davis, California, contains 313 plum accessions, including 154 *P. domestica*, 45 *P. cerasifera*, 63 *P. salicina*, and 39 American plum species (Prunus Crop Germplasm Committee 2010). Several European research institutions have also significant collections of European plums, including the Institute of Plant Genetics and Crop Plant Research, Fruit Genebank, Dresden, Germany; the Swedish University of Agricultural Sciences, Balgard Department of Horticultural Plant Breeding, Kristianstad, Sweden; and the Institut National de la Recherche Agronomique, Bordeaux and Avignon, France (Okie and Hancock 2008). Unfortunately, most of the wild plum species and relatives are poorly represented in these collections.

1.4.3 Apricots

According to different apricot classifications, reported by Bailey (1916), Rehder (1940), and Lingdi and Bartholomew (2003), there are 11 accepted apricot species within the section *Armeniaca* including *P. brigantina* Vill. (alpine apricot), *P. mandshurica* Maxim. (Manchurian apricot), *P. sibirica* L. (Siberian apricot), *P. armeniaca* L. (common apricot), *P. mume* Sieb & Zucc. (Japanese apricot), *P. dasycarpa* Ehrh. (black apricot, a natural plum-apricot hybrid), *P. holosericea* Batal. (Tibetan apricot), *P. hongpingensis* Li., *P. zhengheensis* Zhang & Lu., and *P. hypotrichodes* Cardot. Also, the desert apricot (*P. fremontii* S. Wats.), originated from southern California deserts, is worth mentioning among the listed species, can be freely hybridized with them, and has close morphological traits to other species of apricot (Ledbetter 2008).

Based on precise revising classification of Kostina (Kostina 1946) by Kryukova (1989), there are four major ecogeographical groups of apricots: (1) the Central Asian group composed of six regional subgroups, Fergana, Zeravshan, Khorezm, Shakhrisyabz, Kopet-Dag, and Dzhungar-Zailij; (2) the Iran-Caucasian group with two subgroups, North African and Iran-Caucasian; (3) the European group with two subgroups, Southern European and North American; and (4) the Chinese group.

Self-incompatibility is common among the apricots from Central Asian group. These cultivars are often well suited to a dry climate and susceptible to fungal diseases. Their fruits have a high SSC and low acidity with a wide range of skin color from white to orange and red. Apricot cultivars of Zeravshan and Fergana subgroup are frost resistant. Khorezm's apricots are more resistant to high temperature, spring frost, and soil salinity. Apricot cultivars of Dzhungar-Zailij subgroup are cold hardy (Zhebentyayeva et al. 2012).

Low chilling requirement and early blooming are allocated to the apricots from the Iran-Caucasian group. Most cultivars are mainly self-incompatible, but occurrence of self-compatibility is not rare as well. As compared to those from the Central Asian group, apricot maturation season is not quite long. Fruits are light yellow, white, or creamy in color with sweet kernels. Also, lacking pubescence on the skin of fruits is scarce (Rostova and Sokolova 1992). The apricots in North African subgroup have low chilling requirements, and some are resistant to *Monilia* spp. (Bassi and Pirazzoli 1998).

The European group is considered the youngest in origin and probably the best characterized of the ecogeographical groups (Faust et al. 1998). Apricot cultivars of European group have higher chilling requirements relative to those from the Central Asian group. Most cultivars are self-compatible and more resistant to fungal diseases compared to Central Asian and Iran-Caucasian cultivars. Their fruits are yellow/orange in color, aromatic, rarely glabrous, and mostly with bitter kernel, lower total soluble solids, and higher acidity relative to Central Asian group (Badenes et al. 1998; Ruiz and Egea 2008). By origin, North America apricot cultivars also belong to the European group. Commercial cultivars of North American subgroup are resistant to plum pox virus (PPV), owing to the involvement of Chinese germplasm in diversification of North American apricots (Zhebentyayeva et al. 2008).

The Chinese group of apricot cultivars is not only the oldest but also the most diversified apricot germplasm in the world. Six out of 11 commonly accepted apricot species including *P. armeniaca, P. sibirica, P. mandshurica, P. holosericea, P. mume*, and *P. dasycarpa* are endemic to China (Zhao et al. 2005). Cultivars from the Chinese group are mostly self-incompatible with limited environmental adaptation. Fruits have a short shelf life and are not good enough in quality. In China, apricot production is focused on the development of cultivars for fresh market, kernel production, and ornamental use (Zhebentyayeva et al. 2012). Ornamental apricots developed from the interspecific hybridization of *P. armeniaca* × *P. mume* which have 30–70 petals and varying blooming time (Byrne et al. 2000a).

Seedling apricot is still used and suggested as a first option for new apricot orchards in various growing regions (Khadari et al. 2006). Also, *P. armeniaca* is regarded resistant to root-knot (*Meloidogyne* spp.) nematode and root lesion (*Pratylenchus vulnus*) nematode (Culver et al. 1989). However, lack of reliable vegetative propagation is limiting the apricot rootstocks to only those produced by seed propagation (Reighard et al. 1990).

Considerable amounts of apricot genetic resources are being kept in collections for research and conservation purposes of the species. Over 6000 accessions are held at these institutions in more than 30 countries. Italy (1358), Ukraine (873),

Australia (693), Hungary (472), USA (417), France (406), Slovakia (319), Spain (212), Mexico (200), and Iran (170) are among the countries with the high number of accessions (Ledbetter 2008; Gharaghani et al. 2017).

1.4.4 Cherries

There are over 30 species of cherries, most of which are endemic to Europe and Asia. In the subgenus *Cerasus*, *P. avium*, *P. cerasus*, *P. fruticosa*, and also *P. tomentosa* and *P. pseudocerasus* in China and domesticates of *P. serotina* in South America are grown for their fruits (Webster 1996; Iezzoni 2008).

Sweet cherries can be divided into subgroups, based on fruit color, shape, and texture (Webster 1996). The subgroups include Geans which are heart shaped with tender flesh, black Geans which have dark-colored flesh, amber Geans which have light yellow fruit with translucid flesh and skin, Bigarreaux which has firm and cracking flesh, and Hearts which are dark in color with flesh texture between Geans and Bigarreaux.

Sour cherries have also been further categorized, based on skin and juice color and fruit shape, into either Amarelles (pale red fruits with more or less flattened shape and colorless juice) or Morellos (dark red fruits with globular or cordiform shape and red to dark red in juice color) (Faust and Suranyi 1997). Duke cherries (with dark red skin and semiacid juice) are considered to be a hybrid between sweet and sour cherry and now classified as *P.* × *gondouinii* Rehd. (Tavaud et al. 2004).

Main species in the parentage of sweet and sour cherry rootstocks include *P. avium, P. cerasus, P. canescens, P. fruticosa, and P. mahaleb*. Also, some species like *P. incisa T., P. pseudocerasus L., P. serrulata L., P. concinna K., P. tomentosa T.,* and several interspecific hybrids between these species such as *'Adara'* and *'Myrobalan R1'* have been used in rootstock breeding programs (Iezzoni et al. 1990; Webster and Schmidt 1996; Kappel et al. 2012).

Since cherry is native to West Asia and European countries, significant conservation activities have been done in these countries. The European Cooperative Programme for Plant Genetic Resources (ECPGR, http://www.ecpgr.cgiar.org/) facilitated the long-term ex situ and in situ conservation of cherry genetic resources in Europe. The main focus of this program is the documentation of the accessions kept in the repositories across the continent and also to encourage the plant exchange. The information of the ex situ repositories is available online through the European Internet Search Catalogue (EURISCO, http://eurisco.ipk-gatersleben.de), which currently (Status: 19 August 2015) includes 4667 sweet cherry and 804 sour cherry accretions from 42 institutions in 17 countries (Iezzoni et al. 2017). The Russian Federation is also a member of *Prunus* working group of ECPGR; however, the collection of this country is currently not included to the database.

There are three main collections for cherry germplasm in the USA under the control of the US Department of Agriculture's Agricultural Research Service (USDA-ARS) including the National Clonal Germplasm Repository (NCGR) in Davis, California (57 sweet cherry accessions along with some wild species), the

Plant Genetic Resources Unit in Geneva, New York (81 sour cherry accessions and other tetraploid species), and the National Arboretum in Washington, DC (ornamental cherry accessions). Information of conserved accessions is available through the GRIN-Global database (http://www.grin-global.org/). In Japan, sweet and sour cherry germplasm (57 accessions) are conserved in Morioka Branch of the Fruit Tree Research Station (Iezzoni et al. 2017).

Iran illustrates a major source of germplasm for various fruit species in the *Cerasus* subgenus including 13 species, of which *P. chorassanica* (Pojark.) A.E. Murray, *P. microcarpa* (Boiss.) C.A. Mey. subsp. *diffusa* (Browicz) Schneid., *P. yazdiana* Mozaff., and *P. paradoxa* Dehshiri & Mozaff. are endemic to Iran. These genetic resources may contain useful genes that have not been explored in modern cherry breeding programs. There is a collection of 160 accessions of sweet cherry and 180 accessions of sour cherry at the Kamal Shahr station in Karaj and other affiliated provincial stations of Iran's Horticultural Research Institute (Gharaghani et al. 2017).

1.5 History of Improvement and Worldwide Breeding Programs

1.5.1 General Facts

Major *Prunus* species were domesticated in Central and East Asia and introduced to the West in ancient times. Subsequently, species and technology injections originate from Persia, Greece, Turkey, India, and China. Meanwhile, fruit culture had attained a highly developed level in Greece and Rome by classical time, not exceeded it for more than a millennium (Janick 2005).

The stone fruits which were cultivated at first must be indigenous species that were highly remarkable for humankind. Most stone fruit crops (except peach) are highly cross-pollinated and therefore highly heterozygous. Although these fruit species can be produced from seed, this is usually an inappropriate technique due to long juvenile period and inferior quality of seedling compared to the selected clone. Thus, the basis of most stone fruit improvement has long been related to clonal propagation of special wild seedlings ensuing evolutionary advancement arising from intercrosses of superior clones plus intercrosses with wild species (Zohary and Hopf 2001), resulting in very high seedling diversity. Generally, contemporary stone fruit cultivars have undergone only a few generations and have not diverged from their wild progenitor clones (Janick 2005). The main breeding strategy in these fruit crops was based on continuous selection and combined with the ability to make specific combinations by vegetative propagation. This process was quite successful, given progress in plant breeding; it was not simple to substitute grower-selected clones. The ease of propagation is the key feature to domestication and improvement of fruit crops which is affected by selection. Selection for yield, basic fruit quality attributes including size, shape, color, flavor, and shelf life, which have been practiced by ancient fruit growers for centuries, is still the goal of modern fruit breeding programs (Janick and Moore 1996).

Many domesticated stone fruits differ from their wild progenitors by a few characters that have appeared as mutations. In most cases, owing to a decrease in adaptability, these mutations are not useful for the plant in its natural habitat but would obviously have been automatically preferred by humankind (Forstera and Shub 2011). Mutations associated with stone fruit improvement include some changes such as loss of fruit pubescence in peach and changes in growth habit in many of the stone fruits (Janick and Moore 1996).

Some of stone fruit crops have resulted from spontaneous interspecific hybridization, polyploidy, or both phenomena. This is particularly obvious in cherries and plums (Iezzoni 2008; Okie and Hancock 2008). Spontaneous hybridization between wild races and cultivated clones was important for the early fruit domestication. During the domestication process, it was the dominant force to select from sexual recombinants and is still practiced even in modern fruit breeding (Shulaev et al. 2008).

Fruit breeding as an organized activity is a nineteenth-century innovation, and its origins trace back to mass selection efforts in strawberry and pear. Thomas Andrew Knight was the first who literally initiated the fruit breeding to improve fruits through crossbreeding and selection. He released a number of improved fruit cultivars including some cherry, nectarine, and plum cultivars. Later in the USA and some of the European countries, fruit breeding became a part of research at the public institutions and even the private sector (Janick 2005). Private breeders currently form a significant part of *Prunus* particularly for peach, nectarine, and plum (Byrne 2012).

Although stone fruit breeding has been a major activity since early in the twentieth century and has shown significant advances in the second half of the twentieth century, the results have been uneven and vary from less effectual (in apricot) to extraordinarily successful (in peach and nectarine). Despite numerous breeding programs, extraordinary achievements at *Prunus* were based on growers' selection of seedlings and somatic mutations as well (Potter 2012). In recent decades, developments in molecular genetics may overcome some of the limitations of conventional fruit breeding focused on sexual recombination by increasing selection efficiency utilizing molecular markers and by transgene technology that enables the insertion of individual genes from different sources without disrupting specific genetic combinations (Limera et al. 2017).

1.5.2 Peaches and Nectarines

In Europe, during the Industrial Revolution of the sixteenth century, prominent breeders such as John Rivers released various cultivars. The peach reached Florida, Mexico, and South America in the mid-1500s via Spanish and Portuguese explorers. Before the American Revolution, peaches were grown mainly in rather low-quality seedling stands (Hedrick 1950). In 1850, Charles Downing brought 'Chinese Cling' from China to the USA via England (Scorza and Sherman 1996). After the Civil War, two important cultivars, including 'Belle of Georgia' ('Belle') and 'Elberta', which likely had 'Chinese Cling' as a parent, were released. 'Hiley' (a seedling of 'Belle') and 'J.H. Hale' (a seedling of 'Elberta') were other important early cultivars that released subsequently. This small group of related cultivars constituted the genetic basis of future breeding programs (Scorza et al. 1985). Peach breeding in the USA began at a number of State Experiment Stations in the late 1890s and early 1900s. In 1895, North America developed the first organized breeding program in Geneva, New York. Until the 1920, there were peach breeding programs in Iowa (Ames), Illinois (Urbana), California (Davis), Ontario (Vineland and Harrow), New Jersey (New Brunswick), Virginia (Blacksburg), Massachusetts (Amherst), and New Hampshire (Durham). After this, by 1960, other states including Maryland (College Park), Michigan (East Lansing), Georgia (Fort Valley), Texas (Texas A&M University, College Station), Louisiana (Baton Rouge), Florida (Gainesville), North Carolina (Raleigh), and Arkansas (Fayetteville) started their peach breeding programs, respectively. The 'Redhaven' peach, which was dominant peach cultivar in the eastern USA for decades (Iezzoni 1987) and also recognized as important cultivar worldwide, is developed at Michigan breeding program. Private peach breeding also was established in California (including Grant Merrill, Anderson/Bradford, Luther Burbank Armstrong, Nursery Company, Zaigers Genetics, Metzler and Sons). Most of these programs aimed at improving and developing locally specific types for the fresh market (Okie 1998; Okie et al. 2008).

In Europe, the first peach breeding program was established in Italy (1920s) and much later in France (1960s). After these, additional programs were started in Spain, Romania, Serbia, Greece, Bulgaria, Ukraine, and Poland (Okie et al. 2008; Llácer 2009). Such programs include projects which are supported both privately and publicly. Most of the initial breeding efforts were focused on the peach cultivars produced in the USA, so several European peach cultivars are strongly similar to those of North American cultivars (Faust and Timon 1995).

In Latin America, breeding programs were initiated in Southern Brazil (1950s) at two locations (Pelotas and Sao Paulo) and in Mexico (in the 1980s at Colegio de Postgraduados, Chapingo) aiming to develop cultivars for both the fresh and processing (Byrne et al. 2000b; Byrne and Raseira 2006). Some other programs to develop well-adapted peach types are underway in Chile, Argentina, and Uruguay. Valuable peach breeding activities were also made in Australia, China, Japan, and South Africa (Okie et al. 2008).

The twentieth century was called the "Golden Age of Peach Breeding" by Sansavini et al. (2006), due to significant worldwide breeding activity, with more than 1000 new varieties expected to be released. Although the new clingstone types for canning derive mainly from the public sector, the private sector is developing most of the new peach releases. About half of the cultivars released came from the USA and 30% from Europe, with France and Italy moving ahead (Okie et al. 2008).

1.5.3 Plums

European plums were brought to North America by early settlers, although the plums thrived only in more northern areas. Luther Burbank developed the first set of European plum cultivars, among which only 'Giant', 'Standard', and 'Sugar' were commercially relevant. In 1893, the first public breeding program was founded at Geneva, New York, for European plums. The 'Stanley', which is still an important cultivar in many countries, is released from this program in 1926. More than 100 years ago, improved cultivars of *P. salicina* including 'Kelsey' and 'Abundance' were imported into the USA from Japan. Luther Burbank intercrossed these and other imported materials with *P. simonii* and North American species and released many cultivars including 'Burbank', 'Duarte', 'Beauty', 'Eldorado', 'Formosa', 'Santa Rosa', 'Satsuma', 'Gaviota', 'Shiro', and 'Wickson'. Such plum cultivars established the base for the shipping plum industry worldwide, and some are still commonly cultivated. These cultivars have been crossed in many areas of the world, with the local plums of the specific region. Winter-hardy species like *P. nigra, P. americana,* and *P. besseyi* were crossed to adapted and evolved Japanese plums in the north of USA, to improve the plums that could be grown there. In order to increase disease resistance, the Japanese plums were sometimes crossed with *P. angustifolia* in the southeastern USA, resulting in plums like 'Bruce' (Topp et al. 2008).

The University of California at Davis plum breeding project aims to develop prunes with varied ripening date. Other minor breeding programs are USDA breeding at Prosser and Beltsville. Currently, work at USDA-Kearneysville, W.Va., focused on developing bioengineered plums with high resistance to plum pox virus. The objectives of Japanese plum breeding in California have been and are fruit size and firmness, wider range of skin color, and better eating quality. 'Frontier' (1967), 'Friar' (1968), 'Queen Rosa' (1972), 'Blackamber' (1980), and 'Fortune' (1990) are successful Japanese plum cultivars released by the USDA breeding program at Fresno, California (Okie and Ramming 1999). The main Japanese plum breeding program in southern USA is USDA-ARS at Byron, Georgia, considering the breeding objectives of California plus additional disease resistance. Private breeders and growers (Sun World International, Fred Anderson, John Garabedian, and Floyd Zaiger) in California have also selected many important commercial Japanese plums (Okie and Hancock 2008).

The development of European plums is innately focused in Europe, where breeding efforts have increased in recent years. During the early nineteenth century, plum breeding activities were conducted at research stations in England at Long Ashton and John Innes (Roach 1985). No systematic breeding was performed in other European countries until later, but regional selections of ancient cultivars were produced and developed. In Eastern Europe, breeding activities date back more than 50 years, with many cultivars being released (Okie and Hancock 2008).

At INRA in Bordeaux, France, the breeding targets were to develop dessert plums and drying prunes suited to the French climate. In Italy, at Florence, the breeding programs aim to grow early ripening dessert plums with vigorous productive trees and large, high-quality fruit. In Switzerland, Sweden, Germany, and Norway, breeding projects were begun or restarted with the goal of developing better fresh market plums, with focus on disease resistance, in particular plum pox virus (Okie and Ramming 1999). Japanese plum breeding is becoming progressively significant in Europe, as the market for large fruit of Japanese plums is growing. In Italy, breeding program at Rome and Forli focused on developing cultivars with smaller trees to minimize overall expenses of production combined with large size, dark skin, and good taste, while the target in Florence is to grow late-blooming self-fertile cultivars with high quality, especially yellow-skinned types. In southern France at Avignon, a breeding program was established to develop cultivars adapted to the poor weather during pollination and sharka resistance (Okie and Hancock 2008).

In Canada, Ontario, breeders aimed to produce high-quality dessert plums of varied ripening times around July to October that are cold hardy and productive and have blue color. In Southern Hemisphere, Brazil has three Japanese plum breeding programs aiming to develop low chill red-fleshed cultivars with resistance to bacterial spot and leaf scald. Some other breeding projects in this hemisphere have also been established in Australia, New Zealand, and South Africa. Their objectives include large-fruited, high-quality plums with resistance to bacterial canker and bacterial spot and the storage ability which is crucial to exporting the fruit by ship (Okie and Ramming 1999). There are many breeding programs in Russia and countries derived from former USSR. Cold hardiness, self-fertility, productivity, modest tree size, large fruit size, purple fruit, higher sugar content, and earliness are desired in these regions (Okie and Hancock 2008).

1.5.4 Apricots

Selection of superior apricot genotypes and their clonal propagation initiated around 600 AD in China (Faust et al. 1998) and probably in other regions like Central Asia and ancient Persia. Apricot improvement perhaps began after the development of grafting and budding. Early orchards possibly provided superior selection which was self-incompatible, so fruit inside the orchard must have resulted mainly from cross-pollination. It has now been clearly confirmed that parental choices based on phenotype lead to important genetic achievement in breeding programs of apricot (Couranjou 1995; Bassi et al. 1996); thus, the next generation of seed-propagated trees in ancient era would have made sufficient variations which is worthy for selection and further distribution.

Across the 1600s, many of the so-called apricot cultivars started to appear in the European written record, but apricot was already introduced to these regions several centuries before. Such cultivars seem to have been the result of selection only, from seed-propagated orchards, or by chance seedlings that have grown on their own. Nevertheless, some of these apricots have been relevant in different areas since their discovery and are now commonly used as parents in planned hybridizations (Ledbetter 2008).

In Europe, the number of breeding programs specific to apricot is much lower than those of other stone fruit species. The oldest apricot breeding program began in 1925 at the Nikita Botanical Gardens in Yalta, Crimea, Ukraine, while most other European breeding programs started their research between the 1960s and 1980s (Ledbetter 2008; Zhebentyayeva et al. 2012). The main goals of European apricot breeding programs include resistance to plum pox virus, brown rot, bacterial diseases, chlorotic leaf roll phytoplasma, apricot decline syndrome, adaptability to the environment (water deficit and temperature requirements), extension of the harvest season, productivity, fruit quality, and tree size and structure (Bassi and Audergon 2006). Hybridizations between locally adapted apricot accessions and germplasm from Central Asia are carried out in several projects to achieve those objectives (Benedikova 2006; Ledbetter and Peterson 2004).

Three apricot breeding programs are publicly supported in Italy including the "Dipartimento di Produzione Vegetale" at Milano and Bologna Universities (Pellegrino 2006), the "Dipartimento di Coltivazione e Difesa delle Specie Legnose" at Pisa University (Guerriero et al. 2006), and the "Istituto Sperimentale per la Frutticoltura" at Caserta (Pennone and Abbate 2006). In France, there is an active breeding program by CEP Innovation under the frame of a national agreement with the "Institut National de la Recherche Agronomique" (INRA) and Agri-Obtentions. In Spain, two institutes focused on apricot breeding projects including the "Centro de Edafología y Biología Aplicada del Segura" (CEBAS-CSIC), in Murcia, and the "Instituto Valenciano de Investigaciones Agrarias" (IVIA), in Valencia, attended to produce plum pox virus-resistant cultivars (Badenes and Llácer 2006). A large apricot breeding program has been administered at the National Agricultural Research Foundation, Pomology Institute, at Naoussa, Makedonia, in Greece for the control of sharka disease. In Romania, an apricot breeding program is established within the Agronomic Research Institute in Bucharest, which mainly focused on the modernization of the whole apricot assortment in this country (Cociu 2006). In Bulgaria, a breeding activity was established at the Apricot Research Station in Silistra, focused on the enriching the genetic diversity of this crop (Coneva 2003).

Extraordinary efforts of apricot breeding have also emerged in other parts of the world in regions where apricots are important. There are publicly supported breeding programs in both Australia (South Australian Research and Development Industries, Loxton, South Australia) and New Zealand (Plant and Food Research Institute, Hawke's Bay, NZ) of Oceania, as well as in Tunisia (Institut National de Recherche Agronomiques de Tunisia) and South Africa (Agricultural Research Council of South Africa) of Africa. In Asia, China (Liaoning Institute of Pomology, Xiongyue, Peoples Republic of China), Japan (National Institute of Fruit Tree Science, Tsukuba, Ibaraki, Japan), Turkey (Alata Horticultural Research Institute, Mersin, Turkey), and Iran (Horticultural Research Institute, Karaj, Iran) have publicly funded breeding efforts on apricot. In the USA, Rutgers University, New Brunswick, NJ, and the USDA/Agricultural Research Service, Parlier, CA, have active publicly funded apricot breeding programs. Recently, some breeding activities were started by the University of Santiago in Chile. Many breeding projects have also been privately supported with numerous new cultivars released (Ledbetter 2008; Zhebentyayeva et al. 2012; Gharaghani et al. 2017).

Some of the important apricot cultivars in recorded history include 'Roman' (ancient Rome), 'Shalah' (a landrace from Armenia and progenitor of numerous later cultivars), 'Nancy' (discovered near Nancy, France, in 1755, ancestor of many later cultivars), 'Moor Park' (selected in 1760, Herefordshire, England, preferable to all apricots already produced), 'Royal' (seedling of 'Nancy', discovered in 1808, French), 'Blenheim' (introduced before 1830, Marlborough Blenheim, England, Syn. 'Shipley'), 'Luizet' (chance seedling found in 1838, widely adapted to Europe and N. Africa), 'Hungarian Best' (discovered in 1868, Enying, Hungary), 'Bergeron' (chance seedling of exceptional flavor found in 1820, Saint-Cyr-au-Mont-d'Or, Rh^one, France), 'Stark Earli-Orange' (discovered in 1920, Grandview, Washington, USA, late-blooming apricot used extensively for resistance to sharka), 'Scout' (selected from a seed lot of Manchurian origin and introduced in 1937 by Dominion Experimental Station in Morden, Manitoba, Canada), and 'Perfection' (introduced in 1937, Waterville, Washington, USA, unknown parents, progenitor of many North American cultivars) (Ledbetter 2008).

1.5.5 Cherries

Until the sixteenth century, the details about the improvement of cherries are slightly documented; however, early reports of trace of many ancient sweet cherry varieties date back to German origins. From ancient times to the 1600s, many landraces specific to regions or towns arose and had been popular in different areas across Europe. In European countries with extensive diversity for cherry landrace, the breeding programs began by selecting among landraces to be used directly as cultivar or as parents in hybridization programs (Iezzoni 2008). Almost all of the sour cherry cultivars grown today are either landrace selections themselves or only a generation removed from these landrace selections. Therefore, it can be assumed that the origin of the cultivated sour cherry was derived from few initial genotypes forming the known local cultivar groups including the 'Schattenmorelle' (in Central and Northern Europe), the 'Maraska' (in the Adriatic area), 'Stevnsbaer' (in Denmark), 'Pandy' (Sothern Europe, Hungary, and Romania), the 'Vladimirska' (in Eastern Europe), and the 'Montmorency' and 'Spanish Glaskirsche' (in Western Europe, the Iberian Peninsula, and France) (Iezzoni et al. 1990; Faust and Suranyi 1997; Schuster et al. 2017).

Cherry seeds and budwood were brought to the North America by early settlers and from where pioneers moved the cherries westward. 'Bing', which is still the popular cultivar in North America and even in many parts of the world, was selected by Seth Lewelling in Oregon. Certain significant cultivars that emerged through this selection program included 'Lambert' and 'Democratic' which is still a prominent pollinizer (Iezzoni 2008). A significant progress in sweet cherry improvement arose from the introduction of self-fertility (S4) in this crop. The first self-fertile sweet cherry cultivar, 'Stella', was released in Summerland, British Columbia, Canada

(Lapins 1971). This cultivar had no superior quality to become economically relevant but was commonly used as a source of self-fertility in breeding programs. Later self-compatible releases from this breeding program such as 'Sweetheart' contributed significantly to global sweet cherry production (Iezzoni 2008).

The vast majority of sweet and sour cherry breeding projects belong to Europe which is one of the ancestral homes of the cherries. The most important worldwide sweet and sour cherry breeding programs supported by government or universities across Europe include two programs in the UK (East Malling Research Station, East Malling; John Innes Institute, Norwich), three programs in Italy (Dipartimento di Colture Arboree, University of Bologna; Istituto Sperimentale Frutticoltura, Ministry of Agriculture, Rome; Istituto Sperimentale Frutticoltura, Verona Province), the only program of the Germany (Germany BAZ Institute for Fruit Growing, Dresden), a program in France (INRA, Station de Recherches Fruitieres, Bordeaux), two programs in Hungary (Fruit Research Station, York; Research Institute for Fruit Growing and Ornamentals, Budapest), three programs in Romania (Research Institute of Fruit Growing, Pitesti; Iasi; Bistrita), and breeding activities and efforts in Switzerland (Swiss Federal Research Station for Fruit Growing, Wadensville), Serbia (Fruit and Grape Research Center, Cacak), Ukraine (Institute of Horticulture, Donetsk; Institute of Irrigated Horticulture, Melitopol), Latvia (Latvia State Institute of Fruit-Growing, Dobele), Lithuania (Lithuanian Institute of Horticulture, Babtai), Czech Republic (Research and Breeding Institute of Pomology, Holovousy), Belarus (Research Institute for Fruit Growing, Minsk), and Estonia (Polli Research Center of Horticulture, Karksi) (Iezzoni 2008; Schuster et al. 2017). There are also some active breeding programs in Russia, USA, and Canada (Iezzoni 2008; Kappel et al. 2012). More recently, sweet cherry breeding programs were started in Asia, among which the most important ones are three programs in Japan, two programs in China, two programs in South Korea, and a concise program in Iran (Schuster et al. 2017; Gharaghani et al. 2017).

1.6 General Trends in Stone Fruit Breeding

Breeding of tree fruit species is a long-term process with high costs relative to annual plants, because of the large plant scale and lengthy juvenile cycles. Despite these problems, numerous breeding programs have been developed through world in almost all of important stone fruits. Tree fruit improvement takes at least a decade from the original cross to a released cultivar. Thus, fruit breeders should predict cultivar that requires at least 10 years in advance. However, several cultivars have retained their market interest for many years, e.g., 'Redhaven' peach, 'Bing' cherry, and 'Stanley' plum, but new varieties, especially those of peach and nectarine, possess a fairly short market existence about 10–20 years. Although the new cultivar may have effectively incorporated the desirable traits of interest, the consumer demands for cultivar may have changed during the 15–20-year time frame in which the cultivar was produced. Therefore, the new product may not fulfill the current market demands, when introduced. It is an unavoidable threat in fruit production, but it seems to be a challenge considering the scientific advancements made in the last 50 years (Monet and Bassi 2008).

In the breeding programs, considering these facts, the priorities to be emphasized are quite important. Yield and basic quality attributes have always been and will be important in the future. In addition, some major trends including increased resistance to abiotic and biotic stress, simplified orchard practices, extension of the adaptation zones, extending the harvest window, new fruit types, superfruits or cultivars with high nutritional benefits, eating convenience, and consistently high quality are needs to be considered in developing new cultivars (Byrne 2005).

Integration of intellectual property rights (IP rights) regulations in fruit industry has provided substantial innovative research motivations in both private and public sectors. Improved legislation for plant protection throughout the world has stimulated the fruit breeding to shift from public into the private sectors (Heisey et al. 2001). Crops such as peaches and nectarines, with larger markets and shorter life cycles, are shifting to the private sector more rapidly. For example, approximately 85% of peach and nectarine cultivars have been released by the private sector in the USA in the past decade. While it is still with public agencies to support the development of apricots and cherries, this is evolving as the private sector is getting more interested in producing new cultivars in these crops (Byrne 2012). Another issue in this regard is the shifted philosophy of US and European governments from fully funded programs to partially funded programs which resulted in a dramatically decrease in funding for public fruit breeding programs (Heisey et al. 2001). Although this shift seems to be promising, there are some concerns about the amount of progressing studies into germplasm development, genetics, and new breeding techniques, as private sectors devote less funds and efforts for this purpose (Sansavini 2009). This type of research is very critical for the long-term success and sustainability of the fruit breeding programs worldwide.

Preservation of the environment is among the most important issues affecting the fruit production. These include sustainable fruit industry development, environmental contamination, climate change, and biodiversity. The environmental contamination concerns led to more restrictions on the use of agrochemicals as well as sustainable development of fruit production and marketing systems which is more environmentally friendly. Global warming, which is a real result of climate changes, affects the fruit industry noticeably. In this regard, measurement of the "carbon footprint" is a growing attitude aimed at calculating the carbon cost of fruit production by harvesting, processing, and marketing. Investigations demonstrated that in most cases the carbon footprint of imported fruit is more than that of locally produced fruit in season; nevertheless, this varies widely based on the method of transport, with sea freight becoming less energy-consuming than air freight (Brenton et al. 2009). Short postharvest durability limits the ability of stone fruits to be shipped via sea freight. This fact highlights the need for improved postharvest characteristics, as well as more locally adapted stone fruit cultivars.

From another point of view, 70% of the world's freshwater sources are currently consumed in agriculture (Sansavini 2009); this means water shortage including both of quantity and quality aspects should be considered in the future. Despite works need to be done to develop better water delivery and management techniques, more efforts need to be done to develop more efficient scion and rootstock cultivars that perform well under less or with poorer quality water.

The safety issue is one of the main driving forces of the global fruit industry, though it differs from area to area. As consumers became now conscious of the nutritional value of fruit consumption in human well-being, the demand for this product is increasing (Prior and Cao 2000). Fruits are in the forefront of the food for health movement which are touted to have health benefits such as the prevention of cardiovascular disease, obesity, and high cholesterol (Sloan 2006). Plum, prune, and cherry are stone fruits that ranked among the so-called superfruits with exceptional health benefits. On the other hand, consumer concern over food safety will result in a greater interest in integrated fruit production (IFP) or organic fruit production (OFP) systems which use little or no agrochemicals and minimize the potential pollution of fresh fruits with pesticides and fungicides (Batt and Noonan 2009). Most of the IFP and OFP programs are currently positioned in semiarid environments with conventional cultivars, but economic and environmental benefits have stimulated expanded private and public interest in the growth of disease-resistant cultivars for IFP and OFP systems worldwide (Sansavini 2009).

Consumers now expect to have year-round supply of fruits that are convenient to eat, of a wide variety, and of consistent quality (Byrne 2005; Sloan 2006). Nowadays, long-distance shipment allows stone fruits to be marketed thousands of miles away from the production site in the Southern Hemisphere (Australia, Brazil, Chile, New Zealand, Peru, South Africa) to supply the off-season markets in the Northern Hemisphere. Despite the possibility of long-distance shipment, to sustain the consumer expectation for year-round supply of stone fruits, more efforts need to be done on the varietal diversification of these fruit crops, especially on the harvest season extension and storage ability.

Convenience is another significant moving force in the fruit marketing business. Stone fruits are greatly different in this regard, with apricots and cherries being excellent and peaches and nectarines being not so convenient to consume (Jaeger 2006). Besides important desires like convenience and health, fruits must also be of consistent quality and flavor. Stone fruits are among the fruits that are more difficult to deliver with good quality and flavor than other fruits. Studies have confirmed that lack of constant quality is the main reason for people not to buy peaches (Byrne 2005).

To stay in the fruit business, grower requires to generate high-quality fruit at minimal expense. Skilled labor force, especially in developed countries, and agricultural chemicals are the two largest variable expenses for fruit production (Lucier et al. 2005). Utilization of dwarfing rootstocks made it possible to create commercial orchards with smaller, easier-to-handle trees that usually produce higher yield more precociously. However, in most of stone fruits, dwarfing rootstocks are a fairly new novelty (Webster 2006). This approach needs to be complemented by developing scion cultivars with special tree architecture and growth forms to simplify orchard management or allow the mechanization (Scorza et al. 2006; Schuster et al. 2017). Scion cultivars that set proper fruit load without cross-pollination will also be very advantageous in this regard (Kappel 2008). Outside the environmental and health concerns of using agricultural chemicals, their application would entail considerable economic and management costs. Thus, there is a growing demand for development of stone fruit scion and rootstock cultivars that are tolerant/resistant against pests, diseases, and nutrient problems (Byrne 2005).

1.7 Breeding Objectives

The primary objectives for *Prunus* have been related to the productivity and fruit quality attributes such as fruit size, color, firmness, storage, and processing. The cultivars generally differ in agronomic (i.e., productivity, soil adaptability, biotic and abiotic resistance, etc.) and quality traits. Most fruit breeding programs try to discover the best compromise between agronomic and quality attributes in their new cultivars. In order to breed fruit with the highest quality standards, it is far crucial to find the best combination among environment, agricultural techniques, and genotypes that produce the best performance (Jenks and Bebeli 2011).

1.7.1 Tree and Fruiting Structure

Trends in high productivity, precocity, and low labor cost lead to research emphasis on altering tree growth, simplifying training techniques, and the mechanizing production of fruit tree (Byrne 2005). There are two genetic approaches including the development of dwarfing rootstock and modification of scion tree architecture. Dwarfing rootstocks made an impact on plum, peach, and cherry production not only due to smaller tree size but also to more precocious fruit-bearing. Also, developing new cultivars with unique tree architecture ranging from dwarf to the semi-dwarf, compact, pillar, and weeping had been the other approach to increase tree efficiency (Dosba 2003; Bors 2005; Scorza et al. 2006; Quero-García et al. 2017).

1.7.2 Flower Characteristics

Late blooming seems a solution to avoid spring frost damage in some districts. Alteration of the chilling or the heat accumulation requirements of a cultivar or both could lead to bloom date manipulation (Dirlewanger et al. 2012). In warm zones, increasing the chilling requirement of the cultivar to make flowering delay leads to poor productivity due to lack of chilling. In such a situation, an increase in heat unit accumulation requirement before flowering is preferred to develop late flowering cultivars. Choices of native seedling populations in a region experiencing late frosts frequently provide surely good starting points in hybridization programs. There is a great diversity in average bloom date among Iran-Caucasian apricot germplasm gathered from Anatolia and Turkey that show almost a month of difference between early- and late-blooming cultivars (Asma and Ozturk 2005). Also, large populations of native seedling apricot which provide selection of late-blooming forms would be found in the Erzincan plain of Turkey (Ercisli 2004). Similar variation exists among Iranian apricot landraces (Gharaghani et al. 2017). The tolerance to blossom frost was enhanced in sour cherry by interspecific hybridization with *P. fruticosa* in Russia and Canada (Zhukov and Charitonova 1988; Bors 2005). There is a high heritability (0.86) of full bloom date in European plum (Hansche et al. 1975) that can be modified with a phenotype-based choice. Werner et al. (1988) showed that there was variation in *P. besseyi* for this feature.

Approximately a major goal in most of stone fruit breeding program is self-compatibility specifically, since the molecular markers used for self-fertile seedling selection result in extreme reliability and cost-efficiency (Schuster et al. 2007; Beppu et al. 2010). Cultivars showing entire self-compatibility can be grown as a monoculture system in order to eliminate possible difficulties at blooming and during the harvest time which turns up with two or more self-incompatible varieties developing. However, excessive fruit set is sometimes a problem in a self-compatible orchard, and expenses of thinning may cause the producer a great reduction of profit margin. On the other hand, a self-compatible cultivar may guarantee fruit set in an orchard during bloom periods unlike risk of limited bee pollination due to poor weather conditions. There is a single polymorphic "S" locus in *Prunus* that controls the gametophytic self-incompatibility system and includes genes for pollen and pistil specifics in a way that suppresses growth of pollen tube with the same haplotype in the style (Beppu et al. 2002; Sutherland et al. 2007). The Se-RNase allele admits self-compatibility and can be screened for at the seedling stage using PCR markers (Beppu et al. 2005, 2010) that provides early selection for this trait. 'Stella' allele with a mutated S is ancestor of sweet cherry cultivars with self-fertility trait which was released commercially (Lapins 1971). Nevertheless, the Hungarian cultivar 'Axel' with a mutated S allele (SS) would be an exception (Kappel et al. 2012). Anyway, sources of self-compatibility are diversifying with an increasing interest on self-compatible landraces, as parents in several breeding programs, for instance, 'Cristobalina' or 'Kronio'. In addition, fertilization between plum varieties with special combinations is excluded by cross-incompatibility. For example, a low degree of intersterility has been identified among European plum cultivars, or Tehrani (1990) at Vineland Station in Ontario reported an incompatibility between special varieties. The close relationship and the hexaploidy of the species may give a clarification about it. Three American apricot cultivars were the first cross-incompatible apricot samples, all possessing 'Perfection' apricot in their parentage (Egea and Burgos 1996). The cultivars ('Goldrich', 'Hargrand', and 'Lambertin-1') were the first incompatible group in apricot and received an identical genotype with the allelic designation of S_1S_2. This information provides a starting point for further research to find other self-incompatible alleles.

1.7.3 Tolerance to Abiotic Stresses

Environmental stresses are very important for the productivity, survival, and reproductive biology of stone fruit crops. Therefore, researchers have focused on obtaining genotypes showing high resistance to iron chlorosis, waterlogging, and drought; thus, controlled interspecific crosses have been performed to collect the desired attributes of various *Prunus* species. For example, some *Myrobalan* genotypes were selected as a parent because of their tolerance to waterlogging. In addition, peach, almond, peach × almond, and peach × *P. davidiana* hybrids were used as a different source of tolerance to iron chlorosis, drought, replant problems, and compatibility with peach (Byrne et al. 2012). In addition, breeders consider expanding the environmental ranges of stone fruits, such as chilling requirement reduction to adapt their production into the subtropical climates, increasing cold hardiness through bloom delay in the colder climates as well as improving high heat tolerance during bloom and growing season in the warmer zones (Byrne et al. 2000b; Kozai et al. 2004). Rain-induced fruit cracking and fruit doubling of sweet cherries is one of the most significant annual challenges to fruit quality (Quero-García et al. 2017). Therefore, specific programs are planned for breeding and selecting cultivars resistant to rain-induced cracking.

1.7.4 Tolerance to Biotic Stresses

Resistance to biotic stresses is a considerable factor for *Prunes* breeding. Quarantine pests or pathogens cause diseases infecting stone fruits in different areas worldwide. Plum pox virus (sharka disease) and phytoplasmas (European stone fruit yellows) have made troubles in Europe stone fruit industry. Bacterial leaf spot (caused by *Xanthomonas arboricola* pv *pruni*) and brown rot (caused by *Monilinia fructicola* and *M. laxa*) are examples of worldly spreading quarantine diseases of stone fruits. Therefore, germplasm resisting or tolerant to pathogens, pests, and orchard replant problems are main requirements of breeding programs aiming for new *Prunus* cultivars and, in particular, rootstock development. Active scion breeding programs are pursued for resistance to brown rot (Gradziel et al. 1998), aphids (*Myzus persicae*) (Monet et al. 1998), bacterial leaf spot (Kervella et al. 1998), plum pox virus (Bellini et al. 1996), and powdery mildew (caused by *Sphaerotheca pannosa*) (Byrne et al. 2000b).

To grow *Prunus* fruit using resistant or somehow tolerant rootstocks as an agronomic solution is profitable to increase productivity and efficiency for better tree survival in soils infested with pathogens like fungi, bacteria, virus, and virus-like diseases (Gainza et al. 2015; Dosba 2003). Resistance to peach tree short life (Nyczepir and Beckman 2000), root-knot (*Meloidogyne*) and root lesion (*Pratylenchus*), ring (*Mesocriconema*) and dagger (*Xiphinema* spp.) nematodes (Lu et al. 2000; Pinochet et al. 2000; Claverie et al. 2004), armillaria root rot (*Armillaria* spp.) (Beckman and Pusey 2001), and *Phytophthora* is of paramount importance in *Prunus* rootstock breeding. Genetic resistance is also a solution to pests and predators out of pesticide control; however, this approach is in its infancy and needs to be considered more seriously.

1.7.5 Fruit Quality

In comparison with other crops such as pome fruit, the quality of stone fruit is quite inconsistent. This does not refer to external qualities which are efficiently standardized by packing but rather to internal quality attributes such as firmness, soluble solids, and acidity (Byrne 2005). Thus, the universal goal of all stone fruit breeders is to enhance fruit quality. Measuring genetic gain from planned hybridizations requires fruit quality subdivision, and breeders need to evaluate specific characteristics. Individual characteristics listed as fruit size and the degree of flesh firmness, aroma and flavor characters, skin and flesh color, fruit juiciness, shelf life, and processing quality are parts of fruit quality altogether, which can be evaluated accurately by means of proper instrumentation (Jaeger 2006; Jenks and Bebeli 2011). Incorporation of large fruit size, early ripening, and high soluble solids is the main breeding obstacles regarding the quality improvement of stone fruits in general, and in particular for peach/nectarine, since total soluble solids typically begin to diminish with a lower period of fruit development and larger fruit size (Hansche et al. 1975; Souza et al. 2000). Nevertheless, recent findings have shown it is possible to combine and collect high soluble solids characterized by good fruit size together during a fruit development period of less than 100 days (Byrne 2005). On the other hand, organoleptic fruit quality and shelf life may influence production and marketing, so both the producer and the consumer should think of sensory evaluation of new criteria for fruit quality by employing trained experts or taste panel test (Jenks and Bebeli 2011). Candidate genes involved in the synthesis of specific aromatic compounds must be identified. Some programs are planned to develop cultivars with health-enhancing traits such as increased fiber, antioxidant, vitamin, phenol, and aromatic compound content. Increasing vitamin C content, for instance, is a new possibility for nutrition quality improvement (Wargovich et al. 2012). In addition, both conventional and ecological production systems matter fruit safety objectively (Batt and Noonan 2009). Resistant cultivars decrease pesticide residues, as well as mycotoxin development on non-pesticide cultivars. Both fresh markets and processing industries (juice, dried or canned fruits) demand diversification (Dosba 2003).

1.7.6 Extension of Harvest Period

Stone fruit breeders intend to expand harvesting period and accordingly marketing calendar primarily. Great farm gates and higher market prices are consequences of programs that focus on breeding early ripening genotypes of fruits with good quality (Byrne 2012). According to Hansche et al. (1975), the ripening time is specified by several genes or alleles of additive effect (Dirlewanger et al. 2012). Unlike the early and midseason cultivars, studies recommend to extend the end-of-season calendar using extra-late ripening genotypes especially in the mountain and in the cold summer areas due to their premier pomological attributes, for example, size, firmness, color and taste, and lower susceptibility to biotic stresses such as cracking

(Byrne 2005; Zhebentyayeva et al. 2012). The extension of the harvest season has led to a 1- to 2-month expansion of harvest season for peach (Byrne et al. 2012) and five to six weeks at most for cherries (Kappel et al. 2012). In Northern Hemisphere, the harvest season for European plums is from the middle of June to the middle of October (Topp et al. 2012).

1.7.7 Suitability for Mechanical Harvesting

Modifying tree growth and architecture can facilitate harvest. Additionally, mechanical harvesting is increasing notably due to its influence on labor cost and harvest period reduction (Byrne 2012). Formerly, mechanical harvesting systems were handled for a variety of products specially for processing of crops which are less desirable in appearance. In order to acquire more uniform ripening, easy detachment, non-bruising fruits with suitable firmness, breeding must result in crops properly adaptable to mechanical system or, at least, to a once-over-harvesting method relative to multiple harvesting. The upright tendency for mechanical shaking and low detachment power to pick off fruits and uniformity of ripening within the plant should be mentioned as traits related to harvesting (Kappel et al. 2012; Schuster et al. 2017). Automating the industry of drying apricot through taking advantages of new cultivars capable of mechanical harvest and fruit cutting and with low drying ratios is the goal of Australian breeders (Zhebentyayeva et al. 2012). Mechanical harvesting of cherries supplying fresh market is a proper choice for harvesting sweet cherries due to availability and concern over labor costs. Mechanical harvesting has been used for processing cherries, but higher quality standards should be considered for cherries supplying the fresh markets. The fresh produce trade may require to develop a market for stemless sweet cherries (Quero-García et al. 2017). Tree architecture, stem retention force, and resistance to bruising are issues that matter ideally, without the usage of growth regulators like ethephon. Mechanical harvesting may adapt suitably to dry abscission zone between the fruit and stem which was developed by sweet cherry cultivars including 'Vittoria' (Bargioni 1970) and 'Cristalina' (Kappel et al. 1998), and others have decreased stem retention (e.g., 'Symphony').

1.8 Breeding Program Structure

Breeding of fruit crops has reached a novel season, in which using DNA information to support breeding intentions and activities is now a principle and is growing in applied programs. These tools and techniques have simplified the breeding cycle, allowing the early detection of seedlings carrying traits that are important for growers and consumers, thus increasing the efficiency of new cultivar development (Peace 2017; Laurens et al. 2018). On the other hand, general trends, as discussed before, dictate the breeder to develop new cultivars based on the desires of all involved stakeholders including growers (demand for improved resistance to pests and diseases, high yields, and ease of management), consumers (look for good appeal and taste, eating convenience, enhanced health and safety as well as novelty), and marketing sector (demand for higher transportation ability and better shelf life). These cultivars also should be grown with minimal environmental impact (Byrne 2005; Gallardo et al. 2012). Successful plant breeding at this sophisticated era needs multidisciplinary teams, in which breeding is the central hub, but there are divisions of expertise and labor (Fig. 1.4). In this regard, understanding how to integrate molecular genetics with traditional breeding is key, not the individual components (Ransom et al. 2006). The more organized, frequent, and regular this DNA information used in a breeding plan, the greater the advantages of basic progresses in recognizing genetic diversity, heredity, genomic structure, and phenotypic performance in the development of superior new cultivars (Peace 2017).

1.9 Breeding Strategies

In spite of the difficulties related to tree fruit genetic studies (large plant size and long generation time), a large amount of information has been collected on character inheritance for stone fruits, especially in peach as the model crop of *Prunus* genera, which is the most genetically characterized species of stone fruits. This is because peach has smaller plant size and shorter generation time than other fruit crops. Small chromosome number, self-fertility, tolerance to inbreeding depression, and transmission of many important qualitative traits according to simple Mendelian inheritance also facilitated these studies (Monet and Bassi 2008). In stone fruits, traits such as fruit skin color, fruit size, firmness, and taste are multifactorial properties (both polygenic and affected by environmental factors) and thus more challenging to improve due to their comparatively small degree of heritability.

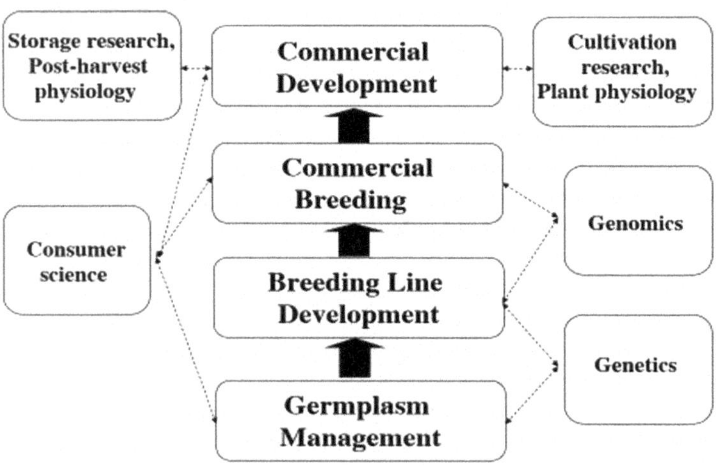

Fig. 1.4 A well-organized structure of a comprehensive fruit breeding program showing the integration of various scientific disciplines. (Adapted from breeding program of Plant and Food Research Institute, New Zealand)

In the last century, many new cultivars have been released and commercialized for major stone fruit crops throughout the world, especially in the Europe and the USA. Most of them come from crossbreeding, either via open pollination or via controlled crosses, and only a few percent from bud sports (Della Strada et al. 1996; Ledbetter 2008; Okie et al. 2008; Quero-García et al. 2017). Mutation breeding and somaclonal variation are among the other possible and more advanced breeding techniques utilized for stone fruits (Predieri and Gatti 2000; Predrieri 2001). Intraspecific hybridization is also a common method for stone fruit breeding, especially for peach and also for creation of new fruits in *Prunus* spp. genera. This approach still continues to provide the overwhelming majority of the world's latest stone fruit cultivars.

Dynamic strategies can be implemented depending on the accessible germplasm and objectives. Very few well-known genotypes were used as progenitors in many of stone fruit breeding, and given the high degree of inbreeding, quality traits have continually developed (Scorza et al. 1985). This constant development is partially attributed to the outcrossing of breeding populations with unrelated genotypes to integrate appropriate traits including fruit quality, diverse chilling requirements, and resistance to biotic stresses (Cantín et al. 2010).

The construction of pre-breeding materials, a time-consuming and costly process, is required if desirable features or variations are not available in current advanced breeding materials, i.e., whether the contributors are wild species or new combinations of desirable attributes produce too much genetic drag (Laurens et al. 2018). For this purpose, usually backcrossing or modified backcrossing strategy is employed to restore the commercially important characteristics of the commercial cultivars. This is the duty of line breeding working group (Fig. 1.4). For example, in peach scion cultivar breeding, the two commonly used species have been *P. davidiana* and *P. dulcis*. *P. davidiana* has been used as a donor for resistance to powdery mildew, green peach aphid, peach leaf curl, and plum pox virus (Foulongne et al. 2003; Rubio et al. 2010), whereas using almonds emphasizes on the introgression of drought tolerance, growth habit (e.g., spur bearing), low bruising, and resistance to certain diseases into peach germplasm (Gradziel 2003; Martínez-Gómez et al. 2004).

Although hybridization between some of the stone fruit species is possible (for instance, peach and plum or plum and apricot), between some others this is practically impossible (cherries with other stone fruits). Previously, "interspecific hybridization" has been exploited to grow new fruits, i.e., 'Pluot', interspecific of plum and apricot; 'Pluerry', interspecific of plum and cherry; and 'Peacharine', interspecific of peach and nectarine (Fig. 1.5), and rootstocks (e.g., 'Viking', 'Atlas', and 'Citation' derived from a combination of apricot, peach, almond, and plum by Zaiger's Inc. Genetics of Modesto, California, USA) within *Prunus* genera. There are several barriers to fertility in the development of such hybrid forms, particularly when breeding populations are established, and several generations of backcrossing are still required to restore fruit quality (Foulongne et al. 2003). Interspecific hybridization with related species is quite common in stone fruit rootstock breeding for

Fig. 1.5 New fruits created by interspecific hybridization

incorporating effective characteristics including tolerance to droughty and calcareous soils (using almond), nematode resistance (using *P. davidiana* and various plum species), waterlogging tolerance (using various plum species), and dwarfing (using various plum species) (Reighard and Loreti 2008).

1.10 Parent Selection

Criteria for selecting the excellent parent are very significant components of the breeding program. Although traits under basic Mendelian inheritance are easily traceable within progenies and over generations, quantitative traits, regulated by polygenic structures, need a specific and more sophisticated approach. Selecting parents having complementary phenotypic traits has been the common practice which results in improving much more of the commercially significant attributes (Monet and Bassi 2008). Parents may be better than standard cultivars which are distinguished by higher efficiency and fruit quality. This method is easy and fast and provides a great possibility to obtain desirable combinations; however, the frequent utilization of conspicuous cultivars as parents may result in high phenotypic homogeneity (Scorza et al. 1985). In other cases, advanced selections carrying one or more beneficial attributes such as those of fruit quality or specific resistance are chosen to introduce the trait of interest into a commercially important cultivar. Heritability estimation of quantitative traits in stone fruits demonstrated that parental materials could be selected based on their phenotype to get achievements rapidly (Souza et al. 2000). It is worth to mention that, if the expression of a trait of interest is determined by superiority or epistasis, a phenotype-based parent's preference may be deceptive and give rise to a worthless offspring. Thus, it is highly recommended that the genetic worth of the parent be evaluated by a progeny test (Monet and Bassi 2008). The best approach is to do a self-pollination (which is very common for peach and nectarine): the more heterozygous the progeny, the more heterozygous the parent. This approach offers valuable knowledge specially on simple characters, which shows recombination as well as recessive traits. However, for quantitative traits under polygenic regulation, prepotency assessment and general or specific combination ability are better approaches to evaluate a given genotype's potential in generating superior progeny. The simple test for progeny would be to examine multiple populations that share a common parent (Cantín et al. 2010). However, more sophisticated statistical designs such as di-allele and half di-alleles could be used for these purposes (Fogle 1974; Ghasemi Soloklui et al. 2018).

Criteria influencing the efficiency and feasibility of the crossing and seedling population development also should be considered. In this regard, self-incompatibility, cross-incompatibility, and blooming time of a given parent are of paramount importance. Recent advances in the detection of S-alleles using molecular methods (PCR based) made it possible to genotype and group parent materials for this purpose in stone fruits (Egea and Burgos 1996; Schuster et al. 2007).

1.11 Crossing and Pollination

The essential operations to develop breeding populations include (1) pollen collection, (2) emasculation of the flowers, (3) pollination, (4) bagging or protecting flowers from foreign pollen, (5) protecting fruit that has set, and (6) growing the seedlings for testing and study of the progeny. For commercial breeding purposes, the size of progeny for selecting new cultivars depends on the prepotency of the parents, goals sought, and commercial cultivars already available (Fogle 1974). Thus, an acceptable progeny size with a high probability to yield a new cultivar can vary from a few hundred to a few thousand seedlings.

Pollen is typically obtained from completely mature flowers that are not fully dehisced ("balloon stage"). The flowers are generally gathered in paper bags; within a few hours of collection, the anthers are isolated by rubbing them over a 4–6 mm mesh sheet. The anthers are most frequently sieved onto absorbent drying paper and permitted to open at ambient room temperature for 12–24 h. The pollen is usually stored in glass shell vials after drying and can be kept for a season at ambient temperature. The pollen is usually stored at −18 °C for longer storage periods (Griggs 1953). Frozen pollen in liquid nitrogen can preserve its longevity for several years. The structure of the flowers of the stone fruits permits easy and rapid emasculation.

The stamens and the single pistil are enclosed under the folded petals (Fig. 1.6). For emasculating, the calyx cup is gently cut with the nails of the thumb and first finger; thus, the entire corolla with its three rows of attached stamens may be lifted from the flower, leaving the pistil undisturbed.

Emasculation happens as the flowers reach anthesis and are not yet open or releasing pollen (balloon stage). To prevent unintended wind pollination, the branches are emasculated from the top to down and should be inspected every other day for 7–10 days after pollination to eliminate any fresh flowers. With varieties of plums, apricots, and cherries that are self-unfruitful or varieties of peach that are pollen-sterile, emasculation is unnecessary in ordinary hybridization. A quick brush of the surface of stigma is all that is required. The entire trees or branches, single shoots, or even individual blossoms that have been pollinated with a single pollen variety should be carefully labeled with full data on a label or tag that will remain until the fruit harvesting.

A 70% alcohol is used after pollination to remove any pollen remaining on the applicator. As a whole, flowers with fewer petals are not visited by pollinators in particular honey bee, such that branches are not isolated for cultivar development crosses. However, in crosses performing for genetic studies, accidental pollination is avoided by isolating the branches with cheesecloth or paper bags. The paper sacs need to be covered with polyethylene sacks, well-aired by making holes in them if rainy weather is anticipated. Plastic houses or parachute coverings fitted with source of heat should be useful for prevent frosting during and after pollination (Werner and Cain 1985).

After fertilization, while the style begins to darken and wither, the protecting paper bag needs to be replaced with an open-mesh bag of heavy net or coarse

Fig. 1.6 Cross section of peach (left) and sweet cherry (right) flower presenting the corolla, ovary, stigma, and anthers

cheesecloth to protect the developing fruit. If the fruit drops off at maturity, it will be held in the bag, preventing the mixture of fruits from different crosses.

1.12 Seed Handling, Germination, and Raising seedling population

There are several differences in the details of *Prunus* seed handling (Grisel 1974), and usually the methods are changed to match the individual program and its resources. The stones (seeds) are removed from the harvested fruit and are allowed to dry in a place free from any pathogen contamination. These are then disinfested with a washing agent on the top and properly rinsed with distilled water to eliminate any contamination. Seeds of stone fruits require an after-ripening period of 1–3 months (depending on the crop and the variety) at low temperatures before they will grow. They are usually soaked for several hours and then placed in moist sand to be held for 1–3 months in a refrigerator or cold storage at about 2–4 ° C or kept out of doors during the winter (Hartmann et al. 2010). Although the seed coats can be removed to facilitate the germination, the more common method of growing the seed is to plant the intact seed directly in pots in the greenhouse or in the nursery (Chao and Walker 1966). A great obstacle in stone fruit breeding is the difficulty in getting the seeds of some crosses (such as early ripening varieties) to resume growth. Although many of such varieties possess desirable characteristics, but they cannot be used as female parents, except in the condition that facilities for embryo rescue is available (Byrne et al. 2012).

Late in the dormant season, seedlings are allocated to be transplanted in the trial orchard at density varying from 1000 (2 m × 5 m) to 33,000 (0.3 m × 1 m) plants per hectare, depending on the fruit crop or evaluation policy (Fig. 1.7). High-density

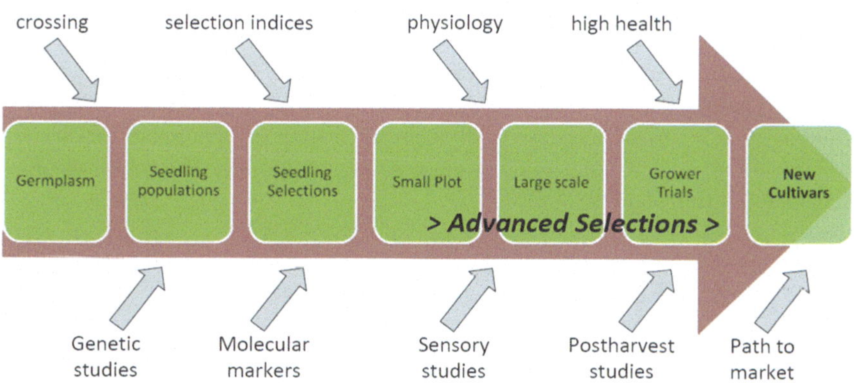

Fig. 1.7 Breeding pipeline presenting different stages of a breeding program for utilization of valuable germplasm and cycling it for new cultivar development as quickly as possible. Consider the order of events (the main arrow) and proper time for the necessary evaluations (outer small arrows). (Adapted from fruit breeding programs of Plant and Food Research Institute, New Zealand as presented by Wendy Cashmore 2010)

planting enables far more seedlings to be assessed in limit areas of field land, even in the same year as hybridization (Sherman et al. 1973), but only the most readily evaluated features such as chilling requirements, the term of fruit development, and fruit quality can be tested effectively (Rodriguez et al. 1994).

1.13 Evaluation

1.13.1 Preselection

Stone fruit crops arc perennial trees with quite long juvenile phase that takes a long time to evaluate the breeding progenies, and also it requires more land for raising an optimum progeny population. Preselection is an efficient approach to reduce the population size in early stages, usually before transplanting to the open field. Easily recognized seedling characters showing strong positive/negative correlation with fruit character in the mature tree would be of considerable value, as they enable the breeders to discard the seedlings with undesirable fruit characters.

In peach, a branching index has been developed for early selection of genotypes with reduced branching (Omar et al. 2010). Correlation between fruit flesh color and foliage color (Connors 1920), pollen sterility and late blooming (Connors 1922), glandular foliage and resistance to leaf curl (Ackerman 1953), and early fruit maturity with solid red or variegated red pattern of leaves in late fall (Sherman et al. 1972) are also reported in peach. In cherry, albinism has been correlated to the isozyme locus Gpi-2, which codes for glucose phosphate isomerase 11 (Tobut and Nicoll 1992). Hartmann and Engelhorn (1992) reported the possibility of screening for precocity of plum, after the second or, better, after the third year by selection of vigorous seedlings with large leaves and many thorns. These characteristics also were positively correlated with fruit size and yield.

Molecular markers linked to interest traits seem to be very promising for this purpose (Bus et al. 2009; Peace 2017). MAS will improve the selection efficiency in traditional fruit breeding, in particular for economically important characteristics that are challenging to identify by phenotype early in the plant life cycle (Laurens et al. 2018). MAS is especially helpful when the evaluation of the character is expensive or time-consuming or the gene expression is recessively controlled (Scorza 2000; Luby and Shaw 2001).

1.13.2 Primary Selection

Selection typically takes place in the first successful crop year, which ranges from the second to the fourth year and even from the fifth year of planting depending on the crop and field management strategy (Fig. 1.7). Normally, 1-year-study is adequate to test most progeny, provided that phenotypes represent a good estimation from the genotype (Hansche et al. 1975). Typically, when choosing new cultivars, only the main characteristics are taken into account (such as leaf and flower traits, bloom and ripening date, fruit type, and yield potential). The rest are easily tossed out without the details being collected. Given the large number of seedlings participating in each of the breeding projects nowadays, data is gathered immediately in a digital template for time-saving to prevent transcription errors.

After a seedling has been selected for further evaluation, the sanitary condition should be tested for the exclusion of intracellular pathogens (e.g., mycoplasmas) and viruses (in particular plum pox virus [PPV]) that may limit efficiency and fruit quality as well as prevent its introduction into the nursery. Diagnostic tests including ELISA, indicator host plants, and PCR-based techniques are available for this purpose (Cambra et al. 2011). When the chosen seedling is shown to be virus-free, any other replications should be founded in an insect-proof screen house to use as the source of clean plant material for later testing and propagation for future release.

1.13.3 Advanced Selection

The advanced selections should be subjected to a check protocol in contrast to commercially established cultivars and other concurrent selections (e.g., from other national or even international breeding programs) based on fruit type and the ripening time. In many breeding projects, this is conducted in cooperation with commercial growers. Of this purpose, trees should be grown on some defined or common rootstocks and in several environments to gather further data with preference to the probable release of a new cultivar (Byrne et al. 2012). Sound statistical plan with replications should be designed to gather objective data to prevent possible biases that resulted from intellectual valuation of the breeder. Plots of 6–8 trees are adequate for yield reports, and 15–30 fruits per tree are appropriate for quality evaluation (Scorza and Sherman 1996). These experimental plots entail at least two or three fruiting seasons of data to ensure a good decision on its commercial potential (Fig. 1.7).

1.13.4 Final Selection

The market value of a possible new cultivar pertains largely to the approval of the farmers and the retail distribution chain. Testing of advanced selections with a commercial grower is incredibly necessary to completely understand the efficiency of a given clone. Thus, the superior candidates from the second stage of the study are then accepted into the final stage of the assessment, i.e., the growers' admission trial. Many growers are willing to try promising selections even at no expense for the breeder. To prevent unintentional or illegitimate propagation of the advanced selections, the experiments are performed under a non-propagation agreement (Byrne et al. 2012). These final trials are typically conducted without a statistically based plan, which yield useful knowledge a breeder has no means to obtain from his structured studies, i.e., the selection's success in different soils and under varied management as well as its fruiting and postharvest behavior under a commercial harvest process. In this final trial, an additional 3–5 fruiting season is necessary to provide enough assurance for introducing an advanced selection as a new cultivar (Fig. 1.7).

In some cases, the trial grower might prefer to topwork the experimental advanced selections into trees in an established orchard of the competitive cultivar. This will provide the opportunity to compare different attributes of both the new advanced selections and the competitive cultivar such as growth habit, blooming seasons, fruit maturity, and fall senescence. The more common practice is to establish new orchards for trialed varieties. Doing so, evaluating can be performed on genotypes grafted on the identical rootstock. Regardless of the composition of the experimental place, it must be large enough, and it is essential for self-incompatible varieties to blossom concurrently with other varieties (Moore and Janick 1983).

1.14 Cultivar Release and Commercialization

The creation of a new sound cultivar is very expensive and time-consuming; thus, to return the investment, legal varietal protection is necessary. In the past, cultivar protection was sought mainly by private breeders, but also cultivars obtained from public projects are today being preserved. The requirements and procedures for protection and patenting may differ from one country to another. In the European Union, the Community Plant Variety Office (CPVO) covering the 27 member states (http://www.cpvo.europa.eu) manages a uniform system of plant cultivar rights, taking the necessary measures for technical examination of the candidate cultivar. The purpose of this is to confirm that the cultivar is distinct from others, uniform in its characteristics, and stable in the long term (DUS). In the USA, patenting a cultivar is quite similar to patenting an industrial process. For a cultivar to be patented in the USA, it must be original and healthy (virus-free) and legally protected for 20 years and comprises its phenotype only, even the fruit. In Europe, a new cultivar gets a license for a term of 30 years that has about the same meaning as a patent in the USA. Integration of trade labels into the plant conservation and commercialization policy is among the newest movements in intellectual property of the new cultivars (Clark et al. 2012).

1.15 Mutation Breeding

The term mutation breeding applies to the intentional induction and formation of mutant lines for crop improvement. However, this concept has often been used in a broader context to describe the utilization of both natural and engineered mutants. Mutation breeding includes the development of new varieties through producing and using genetic variation by chemical and physical mutagenesis. At the DNA level, mutations are either DNA base pair changes (point mutation) or small to large DNA fragment insertions, deletions, and translocation. At the phenotypic level, mutations can result in a wide spectrum of morphologies (Forstera and Shub 2011).

Since it takes several years for seedlings to produce flowers in stone fruits and in many cases the flowers often exhibit self-incompatibility, which restricts pure line selection, i.e., sweet cherry and Japanese plum cultivars, mutation breeding has an advantage over hybridization in these species.

Multiple forms of plant propagules, such as seed, bud, shoot tip, leaf, ovules, and more recently single cells and other forms of in vitro cultured plant tissues (Predieri and Gatti 2000), have been used for mutation treatment. Selection of a particular targeted variety should be done according to the needs of consumers, grower, and processors. Physical (X-ray or gamma-ray irradiation) and chemical (such as ethyl methanesulfonate) mutagen can be used to induce mutation, although the frequency and types of mutations are direct effects of the dose and degree of exposure of the mutagen rather than its type (Mba et al. 2012). Deployment of efficient and affordable selection methods is of paramount importance in designing an effective mutation breeding program (Predieri and Gatti 2000). Following selection, it is imperative to confirm if the selected character is inheritable. Since many of the stone fruits are strongly heterozygous, the principal mechanism of induced mutation is assumed to be the detection of recessive traits by knocking out the dominant allele at a heterozygous locus of the wild type. The heritability of a newly discovered mutation must be evaluated in multilocation experiments over several years (Shua et al. 2011).

Natural and induced mutations have produced valuable stone fruit cultivars. Among the bud sports, the most common and wide spread mutations in stone fruits are color alterations (Moore and Janick 1983; Walker et al. 2006). The induced mutation has also been used to introduce many beneficial properties influencing plant growth, flowering period, fruit ripening, fruit color, self-compatibility, self-thinning, and tolerance to pathogens (Janick and Moore 1996). Gamma irradiation has been successfully used in developing plum mutants (Predrieri and Gatti 2000) and late-ripening mutants of 'Fairhaven' and 'Elberta' peaches (Scorza and Sherman 1996). 'Galaxy' is an irradiation-induced spur-type mutant of 'Montmorency' cherry (Okie and Weinberger 1996). Induced mutations have also produced self-fertile clones and compact tree types in cherry (Brown et al. 1996). Thermal neutron irradiation was used to create the early ripening apricot mutant cultivar 'Early Blenheim' (Layne et al. 1996).

Similar to many modern breeding systems, mutation breeding often uses genomic advancements to promote the development of the desirable lines through genotyping rather than phenotyping. Molecular mutation breeding has the potential to improve the power and efficiency of plant breeding comparison to conventional methods (Forstera and Shub 2011).

1.16 Application of Cell and Tissue Culture in Stone Fruit Breeding

Cell and tissue culture are primarily used for mass propagation of *Prunus* spp. cultivars/rootstock and recently also employed for some breeding purposes.

In *Prunus* species, the crop breeding cycle is too long and can take up to 15 years. By combining micropropagation and induced mutation protocols in vitro, breeders try to improve the recovery of the somaclones (Predrieri 2001; Afrasiabi and Iqbal 2010). In vitro screening has been used to select somaclonal variants of peach with resistance to bacterial diseases; however, the frequency of somaclonal variation is proved to be genotype-dependent (Hammerschlag and Ognjanov 1990).

Embryo rescue technique has been used to germinate seed from early ripening genotypes, particularly in peach and nectarine, where the flesh matures before the embryo is fully developed. The embryo is then excised from the seed and cultured on 0.6–0.7% agar containing 2–4% sucrose and nutrients (Ramming 1990). Recent developments in breeding for early ripening apricots have contributed to a substantial decline in the viability of seed from very early ripening cultivars. The in vitro embryo culture is the answer to this issue (Burgos and Ledbetter 1993).

Haploid and double haploid production enable breeders to develop the homozygous clones from the heterozygous parent increasing the speed and efficiency of breeding in perennial crops such as stone fruit species (Germana 2006). Successful production of F1 hybrids of double haploid has been reported in peach (Scorza and Pooler 1999).

Somatic hybridization technology has been successfully applied in some stone fruit trees. Somatic hybrids between two sexually incompatible rootstocks including wild pear (*Pyrus communis* var. *pyraster* L.) and 'Colt' cherry (*P. avium* × *pseudocerasus*) have been reported (Ochatt et al. 1989). Mesophyll protoplasts of wild pear and protoplasts of 'Colt' cherry derived from suspension cultures were electroporated as separate populations and fused chemically. This strategy resulted in a unique heterokaryon and was the basis for selection of hybrids. All somatic hybrid plants had 58 chromosomes, equal to the overall complement of somatic chromosome numbers of the parents. The hybrid plants were intermediate for most morphological characters compared to the parents. Successful production of protoplast-derived plants of two forms of prune/plum, the ornamental species *P. spinosa* and the fruit-bearing and rootstock genotype *P. cerasifera*, Myrobalan, has also been reported (Ochatt 1992).

Stone fruit germplasm are conserved in field collections, in which plant materials can be easily accessed and evaluated. This type of gene banks is very costly to establish and maintain and also require considerable inputs such as land, labor, and facilities to ensure sufficient diversity (Engelmann and Engels 2002; Hofer and Hanke 2017). Cryopreservation techniques with or without cryoprotectants and vitrification agents proved to be efficient in long-term storage and maintenance of stone fruit genetic resources. Dormant buds, shoot tips, and zygotic embryos derived from in vitro culture as well as embryogenic cultures have been used for this purpose (Michalak et al. 2015; Hofer and Hanke 2017). Shoot tips of *Prunus* hybrids have been cryopreserved using a two-step procedure, slow cooling prior to storage in liquid nitrogen, with recovery rate of 74% after thawing (De Boucaud et al. 2002). A procedure without cryoprotectants has been used for dormant vegetative buds of sour cherry. The shoots cooled at 1 °C/h to −30 °C before immersion in −160 °C. Approximately 40% of five sour cherry scions survived as evaluated by chip budding to rootstocks (Towill and Forsline 1999). Shoot-tip vitrification and encapsulation-based procedures have been developed for plum cryopreservation (De Carlo et al. 2000).

1.17 Genetic Transformation

Conventional breeding of stone fruit species is hindered by limitations imposed by their large tree size, long juvenility, high heterozygosity, and self-incompatibility. Genetic transformation is a promising tool that enables the breeder to reduce the time and spaces required for the improvement of these fruit crops. Genetic engineering is key to the improvement of the existing cultivars with traits that cannot be readily found in the existing germplasm (Martínez-Gómez et al. 2005). The major barrier to the genetic transition of *Prunus* species is the regeneration of transformed tissues/plantlets. Reliable protocols for transformation, selection, and regeneration of transgenic plantlets are therefore necessary for the use of genetic engineering techniques for improvement of stone fruit cultivars (Byrne et al. 2012).

Agrobacterium-mediated transformation is the common method for the transformation of *Prunus* species; however, there are a limited number of reports (Ye et al. 1994) on microprojectile bombardment-based protocols. Poor regeneration efficiency is an obstacle for *Prunus* spp., which could be influenced by transformation method and environment, agrobacterium strain, and antibiotic concentration used to select the transforms (Scorza et al. 1994).

Major objectives of *Prunus* genetic transformation include alteration of growth habit, resistance to diseases and pests, tolerance of abiotic stress, and regulation of fruit ripening (Callahan et al. 1991; Scorza and Hammerschlag 1992).

Cytokinin-overexpressing transgenic 'Redhaven' peach have been regenerated using *Agrobacterium*-based protocol, which are dwarf in stature, produce more branches, and have delayed senescence of leaves (Hammerschlag and Smigocki 1998). Cherry rootstock 'Colt' was transformed by the introduction of tumor-inducing (T)-DNA from *Agrobacterium rhizogenes* and produced trees with reduced plant height and internode length (Gutiérrez-Pesce et al. 1998). Transgenic sour cherry lines have been produced that contain the antifreeze protein gene to reduce ice-crystal formation at freezing temperatures (Dolgov 1999). Apricot and European plum were transformed with virus coat protein genes for tolerance to PPV disease (Rugini and Gutierrez-Pesce 1999; Scorza et al. 1994). The development of PPV-resistant plums could serve as a model to study virus resistance strategies in transgenic fruit trees and presents the potential benefit of transformation technology (Ravelonandro et al. 2000).

Currently, the approval of genetically modified fruit cultivars by the consumer is relatively slight, especially in the European Union. These barriers may be circumvented by new technologies. A variety of stone fruit breeding projects rely on cisgenesis, described as the genetic alteration of plants injecting genes from the gene pool of the plant itself (including crossable relatives), which is a focus of several stone fruit breeding programs. Others often use marker-free methods where recombinase, which is chemically inductive, can excrete marker genes and generate plants that cannot easily be identified as transgenic. These plants can be designed to produce an optimal transgenic structure targeting a feature of interest but without an abundance of unacceptable molecular luggage (Nocker and Gardiner 2014; Terns and Terns 2014; Limera et al. 2017).

1.18 Application of Biotechnology in Stone Fruit Breeding

The most important goals for genetic improvement of stone fruits (*Prunus* spp.) include reduced cost of production by increasing productivity, low-priced, high-quality fruits with long storage life, and reduced pest and disease damage. Cold hardiness, late blooming to escape spring frosts, extension of maturity season, greater diversity of fruit types, modified tree architecture, and improved fruit storage life are also important (Callahan et al. 1991). To achieve these objectives, the integration of the new genomics and molecular tools into the breeding programs will be important (Scorza 2000). These modern methods improve the efficiency of breeding systems by detecting essential genes at the molecular stage.

The existence of whole-genome sequences and expressed sequence tag (EST) libraries for essential crops accelerates the gene discovery process. A worldwide consortium of laboratories as well as international collaborations has attempted to build genomic resources for peach to be used as a model for the identification, cloning, and characterization of genes in *Prunus*. The first and second versions of the integrated peach genome sequence newly published, *peach v1.0* (http://www.rosaceae.org/peach/genome) and *peach v2.0* (http://www.rosaceae.org/species/prunus_peesica/genome_v2.0.al), together with the already accessible *Prunus* Genome (http://www.rosaceae.org/) and ESTree databases (http://www.itb.cnr.it/estree/index.php), enable entry to genomic data for all of *Prunus* species, in particular peach, and represent a very valuable source of knowledge for genome comparative studies and detection of essential genes (Byrne 2012).

A major project called "FruitBreedomics" has recently been financed in collaboration between many institutions and countries in the European Union, with the aim of combining breeding with genomics (Laurens et al. 2018). The RosBREED project in the USA (http://www.rosbreed.org) seeks quite the same goal, providing MAS in the Rosaceae family (Peace 2017). In the USA, the Genome Database for Rosaceae (GRD), including stone fruit genomics and genetics data (http://www.bioinfo.wsu.edu/gdr/), is a unit web database for Rosaceae. The establishment of GDR was an important step toward integrating all Rosaceae, including *Prunus*, structural and functional genomics initiatives (Jung et al. 2008).

Molecular markers are valuable tools to locate the economically important *Prunus* genes using linkage maps. Undesirable plants can be eliminated from progeny populations using MAS as early as the seedling stage (Scorza 2000). Several molecular markers such as randomly amplified polymorphic DNA (RAPD), amplified fragment length polymorphisms (AFLPs) and simple sequence repeat (SSR), and more recently single nucleotide polymorphism (SNP) are widely used to characterize genetic resources, identify cultivars, study traits, and generate linkage maps in stone fruit crops (Bus et al. 2009). Molecular markers associated with features of interest are important for marker-assisted selection (MAS) to increase selection efficiency in typical fruit breeding methods, particularly for economic features that are so difficult to identify them early in the plant life cycle by phenotype. MAS is especially profitable when the evaluation of the character is expensive or time-consuming, and the gene expression is recessively controlled (Scorza 2000; Luby and Shaw 2001). The advent of genomics and the continually advancing in technologies of DNA sequencing and whole-genome sequences have made new possibilities for the development of new markers and the identification and the comprehension of the functional genes which regulate essential phenotypes in fruit breeding (Laurens et al. 2018).

Both intra- and interspecific genetic association maps are present in almost all stone fruits and were crucial in recognizing and selecting the target genes or markers associated with them (Shulaev et al. 2008). Although various linkage maps of *Prunus* species have been released, the interspecific linkage map produced from an interspecific cross between almond ('Texas') and peach ('Earlygold') is the most accurate of all such linkage maps (Dirlewanger et al. 2004b). Several associations have been identified between molecular markers and simply inherited traits (Fig. 1.8)

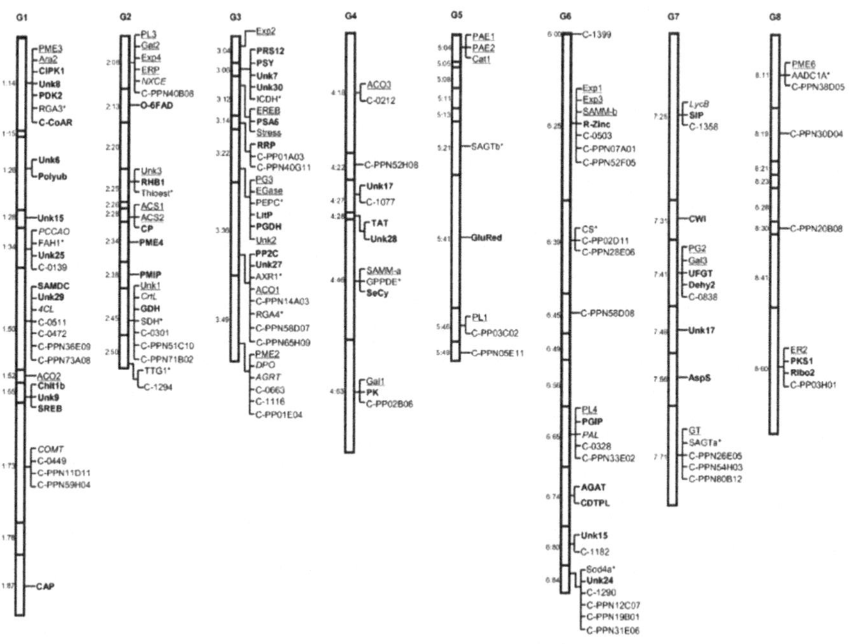

Fig. 1.8 Candidate genes (CGs) and novel *Prunus* EST-SSR markers bin-mapped to the *Prunus* T × E reference map. Chilling injury (CI) resistance CGs are in bold fonts, texture CGs are underlined, CGs related to fruit pigmentation are italicized, other CGs are asterisked, and new *Prunus* EST-SSRs are in normal font. (Adapted from Ogundiwin et al. 2009)

such as compact growth, bud dormancy, fruit quality, flesh color, and pathogen resistance (Byrne et al. 2012). Linkage relationships between molecular markers and quantitative trait locus (QTL) analyses have been reported for fruit quality, powdery mildew resistance, internode length, blooming time, ripening time, skin color, soluble solid content, and disease resistance (Cao et al. 2016; Laurens et al. 2018).

A resistance map for *Prunus* using a candidate gene approach with more than 40 RGAs (resistance gene analogs) and RAGs (resistance-associated genes) is mapped in regions reported to contain resistance to plum pox virus, powdery mildew, and parasitic nematodes (Lalli et al. 2005; Abbott et al. 2007). More recently, gene mapping, using crossed populations from known resistance source, and high-throughput analysis by microarray profiling and RNA-seq have employed to determine genes associated with pest and disease resistance of stone fruits (Cao et al. 2016; Gao et al. 2016).

These advances allowed identification of regions linked to the traits of interest; however, the process needs phenotyping of large families and development of the markers that cover whole genome, a costly and time-consuming procedure that addressed only a few traits, until now. Furthermore, all of the agronomically significant characters are inherited quantitatively, and while numerous QTLs have been detected in *Prunus*, more efforts are required until QTL-associated markers can be systematically incorporated into selection programs. Despite the growing accessibility of genomic resources in *Prunus* plant species, the presence of a highly saturated reference map in many of the stone fruit crops, and most of the basic traits being adequately identified for molecular selection, the usage of molecular markers for commercial breeding is still in its infancy, and only a few molecular markers are used for marker-assisted selection in stone fruit breeding (Peace 2017; Laurens et al. 2018).

1.19 Fast Breeding

Stone fruits typically have long breeding cycles; thus, developing a new cultivar through traditional breeding will take several breeding cycles and dozens of years to complete. However, latest progresses in genomics and biotechnology have the ability to greatly revolutionize and improve cultivar development in these crops. Such methods include genomics-assisted detection of useful germplasm and cycling across generations as rapidly as possible (Scorza et al. 2014).

Embryo rescue or chemical treatment of seed to break seed dormancy provides a solution to flowering. Maintaining seedlings under optimum growth conditions in the greenhouse is another way to greatly minimize the juvenile phase. Biotechnological engineering of endogenous, genetic flowering networks is very promising in reducing the duration of the breeding period (Nocker and Gardiner 2014).

Genome-wide selection (GWS) is a rapidly developing approach, which uses the genomic estimated breeding values (GEBVs) as selection parameters, instead of traditionally using the estimated breeding values (EBVs) by fruit breeders (Kumar et al. 2012a; Limera et al. 2017). GEBVs of seedlings in breeders' selection population are determined on the basis of genetic marker information in order to recognize remarkable "elite" individuals that would be used to advance generations or assessed in the field as prospective cultivars. Marker-assisted selection (MAS) for attributes regulated by major genes or quantitative trait loci is now widely used in certain outbreeding fruit crops (Bus et al. 2009). As GWS is quite expensive, it is recommended to use a fairly inexpensive MAS prescreen with a few (less than 10) markers and only those seedlings able to pass through this filter to be exposed to a far more costly screen with a few thousand genetic markers needed to allow GWS to be applied (Nocker and Gardiner 2014; Peace 2017). Fast breeding by incorporating the juvenility abbreviation and GWS has the potential to shorten the primary selection phase into only 2 years, instead of 7 years in standard breeding programs. Thanks to the high accuracy of GWS proved in the initial studies (Kumar et al. 2012b), pollen from several of these elites could be used to advance to the next generation, prior to phenotyping, thus reducing the length of each cycle of breeding.

Recent advances in gene manipulation, including cis-genesis, marker-free transformation, and targeted genome engineering (such as the CRISPR/Cas system), have also recently emerged as powerful tools in the context of fast fruit breeding (Terns and Terns 2014; Limera et al. 2017). Despite some challenges, these tools have a broad potential relative to conventional fruit breeding and attract the attention of governments, entrepreneurs, and citizens due to the revolutionary effect they may have on fruit breeding (Dalla Costa et al. 2017).

1.20 Conclusion

Although stone fruit breeding has been major efforts since early in the twentieth century and has shown major progresses in the second half of the twentieth century, the results have been uneven and differ from less effectual (in apricot) to extraordinarily successful (in peach and nectarine).

Improved legislation for plant protection throughout the world has stimulated the fruit breeding to shift from public into the private sectors. Efficiency and fundamental quality attributes have often been significant breeding objective of stone fruits and will be in the future. In addition, some major trends including increased tolerance to biotic and abiotic stress, simplified orchard practices, extension of the adaptation zones, extending the harvest window, new fruit types, superfruits or cultivars with improved nutritional benefits, eating convenience, and constantly high quality are needs to be considered in developing new cultivars.

Breeding of fruit crops has entered a new era, in which using DNA information to support breeding decisions and operations is facilitated. This approach includes genomics-assisted identification of valuable germplasm and cycling it through generations as quickly as possible. In this regard, the availability of genomic resources is growing, highly saturated reference map is developed, and most of the simple traits are sufficiently marked for molecular selection in many of stone fruit crops.

Advances in gene manipulation, including cis-genesis, marker-free transformation, and targeted genome engineering (such as the CRISPR/Cas system), have also recently emerged as powerful tools in the context of fast fruit breeding. Successful plant breeding at this sophisticated era needs multidisciplinary teams, in which breeding is the central hub, but there are also well-structured divisions of expertise and labor. In this regard, understanding how to integrate molecular genetics with traditional breeding is key, not the individual components.

Despite significant progresses, the use of genomics and biotechnological tools for commercial breeding is still in its infancy, and further advances and efforts need to sustain the involvement of these tools in the stone fruit improvement activity in the future.

References

Abbott, A., Arús, P., & Scorza, R. (2007). Peach. In C. Kole (Ed.), *Genome mapping and molecular breeding in plants, fruits and nuts* (pp. 137–156). Heidelberg: Springer.

Ackerman, W. L. (1953). *The evaluation of peach leaf curl in foreign and domestic peaches and nectarines grown at the U.S. Plant Introduction Garden, Chico, California* (pp. 1–31). Div. Plant Exploration and Introduction. Bur. Plant Ind. Soils and Agric. Eng., USDA, Mimeo.

Afrasiabi, H., & Iqbal, J. (2010). In vitro techniques and mutagenesis for the genetic improvement of potato cvs. desiree and diamant. *Pakistan Journal of Botany, 42*(3), 1629–1637.

Ashton, R. W. (2008). *Plums of north America*. Temple: Third Millennium Publishing.

Asma, B. M., & Ozturk, K. (2005). Analysis of morphological, pomological and yield characteristics of some apricot germplasm in Turkey. *Genetic Resources and Crop Evolution, 52*, 305–313.

Badenes, M. L., & Llácer, G. (2006). Breeding for resistance: Breeding for Plum pox virus resistant apricots in Spain. *Bulletin OEPP/EPPO Bulletin, 36*, 323–326.

Badenes, M. L., Martínez-Calvo, J., & Llácer, G. (1998). Analysis of apricot germplasm from the European ecogeographical group. *Euphytica, 102*, 93–99.

Bailey, L. H. (1916). *Prunus. The standard cyclopedia of horticulture* (Vol. V, pp. 2822–2845). Harrisburg: J. Horace McFarland/P–R. Mount Pleasant Press.

Bargioni, G. (1970). 'Vittoria', a new sweet cherry cultivar (in Italian). *Revistadella Otro. Italiana, 63*, 3–12.

Bassi, D., & Audergon, J. M. (2006). Apricot breeding: Update and perspectives. *Acta Horticulturae, 701*, 279–294.

Bassi, D., & Pirazzoli, C. (1998). The stone fruit industry in the Mediterranean region: agronomical and commercial overview. In B. Di Terlizzi, A. Myrta, & V. Savino (Eds.), *Options Méditerranéennes, Série B/n°19, Stone fruit viruses and certification in the Mediterranean: problems and prospects* (pp. 3–38). Bari: CIHEAM.

Bassi, D., Bartolozzi, F., & Muzzi, E. (1996). Patterns and heritability of carboxylic acids and soluble sugars in fruits of apricot (*Prunus armeniaca* L.). *Plant Breeding, 115*(1), 67–70.

Batt, P. J., & Noonan, J. (2009). Global trends in food quality: An exploratory study in fresh produce supply chains. *Acta Horticulturae, 831*, 95–103.

Beckman, T. G., & Pusey, P. L. (2001). Field testing peach rootstocks for resistance to *Armillaria* root rot. *HortScience, 36*, 101–103.

Bellini, E., Natarelli, L., Nencetti, V., Surico, G., & Mugnai, L. (1996). Peach genetic improvement: Results achieved with a breeding programme for resistance to *Taphrina deformans* (BERK. TUL.). *Acta Horticulturae, 374*, 33–37.

Benedikova, D. (2006). Gene pool utilisation in apricot breeding in Slovak Republic. *Acta Horticulturae, 717*, 173–174.

Beppu, K., Yamane, H., Yaegaki, H., Yamaguchi, M., Kataoka, I., & Tao, R. (2002). Diversity of S-RNase genes and S-haplotypes in Japanese plum (*Prunus salicina* Lindl.). *Journal of Horticultural Science and Biotechnology, 77*, 658–664.

Beppu, K., Komatsu, N., Yamane, H., Yaegaki, H., Yamaguchi, M., Tao, R., & Kataoka, I. (2005). Se-haplotype confers self-compatibility in Japanese plum (*Prunus salicina* Lindl.). *Journal of Horticultural Science and Biotechnology, 80*, 760–764.

Beppu, K., Syogase, K., Yamane, H., Tao, R., & Kataoka, I. (2010). Inheritance of self-compatibility conferred by the Se-haplotype of Japanese plum and development of Se-RNase gene specific PCR primers. *Journal of Horticultural Science and Biotechnology, 85*, 215–218.

Bors, R. H. (2005). Dwarf sour cherry breeding at the University of Saskatchewan. *Acta Horticulturae, 667*, 135–140.

Brenton, P., Edwards-Jones, G., & Jensen, M. F. (2009). Carbon labeling and low-income country exports: A review of the development issues. *Development Policy Review, 27*(3), 243–267.

Brown, S. K., Iezzoni, A. F., & Fogle, H. W. (1996). Cherries. In J. Janick & J. N. Moore (Eds.), *Fruit breeding. Volume 1: Tree and tropical fruits*. New York: Wiley.

Burgos, L., & Ledbetter, C. A. (1993). Improved efficiency in apricot breeding—Effects of embryo development and nutrient media on in-vitro germination and seedling establishment. *Plant Cell Tissue and Organ Culture, 35*, 217–222.

Bus, V. G. M., Esmenjaud, D., Buck, E., & Laurens, F. (2009). Application of genetic markers in rosaceous crops. In *Genetics and Genomics of the Rosaceae. Plant genetics and genomics: Crops and models* (Vol. 6, pp. 563–599). New York: Springer.

Byrne, D. H. (2005). Trends in stone fruit cultivar development. *HortTechnology, 15*(3), 494–500.

Byrne, D. H. (2012). Trends in fruit breeding. In M. L. Badenes & D. H. Byrne (Eds.), *Fruit breeding, handbook of plant breeding 8* (pp. 3–36). New York: Springer.

Byrne, D. H., & Raseira, M. C. B. (2006). Inbreeding of the major commercial fresh market peach cultivars grown in Southern Brazil. *Acta Horticulturae, 713*, 99–101.

Byrne, D. H., Ramming, D. W., & Topp, B. (2000a) China germplasm collection trip report, August 7-August 25, 2000.

Byrne, D. H., Sherman, W. B., & Bacon, T. A. (2000b). Stone fruit genetic pool and its exploitation for growing under warm winter conditions. In A. Erez (Ed.), *Temperate fruit crops in warm climates* (pp. 157–230). Boston: Kluwer Academic Publishers.

Byrne, D. H., Raseira, M. B., Bassi, D., Piagnani, M. C., Reighard, G. L., Moreno, M. A., & Pérez, S. (2012). Peach. In M. L. Badenes & D. H. Byrne (Eds.), *Fruit breeding* (Handbook of Plant Breeding) (Vol. 8, pp. 505–569). New York: Springer.

Callahan, A., Scorza, R., Morgens, P., Mante, S., Cordts, J., & Cohen, R. (1991). Breeding for cold hardiness: searching for genes to improve fruit quality in cold hardy peach germplasm. *HortScience, 26*, 522–526.

Cambra, M., Boscia, D., Gil, M., Bertolini, E., & Olmos, A. (2011). Immunology and immunological assays applied to the detection, diagnosis and control of fruit tree viruses. In A. Hadidi, M. Barba, T. Candresse, & W. Jelkmann (Eds.), *Virus and virus-like disease of pome and stone fruits* (pp. 303–313). St. Paul: APS Press.

Cantín, C. M., Gogorcena, Y., & Moreno, M. A. (2010). Phenotypic diversity and relationships of fruit quality traits in peach and nectarine [*Prunus persica* (L.) Batsch] breeding progenies. *Euphytica, 171*, 211–226.

Cao, K., Zhou, Z., Wang, Q., et al. (2016). Genome-wide association study of 12 agronomic traits in peach. *Nature Communications, 7*, 13246. https://doi.org/10.1038/ncomms13246.

Cashmore, W. (2010). *Plant breeding for the future: Fruit breeding for the twenty-first century*. Retrieved from https://www.upov.int/edocs/mdocs/upov/en/upov_sym_ge_11/upov_sym_ge_11_13.pdf

Chao, L., & Walker, D. R. (1966). Effect of temperature, chemicals and seed coat on apricot and peach seed germination and growth. *Proceedings of the American Society for Horticultural Science, 88*, 232–238.

Clark, J. R., Aust, A. B., & Jondle, R. (2012). Intellectual property protection and marketing of new fruit cultivars. In M. L. Badenes & D. H. Byrne (Eds.), *Fruit breeding. Handbook of*

plant breeding (Vol. 8). New York: © Springer Science+Business Media, LLC. https://doi.org/10.1007/978-1-4419-0763-9_3.
Claverie, M., Bosselut, N., Lecouls, A. C., Voisin, R., Lafargue, B., Poizat, C., Kleinhentz, M., Laigret, F., Dirlewanger, E., & Esmenjaud, D. (2004). Location of independent root-knot nematode resistance genes in plum and peach. *Theoretical and Applied Genetics, 108*(4), 765–773.
Cociu, V. (2006). 50 years of apricot breeding in Romania. *Acta Horticulturae, 701*, 355–358.
Coneva, E. (2003). New apricot germplasm selected by ten characteristics. *Acta Horticulturae, 622*, 465–472.
Connors, C. H. (1920). Some notes on the inheritance of unit characters in the peach. *Proceedings of the American Society for Horticultural Science, 16*, 24–36.
Connors, C. H. (1922). Inheritance of foliar glands of the peach. *Proceedings of the American Society for Horticultural Science, 18*, 20–26.
Couranjou, J. (1995). Genetic-studies of 11 quantitative characters in apricot. *Scientia Horticulturae, 61*, 61–75.
Crane, M. B., & Lawrence, W. J. C. (1956). *The genetics of garden plants* (4th ed.). London: Macmillan.
Culver, D. J., Ramming, D. W., & McKenry, M. V. (1989). Procedures for field and greenhouse screening of *Prunus* genotypes for resistance and tolerance to root-lesion nematode. *Journal of the American Society for Horticultural Science, 114*, 30–35.
Dalla Costa, L., Malnoy, M., & Gribaudo, I. (2017). Breeding next generation tree fruits: Technical and legal challenges. *Horticulture Research, 4*, 17067. https://doi.org/10.1038/hortres.2017.67.
De Boucaud, M. T., Brison, M., Helliot, B., & Paulus, V. H. (2002). Cryopreservation of prunus. In L. E. Towill & Y. P. S. Bajaj (Eds.), *Cryopreservation of plant germplasm II—Biotechnology in agriculture and forestry* (Vol. 50, pp. 287–311). Berlin: Springer-Verlag.
De Carlo, A., Benelli, C., & Lambardi, M. (2000). Development of a shoot-tip vitrification protocol and comparison with encapsulation-based procedures for plum (*Prunus domestica* L.) cryopreservation. *CryoLetters, 21*, 215–222.
Della Strada, G., Fideghelli, C., & Grassi, F. (1996). Peach and nectarine cultivars introduced in the world from 1980 to 1992. *Acta Horticulturae, 374*, 43–51.
Dirlewanger, E., Cosson, P., Howad, W., Capdeville, G., Bosselut, N., Claverie, M., Voisin, C., Pozat, R., Lafargue, B., Baron, O., Laigret, F., Kleinhentz, M., Arús, P., & Esmenjaud, D. (2004a). Microsatellite genetic linkage maps of myrobalan plum and an almond-peach hybrid location of root-knot nematode resistance genes. *Theoretical and Applied Genetics, 109*, 827–838.
Dirlewanger, E., Graziano, E., Joobeur, T., Garriga-Caldere, F., Cosson, P., Howard, W., & Arús, P. (2004b). Comparative mapping and marker assisted selection in Rosaceae fruit crops. *Proceedings of the National Academy of Sciences, 101*, 9891–9896.
Dirlewanger, E., Quero-Garcia, J., Le Dantec, L., Lambert, P., Ruiz, D., Dondini, L., et al. (2012). Comparison of the genetic determinism of two key phenological traits, flowering and maturity dates, in three *Prunus* species: Peach, apricot and sweet cherry. *Heredity, 109*(5), 280–292.
Dolgov, S. V. (1999). Genetic transformation of sour cherry (*Cerasus vulgaris* Mill.). In Y. P. S. Bajaj (Ed.), *Biotechnology in agriculture and forestry, transgenic trees* (Vol. 44, pp. 29–38). Berlin: Springer Verlag.
Dosba, F. (2003). Progress and prospects in stone fruit breeding. In *XXVI International Horticultural Congress: Genetics and Breeding of Tree Fruits and Nuts* (Vol. 622, pp. 35–43).
Edin, M., & Garcin, A. (1994). Un nouveau porte-greffe du pêcher Cadaman®-Avimag. *L'Arboriculture Fruitière, 475*, 20–23.
Egea, J., & Burgos, L. (1996). Detecting cross-incompatibility of three North American apricot cultivars and establishing the first incompatibility group in apricot. *Journal of the American Society for Horticultural Science, 121*, 1002–1005.
Engelmann, F., & Engels, J. M. M. (2002). Technologies and strategies for *ex situ* conservation. In J. M. M. Engels, V. R. Rao, A. H. D. Brown, & M. T. Jackson (Eds.), *Managing plant genetic diversity* (pp. 89–104). Wallingford/Rome: CABI/IPGRI.

Ercisli, S. (2004). A short review of the fruit germplasm resources of Turkey. *Genetic Resources and Crop Evolution, 51*, 419–435.

Erturk, Y., Ercisli, S., Maghradze, D., Orhan, E., & Agar, G. (2009). An assessment of genetic variability and relationships among wild-grown blackthorn (*Prunus spinosa* L.) plants based on RAPD markers. *Genetics and Molecular Research, 8*, 1238–1244.

Faust, M., & Suranyi, D. (1997). Origin and dissemination of cherry. *Horticulture Review, 19*, 263–317.

Faust, M., & Timon, B. (1995). Origin and dissemination of the peach. *Horticulture Review, 17*, 331–379.

Faust, M., Surányi, D., & Nyujtó, F. (1998). Origin and dissemination of apricot. In J. Janick (Ed.), *Horticultural reviews* (Vol. 22, pp. 225–266). New York: Wiley.

Felipe, A. (2009). 'Felinem', 'Garnem', and 'Monegro' almond x peach hybrid rootstocks. *HortScience, 44*, 196–197.

Fogle, H. W. (1974). Evaluating combining ability in peach and nectarine. *HortScience, 9*, 334–335.

Forstera, B. P., & Shub, Q. Y. (2011). Plant mutagenesis in crop improvement: Basic terms and applications. In Q. Y. Shu, B. P. Forster, & H. Nakagawa (Eds.), *Plant mutation breeding and biotechnology*. Vienna: Plant Breeding and Genetics Section, Joint FAO/IAEA Division of Nuclear Techniques in Food and Agriculture International Atomic Energy Agency.

Foulongne, M., Pascal, T., Pfeiffer, F., & Kervella, J. (2003). QTLs for powdery mildew resistance in peach × *Prunus davidiana* crosses, consistency across generations and environments. *Molecular Breeding, 12*, 33–50.

Gainza, F., Opazo, I., Guajardo, V., Meza, P., Ortiz, M., Pinochet, J., & Muñoz, C. (2015). Rootstock breeding in *Prunus* species: Ongoing efforts and new challenges. *Chilean Journal of Agricultural Research, 75*, 6–16.

Gallardo, R. K., Nguyen, D., McCracken, V., Yue, C., Luby, J., & McFerson, J. R. (2012). An investigation of trait prioritization in rosaceous fruit breeding programs. *HortScience, 47*(6), 771–776.

Gao, L., Wang, Y., Li, Z., Zhang, H., Ye, J., & Li, G. (2016). Gene expression changes during the gummosis development of peach shoots in response to *Lasiodiplodia theobromae* infection using RNA-Seq. *Frontiers in Physiology, 7*, 170. https://doi.org/10.3389/fphys.2016.00170.

Germana, M. A. (2006). Doubled haploid production in fruit crops. *Plant Cell Tissue Organ Culture, 86*, 131. https://doi.org/10.1007/s11240-006-9088-0.

Gharaghani, A., Solhjoo, S., & Oraguzie, N. A. (2017). Review of genetic resources of almonds and stone fruits (*Prunus* spp.) in Iran. *Genetic Resources and Crop Evolution, 64*, 611–640.

Ghasemi Soloklui, A. A., Gharaghani, A., Oraguzie, N., & Saed-Mucheshi, A. (2018). Heritability and combining ability for cold hardiness from partial dialleles in Iranian pomegranate cultivars. *HortScience, 53*(4), 1–5.

Gradziel, T. M. (2003). Interspecific hybridizations and subsequent gene introgression within *Prunus* Subgenus Amygdalus. *Acta Horticulturae, 622*, 249–255.

Gradziel, T. M., Thorpe, M. A., Botstock, R. M., & Wilcox, S. (1998). Breeding for brown rot (*Monilinia fructicola*) resistance in clingstone peach with emphasis on the role of fruit phenolics. *Acta Horticulturae, 465*, 161–170.

Griggs, W. H. (1953). *Pollination requirements of fruits and nuts* (California Agricultural Experiment Station Circular, 424).

Grisel, T. J. (1974) *Prunus* L. Cherry, peach and plum. In C. S. Schopmeyer (Technical Coordinator), *Seeds of woody plants in the United States* (pp. 658–673) (Agriculture Handbook No. 450). Washington, DC: Forest Service, USDA.

Guerriero, R., Viti, R., Monteleone, P., Iacona, C., & Gentili, M. (2006). Pisa University's contribution to the national apricot breeding programme: Three new apricot cultivars for Tuscan fruit growers. *Acta Horticulturae, 717*, 137–140.

Gutiérrez-Pesce, P., Taylor, K., Muleo, R., & Rugini, E. (1998). Somatic embryogenesis and shoot regeneration from transgenic roots of the cherry rootstock Colt (*Prunus avium* × *Prunus pseudocerasus*) mediated by pRi 1855 T-DNA of *Agrobacterium rhizogenes*. *Plant Cell Reports, 17*, 574–580.

Hammerschlag, F. A., & Ognjanov, V. (1990). Somaclonal variation in peach: screening for resistance to *Xanthomonas campestries* pv. *pruni* and *Pseudomonas syringae* pv. *syringae*. *Acta Horticulturae, 280*, 403–408.
Hammerschlag, F. A., & Smigocki, A. C. (1998). Growth and *in vitro* propagation of peach plants transformed with the shoot mutant strain of *Agrobacterium tumefaciens*. *HortScience, 33*, 897–899.
Hancock, J. F., Scorza, R., & Lobos, G. A. (2008). Peaches. In J. F. Hancock (Ed.), *Temperate fruit crop breeding: germplasm to genomics* (pp. 265–298). New York: Springer.
Hansche, P. E., Hesse, C. O., & Beres, V. (1975). Inheritance of fruit size, soluble solids and ripening date in *Prunus domestica* cv. Agen. *Journal American Society for Horticultural Science, 100*, 522–524.
Hartmann, W., & Engelhorn, E. (1992). Some characteristics of young seedling for pre-selection for precocity and fruit size in plum breeding. *Acta Horticulturae, 317*, 125–132.
Hartmann, W., & Neumüller, M. (2009). Plum. In S. M. Jain & P. M. Priyadarshan (Eds.), *Breeding plantation tree crops: Temperate species* (pp. 161–222). New York: Springer.
Hartmann, H. T., Kester, D. E., Davies, F. T., Jr., & Geneve, R. L. (2010). *Plant propagation: Principles and practices* (8th ed.). New York: Pearson Publishing.
Hedrick, U. P. (1950). *A history of horticulture in America to 1860*. New York: Oxford University Press.
Heisey, P. W., Srinivasan, C. S., & Thirtle, C. (2001). *Public sector plant breeding in a privatizing world* (Agriculture Information Bulletin No. 72). Resource Economics Division, Economic Research Service, U. S. Department of Agriculture.
Hofer, M., & Hanke, M. (2017). Cryopreservation of fruit germplasm. *In Vitro Cell. Dev.Biol.-Plant, 53*, 372–381. https://doi.org/10.1007/s11627-017-9841-6.
Iezzoni, A. F. (1987). The 'Redhaven' peach. *Fruit Varieties Journal, 41*, 5–52.
Iezzoni, A. F. (2008). Cherries. In J. F. Hancock (Ed.), *Temperate fruit crop breeding: Germplasm to genomics* (pp. 153–175). New York: Springer.
Iezzoni, A., Schmidt, H., & Alberini, A. (1990). Cherries (*Prunus*). *Acta Horticulturae, 290*, 111–173.
Iezzoni, A., Wunsch, A., Hofer, M., Giovannini, D., Jensen, M., Quero-García, J., Campoy, J. A., Vokurka, A., & Barreneche, T. (2017). Biodiversity, germplasm resources and breeding methods. In J. Quero-García, A. Iezzoni, J. Pulawska, & G. A. Lang (Eds.), *Cherries: Botany, production and uses*. Wallingford: CABI.
Jacob, H. B. (2007). Experience with new mirabelle cultivars from Geisenheim: 'Bellamira'(R) and 'Miragrande'(R) as fruit for the fresh market and for distillation. *Acta Horticulturae, 734*, 347–351.
Jaeger, S. R. (2006). Non-sensory factors in sensory science research. *Food Quality and Preference, 17*, 132–144.
Janick, J. (2005). The origins of fruits, fruit growing, and fruit breeding. *Plant Breeding Reviews, 25*, 255–320.
Janick, J., & Moore, J. N. (1996). *Fruit breeding. Volume 1: Tree and tropical fruits*. New York: Wiley.
Jenks, M. A., & Bebeli, P. (Eds.). (2011). *Breeding for fruit quality*. New York: Wiley.
Jung, S., Staton, M., Lee, T., Blenda, A., Svancara, R., Abbott, A., & Main, D. (2008). GDR (Genome Database for Rosaceae): Integrated web-database for Rosaceae genomics and genetics data. *Nucleic Acids Research, 36*, D1034–D1040.
Kappel, F. (2008). Breeding cherries in the "New World". *Acta Horticulturae, 795*, 59–69.
Kappel, F., Lane, W. D., MacDonald, R., Lapins, K., & Schmid, H. (1998). 'Santina', 'Sumpaca Celeste', and 'Sumnue Cristalina' sweet cherries. *HortScience, 33*, 1087–1089.
Kappel, F., Granger, A., Hrotkó, K., & Schuster, M. (2012). Cherries. In M. L. Badenes & D. H. Byrne (Eds.), *Fruit breeding. Handbook of plant breeding* (pp. 459–504). New York: Springer.
Kervella, J., Pascal, T., Pfeiffer, F., & Dirlewanger, E. (1998). Breeding for multiresistance in peach tree. *Acta Horticulturae, 465*, 177–184.

Kester, D. E., & Assay, R. N. (1986). 'Hansen 2168' and 'Hansen 536', two Prunus rootstock clones. *HortScience, 21*, 331–332.
Khadari, B., Krichen, L., Lambert, P., Marrakchi, M., & Audergon, J. M. (2006). Genetic structure in Tunisian apricot, *Prunus armeniaca* L., populations propagated by grafting: A signature of bottleneck effects and ancient propagation by seedlings. *Genetic Resources and Crop Evolution, 53*, 811–819.
Knight, R. L. (1969). *Abstract bibliography of fruit breeding and genetics*. London: Easter Press.
Kole, C., & Abbott, A. G. (2012). *Genetics, genomics and breeding of stone fruits*. New York: CRC Press.
Kostina, K. F. (1946). The origin and evolution of cultivated apricot. *Proceedings (Trudi) of the Nikita Botanical Garden, 24*, 25–31. (in Russian).
Kozai, N., Beppu, K., Mochioka, R., Boonprakob, U., Subhadrabandhu, S., & Kataoka, I. (2004). Adverse effects of high temperature on the development of reproductive organs in 'Hakuho' peach trees. *Journal of Horticultural Science and Biotechnology, 79*, 533–537.
Kryukova, I. V. (1989). Botanical classification and geographical distribution. In V. K. Smykov (Ed.), *Apricot* (pp. 9–23). Moscow: Agropromizdat. (in Russian).
Kumar, S., Chagné, D., Bink, M. C., Volz, R. K., Whitworth, C., & Carlisle, C. (2012a). Genomic selection for fruit quality traits in apple (Malus×domestica Borkh.). *PLoS one, 7*(5), e36674. https://doi.org/10.1371/journal.pone.0036674.
Kumar, S., Bink, M. C. A. M., Volz, R. K., et al. (2012b). Towards genomic selection in apple (*Malus* × *domestica* Borkh.) breeding programmes: Prospects, challenges and strategies. *Tree Genetics & Genomes, 8*, 1–14. https://doi.org/10.1007/s11295-011-0425-z.
La Rue, J. H., & Johnson, R. S. (1989). *Peaches, plums and nectarines—Growing and handling for fresh market*. Oakland: University of California.
Lalli, D. A., Decroocq, V., Blenda, A. V., Schurdi-Levraud, V., Garay, L., Le Gall, O., Damsteegt, V., Reighard, G. L., & Abbott, A. G. (2005). Identification and mapping of resistance gene analogs (RGAs) in *Prunus*: A resistance map for *Prunus*. *Theoretical and Applied Genetics, 111*, 1504–1513.
Lapins, K. O. (1971). Stella, a self-fruitful sweet cherry. *Canadian Journal of Plant Science, 51*, 252–252.
Larsen, H. J. (2010). *Evaluating tree fruit bud and fruit damage from cold. Colorado State University Extension* (Fact sheet No. 7.426). Retrieved from www.ext.colostate.edu
Laurens, F., Aranzana, M. J., Arus, P., Bassi, D., Bink, M., Bonany, J., Caprera, A., Corelli-Grappadelli, L., Costes, E., Durel, C. E., Mauroux, J. B., Muranty, H., Nazzicari, N., Pascal, T., Patocchi, A., Peil, A., Quilot-Turion, B., Rossini, L., Stella, A., Troggio, M., Velasco, R., & Van de Weg, E. (2018). An integrated approach for increasing breeding efficiency in apple and peach in Europe. *Horticulture Research, 5*, 11. https://doi.org/10.1038/s41438-018-0016-3.
Layne, R. E. C., Bailey, C. H., & Hough, L. F. (1996). Apricots. In J. Janick & J. N. Moore (Eds.), *Fruit breeding, vol. 1: Tree and tropical fruits* (pp. 79–111). New York: John Wiley & Sons.
Ledbetter, C. A. (2008). Apricot. In J. F. Hancock (Ed.), *Temperate fruit crop breeding: Germplasm to genomics* (pp. 38–82). New York: Springer.
Ledbetter, C. A., & Peterson, S. J. (2004). Utilization of Pakistani apricot (*Prunus armeniaca* L.) germplasm for improving Brix levels in California adapted apricots. *Plant Genetic Resources Newsletter, 140*, 14–22.
Lespinasse, Y., Durel, C. E., Eskes, A., Esmenjaud, D., & Poessel, J. L. (2003). Resistance to biotic stress in fruit trees. *Acta Horticulturae, 622*, 303–315.
Limera, C., Sabbadini, S., Sweet, J. B., & Mezzetti, B. (2017). New biotechnological tools for the genetic improvement of major woody fruit species. *Frontiers in Plant Science, 8*, 1418. https://doi.org/10.3389/fpls.2017.01418.
Lingdi, L., & Bartholomew, B. (2003). Armeniaca. In C. Y. Wu & P. H. Raven (Eds.), *Flora of China, vol 9 (Pittosporaceae through Connaraceae)* (pp. 396–401). Beijing/St. Louis: Science Press/Missouri Botanical Garden Press.
Llácer, G. (2009). Fruit breeding in Spain. *Acta Horticulturae, 814*, 43–56.

Lu, Z.-X., Reighard, G. L., Nyczepir, A. P., Beckman, T. G., & Ramming, D. W. (2000). Inheritance of resistance to root-knot nematodes (*Meloidogyne* sp.) in *Prunus* rootstock. *HortScience, 35*, 1344–1346.

Luby, J. J., & Shaw, D. V. (2001). Does marker-assisted selection make dollars and sense in a fruit breeding program? *HortScience, 35*, 872–879.

Lucier, G., Pollack, S., Ali, M., & Perez, A. (2005) *Fruit and vegetable backgrounder* (USDA. ERS. VGS-313-01). Retrieved from www.ers.usda.gov

Martínez-Gómez, P., Rubio, M., Dicenta, F., & Gradziel, T. M. (2004). Resistance to plum pox virus (Dideron isolate RB3.30) in a group of California almonds and transfer of resistance to peach. *Journal of the American Society for Horticultural Science, 129*, 544–548.

Martínez-Gómez, P., Sánchez-Pérez, R., Rubio, M., Dicenta, F., Gradziel, T., & Sozzi, G. (2005). Application of recent biotechnologies to prunus tree crop genetic improvement. *International Journal of Agriculture and Natural Resources, 32*(2), 55–78. https://doi.org/10.7764/rcia.v32i2.308.

Mba, C., Afza, R., & Shu, Q. Y. (2012). Mutagenic radiations: X-rays, ionizing particles and ultraviolet. In Q. Y. Shu, B. P. Forster, & H. Nakagawa (Eds.), *Plant mutation breeding and biotechnology*. Vienna: Plant Breeding and Genetics Section, Joint FAO/IAEA Division of Nuclear Techniques in Food and Agriculture International Atomic Energy Agency.

Michalak, M., Plitta-Michalak, B. P., & Chmielarz, P. (2015). A new insight in desiccation tolerance and cryopreservation of mazzard cherry (*Prunus avium* L.) seeds. *Open Life Science, 10*, 354–364. https://doi.org/10.1515/biol-015-0036.

Minev, I., & Balev, M. (2002). Interspecific hybrids of the *Prunus* genus bred at RIMSA, Troyan. *Acta Horticulturae, 577*, 195–198.

Monet, R., & Bassi, D. (2008). Classical genetics and breeding. In D. R. Layne & D. Bassi (Eds.), *The peach. Botany, production and uses* (pp. 61–84). Wallingford: CAB International.

Monet, R., Guye, A., & Massonie, G. (1998). Breeding for resistance to green peach aphid, *Myzus persicae* Sulzer in the peach. *Acta Horticulturae, 465*, 171–175.

Moore, J. N., & Janick, J. (1983). *Methods in fruit breeding* (1st ed.). West Lafayette: Purdue University Press.

Neumüller, M., Lanzl, S., Hartmann, W., Feucht, W., & Treutter, D. (2009). Towards an understanding of the inheritance of hypersensitivity resistance against the sharka virus in European plum (*Prunus domestica* L.): Generation of interspecific hybrids with lower ploidy levels. *Acta Horticulturae, 814*, 721–726.

Nocker, S. V., & Gardiner, S. E. (2014). Breeding better cultivars, faster: Applications of new technologies for the rapid deployment of superior horticultural tree crops. *Horticulture Research, 1*, 14022. https://doi.org/10.1038/hortres.2014.22.

Nyczepir, A. P., & Beckman, T. G. (2000). Host status of guardian peach rootstock to *Meloidogyne* sp. and *M. javanica*. *HortScience, 35*(4), 772.

Ochatt, S. J. (1992). The development of protoplast-to-tree systems for *Prunus cerasifera* and *P. spinosa*, Rosaceae. *Plant Science, 81*, 253–259.

Ochatt, S. J., Patat-Ochatt, E. M., Rech, E. L., Davey, M. R., & Power, J. B. (1989). Somatic hybridization of sexually incompatible top-fruit tree rootstocks, wild pear (*Pyrus communis* var pyraster L.) and Colt cherry (*Prunus avium* × *P. pseudocerasus*). *Theoretical and Applied Genetics, 78*, 35–41.

Ogundiwin, E., Peace, C., Gradziel, T., Parfitt, D., Bliss, F., & Crisosto, C. (2009). A fruit quality gene map of Prunus. *BMC Genomics, 10*, 587. https://doi.org/10.1186/1471-2164-10-587.

Okie, W. R. (1987). Plum rootstocks. In R. C. Rom & R. F. Carlson (Eds.), *Rootstocks for fruit crops* (pp. 321–360). New York: Wiley.

Okie, W. R. (1998). *Handbook of peach and nectarine varieties* (Agriculture Handbook Number 714). United States Department of Agriculture.

Okie, W. R., & Hancock, J. F. (2008). Plums. In J. F. Hancock (Ed.), *Temperate fruit crop breeding: Germplasm to genomics* (pp. 337–358). New York: Springer.

Okie, W. R., & Ramming, D. W. (1999). Plum breeding worldwide. *HortTechnology, 9*, 162–176.

Okie, W. R., & Weinberger, J. H. (1996). Plums. In J. Janick & J. N. Moore (Eds.), *Fruit breeding, Vol. 1. Tree and tropical fruits*. New York: Wiley.

Okie, W. R., Bacon, T., & Bassi, D. (2008). Fresh market cultivar development. In D. R. Layne & D. Bassi (Eds.), *The peach—Botany, production and uses* (pp. 139–174). Wallingford: CAB International.

Omar, C. M., Sherman, W. B., & Chaparro, J. (2010). Development of a branching index for evaluation of peach seedlings using interspecific hybrids. *HortScience, 45*(6), 852–856.

Peace, C. P. (2017). DNA-informed breeding of rosaceous crops: Promises, progress and prospects. *Horticulture Research, 4*, 17006. https://doi.org/10.1038/hortres.2017.6.

Pellegrino, S. (2006). *Apricot industry in Italy (in Spanish) 'Updating the apricot production technology course*. Zaragoza: Escuela Agraria de Cogullada.

Pennone, F., & Abbate, V. (2006). Apricot breeding in Caserta: New perspectives of apricot growing in Southern Italy. *Acta Horticulturae, 717*, 157–161.

Pinochet, J., Fernandez, C., Calvet, C., Hernandez-Dorrego, A., & Felipe, A. (2000). Selection against *Pratylenchus vulnus* population attacking *Prunus* rootstocks. *HortScience, 35*, 1333–1337.

Potter, D. (2012). Basic information on the stone fruit crops. In C. Kole & A. G. Abbott (Eds.), *Genetics, genomics and breeding of stone fruits* (pp. 1–21). New York: CRC Press.

Potter, D., Eriksson, T., Evans, R. C., Oh, S., Smedmark, J. E. E., Morgan, D. R., Kerr, M., Robertson, K. R., Arsenault, M., Dickinson, T. A., & Campbell, C. S. (2007). Phylogeny and classification of *Rosaceae*. *Plant Systematic and Evolution, 266*, 5–43.

Predrieri, S., & Gatti, E. (2000). Effects of gamma radiation on plum (*Prunus salicina* Lindl.) 'Shiro'. *Advances in Horicultural Science, 14*, 215–223.

Predrieri, S. (2001). Mutation induction and tissue culture in improving fruits. *Plant Cell Tissue Organ Culture, 64*, 185–210.

Prior, R. L., & Cao, G. (2000). Antioxidant phytochemicals in fruits and vegetables: Diet and health implications. *HortScience, 35*, 588–592.

Prunus Crop Germplasm Committee. (2010). *Prunus* vulnerability statement. USDA - ARS National Plant Germplasm System [Online]. Retrieved from http://www.ars-grin.gov/npgs/cgc_reports/prun2010.doc

Quero-García, J., Shuster, M., Lopez-ortega, G., & Charlot, G. (2017). Sweet cherry varieties and improvement. In J. Quero-García, A. Iezzoni, J. Pulawska, & G. A. Lang (Eds.), *Cherries: Botany, production and uses*. Wallingford: CABI.

Ramming, D. W. (1990). The use of embryo culture in fruit breeding. *HortScience, 25*, 393–398.

Ransom, C., Particka, C., Ando, K., & Olmstead, J. (2006). Report of breakout group 1. What kind of training do plant breeders need, and how can we most effectively provide that training? *HortScience, 41*, 53–54.

Ravelonandro, M., Scorza, R., Callahan, A., Levy, L., Jacquet, C., Monsion, M., & Damsteegt, V. (2000). The use of transgenic fruit trees as a resistance strategy for virus epidemics: The plum pox (sharka) model. *Virus Research, 71*, 63–69.

Rehder, A. (1940). *Manual of cultivated trees and shrubs* (2nd ed., pp. 452–481). New York: Macmillan.

Reighard, G. L., & Loreti, F. (2008). Rootstock development. In D. Layne & D. Bassi (Eds.), *The peach, botany, production and uses* (pp. 193–220). Wallingford: CAB International.

Reighard, G. L., Cain, D. W., & Newall, W. C. (1990). Rooting and survival potential of hardwood cuttings of 406 species, cultivars and hybrids of *Prunus*. *HortScience, 25*(5), 517–518.

Reynders-Aloisi, S., & Grellet, F. (1994). Characterization of the ribosomal DNA units in two related *Prunus* species (*P. cerasifera* and *P. spinosa*). *Plant Cell Reports, 13*, 641–646.

Roach, F. A. (1985). *Plums. Cultivated fruits of britain, their origin and history* (pp. 142–460). Oxford: Basil Blackwell.

Rodriguez, J., Sherman, W. B., Scorza, R., Wisniewski, M., & Okie, W. R. (1994). 'Evergreen' peach, its inheritance and dormancy behavior. *Journal of the American Society for Horticultural Science, 119*, 789–792.

Rostova, I. S., & Sokolova, E. A. (1992). Variability of anatomical and morphological leaf characters in apricot (*Armeniaca* Scop.) species and varieties. *Bulletin of Applied Botany, Genetics and Plant Breeding, 146*, 74–86.
Rubio, M., Pascal, T., Bachellez, A., & Lambert, P. (2010). Quantitative trait loci analysis of Plum pox virus resistance in *Prunus davidiana* P1908, new insights on the organization of genomic resistance regions. *Tree Genetics & Genomes, 6*, 291–304.
Rubio-Cabetas, M. J., Lecouls, A. C., Salesses, G., Bonnet, A., Minot, J. C., Voisin, R., & Esmenjaud, D. (1998). Evidence of a new gene for high resistance to *Meloidogyne* spp. in Myrobalan plum (*Prunus cerasifera*). *Plant Breeding, 117*(6), 567–571.
Rugini, E., & Gutierrez-Pesce, P. (1999). Transgenic *Prunus* fruit species (almond, apricot, cherry rootstocks, sour cherry and sweet cherry, peach and plum). In Y. P. S. Bajaj (Ed.), *Biotechnology in agriculture and forestry, Vol. 44, Transgenic trees* (Vol. 44, pp. 245–262). Berlin: Springer-Verlag.
Ruiz, D., & Egea, H. (2008). Phenotypic diversity and relationships of fruit quality traits in apricot (*Prunus armeniaca* L.) germplasm. *Euphytica, 163*, 143–158.
Sansavini, S. (2009). Horticulture in Europe: From history to innovation. *Acta Horticulturae, 817*, 43–58.
Sansavini, S., Gamberini, A., & Bassi, D. (2006). Peach breeding, genetics and new cultivar trends. *Acta Horticulturae, 713*, 23–48.
Schuster, M., Flachowsky, H., & Köhler, D. (2007). Determination of self-incompatible genotypes in sweet cherry (Prunus avium L.) accessions and cultivars of the German Fruit Gene Bank and from private collections. *Plant Breeding, 126*, 533–540.
Schuster, M., Apostol, J., Iezzoni, A., Jensen, M., & Milatovic, D. (2017). Sour cherry varieties and improvement. In J. Quero-García, A. Iezzoni, J. Pulawska, & G. A. Lang (Eds.), *Cherries: Botany, production and uses*. Wallingford: CABI.
Scorza, R. (2000). Progress in tree fruit improvement through molecular genetics. *HortScience, 36*, 855–858.
Scorza, R., & Hammerschlag, F. A. (1992). Stone fruits. In F. A. Hammerschlag & R. E. Litz (Eds.), *Biotechnology of perennial fruit crops* (pp. 277–301). Wallingford: CAB International.
Scorza, R., & Pooler, M. (1999). Growth and yield of F-1 hybrid peaches developed from doubled haploids. *HortScience, 34*, 928–931.
Scorza, R., & Sherman, W. B. (1996). Peaches. In J. Janick & J. N. Moore (Eds.), *Fruit breeding Vol. I. Tree and tropical fruits* (pp. 325–440). New York: Wiley.
Scorza, R., Mehlenbacher, S. A., & Lightner, G. W. (1985). Inbreeding and coancestry of freestone peach cultivars of the eastern United States and implications for peach germplasm improvement. *Journal of American Society for Horticultural Science, 110*, 547–552.
Scorza, R., Ravelonandro, M., Callahan, A., Cordts, J. M., Fuchs, M., Dunez, J., & Gonsalves, D. (1994). Transgenic plums (*Prunus domestica* L.) express the plum pox virus coat protein gene. *Plant Cell Reports, 14*, 18–22.
Scorza, R., Miller, S., Glenn, D. M., Okie, W. R., & Tworkoski, T. (2006). Developing peach cultivars with novel tree growth habits. *Acta Horticulturae, 713*, 61–64.
Scorza, R., Dardick, C., Callahan, A. M. et al. (2014). 'FasTrack' a revolutionary approach to long generation cycle tree fruit breeding. In *Proceedings of the 1st International Rapid Cycle Crop Breeding Conference*; 7–9 January 2014; Leesburg, VA, USA.
Sherman, W. B., Sharpe, R. H., & Prince, V. E. (1972). Two red leaf characters associated with early ripening peaches. *HortScience, 7*, 502–503.
Sherman, W. B., Sharpe, R. H., & Janick, J. (1973). The fruiting nursery, ultrahigh density for evaluation of blueberry and peach seedlings. *HortScience, 8*, 170–172.
Shua, Q. Y., Forsterb, B. P., & Nakagawac, H. (2011). Principles and applications of plant mutation breeding. In Q. Y. Shu, B. P. Forster, & H. Nakagawa (Eds.), *Plant mutation breeding and biotechnology*. Vienna.: Plant Breeding and Genetics Section, Joint FAO/IAEA Division of Nuclear Techniques in Food and Agriculture International Atomic Energy Agency.
Shulaev, V., Korban, S. S., Sosinski, B., Abbott, A. G., Aldwinckle, H. S., Folta, K. M., Iezzoni, A., Main, D., Aru's, P., Dandekar, A. M., Lewers, K., Brown, S. K., Davis, T. M., Gardiner, S. E.,

Potter, D., & Veilleux, R. E. (2008). Multiple models for Rosaceae genomics. *Plant Physiology, 147*, 985–1003. https://doi.org/10.1104/pp.107.115618.
Sloan, E. (2006). Top 10 functional food trends. *Food Technology, 04*(6), 23–34.
Souza, V. A. B., Byrne, D. H., & Taylor, J. F. (2000). Predicted breeding values for nine plants and fruits characteristics of 28 peach genotypes. *Journal of American Society for Horticultural Science, 125*(4), 460–465.
Sutherland, B. G., Tobutt, K. R., & Robbins, T. P. (2007). Molecular genotyping of self-incompatible plum cultivars. *Acta Horticulturae, 734*, 47–51.
Tavaud, M., Zanetto, A., David, J. L., Laigret, F., & Dirlewanger, E. (2004). Genetic relationships between diploid and allotetraploid cherry species (*Prunus avium*, *Prunus* x *gondouinii* and *Prunus cerasus*). *Heredity, 93*, 631–638.
Tehrani, G. (1990). Seventy-five years of plum breeding and pollen compatibility studies in Ontario. *Acta Horticulturae, 283*, 95–103.
Terns, R. M., & Terns, M. P. (2014). CRISPR-based technologies: Prokaryotic defense weapons repurposed. *Trends in Genetics, 30*, 111–118.
Tobut, K. R., & Nicoll, R. J. (1992). Joint meeting of the European Primus and apple genome mapping groups. East Mailing, UK. May 1991. *Fruit Variety Journal, 46*, 263–265.
Topp, B. L., Sherman, W. B., & Raseira, M. C. B. (2008). Low-chill cultivar development. In D. R. Layne & D. Bassi (Eds.), *The peach. Botany, production and uses* (pp. 106–138). Wallingford: CAB International.
Topp, B. L., Russell, D. M., Neumüller, M., Dalbó, M. A., & Weisheng, L. (2012). Plum. In M. L. Badenes & D. H. Byrne (Eds.), *Fruit breeding. Handbook of plant breeding* (pp. 571–621). New York: Springer.
Towill, L. E., & Forsline, P. L. (1999). Cryopreservation of sour cherry (*Prunus cerasus* L.) using a dormant vegetative bud method. *Cryo-Letters, 20*, 215–222.
Vavilov, N. I. (1951). The phytogeographic basis of plant breeding. In F. Verdoorn (Ed.), *Chronica Botanica, an international collection of studies in the method and history of biology and agriculture* (K. S. Chester, Trans. from Russian, Vol. 13, No. 1/6, pp. 13–54).
Walker, A. R., Lee, E., & Robinson, S. P. (2006). Two new grape cultivars, bud sports of Cabernet Sauvignon bearing pale-coloured berries, are the result of deletion of two regulatory genes of the berry colour locus. *Plant Molecular Biology, 62*, 623–635. https://doi.org/10.1007/s11103-006-9043-9.
Wang, L., Zhu, G., & Fang, W. (2002). Peach germplasm and breeding programs at Zhengzhou in China. *Acta Horticulturae, 592*, 177–182.
Wargovich, M. J., Morris, J., Moseley, V., Weber, R., & Byrne, D. H. (2012). Developing fruit cultivars with enhanced health properties. In M. L. Badenes & D. H. Byrne (Eds.), *Fruit breeding. Handbook of plant breeding* (pp. 37–69). New York: Springer.
Webster, A. D. (1996). The taxonomic classification of sweet and sour cherries and a brief history of their cultivation. In W. D. Webster & N. E. Looney (Eds.), *Cherries: Crop physiology, production and uses*. Wallingford: CAB International.
Webster, T. (2006). Control of growth and cropping of temperate fruit trees. *Chronica Horticulturae, 46*(3), 20–26.
Webster, A. D., & Schmidt, H. (1996). Rootstocks for sweet and sour cherries. In W. D. Webster & N. E. Looney (Eds.), *Cherries: Crop physiology, production and uses*. Wallingford: CAB International.
Werner, D. J., & Cain, D. W. (1985). Cages for protection of tree fruit hybridizations. *HortScience, 20*, 450–451.
Werner, D. J., Mowrey, B. D., & Young, E. (1988). Chilling requirement and post-rest heat accumulation as related to difference in time of bloom between peach and western sand cherry. *Journal of the American Society for Horticultural Science, 113*, 775–778.
Ye, X., Brown, S. K., Scorza, R., Cordts, J., & Sanford, J. C. (1994). Genetic transformation of peach tissues by particle bombardment. *Journal of the American Society for Horticultural Science, 119*, 367–373.

Zarrouk, O., Gogorcena, Y., Moreno, M. A., & Pinochet, J. (2006). Graft compatibility between peach cultivars and *Prunus* rootstocks. *HortScience, 41*, 1389–1394.
Zhao, F., Liu, W., Liu, N., Yu, X., Sun, M., Zhang, Y., & Zhou, Y. (2005). Reviews of the apricot germplasm resources and genetic breeding in China. *Journal of Fruit Science, 22*, 687–690. (in Chinese).
Zhebentyayeva, T. N., Reighard, G. L., Lalli, D., Gorina, V. M., Krška, B., & Abbott, A. G. (2008). Origin of plum pox virus resistance in apricot: What new AFLP and targeted SSR data analyses tell. *Tree Genetics & Genomes, 4*, 403–417.
Zhebentyayeva, T., Ledbetter, C., Burgos, L., & Llácer, G. (2012). Apricot. In M. L. Badenes & D. H. Byrne (Eds.), *Fruit breeding. Handbook of plant breeding* (pp. 415–458). New York: Springer.
Zhukov, O. S., & Charitonova, E. N. (1988). *Selektzia vishi*. Moscow: Agroproizdat.
Zohary, D., & Hopf, M. (2001). *Domestication of plants in the Old World* (3rd ed.). Oxford: Oxford University Press. 334 pp.

Nutrient Management in Stone Fruits

2

Shabnam Ahad, Mohammad Maqbool Mir, Umar Iqbal, Gh. Hassan Rather, M. U. Rehman, Shamim A. Simnani, Aroosa Khalil, Amarjeet S. Sindouri, Shafat A. Banday, and I. A. Bisati

Abstract

Fruit crops are nutritionally more effective than the field crops collectively by the virtue of their permanent nature of tree morphology (nutrients locked therein), prolonged growth stages, differences in root distribution patterns, developmental stages from the point of view of nutrient requirement, and preferential necessity of certain nutrients by particular cultivars. Perennial fruit crops have huge commercial status in world trade and in the economy of various regions, even if they cover only 1% of global agricultural land. Due to huge distinction in the nutrient use efficiency of perennial fruit crops, their nutrient management-based production system is characteristically intricate to understand. Modern orchardists operate to rising standards of production using a range of best management practice systems, and orchardists are becoming increasingly aware that an ecologically stable soil system is vital for sustaining healthy crops. Fertile soils normally hold all the nutrition required for healthy crop growth but rely on the right combination and volume of microbial populations to digest and transform these minerals to compounds readily available for plant uptake. Thus, adequate mineral nutrition is a preharvest factor affecting fruit quality. While developing fertilization program in any stone fruit orchard, available nutrient status of the soil and plant needs to be taken into contemplation along with the actual plant nutrient requirements. For understanding the significance of balanced supply of nutrients, sound knowledge about nutrient interactions is essential.

Keywords

Nutrients · Management · Stone fruits · Requirements · Fertigation

S. Ahad (✉) · M. M. Mir · U. Iqbal · G. H. Rather · M. U. Rehman · S. A. Simnani · A. Khalil · A. S. Sindouri · S. A. Banday · I. A. Bisati
Division of Fruit Science, Sher-e-Kashmir University of Agricultural Sciences & Technology of Kashmir, Srinagar, Jammu and Kashmir, India

M. M. Mir et al. (eds.), *Production Technology of Stone Fruits*,
https://doi.org/10.1007/978-981-15-8920-1_2

2.1 Introduction

The term stone fruits also known as drupes is used to denote fruits of the *Prunus* species, viz., peach, nectarine, plum, cherry, almond, an d apricot, which contain hard stony seed in the flesh. These are the most important fruits worldwide among fruiting tree species, which plays a significant role for small horticulturists and rural economies (Jacob 2010). According to a report, global stone fruit production amounted to 829.57 thousand metric tons in 2017 (Shahbandeh 2020). Nearly half of worldwide production of stone fruit comes from China. The area under stone fruits is speedily rising in temperate parts of world due to its rich nutritional and economic value. In India states of Jammu and Kashmir (J&K), hills of Himachal Pradesh (HP), Uttarakhand, and to some extent North Eastern region are the major stone fruit-producing areas. The attention toward growing these crops under subtropical regions during the last three decades has also increased due to the availability of high-quality and low-chilling cultivars (Sharpe et al. 1990). Nutritionally, stone fruits are rich sources of vitamins and minerals, and there is increasing interest in their potential value as nutraceuticals owing to the good amount of antioxidants like anthocyanins and phenolic compounds. Almost all the stone fruits are ideal for alcoholic fermentation due to the lower sugar content. These fruits are cultivated both in hills and submountainous provinces of North India, and in the hills, the orchards are rainfed and are confined to the slopes where moderate- and high-chilling varieties of high dessert quality are predominantly suitable. The soils in the hills are poor unlike in the plains and experience frequent dry spells during the summer and excessive leaching of nutrients during the heavy rains. Thus, nutrient management in such orchards deserves greater attention in the hills. For proper growth and metabolism, plants require various chemical compounds called as nutrients, and their supply and absorption is termed as nutrition. Based on the criteria proposed by Arnon and Stout (1939), 20 elements required for higher plants are known. Some nutrients are required by plants in larger quantities, called as macro- or major nutrients, while some are required by plants in small quantities, called as minor or trace elements (Devlin 1975). For the proper vegetative and reproductive growth of plants, both macro- and micronutrients are equally essential. Each nutrient provides a unique function in the overall metabolism of the tree, and good tree growth and crop and fruit quality depend not only on the proper supply of each nutrient but also on the proper balance between them. The permanent nature of the wood profile of fruit crops (nutrients are locked), their extended physical condition, differential root development (root volume distribution), developmental stages from the point of view of nutrient requirements, and the priority of certain nutrients by particular species collectively make them more proficient than annual crops. The perennial nature of woody framework of fruit crops (nutrients locked therein), their extended physiological stages of growth, differential root growth (root volume distribution), growth stages from the point of view of nutrient requirement, and preferential requirement of some nutrients by definite fruit crop collectively make them nutritionally more proficient than the annual or yearly crops (Scholberg et al. 2012; Srivastava et al. 2008). The response of fruit trees to disease resistance is significantly affected by the supply and balance of nutrients by affecting the total plant metabolism, changes in cell wall strength and composition, synthesis of various defense compounds, and the profusion of metabolites (sugars and amino acids). A proper rise in concentration of some nutrients up to the threshold level promotes the absorption of other nutrients (synergism), while in case of antagonism excess level of a specific nutrient may inhibit the accumulation of other particular nutrients.

Thus, maintaining the right amount, source, time, and method of nutrient application with the goal of boosting productivity while curtailing those losses creating ecological harms is called as nutrient management (Johnson 2011). Practically, it takes into account the development of nutrient budgets that entails the knowledge about the quantity of nutrients present inside the soil, determining the nutrient needs of the crop, thereby calculating all the sources of nutrients so that the nutrient demand of the crop will be met. In nutrient management for enhancing crop production and maintaining soil quality, it is critical to maintain adequate but not disproportionate nutrient concentrations. So apart from the environmental remunerations, there are economic aids of proper orchard nutrient management. Right applications of water and nutrients allow the temporal tuning of fertigation requirements, as well as provide a resourceful use of water and the diminution of pollution (Neilsen and Neilsen 2002). Proper application of water and nutrients allows for temporary tuning of regenerative needs, as well as reduce water resource utilization and pollution (Neilsen and Neilsen 2002). In addition to optimizing the overall nutrition of the tree, there is a growing concern with fertilizing specific organs of the plant. For promising standard yields of fruit orchards, our main aim should be to promote flowering and fruit set. The concept of integrated nutrient management augments complete performance of fruit trees in terms of growth, production, and quality of fruits by facilitating accessibility of nutrients like N, P, K, Ca, Mg, and B (Merwe 2012). Globally, the fertilizer costs are continuously mounting as the world food and fiber demand grows; thus, farmers will definitely follow the 4Rs in orchard nutrition management practice.

Efficient nutrient management calls for supplying the exact nutrient to the particular plant at the correct time. Therefore, the practice of orchard nutrition significantly affects productivity and fruit quality and has to be performed very vigilantly both before and after harvest (Crisosto et al. 1997).

2.2 Availability and Uptake of Nutrients

It is quite intricate to determine nutrient uptake patterns of large-sized and perennial deciduous fruit tree. As compared to the annual crops, the farming system of perennials might be more sustainable owing to the comparatively low nutrient removal through yield, high remobilization, and storage in the woody morphology (Wu et al. 2008). The various factors affecting availability and uptake of nutrients are as under:

(a) Soil texture.
(b) Soil structure.

(c) Soil moisture.
(d) Soil temperature.
(e) Soil pH.

To determine the fertilization management of any orchard, primary knowledge about nutrient availability is a priority. Soils with high amount of clay and loam tend to hold most of the essential nutrients more effectively than the sandy soils as high clay content increases the cation exchange capacity. Similarly, good soil structure is necessary for proper movement, penetration, and retention of nutrients. The process of nitrification is carried out under aerobic conditions to convert ammonia into nitrate; however, the fraction of nitrogen not available to root absorption is lost through leaching. In a sustainable fertilization program, a proportion of nutrients offered can be calculated by multiplying concentration by the mass of soil explored by roots, i.e., soil unit which on a hectare (1 ha ¼10,000 m^2) basis depends on root depth and soil bulk density. Approximately a soil volume of 8000 m^3/ha and 10,400 mg/ha mass explored by roots has been accounted for a root depth of 0.8 m and an apparent soil weight of 1.3 Mg/m^3, respectively. This is a hypothetical value in case the root system explores all the volume of soil in which it grows. According to our knowledge, root density of fruit crops ranges between less than 1 and 10 kg/m^3, meaning roughly from 0.1% to 1% of soil mass and lower probability to intercept the nutrient. According to Bengough et al. (2011), the growth of root system is directly affected by soil structure, and pH affects nutrient availability by changing the nutrient form, ultimately affecting the crop productivity level. In case of nitrogen, at various pH levels, different forms of N have different leaching capabilities; other nutrients may become adsorbed or desorbed, precipitated, mineralized, or immobilized at different pH values. Many nutrients are more available in slightly acidic soils; P is most available at neutral pH (about 6.5); Mb is available at high pH and can be toxic to plants. Generally, a soil pH of 6.0–7.5 is satisfactory for most of the fruit crops as most of the nutrients are available at this pH range.

2.3 Tissue Analysis and Nutrient Status

In contrast to soil analysis, which merely shows what is in the ground, leaf analysis shows what the trees actually absorb; thus, it is used as an accurate guide for determining the nutritional status of stone fruit trees. The accurate time for leaf collection starts around mid-July until mid-August due to the most stable nutrient levels in fruit trees at that particular time. During spring, trees are vigorously growing and transporting nutrients up into the leaves, while in fall, senescence is beginning and nutrients are being transported out of leaves. In case of peach and plum trees, leaf sample should be collected from the midportion of current season in nonbearing shoots during the month of June to August (Khera et al. 1981). In case of apricots, August is a suitable period for leaf sampling as least changes occurred in the nutrient levels of various elements in that month. Status of one nutrient in leaves affects the status of other nutrients. For N compounds, leaves act as the main sink, and there

Table 2.1 Nutrient levels in leaves of stone fruit (peach and plum) (Johnson and Uriu 1989)

Nutrient	Deficient	Low	Normal	High	Excessive
Dry weight (%)					
N	1.7	1.7–2.3	2.4–3.0	3.1–4.0	4.0
P	0.09	0.09–0.13	0.14–0.25	0.26–0.40	0.40
K	1.0	1.0–1.5	1.6–3.0	3.1–4.0	4.0
Ca	1.0	1.0–1.4	1.5–3.0	3.1–4.0	4.0
Mg	0.20	0.20–0.29	0.30–0.80	0.81–1.10	1.10
Dry weight (ppm)					
Fe	60	60–90	100–250	251–500	500
Cu	4	4–5	6–16	17–30	30
Mn	20	20–39	40–160	161–400	400
Zn	15	15–19	20–50	51–70	70
B	20	20–24	25–60	61–80	80

is a close relation between the nitrogen content of leaves, size, and color of fruits (Williams and Billingsley 1974) The range of nutrient levels in leaves of stone fruits like peach and plum is shown in Table 2.1.

As a rule of thumb, every 10% increase in nitrogen fertilizer application results in an increase of 0.1% leaf nitrogen. However, if nitrogen is delivered via fertigation, nitrogen needed is less, as uptake efficiency is higher in fertigation than in regular soil application (Lehnert 2010).

2.4 Nutrient Needs by Stone Fruits

The various ways used for determining the nutrient needs of perennial fruit crops include surveying in orchards, knowledge about experience of orchardists, following nutrient management program of high-yielding orchards, determining the nutrient removal by fruits, visualizing deficiency symptoms, and leaf/soil analysis (Tagliavini and Scandellari 2012). To quantify the annual nutrient demand of a tree, the paramount approach is to determine the amount of nutrients absorbed by plant which involves the fraction amassed in fruits and permanent organs like roots, trunk, and branches and the fraction which returns back to the soil after harvest in the form of fallen leaves, pruned wood, and thinned fruits. Soil nutrient availability and tree nutrient status need to be taken into contemplation along with nutrient needs of tree before developing a fertilization program in any stone fruit orchard. In case of peaches, nutritional requirements are comparatively higher than other stone fruits as it bears on previous year's growth making the production of annual growth imperative for fruit production. Application of inorganic fertilizers may be regulated on the basis of soil test report and leaf analysis. As proposed by Stassen (1987), the nitrogen requirements of peach orchards are as follows:

About 10.5 g N/kg of fruit produced is required by young, nonbearing trees, while only 5.6 g N/kg of fruit produced is required by mature trees. This was attributed to a larger leaf-to-fruit ratio as well as a higher fixation by tree morphology of young trees. In another study, Stassen (2001) stated that for every ton of peaches produced, about 4.0 kg of N is required. Woolridge (2007) stated that in unproductive and sandy soils, fruit produce, N utilization, and overall behavior of trees in 'Keisie' peach orchards are likely to be optimized by the application of 8.4 g N/kg of fruit produced. This was based on the field observations taken during fourth to seventh leaf of the orchard. Leaching, volatilization, and ineffective placement lead to about 30% of the applied nitrogen loss. Such kind of losses may be greater than 30% in sandy and unproductive sites. After subtracting 30% from 8.4 g of N per kg of fruit produced, it would be 5.9 grams of N per kg of fruit, very near to 5.6 g/kg, the value reported by Stassen (1987), but greater than 4.0 kg N/kg, suggested later by Stassen in 2001. To overcome this problem of nutrient losses and to ensure the better uptake, timing of fertilizer application should match the sink demand of plant (Klein and Weinbaum 2000). Therefore, to have profitable effects on tree physiology, time of application should match ample nutrient absorption. The full applications of autumn nitrogen on peach trees resulted in precocious flowering and better fruit set than where the fall nitrogen applications were reduced. Moreover, they concluded that fresh growth throughout the spring was mainly dependent on stored nitrogen which in turn was dependent on the fall application during the preceding year. As per Stassen et al. (1983), a full bearing peach tree must take up at least 48% of the total yearly N-requirement during the postharvest period in order to bring the tree to the same nitrogen level as last year. Stassen et al. (1981) stated that stored N accounted for nearly 65% of the N increase in the fresh growth, while Stassen et al. (1983) stated a value of 80%. This stresses the importance of satisfactory fall applications of nitrogen. Thus, from bud movement during early spring and from the cessation of shoot elongation up to leaf fall where each of these periods should receive around 50% of the annual N requirement are the two main stages when nitrogen needs to be applied to the peach (Stassen 1987). Woolridge (2007), however, for the same 'Keisie' peach orchard as described above, focuses on split applications, i.e., 60% of the total N to be given at full bloom, 30% approximately 42 days after the first application, and the remaining 10% after the cessation of shoot growth during fall.

Similarly, in case of mature plum trees, biologically sound fertilizer scheduling can be developed based on the quantity of yearly removal of nutrients (Weinbaum et al. 1992; Peng et al. 2003). The overall quantity of each nutrient present in the fruit at harvest is directly proportional to fruit yield, and permanent removal of nutrients takes place at every harvest as shown in Table 2.2. Nitrogen is applied in two split doses, one half prior to flowering and the remaining half after fruit set, whereas the entire dose of P and K along with farmyard manure is applied in December to January. The heavy application of K fertilizer is necessary for prune trees during the years of cropping to prevent collapse of trees. Under rainfed conditions, single dose of N fertilizers is applied, i.e., 15 days before bud break. Excessive nitrogen supply has deleterious influence on the stone fruit quality as it significantly reduces fruit firmness and sugar content (Rettke et al. 2006) and leads to fading color (Crisosto et al. 1997) and exposure to various diseases developing after harvest (Daane et al. 1995). However, insufficiency of nitrogen leads to small-sized

Table 2.2 Amount of the yearly macronutrients removed by several stone fruits

Species	N	P	K	Ca	Mg	Note	Literature
Peach	49–170	6–26	60–212	94–259	14–37	Peach BabyGold 5/GF677 Catherina/GF677, on calcareous, clay-loam soil	Rufat and DeJong (2001), El-Jendoubi et al. (2013)
Plum	117–154	10	112	95	17	Dabrowica Plume and Stanley/ Myrobalan on sandy loam soil; 4-year-old Red-Beaut	Plich and Wójcik (2002), Alcaraz-Lopez et al. (2003)
Cherry	39–65	6–11	16–47	26–55	5–9	Unbearing Bing/ Gisela 6; 12-year-old Bigarreau Moreau, Ferrovia, Stella, Colafemmina, Droganova, Montagnola/ Mazzard	Bonomelli et al. (2010), Roversi and Monteforte (2005)

fruits with poor flavor (Taylor 2009). There is a negative impact of potassium deficiency on stone fruit productivity, quality (Chatzitheodorou et al. 2004), and post-harvest life (Ruiz 2006). Johnson and Uriu (1989) stated that peaches under potassium deficiency either lack color or have dull-looking surfaces. The acidic content of fruits is highly affected by the potassium level (Kader and Rolle 2004). In many cases, the nutrient interactions and balanced supply by fertilizer application are more vital than to analyze individually. For example, a balanced level of nitrogen and potassium is responsible for red color development of fruits, and in case of stone fruits, these levels depend on other factors such as cultivar, tree status, soil characteristics, ground cover, and irrigation method.

In cherry orchards, management of potassium (K) supply is imperative (Hanson and Proebsting 1996), as performing vital function in translocation of various metabolites. Therefore, inadequate supply of potassium to stone fruits may result in lower yield and poor quality of fruits. In modern high-density planting systems, the young trees planted are heterografts, and the main part controlling the mineral uptake from the soil is the rootstock. The differential nutrient concentrations transporting to the leaves and fruits may be due to the scion influence on the xylem. For example, Ystaas and Froynes (1998) observed the lower leaf N and K content and considerably higher levels of calcium (Ca) and magnesium (Mg) on Colt than those on seedling Mazzard rootstock. Hrotko et al. (1997) stated that trees on moderately vigorous M × M 14, M × M 97 rootstocks had greater N, P, and K contents in leaves as compared to vigorous Mahaleb SL 64 and Colt. This confirms the influence of rootstocks on leaf mineral content. Thus, in an orchard established with trees on clonal rootstocks, nutrient utilization is much efficient.

2.5 Requirement of Organic Manures in Stone Fruits

Organic manures are valuable by-products of agriculture and allied industries, derived from plant and animal sources. Organic manures which supply plant nutrients in small quantities and are bulky in nature are known as bulk organic manures, e.g., FYM, night soil, compost, etc., and those containing higher percentage of major nutrients are concentrated organic manures, e.g., oil cakes, bone meals, poultry manure, etc. Organic manures influence plant growth and yield favorably through supplying plant nutrients; improving soil fertility like physical, chemical, and biological properties of soil; and enhancing uptake of humic substances and their decomposition products. The optimum growth and development of fruit trees is ensured by the combined use of organic and inorganic fertilizers. Chauhan (2008) suggested that for enhancing the growth of plum trees, integrated application of organic manures along with inorganic fertilizers is a better option. She reported that the application of 80% recommended dose of fertilizer +20 kg vermicompost +60 g biofertilizers gave better results. The application of organic manures should be at the following rates as per the age of the fruit trees. The manure can be applied at any time, but the most favorable time is February to March as shown in Table 2.3.

- Mix biofertilizers, viz., *Azotobacter*, PSB, or AMF (arbuscular mycorrhizal fungi mixed culture), at 1.0 g each/kg of fully decomposed FYM to improve fertilizer use efficiency.
- For nursery plantation, prepare the slurry of Gur at 450 g in 1 L of water, add *Azotobacter* or *Azospirillum* at 100 g in this slurry, dip the seedlings for 1–2 min, plant them immediately, and dry them in shade before planting. Add AMF or PSB 2 1.0 g per kg of FYM.

2.6 Improving Nutrient Use Efficiency

The term nutrient use efficiency refers to yield per unit input, and in horticulture this is related to the input of fertilizer. NUE expresses the ability of plants to make efficient utilization of nutrients for maximum yields. Improvement of nutrient use efficiency is a critical prerequisite for increasing the crop production in marginal lands with limited accessible nutrients. NUE relies not merely on the efficient uptake of nutrients from soil but also on the transportation, storage, mobilization, assimilation, and even on the environmental conditions. It can be enhanced in many ways:

Table 2.3 Application of organic manures to stone fruits

Age of trees (years)	Quantity(kg)	Time of application
1–5	10–20	February to March
6–10	20–30	
11 and above	30–50	

(a) Use of controlled release fertilizers (CRFs).
(b) Application of biofertilizers.
(c) Application of organic matter.
(d) Foliar fertilization.
(e) Fertigation.

(a) **Controlled release fertilizers**

As per the Association of American Plant Food Control Officials, CRFs are considered as those fertilizers holding nutrients in that form which cannot be absorbed by the plants instantly. The coating or encapsulation with inorganic or organic ingredients actually controls the rate, pattern, and length of nutrient release by CRFs. Thus, CRFs control the nutrient release and make them available gradually for longer period than the quick release fertilizers like urea, and release rate should synchronize with the changing crop requirements. Polymer-coated urea exemplifies controlled release fertilizers (Du et al. 2006; Loper and Shober 2012). The mechanism behind the controlled release of nutrients may be either the encapsulation or coating is semipermeable; water after passing into the prill solubilizes the fertilizer to release the nutrients slowly from the prill (Trenkel 2010).

(b) Biofertilizer

The term biofertilizers entitled as microbial inoculants are the carrier-based preparations holding beneficial microbes whose main function is to enrich the soil fertility as well as plant growth by their increased population and biological activity in the root zone (SubbaRao 1998). In other words, biofertilizers are ecofriendly compared to commercial fertilizers as based on renewable energy sources (Verma and Bhattacharyya 1994) and reduce the use of chemical fertilization. Biofertilization provides a great assistance in stimulating production and quality of pome and stone fruits (Thakur and Thakur 2014). Various types of microbes like *Glomus fasciculatum*, *Glomus mosseae*, *Azospirillum*, *Azotobacter*, and phosphate solubilizing bacteria are found beneficial for various fruit trees. The absorption of mobile elements like nitrogen is increased in association with VAM fungi. The higher status of N and organic C showed optimistic association with microbial population; however, more soil phosphorous and zinc content results in the lower fraction and intensity of mycorrhizal colonization. The application of *Azotobacter* encourages nitrogen fixation and synthesis of plant growth-promoting hormones like auxins and gibberellins (Khalid et al. 2004; Singh et al. 2017).

(c) **Organic matter**

The use of organic manures has a good scope for improving soil structure, range of microbial population, soil water-holding capacity, capacity to hold exchangeable cations, and subsequently crop yields (Zink and Allen 1998). Organic matter may be farmyard manure and vermicompost (excreta of earthworms), which are rich in humus and nutrients and are advantageous in mutually with microbes such as *Azotobacter*, *Azospirillum*, phosphate solubilizing bacteria, etc. (Singh et al. 2017). Vermicompost is a microbial decomposition product with a high porosity, aeration, drainage, water-holding capacity, and microbial activity, which are stabilized by a non-thermophilic interaction between earthworms and microorganisms. Vermicompost covers the majority of nutrients in plant-available forms, and in addition to nutrients different microbial secretions act as growth-promoting substances. In case of green manuring, undecomposed green plant tissues are plowed or turned into the soil with the objective of refining soil physical properties as well as overall soil productivity. Apart from the environmental conditions, the quantity and composition of material determines the rate of decay of organic matter (Lourenzi et al. 2011). Organic matter also increases various endogenous hormone levels (viz., Auxin and GA3), accountable for better pollen germination and pollen tube growth, and finally increases the fruit set as well as yield (Sumner 1990; Mahendra and Singh 2009; Bhat et al. 2017). Swierczynski and Stachowiak (2010) reported higher yield from plum and sour cherry orchards inoculated with a mycorrhizal fungi.

The response of fruit quality characteristics and their nutraceutical value to the compost application is still unclear with distinct results observed in different species applied with different types of compost. The nutrient availability is limited as a result of organic fertilization thus stances more stress on plants that divert resources toward synthesis of antioxidants like polyphenols. There is a stimulation of microbial populations due to constant application of compost that allows the degradation of organic C according to temperature fluctuations (Baldi et al. 2010).

(d) **Foliar fertilization in stone fruits**

To meet tree nutrient demand rightly, the use of foliar sprays has been a suggestive means for applying nutrients in controlled quantities, thus being more target-oriented and ecofriendly method of fertilization (Fernandez and Eichert 2009). Considering the limitations of soil application, foliar fertilization is an efficient way for meeting the plant nutrient needs (Wojcik 2004). In case of micronutrients, foliar application is 10–20 times effectual than soil application (Zaman and Schumann 2006). Fruit trees vary significantly in their nutritional needs depending on age, rootstock, and type of cultivars. Usually stone fruits have much higher nutritional needs (N, P, B, Zn, and Mn) than pome fruits. In contrast to pome fruits, stone fruits develop flowers prior to leaves, which means early spring growth is dependent on the nutrients contained in shoots during the summer or autumn of the last year. Thus, it is suggested to go for late summer or fall applications of nutrients in stone fruit trees, especially with foliar fertilizers.

The most common macronutrients applied as foliar fertilizers are nitrogen as urea, ammonium nitrate, and ammonium sulfate; phosphorous as H_3PO_4, KH_2PO_4, $NH_4H_2PO_4$, $Ca(H_2PO_4)_2$, and phosphites; potassium as K_2SO_4, KCl, KNO_3, K_2CO_3, and KH_2PO_4; magnesium as $MgSO_4$, $MgCl_2$, and $Mg(NO_3)_2$; and calcium as $CaCl_2$, Ca propionate, and Ca acetate. Among the micronutrients, boron is applied as boric acid ($B(OH)_3$, borax ($Na_2B_4O_7$), Na-octaborate ($Na_2B_8O_{13}$), and B-polyols; iron as $FeSO_4$, Fe(III)-chelates, and Fe-complexes; manganese as $MnSO_4$ and Mn(II) chelates; and zinc as $ZnSO_4$, Zn(II)-chelates, ZnO, and Zn-organic complexes (Fernandez et al. 2013). Foliar nutrition significantly controls flowering, fruit yield, and fruit quality foliar sprays. Boron sprays increase pollen tube germination and fruit set in most of fruit crops

(Wang et al. 2015). As a result of a field trial in apricot New Castle, it was reported that 400 g N, supplemented with two foliar sprays of urea in April, resulted in higher yield, larger fruit, optimum vegetative growth, and improved N status of leaves (Sud and Bhutani 2015). Zinc nutrition is a key economic factor in cultivation of fruit trees, chiefly in peaches, as it is considered as sensitive to Zn deficiency (Chapman 1966).

(e) **Fertigation in Stone Fruits**

Fertigation is one of the recent techniques of applying fertilizers through drip irrigation systems, which permits the application of various fertilizer formulations directly at the site of roots and thus improves fertilizer use efficiency (Shirgure and Srivastava 2014). This technique had come into existence in the late 1960s in Israel (Goldberg and Shmueli 1970; Magen 1995); however, the earliest study on fertigation was probably proposed by Bryan and Thomas (1958). In recent past few decades, many researchers after using this technique in different stone fruits have pointed out the advantages of fertigation than other means of conventional fertilization. In peach, fertigation significantly improved the tree growth, yield, and physical and chemical properties of fruits (Banyal et al. 2014). The valuable effects of fertigation on yield attributes, leaf and fruit nutrient content, tree growth, nutrient use efficiency, and chemical traits were observed in nectarines (Singh et al. 2015). The fertigation treatments have been found significantly superior in terms of vegetative and reproductive growth, time of application, and yield, nutrient, and quality characteristics in cherry (Neilsen et al. 2010; Salgado et al. 2012). Under drip fertigation, where a portion of soil is wetted, water use efficiency is found to be increased up to 90% (Manohar et al. 2001), and fertilizer use efficiency is also higher which helps to save nutrients up to 80%. Under fertigation, water and fertilizers are supplied at the right time and required levels. Thus, overfeeding is totally avoided and also helps to meet the physiological needs of the trees at different stages of growth. Cultivation of stone fruit trees under light soil condition always possesses problem, and hence fertigation offers growing of these crops under such soil condition and also minimizing soil compaction by avoiding involvement of heavy traffic of equipment as in conventional method of fertilizer application and thereby maintaining and improving the physical, chemical, and biological nature of soil (Haynes and Swift 1987). In conventional approaches, overfertilization and irrigation result in high intensity of weeds and pathogens, whereas fertigation facilitates reduced weed population and contact time of pathogen with the tree (Yarwood 1978; Battilani 2008). Fertigation can be referred as spoon feeding approach of fertilization, where fertilizer requirement of crop is calculated on the basis of individual tree demand for NPK on daily basis over the entire growing period of crop (Kabirigi et al. 2017).

2.7 Nutrient Interactions

Just as humans require balanced proportion of each and every nutrient, plants too demand for the environments of stable plant nutrition. Nutrient balancing among trace elements is as important and yet more challenging than balancing between major nutrients. Interaction among nutrients occurs when application of single nutrient influences the uptake and assimilation of other nutrients, when excess concentration of any nutrient occurs in the growing media, and when others get affected either positively or negatively (Frageria et al. 1997). These interactions occurring either at the root surface or within the plant are divided into two groups. In the first group, the interactions are due to formation of precipitates or complexes which occur between ions which form chemical bonds. The second form of interaction is among ions having competition for the site of adsorption, absorption, transport, and function on plant root surfaces or within plant tissues. Such interactions are more common among the nutrients having alike chemical properties like charge, size, electronic configuration, and geometry of coordination, e.g., Ca^{2+}, Mg^{2+}, K^{+}, and Na^{+}.

Nutrient interaction may be synergistic or antagonistic and also possible to have no interactions. When nutrients in combination result in a growth response that is greater than the sum of their individual effects, the interaction is positive, whereas the negative interactions lead to the lower combined effect. In the former case, the nutrients are synergistic, whereas in the latter they are antagonistic. In case of absence of interaction, there is no deviation from two nutrients' additive response when applied separately (Sumner and Farina 1986). Nitrogen and phosphorous are positively correlated resulting in yield improvement (Terman et al. 1977; Adams 1980). Better understanding of these nutrient interactions in most of the stone fruit crops can lead to more efficient fruit production predominantly in the modern high-density orchards.

2.8 Challenges for Plant Nutrition Management

To continue and wherever feasible to increase the sustainable crop productivity in order to meet the rising demand for food and to improve the quality of land and water resources are the main challenges for plant nutrition management. To maintain the sustainability of any fruit-based cropping system, the nutrients removed from the soil have to be replaced by whatever sources are available. For an orchardist, nutrient losses lead to a huge economic loss. In many developing countries, horticultural production is under grave threat due to inadequate replacement combined with imbalanced nutrient management practices. In industrialized countries, further intensification is limited by ecological concerns and international trade agreements checking surplus food production, while in developing countries, it is limited by the high cost of external sources of nutrients and their scarce accessibility. Taking into account the significance of plant nutrients to horticultural production, it is imperious to set up the correlation among yield, use of nutrients, economic viability, and environmental quality. Sound devotion is now being paid to Integrated Plant Nutrition Systems (IPNS) increasing both soil productivity and crop yields being ecologically, socially, and economically viable. Over the past three decades, in developing countries, about 55% of the yield increase is due to additional nutrients applied as fertilizer.

2.9 Conclusion

In perennial fruit orchards, correct nutrient management is crucial if generative and vegetative progression and fruit quality are to be influenced in an effective way. To avoid either insufficiency or toxicity conditions that might hamper in the typical plant functions is the primary objective of nutrient management. Nutrient management is the practice of scheduling the right amount, source, timing, and method of nutrient application. It mainly takes into account the nutrient requirements of plant, various supply sources, actual nutrient status of the tree, and different management strategies. The flexibility in the timing and precision of nutrient supply is achieved through fertigation which gives greater nutrient mobility (P and K) than broadcasting. Foliar nutrition of micronutrients has a benefit of low application rates, uniform delivery of fertilizer materials, and rapid responses to applied nutrients. Advancing the nutrient use efficiency is very vital mutually from commercial as well as ecological point of view. The most common practice adopted by the farmers to improve the yield and quality of fruit crops is the incorporation of organic fertilizers that limits chemical involvement and finally reduces the opposing influence on the wider environment.

References

Adams, F. (1980). Interactions of phosphorous with other elements in soil and plants. In R. C. Dinaeur (Ed.), *The role of phosphorous in agriculture* (pp. 650–680). Madison: American Society of Agronomy.

Alcaraz-Lopez, C., Botia, M., Alcaraz, C. F., & Riquelme, F. (2003). Effects of foliar sprays containing calcium, magnesium and titanium on plum (Prunus domestica L.) fruit quality. *Journal of Plant Physiology, 160*, 1441–1446.

Arnon, D. I., & Stout, P. R. (1939). The essentiality of certain elements in minute quantity for plants with special reference to copper. *Plant Physiology, 14*, 371–375.

Baldi, E., Toselli, M., Marcolini, G., Quartieri, M., Cirillo, E., & Innocenti, A. (2010). Compost can successfully replace mineral fertilizers in nutrient management of commercial peach orchard. *Soil Use Management, 26*, 346–353.

Banyal, S. K., Sharma, D., & Jarial, K. (2014). Effect of nitrogen fertigation on yield and fruit quality of peach (Prunus persica L.) under low hill conditions of Himachal Pradesh. *Current Advances in Agricultural Sciences, 6*, 158–160.

Battilani, A. (2008). Manipulating quality of horticultural crops with fertigation. *Acta Horticulturae, 792*, 47–60.

Bengough, A. G., McKenzie, B. M., Hallettand, P. D., & Valentine, T. A. (2011). Root elongation, water stress, and mechanical impedance: A review of limiting stresses and beneficial root tip traits. *Journal of Experimental Botany, 62*, 59–68.

Bhat, S., Lal, R. L., Misra, K. K., & Singh, D. K. (2017). Response of organic formulations on fruiting and yield of litchi (Litchi chinensis Sonn.) cv. Rose Scented. *Journal of Pharmacognosy and Phytochemistry, 6*, 2092–2095.

Bonomelli, C., Gilabert, H., & Ayala, M. (2010). Nitrogen uptake, growth and biomass accumulation in the first growing season of cherry trees on Gisela®6. *Acta Horticulturae, 868*, 177–183.

Bryan, B. B., & Thomas, E. L.(1958). *Distribution of fertilizer materials applied through sprinkler irrigation systems*. University of Arkansas Agricultural Experimental Station 59812, Fayetteville.

Chapman, H. D. (1966). Zinc. In H. D. Chapman (Ed.), *Diagonostic criteria for plants and soils* (pp. 484–499). Brekley: University of California's Division of Agriculture Science.

Chatzitheodorou, I. T., Sotiropoulos, T. E., & Mouhtaridou, G. I. (2004). Effect of nitrogen, phosphorus, potassium fertilization and manure on fruit yield and fruit quality of the peach cultivars 'Spring Time' and 'Red Haven'. *Agronomy Re Agronomy Research, Madison, 2*, 135–143.

Chauhan, A. (2008). *Studies on integrated nutrient management in plum cv. Santa Rosa*. Ph.D. thesis, Dr Y. S. Parmar University of Horticulture and Forestry, Nauni, Solan, H.P., India.

Crisosto, C. H., Johnson, R. S., Dejong, T., & Day, K. R. (1997). Orchard factors affecting postharvest stone fruit quality. *HortScience, 32*, 820–823.

Daane, K. M., Johnson, R. S., Michailides, T. J., Crisosto, C. H., Dlott, J. W., Ramirez, H. T., Yokota, G. T., & Morgan, D. P. (1995). Excess nitrogen susceptibility to raises nectarine disease and insects. *California Agriculture, Berkeley, 49*(4), 13–17.

Devlin, R. (1975). *Plant physiology* (3rd ed.). New York: D. Van Nostrand, 600p.

Du, C., Zhou, Z., & Shaviv, A. (2006). Release characteristics of nutrients from polymer-coated compound controlled release fertilizers. *Journal of Polymers and the Environment, 14*(3), 223–230.

El-Jendoubi, H., Abadía, J., & Abadía, A. (2013). Assessment of nutrient removal in bearing peach trees (Prunus persica L. Batsch) based on whole tree analysis. *Plant and Soil, 369*(1–2), 421–437.

Fernandez, V., & Eichert, T. (2009). Uptake of hydrophilic solutes through plant leaves: current state of knowledge and perspectives of foliar fertilization. *Critical Reviews in Plant Sciences, 28*, 36–68.

Fernandez, V., Sotiropoulos, T., & Brown, P. H. (2013). *Foliar fertilisation: Principles and practices*. Paris: International Fertilizer Industry Association (IFA).

Frageria, N. K., Baligar, V. C., & Jones, C. A. (1997). *Growth and mineral nutrition of crop plants* (2nd ed.). New York: Marcel Dekker.

Goldberg, D., & Shmueli, M. (1970). Drip irrigation a method used under arid and desert conditions of high water and soil salinity. *American Society of Agricultural Engineers, 13*, 38–41.

Hanson, E. J., & Proebsting, E. I. (1996). Cherry nutrient requirements and water relations. In A. D. Webster & N. E. Looney (Eds.), *Cherries: Crop physiology, production and uses* (pp. 243–257). Wallingford: CAB International.

Haynes, R. J., & Swift, R. S. (1987). Effect of trickle fertigation with three forms of nitrogen on soil pH, levels of extractable nutrients below the emitter and plant growth. *Plant and Soil, 102*, 211–221.

Hrotko, K., Hanusz, B., Papp, J. & Simon. G. (1997). Effect of rootstocks on leaf nutrient status of sweet cherry trees. In *Third International Cherry Symposium 1997*, July 23–29, Norway-Denmark. Programme and Abstracts (p. 103).

Johnson, J. (2011). *4Rs right for nutrient management*. Natural resources conservation. Retrieved from service.www.nrcs.usda.gov

Johnson, R. S., & Uriu, K. (1989). Mineral nutrition. In J. H. La Rue & R. S. Johnson (Eds.), *Peach, plums and nectarines: Growing and handling for fresh market*. Berkeley: University of California, 252p. (Cooperative Extension, 3331).

Kabirigi, M., Shrestha, O. P., Prescella, B. V., Niamwiza, C., Quintin, S. P., & Mwamjengwa, I. A. (2017). Fertigation for environmentally friendly fertilizers application: Constraints and opportunities for its application in developing countries. *Agricultural Sciences, 8*, 292–301.

Kader, A. A., & Rolle, R. S.(2004). *The role of post-harvest management in assuring the quality and safety of horticultural produce* (FAO Bulletin, Vol. 152). Washington, 52p.

Khalid, A., Arshad, M., & Zahir, Z. A. (2004). Screening plant growth promoting rhizobacteria for improving growth and yield of wheat. *Journal of Applied Microbiology, 96*, 473–480.

Khera, A. P., Makhija, M., Chitkara, S. D., & Chauhan, K. S. (1981). Development of leaf nutrient concentration standards for subtropical peach (Prunus persica). *Haryana Agriculture University Journal of Research, 11*, 181–184.

Klein, I., & Weinbaum, S. A. (2000). Fertilization of temperate-zone fruit trees in warm and dry climates. In E. Amnon (Ed.), *Temperate fruit crops in warm climates*. Amsterdam: Kluwer Academic Publishers.

Lehnert, R. (2010). Nitrogen fertilization in apples. Apples, August 2010 Issue, Production.

Loper, S., & Shober, A. L. (2012). *Soils & fertilizers for master gardeners: Glossary of soil*.

Lourenzi, C. L., Ceretta, C. A., Silva, L. S., Trentin, G., Girotto, E., Lorensini, F., Tiecher, T. L., & Brunetto, G. (2011). Soil chemical properties related to acidity under successive pig slurry application. *Revista Brasileira de Ciência do Solo, 35*, 1827–1836.

Magen, H. (1995). Fertigation: An overview of some practical aspects. *Fertilizer News, 40*, 97–100.

Mahendra, S. H. K., & Singh, J. K. (2009). Studies on integrated nutrient management on vegetative growth, fruiting behavior and soil fertilizer status of ber (Zizypus mauritiana Lamk.) orchard cv. Banarasi Karaka. *The Asian Journal of Horticulture, 4*(1), 230–232.

Manohar, K., Khan, R., Kariyanna, M. M., & Sreerama, R. (2001). An overview of status, potential and research accomplishment of drip irrigation in Karnataka. In *Proceedings of International Conference on Micro and Sprinkler Irrigation System* (pp. 69–79).

Merwe, P. D. J. (2012). *The effects of organic and inorganic mulches on the yield and fruit quality of Cripp's Pink apple trees*. M.Sc. thesis, Faculty of Agriculture, Stellenbosch University, p. 126.

Neilsen, D., & Neilsen, G. H. (2002). Efficient use of nitrogen and water in high-density apple orchards. *HortTechnology, 12*, 19–25.

Neilsen, G. H., Neilsen, D., Kappel, F., Toivonen, P., & Herbert, L. (2010). Factors affecting establishment of sweet cherry on Gisela 6 rootstock. *HortScience, 45*(6), 939–945.

Peach, J. D. (2010). *Temperate horticulture current scenario* (pp. 73–88). New Delhi: Oxford Book Company.

Peng, F. T., Jiang, M. Y., Gu, M. R., & Shu, H. R. (2003). Advances in research on nitrogen nutrition of deciduous fruit crops. *Journal of Fruit Science, 20*, 54–58.

Plich, H., & Wójcik, P. (2002). The effect of calcium and boron foliar application on postharvest plum fruit quality. *Acta Horticulturae, 594*, 445–451.

Rettke, M. A., Pitt, T. R., Maier, N. A., & Jones, J. A. (2006). Quality of fresh and dried fruit of apricot (cv. Mooprark) in response to soil applied nitrogen. *Australian Journal of Experimental Agriculture, Melbourne, 46*(1), 123–129.

Roversi, A., & Monteforte, A. (2005). Preliminary results on the mineral uptake of six sweet cherry varieties. *Acta Horticulturae, 721*, 123–128.

Rufat, J., & DeJong, T. M. (2001). Estimating seasonal nitrogen dynamics in peach trees in response to nitrogen availability. *Tree Physiology, 21*, 1133–1140.

Ruiz, R. (2006). Effects of different potassium fertilizers on yield, fruit quality and nutritional status of 'Fairlane' nectarine trees and on soil fertility. *Acta Horticulturae, The Hague, 721*, 185–190.

Salgado, E., Livellara, N., & Pinilla, J. (2012). Programmed fertigation effects on the growth and production of young cherry trees in central Chile. *Journal of Soil Science and Plant Nutrition., 121*, 15–22.

Scholberg, J., & Morgan, K. T. (2012). Nutrient use efficiency in citrus. In A. K. Srivastava (Ed.), *Advances in citrus nutrition* (pp. 205–229). The Netherland: Springer Verlag.

Shahbandeh, M. (2020). *Global stone fruit production 2000–2017*. Retrieved from https://www.statista.com/statistics/577598/world-stone-fruit-production

Sharpe, R. H., Sherman, W. B., & Martsolf, J. D. (1990). Peach cultivars in Florida and their chilling requirements. *Acta Horticulturae, 279*, 191–197.

Shirgure, P. S., & Srivastava, A. K. (2014). Fertigation in perennial fruit crops: Major concerns. *Agrotechnology, 3*(1), 1–2.

Singh, D., Sharma, S. D., & Kumar, P. (2015). Nitrogen fertigation for nectarines (Prunus persica var. nucipersica): Lateral and vertical nutrient acquisition and cropping behaviour in rainfed agro-ecosystem. *Indian Journal of Agricultural Sciences, 85*(11), 1440–1447.

Singh, Y., Satya, P., Om, P., & Dharmendra, K. (2017). Effect of integrated nutrient management on fruit yield and quality of Amrapali mango (Mangifera indica L.) under high density planting. *International Journal of Pure and Applied Bioscience, 5*(3), 6773.

Srivastava, A. K., Singh, S., & Albrigo, L. G. (2008). Diagnosis and remediation of nutrient constraints in citrus. *Horticultural Reviews, 34*, 277–364.

Stassen, P. J. C. (1987). Macro-element content and distribution in peach trees. *Deciduous Fruit Grower, 37*, 245–249.

Stassen, P. J. C. (2001). Is voeding–en waterbestuurvolgens 'n oophidroponiesebenadering lewensvatbaarvirtafeldruifverbouing? SuidAfrikaanseWingerd –en WynkundeVereniging Tafel – en DroogdruifKortkursus.

Stassen, P. J. C., Terblanche, J. H., & Strydom, D. K. (1981). The effect of time and rate of nitrogen application on development and composition of peach trees. *Agroplantae, 13*, 55–61.

Stassen, P. J. C., Du Preez, M., & Stadler, J. D. (1983). Reserves in full bearing peach trees. Macro element reserves and their role in peach trees. *Deciduous Fruit Grower, 33*, 200–206.

SubbaRao, N. S. (1998). *Biofertilizer in agriculture and forestry* (3rd ed., p. 242). New Delhi: Oxford and IBH.

Sud, G., & Bhutani, V. P. (2015). Effects of soil and foliar applications of urea on productivity and the tree nitrogen status of apricot. *Journal of Horticultural Science, 63*(4), 585.

Sumner, M. E. (1990). Crop response to Azospirillium inoculation. *Advances of Soil Science, 12*, 53–123.

Sumner, M. E., & Farina, M. P. W. (1986). Phosphorous interactions with other nutrients and lime in field cropping systems. *Advances of Soil Science, 5*, 201–236.

Swierczynski, S., & Stachowiak, A. (2010). The influence of mycorrhizal fungi on the growth and yielding of plum and sour cherry trees. *Journal of Fruit and Ornamental Plant Research, 18*, 71–77.

Tagliavini, M. & Scandellari, F. (2012). Methodologies and concepts in the study of nutrient uptake requirements and partitioning in fruit trees. In *Proceedings of the Seventh International Symposium on Mineral Nutrition of Fruit Crops*, Chanthaburi, Thailand, 19–25 May (pp. 19–25).

Taylor, K. C. (2009). *Cultural management of the bearing peach orchards*. Athens: University of Georgia. 4p. (circular, 879).

Terman, G. L., Noggle, J. C., & Hunt, C. M. (1977). Growth rate nutrient concentration relationships during early growth of corn as affected by applied N, P and K. *Soil Science Society of America, 41*, 363–368.

Thakur, N., & Thakur, B. S. (2014). Studies on the effect of integrated nutrient management on growth and yield of plum cv. Santa Rosa. *The Asian Journal of Horticulture, 9*(1), 112–115.

Trenkel, M. E. (2010). *Slow- and controlled-release and stabilized fertilizers: An option for enhancing nutrient use efficiency in agriculture*. Paris: International Fertilizer Industry Association (IFA).

Verma, L. N., & Bhattacharyya, P. (1994). *Production distribution and promotion of biofertilizers. Fertilizers, organic manures.recyclable wastes and biofertilizers* (pp. 132–147). New Delhi: FD Co.

Wang, N., Yang, C., Pan, Z., Liu, Y., & Peng, S. (2015). Boron deficiency in woody plants: Various responses and tolerance mechanisms. *Frontiers in Plant Science, 6*, 916.

Weinbaum, S. A., Johnson, R. S., & DeJong, T. M. (1992). Causes and consequences of over fertilization in orchards. *Hort Technology, 2*, 112–121.

Williams, M. W., & Billingsley, H. D. (1974). Effect of nitrogen fertilizers on yield size and color of golden delicious apples. *Journal of American Society of Horticultural Sciences, 2*, 144–145.

Wojcik, P. (2004). Uptake of mineral nutrients from foliar fertilization. *Journal of Fruit and Ornamental Plant Research, 12*, 201–218.

Woolridge, J. (2007). Effects of early season and autumn nitrogen applications on Keisie canning peach. *SA Fruit Journal, 6*(3), 22–23.

Wu, F. Q., Liu, H. B., Sun, B. S., Wang, J., & Gale, W. J. (2008). Net primary production and nutrient cycling in an apple orchard-annual crop system in the Loess Plateau, China: A comparison of Qinguan apple, Fuji apple, corn and millet production subsystems. *Nutrient Cycling in Agroecosystems, 81*, 95–105.

Yarwood, C. E. (1978). Water and the infection process. In T. T. Kozlowski (Ed.), *Water deficit and plant growth* (Vol. 5, pp. 141–165). New York: Academic Press.
Ystaas, J., & Froynes, O. (1998). The influence of eleven cherry rootstocks on the mineral leaf content of major nutrients in "stella" and "ulster" sweet cherries. *Acta Horticulturae, 468*, 367–372.
Zaman, Q., & Schumann, A. W. (2006). Nutrient management zones for citrus based on variation in soil properties and tree performance. *Precision Agriculture, 7*, 45–63.
Zink, T. A., & Allen, M. F. (1998). The effect of organic amendments on the restoration of a disturbed coastal sage scrub habitat. *Restoration Ecology, 6*, 52–58.

Pollination Management in Stone Fruit Crops

3

Sara Herrera, Jorge Lora, José I. Hormaza, and Javier Rodrigo

Abstract

Pollination and subsequent fertilization are needed to ensure fruit set in stone fruit crops. Knowing the pollination requirements for particular cultivars is a necessity for an appropriate orchard design. Pollination management is receiving increasing importance due to the release of a number of new cultivars of unknown pollination requirements in most *Prunus* species. In addition, variable environmental conditions due to climate change and the interest in expanding stone fruit crops to new production areas with different climatic conditions are leading to a lack of coincidence in flowering time between pollinating and pollinated cultivars in many situations. In this work, the available information on pollination requirements of cultivars of the most cultivated stone fruit crops (almonds, apricots, cherries, peaches, and plums) is reviewed, paying special attention to the reproductive process, the self-(in)compatibility of each cultivar, the inter-incompatibility *S*-alleles and incompatibility groups, and the external factors affecting flowering and pollination. Results obtained in the last decades have allowed establishing the pollination requirements of most commercial cultivars in the main stone fruit crops. However, for the coming years, it will be necessary to continue determining the self-(in)compatibility and the inter-incompatibility relationships of the new releases.

S. Herrera · J. Rodrigo
Unidad de Hortofruticultura, Centro de Investigación y Tecnología Agroalimentaria de Aragón (CITA), Zaragoza, Spain

Instituto Agroalimentario de Aragón – IA2 (CITA-Universidad de Zaragoza), Zaragoza, Spain

J. Lora · J. I. Hormaza (✉)
Instituto de Hortofruticultura Subtropical y Mediterránea La Mayora (IHSM La Mayora-UMA-CSIC), Málaga, Spain
e-mail: ihormaza@eelm.csic.es

M. M. Mir et al. (eds.), *Production Technology of Stone Fruits*,
https://doi.org/10.1007/978-981-15-8920-1_3

Keywords

Almond · Apricot · Cherry · Fertilization · Flowering · Inter-incompatibility · Peach · Plum · Pollination · Self-(in)compatibility

3.1 Introduction

Stone fruit crops include different plant species that produce a type of indehiscent fruit called drupe that contains a single large seed covered by an outer shell that constitutes the endocarp. The endocarp together with the fleshy mesocarp (pulp) and exocarp (skin) constitute the pericarp. The pericarp and the seed coat are of maternal origin. Most stone fruit tree crops belong to the genus *Prunus*, and examples include peach (*P. persica* L. Batsch), European (*P. armeniaca* L.) and Japanese (*P. mume* Siebold & Zucc.) apricots, sweet (*P. avium* L.) and sour (*P. cerasus* L.) cherries, almond (*P. dulcis* (Mill.) D.A. Webb), or different species of plums such as the European (*P. domestica* L.) and the Japanese (*P. salicina* Lindl.) plums (Byrne 2005; Badenes and Byrne 2012).

One of the main factors that limit production in crops in which the main product is fruits is the reproductive phase. During this phase, a critical stage is the days following flower opening, since pollination and subsequent fertilization are needed to ensure fruit set because *Prunus* species are unable to set fruit parthenocarpically (Sedgley and Griffin 1989). Indeed, some efforts to induce artificial parthenocarpy using gibberellins have been unsuccessful (Bukovac 1963; Webster and Goldwin 1978). This period between pollination and fertilization is the progamic phase and takes place in a few days, but the development of the flower buds occurs for several months, and a number of external and internal factors during this period may condition the success of the reproductive process (Hedhly et al. 2009; Hedhly 2011). Knowing the pollination requirements for particular cultivars is a requirement for an appropriate orchard design and management.

Although the fruit crops considered as stone fruits belong to the same genus, *Prunus*, in the Rosaceae family, there is a high variability among species in flower characteristics, chilling and heat requirements, pollination behavior, and flowering times. However, the main biological processes are conserved although with different timing of the different processes among species.

In recent years, pollination management is emerging as a field with increasing importance, due to the release, in most stone fruit crops, of a number of new cultivars with unknown pollination requirements. In addition to understanding incompatibility relationships among cultivars, variable environmental conditions can lead to a lack of coincidence in flowering time between pollinating and pollinated cultivars (Atkinson et al. 2013). Thus, the increasing interest in expanding the marketing season in many crops is resulting in the cultivation of different fruit crop species in new production areas with different climatic conditions (Hedhly et al. 2009; Fadón and Rodrigo 2018). On the other hand, the decrease in winter chilling in many regions caused by climate change is resulting in alterations in flowering time (Luedeling et al. 2011; Atkinson et al. 2013) and, as a consequence, in the advance of the flowering season in low-chilling cultivars and the delay in high-chilling cultivars (Luedeling 2012). In this work, the available information on pollination requirements of cultivars of the most cultivated stone fruits (almonds, apricots, cherries, peaches, and plums) is reviewed, paying special attention to the reproductive process and the factors affecting flowering and pollination.

3.2 Reproductive Biology in *Prunus*

Flower bud development in *Prunus* species requires several months from the summer of the previous year to flowering in late winter or early spring of the following season (Fadón et al. 2015). Flower induction occurs during the summer, and flower differentiation starts before dormancy in late summer and autumn (Fadón et al. 2018a, b). After breaking dormancy, flowers complete differentiation when temperatures rise.

In autumn, prior to the exposure to low temperatures, flower buds cease development and progressively enter in a dormant stage, endodormancy (Lang et al. 1987), during which they are completely closed and covered by brown scales, corresponding to phenological stage A in the Baggiolini code (Baggiolini 1952) or to 50/00 in the BBCH scale (Fadón et al. 2015). Each flower bud can contain one (almond, apricot, peach) (Herrero and Arbeloa 1989; Kester and Gradziel 1996; Julian et al. 2010) or several (sweet and sour cherry, plum) (Fadón et al. 2015; Guerra and Rodrigo 2015; Herrero et al. 2017) flower primordia. In each flower primordium, the different whorls (sepals, petals, stamens and carpels) start to differentiate before endodormancy (Fadón et al. 2018a, b). The carpel and stamens are surrounded by the petals, which, in turn, are surrounded by the sepals (Diaz et al. 1981; Lamp et al. 2001; Julian et al. 2010; Fadón et al. 2015).

Flowers of *Prunus* are hermaphroditic, with numerous stamens surrounding a single pistil, enclosed by five petals and five sepals (Sterling 1964). Flowering time depends on the cultivar and the weather conditions that may cause high year-to-year variations. Pollination, the transfer of pollen from the anthers to the stigma, is usually entomophylous, mainly by bees (Griggs 1953, Fig. 3.1). The sequence of events from pollination to fertilization is consistent in different species of the genus (Herrero 1992, 2000).

Pollination can also be hindered by a short period of stigmatic receptivity that reduces the effective pollination period (EPP) in which pollination is effective to produce a fruit. The length of the EPP in stone fruits varies depending on location, season, cultivar, and environmental conditions (mainly temperature and humidity), ranging from 2 days to more than 1 week (Sanzol and Herrero 2001). Usually, stigmas are receptive at anthesis, as it has been shown in apricot (Egea et al. 1991), peach (Martínez-Tellez and Crossa-Raynaud 1982), or sweet cherry (Guerrero-Prieto et al. 1985), although immature stigmas at anthesis have been observed in some almond (Yi et al. 2006) and peach cultivars (Sanzol and Herrero 2001).

Fig. 3.1 Honeybee pollinating apricot flowers

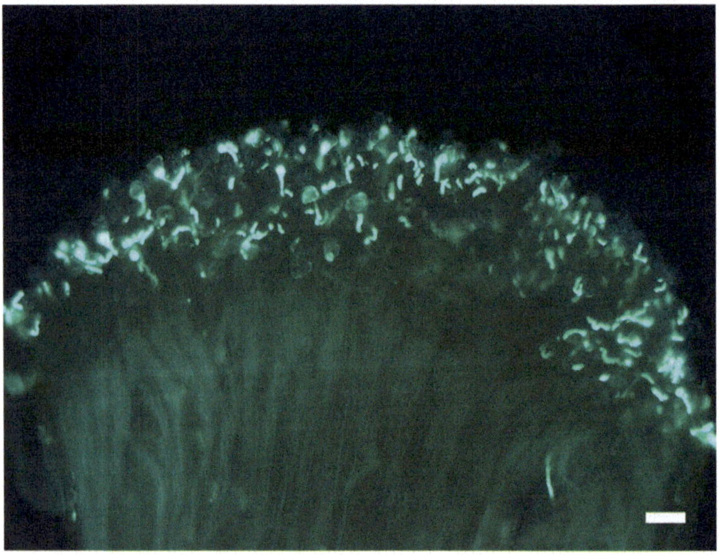

Fig. 3.2 Pollen grains germinating on the stigma surface of an apricot flower. Bar = 100 μm

The EPP can be shortened by warm temperatures that can accelerate both stigma and ovule degeneration (Ortega et al. 2004; Hedhly et al. 2005; Kodad and Socias i Company R 2013; Zhang et al. 2018).

The pollen grain germinates on the stigma within few hours after pollination (Fig. 3.2), producing a pollen tube that grows along the closed style and reaches the ovary in the following days, where it goes through one of the two ovules reaching the embryo sac and achieving double fertilization. Intense competition among pollen tubes occurs in the style, reducing the number of more than 100 germinating pollen grains on the stigma to just one to three pollen tubes reaching the ovary (Hormaza and Herrero 1996). The duration of this progamic phase may vary between species, cultivars, and years (Williams 1970; Sedgley and Griffin 1989; Guerra and Rodrigo 2015). Thus, this phase can last for 4–8 days in sweet and sour cherry and Japanese plum (Cerovic and Micic 1999; Hedhly et al. 2007; Jia et al. 2008), 5–8 days in apricot and almond (Pimienta and Polito 1983; Rodrigo and Herrero 2002a), and more than 2 weeks in peach and European plum (Thompson and Liu 1973; Herrero and Arbeloa 1989).

Only a reduced number of the pollinated flowers develop into fruits. The percentage of fruit set varies among species and cultivars: 5–15% in Japanese plum, (Dorsey 1919; Beppu et al. 2005; Jia et al. 2008; Guerra et al. 2010), 18–33% in apricot and sweet cherry (Rodrigo and Herrero 2002a; Hedhly et al. 2007), 30–40% in almond and peach (Harrold 1935; Kester and Griggs 1959), and 25–50% in sour cherry (Bradbury 1929), although the final fruit set is highly dependent on pollination and weather conditions.

3.3 Pollen-Pistil Incompatibility

Different barrier mechanisms to avoid self-fertilization and promote outcrossing have arisen in seed plants. Some are physical barriers such as dichogamy, in which the female and male parts do not mature at the same time, or heterostyly in which populations are composed of two (distyly) or three (tristyly) different floral morphs with reciprocal arrangements of height of anthers and stigmas (reciprocal herkogamy). However, the most common mechanism in evolutionary-derived angiosperms, such as the Rosaceae, is a genetic self-incompatibility system based in cell-to-cell recognition. There are different systems of genetic self-incompatibility, and the most widespread in flowering plants is gametophytic self-incompatibility (GSI) in which the cell-to-cell recognition occurs in the style (Kao and McCubbin 1996) and is controlled by a multiallelic locus called *S*-locus (Kao and Huang 1994) (Fig. 3.3). GSI has been found in Solanaceae, Plantaginaceae, and Rosaceae, including stone fruit species of *Prunus* such as apricot, peach, almond, European plum, Japanese plum, sweet cherry, and sour cherry. Since fertilization and seed formation are essential for fruit production, the knowledge of the genetic basis of GSI is essential for the development of breeding programs and efficient design of stone fruit orchards (Herrera et al. 2018a, b).

GSI has been extensively studied in Solanaceae. Indeed, the first protein associated with self-incompatibility, a stylar glycoprotein, was isolated in *Nicotiana alata* (Anderson et al. 1986). A few years later, several S-glycoproteins were shown to be RNases (McClure et al. 1989, 1990). The S-glycoprotein is secreted in the extracellular matrix and enters the pollen tube, where the incompatibility reaction, as hypothesized in the inhibitor model, has been supported by immunocytochemical evidences in *Solanum chacoensis* (Luu et al. 2000). Similar evidences are still lacking in Rosaceae, but the inhibitor model has also been generally accepted in the

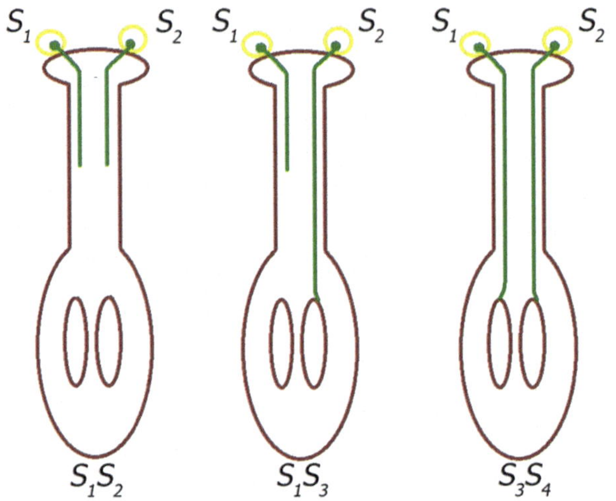

Fig. 3.3 Diagram of the gametophytic self-incompatibility system (GSI) present in stone fruit crops. Pollen tubes containing an *S*-allele present in the pistil will not reach the ovule, avoiding fertilization

cell-to-cell recognition in this family (Tao and Iezzoni 2010). Thus, the recognition between the female (RNase) and the male S-determinants occurs in the pollen tube, in which only the self-RNase shows cytotoxic activity. In Rosaceae, the first *S-RNase* DNA sequence was isolated from apple (*Malus* × *domestica*) (Broothaerts et al. 1995), although previously an S-RNase protein had been isolated from *Pyrus* (Sassa et al. 1993). Later on, *S-RNases* were also identified and isolated in almond (Tao et al. 1997), sweet cherry (Tao et al. 1999; Sonneveld et al. 2001), sour cherry (Yamane et al. 2001), Japanese plum (Yamane et al. 1999), apricot (Romero et al. 2004), and peach (Tao et al. 2006). In other cases, such as European plum, self-incompatibility is relatively rare probably due to its hexaploid ($2n = 6x = 48$) genome (Hegedűs and Halász 2006) although *S-RNase* alleles have been characterized in this species (Sutherland et al. 2004). This breakdown of SI in polyploid species has also been observed in tetraploids in Solanaceae and Plantaginaceae (Tao and Iezzoni 2010), but, interestingly, this is not the case in tetraploid sour cherry. This has been explained in terms of the disfunction of some *S*-haplotypes caused by polyploidization and gene duplication in sour cherry that has not been observed in diploid sweet cherry (Tsukamoto et al. 2006).

During the first years of the twenty-first century, several studies in *Antirrhinum* (Plantaginaceae) (Lai et al. 2002), *Prunus* (Rosaceae) (Entani et al. 2003; Ushijima et al. 2003, 2004), and *Petunia* (Solanaceae) (Sijacic et al. 2004) identified the pollen S-determinant as an F-box protein. In almond, the *F-box* gene named *SFB* (*S* haplotype-specific F-box protein) was linked to the *S-RNase* and showed a specific pollen expression (Ushijima et al. 2003). Studies of the S_1 and S_7 haplotypes of *Prunus mume* allowed the identification of an *F-box* in this species that was named *SLF* (Entani et al. 2003). The linkage between the *F-box* and the *S-RNase* genes was also confirmed in *Prunus armeniaca* in which three *F*-box allelic variants (SFB_1, SFB_2, and SFB_4) were linked to the three *S-RNase* alleles S_1, S_2, and S_4, respectively (Romero et al. 2004).

The F-box protein is a component of the Skp1/Cullin or CDC53/F-box (SCF) complex that regulates protein degradation, expected to be involved in detoxification of non-self S-RNases by the ubiquitin-proteasome pathway in self-compatible interactions (del Pozo and Estelle 2000; Qiao et al. 2004). According to this model, self-compatible cultivars with mutations in the *F-box* gene have not been observed in Solanaceae and Plantaginaceae since it is believed that these mutations would result in self-incompatibility or lethality (Tao and Iezzoni 2010). However, mutations in the *F-box* gene result in self-compatibility in *Prunus*. That is the case for most traditional apricot cultivars in Western Europe (Vilanova et al. 2006) and Hungary (Halász et al. 2007b). *SFB* mutations have also been reported in self-compatible cultivars of peach, sweet cherry, and Japanese apricot (Tao et al. 2006; Abdallah et al. 2020). Thus, in *Prunus*, an *S*-RNase inhibitor that interacts with F-box has been proposed (Tao and Iezzoni 2010). Indeed, recent experimental evidence suggested an *S*-locus F-box like-2 protein as the "general inhibitor" that detoxifies S-RNase in a nonspecific manner if it is not affected by the F-box protein (Matsumoto and Tao 2016).

The disruption of the female *S*-determinant can also result in self-compatibility. Mutations in the *S-RNase* gene have been reported in several species of the Rosaceae, such as pear (Sassa et al. 1997), peach (Tao et al. 2006), or almond (Bošković et al. 2007). In addition to the disruptions in the male or female *S*-determinant genes, the study of self-compatible cultivars in the Rosaceae has revealed additional factors not linked to the *S*-locus involved in cell-cell recognition. Thus, the *M*-locus, localized in chromosome 3 (Zuriaga et al. 2012), appears to be involved in self-compatibility in the apricot cultivars Katy and Canino (Zuriaga et al. 2013; Muñoz-Sanz et al. 2017). In sweet cherry, in which natural self-compatibility is rare, self-compatibility was also found associated with a modifier gene in the Spanish cultivar Cristobalina (Wünsch and Hormaza 2004a; Cachi and Wünsch 2011).

3.4 Pollination Requirements

Although most of the species of the *Prunus* genus exhibit GSI, mutations in both the *F-box* and *SFB* genes have resulted in a variable number of self-compatible genotypes in different species (Tao et al. 2006; Vilanova et al. 2006; Bošković et al. 2007; Halász et al. 2007a). Both compatible and incompatible pollen grains are able to germinate on the stigma producing a pollen tube that grows along the style (Fig. 3.4). However, pollen tube growth is arrested in the style (Fig. 3.5), preventing fertilization in incompatible relationships, whereas compatible pollen tubes may reach the ovary and fertilize the ovules (Fig. 3.3). Self-(in)compatibility affects orchard management, since self-incompatible cultivars need compatible pollen from other cultivars, and, therefore, it is necessary to design the orchards with trees

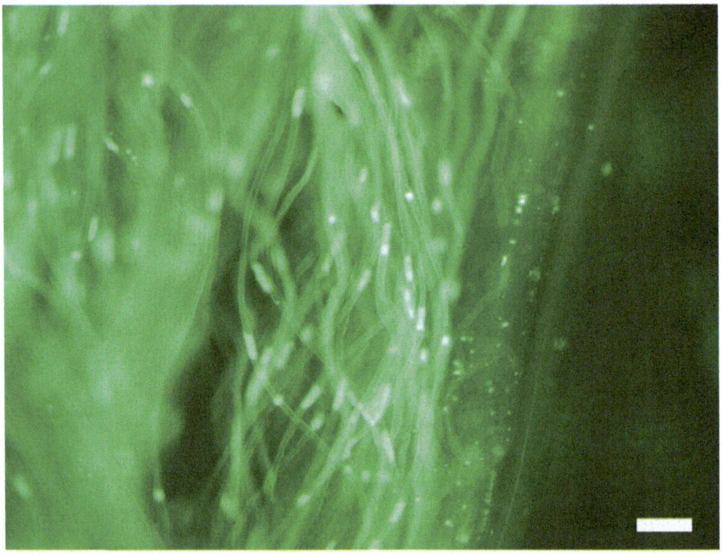

Fig. 3.4 Pollen tubes growing along the style of an apricot flower. Bar = 100 μm

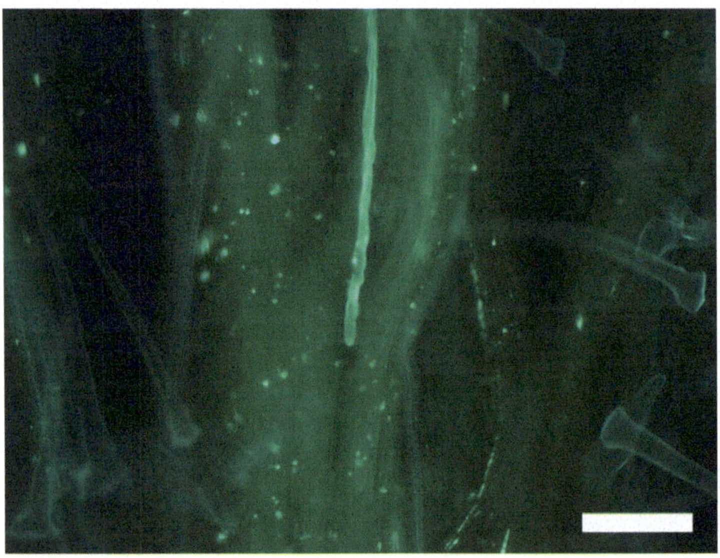

Fig. 3.5 Pollen tube arrested in the style of an apricot flower. Bar = 100 μm

of several inter-compatible cultivars. On the other hand, the ovules of self-compatible cultivars can be fertilized with their own pollen and, therefore, can be grown without the presence of trees of other cultivars.

Peach commercial cultivars are self-compatible (Abdallah et al. 2020), but in the other cultivated *Prunus* species, both self-incompatible and self-compatible cultivars are usually found. In almond, sweet cherry, and Japanese and European plums, most commercial cultivars are self-incompatible and need cross-pollination, whereas most apricot cultivars are self-compatible. However, the release of new cultivars from breeding programs in which self-compatibility is one of their main objectives is resulting in an increasing number of self-compatible cultivars in almond (Table 3.1), sweet cherry (Table 3.2), and Japanese plum (Table 3.3). On the other hand, the use of self-incompatible cultivars as parents in apricot breeding programs, with the purpose of introducing a source of resistance to sharka, a common disease among stone fruit crops caused by the plum pox virus (PPV), is resulting in the release of an increasing number of self-incompatible cultivars (Table 3.4).

3.5 Inter-incompatibility *S*-Alleles and Incompatibility Groups

The sequences of the *S-RNase* gene reveal a high diversity that results in numerous *S*-alleles, although all showing five highly conserved regions (C1–C5). The conserved regions C1, C2, C3, and C5 are similar in Plantaginaceae, Solanaceae, and Rosaceae, but the fourth region differs in Rosaceae and is called RC4. In Rosaceae, one hypervariable region (RHV) is found, whereas two (HVa and HVb) are found in Plantaginaceae and Solanaceae (Fig. 3.6) (Tao and Iezzoni 2010). In addition to the high diversity of the encoding gene sequences, a higher diversity is found in the introns, and this diversity is essential for *S*-allele identification in addition to the general use of specific primers from the exon sequences (Herrera et al. 2018a, b) (Fig. 3.7).

Knowing the *S*-alleles of each cultivar is necessary to select compatible pollinizers in the design of new orchards and also to identify and solve situations of low yield related to lack of pollination. In different *Prunus* species, the identification of a variable number of *S-RNase* and *SFB* alleles has allowed allocating the self-incompatible cultivars in incompatibility groups: almond (Table 3.1), sweet cherry (Table 3.2), Japanese plum (Table 3.3), apricot (Table 3.4), European plum (Table 3.5), and sour cherry (Table 3.6). Those cultivars with the same *S*-alleles are genetically inter-incompatible and are included in the same incompatibility group. On the other hand, those cultivars with at least one different *S*-allele are inter-compatible and are assigned to different incompatibility groups. Those cultivars whose *S*-genotype has not been found in any other cultivar until now are included in group 0. Self-compatible cultivars are grouped in the last row of each table as group SC. Cultivars from both groups 0 and SC may be considered as universal pollinizers. Adequate pollinizers must be inter-compatible and coincident in flowering time.

Table 3.1 Incompatibility groups and *S*-genotype of 219 almond cultivars

I.G.	*S*-RNase genotype	Cultivars
I	S_7S_8	Galaxy, Golden state, Grace, IXL, Long IXL, Mckinlay's Magnificent, Nonpareil, Riedenhoure, Tardy Nonpareil, West Steyn
II	S_1S_5	Ballico, Bulbuente, Garbi, Glorieta, Languedoc, Texas (=Mission), Wawona
III	S_5S_7	Baxendale, Durango, Granada, Harvey, Le Grand, Mono, Robson, Sauret no. 2, Thompson, Wassum, Wood Colony
IV	S_1S_7	Alrich, Jenette, Merced, Ne Plus Ultra, Norman, Pearl, Price, Ripon, Rosetta, Sano
V	S_5S_8	Carmel, Jubilee, Livingston, Monarch, Reams, Sauret no. 1, Tioga
VI	S_1S_8	Avalon, Bigelow, Blue Gum, Butte, Dottie Won, Duro Amarelo, Folsom, Grace, Kutsch, Monterey, Northland, Plateau, Rivers Nonpareil, Sultana
VII	S_8S_{13}	Eureka, Kapareil, Solano, Sonora, Vesta
VIII	S_1S_3	Ferragnes, Ferralise, Mourisca
IX	S_7S_{14}	Harpareil, Jordanolo
X	S_6S_8	Drake, Kochi, Smith XL
XI	S_1S_6	Abizanda, Fritz, Peerless, Ruby, Rumbeta-2
XII	S_2S_9	Anxaneta, Tarragones
XIII	S_1S_{10}	Ardechoise, Coop, Desmayo Largueta, Pep de Juneda, Zahaf
XIV	S_2S_{10}	Achaak, Alnem88, Ferrastar
XV	$S_{12}S_{23}$	Pajarera-2, Pestañeta (=Pestanhieta)
XVI	$S_{22}S_{23}$	Malagueña, Muel, Pau, Planeta Fina, Planeta Roja, Verdeta
XVII	$S_{13}S_{27}$	Garrigues, Pajarera-1
XVIII	S_8S_{12}	AS1, Marcona Flota
XIX	S_3S_{23}	Belle d'Aurons, Peraleja
XX	S_3S_5	Bartre, Castañera
XXI	S_1S_9	Masbovera, Moncayo, Tarraco
XXII	S_1S_4	Casanova, Coelhinha, Ferraduel
XXIII	S_7S_{13}	Chellastone, Milow
XXIV	$S_{11}S_{22}$	Parada, Pestañeta Menuda
XXV	S_8S_{23}	Pierce
XXVI	$S_{fa}S_{36}$	Alzina, Garondes
XXIX	S_1S_2	Cristomorto
XXX	S_3S_4	Aï
XXXI	S_6S_7	Tokyo
XXXII	S_4S_{12}	Carretas Bajas
XXXVIII	$S_{62}S_{63}$	Safari
XL	S_4S_{13}	CEBAS I, Verdeal
XLI	S_3S_9	Tardive de la Verdiere
XLII	S_1S_{21}	Bonita
XLIII	$S_{10}S_{22}$	Angones
XLV	$S_{12}S_{22}$	Del Cid, Esperanza Forta
XLVI	$S_{12}S_{28}$	Colorada, Nano

(continued)

Table 3.1 (continued)

I.G.	*S*-RNase genotype	Cultivars
XLVII	S_6S_{10}	Pauet
XLVIII	S_1S_{12}	Mollar de Tarragona
XLIX	$S_{12}S_{27}$	Carreró
L	S_5S_{12}	Caima, Tardaneta
0 (cultivars with unique *S*-genotype)		Aldrich (S_7S_{17}), Asperilla ($S_{10}S_{27}$), Atascada (S_5S_{22}), Atocha ($S_{13}S_{22}$), Avellaneda Gruesa ($S_{22}S_{26}$), Belardino (S_2S_{11}), Bertina (S_6S_{11}), Biota (S_5S_{13}), Boa Casta (S_8S_{21}), Bonita de São Brás (S_8S_{22}), Carrion (S_5S_{14}), Castilla (S_6S_{22}), Colossal (S_7S_{11}), Fascionello ($S_{18}S_{52}$), Fina del Alto ($S_{28}S_{29}$), Fournat de Brezenaud ($S_{24}S_{27}$), Fura Saco (S_4S_{23}), Gabaix ($S_{10}S_{24}$), Harriot (S_6S_{14}), Jiménez Salazar ($S_{21}S_{26}$), Jordi (S_5S_6), La Mona ($S_{23}S_{25}$), Liso ($S_{10}S_{23}$), Marcona ($S_{11}S_{12}$), Marcona de San Joy ($S_{22}S_{27}$), Menut ($S_{10}S_{13}$), Mollar (S_8S_{24}), Mollar de la Princesa ($S_{24}S_{53}$), Padre (S_1S_{18}), Padre Santo (S_3S_{10}), Pané-Barquets (S_1S_{34}), Parque Samá (S_1S_{35}), Planeta de les Garrigues ($S_{22}S_{35}$), Ponç ($S_{fa}S_{27}$), Pou d'Establiments ($S_{12}S_{33}$), (S_5S_9), Ramillete (S_6S_{23}), Redonda de Palma (S_3S_{25}), Retsou (S_2S_3), Rof (S_5S_{23}), Rumbeta ($S_{11}S_{21}$), Somerton (S_1S_{23}), Tejeda-2 (S_1S_{22}), Tendra Amarga ($S_{21}S_{23}$), Tío Martín ($S_{23}S_{27}$), Titan (S_8S_{14}), Totsol (S_8S_{31}), Verruga (S_5S_{10}), Vivot ($S_{fa}S_{23}$), Winters (S_1S_{14}). Yosemite (S_8S_{10})
Self-compatible cultivars		All-in-One, Almenara, Antoñeta, Aylés, Belona, Blanquerna, Cambra, Constanti, Falsa Barese, Felisia, Filippo Ceo, Francoli, Garden Princess, Genco, Guara, Independence, Lauranne, Mandaline, Mardía, Marinada, Marita, Marta, Matan, Occhiorosso, Penta, Soleta, Steliette, Supernova, Sweetheart, Tardona, Teresa, Tuono, Vairo, Vialfas
References		Kester and Gradziel (1996), Bošković et al. (1997, 1998, 2003) Boutard (1999), Certal et al. (2002), Channuntapipat et al. (2003), Martínez-Gómez et al. (2003), Sánchez-Pérez et al. (2004), Ortega et al. (2005, 2006), Barckley et al. (2006), Kodad et al. (2008a, b), Marchese et al. (2008), Kodad and Socias i Company R (2009), Kodad et al. (2010), Gómez et al. (2019)

3.6 External Factors Affecting Flowering and Pollination

Alterations in flower biology may lead to failure in pollination and subsequent lack of fertilization and fruit set. These may occur at flower bud development or at flowering. During the onset of flower differentiation, high temperatures may cause various floral abnormalities, as double pistil formation and appendices with pistil or petal appearance at the end of the stamen filaments instead of the anthers or with a leafy appearance in different floral organs (Herrero et al. 2017).

As other temperate fruit trees, *Prunus* spp. need to accumulate chilling during winter dormancy to break endodormancy. Once the cultivar-specific chilling requirements are fulfilled, flower buds accumulate heat and enter ecodormancy (Lang et al.

Table 3.2 Incompatibility groups and *S*-genotype of 135 sweet cherry cultivars

I.G.	*S*-RNase genotype	Cultivars
I	S_1S_2	Canada Giant (Sumgita), Ferdouce, Starking Hardy Giant, Summit, Tulare
II	S_1S_3	Areko, Black Star, Coral, Cristalina, Early Robin (Doty), Early Van Compact, Lala Star, Prime Giant, Royal Lee, Rosie, Royal Ansel (Royal Bailey), Redstone, Regina, Samba (Sumste), Satin (Sumele), Sonnet, Sumbola, Van, Vera
III	S_3S_4	Belge, Bing, Karina, Lambert, Napoleon (Monzón, Royal Ann), Somerset, Sweet Lorenz, Sweet Valina, Ulster
IV	S_2S_3	Coralise (Gardel), Nimba, Sue, Vega
VI	S_3S_6	Ambrunés, Duroni 3, Ferdiva, Fertard, Fertille, Kordia, Pico Negro, Satonishiki, Stark's Gold (Dönissens Gelbe, Gold), Techlovan
VII	S_3S_5	Hedelfinger
IX	S_1S_4	Ebony Pearl, King, Rainier, Royal Brynn, Royal Lynn, Sweet Gabriel, Sylvia
X	S_6S_9	Folfer, Penny, Ramón Oliva
XIII	S_2S_4	Royalton, Sam, Vic
XIV	S_1S_5	Blanca de Provenza
XV	S_5S_6	Colney
XVI	S_3S_9	Burlat, Chelan, Moreau, Precoce Bernard, SMS-280, Tieton
XVII	S_4S_6	Larian, Royal Hazel, Royal Tenaya (Royal Marie)
XVIII	S_1S_9	Bigisol (Early Bigi), Brooks, Earlise (Rivedel), Marvin (Niran, 4–70), Rocket, Sweet Early (Panaro 1), Tamara
XIX	S_3S_{13}	Reverchon
XX	S_1S_6	Vanda
XXI	S_4S_9	Cashemire, Merchant
XXII	S_3S_{12}	0900-Ziraat, Ferrovia, Schneiders
XXIV	S_6S_{12}	Aida
XXV	S_2S_6	Fercer (Arcina)
XXVII	S_4S_{12}	Margit, Kavics
XXXIII	S_1S_{14}	Fermina
XXXIV	S_4S_{13}	Black Pearl
XL	S_6S_{22}	Pico Colorado
XLIII	S_2S_9	Primulat (Ferprime)
0 (cultivar with unique *S*-genotype)	S_5S_{22}	Rita
Self-compatible cultivars		Alex, Blackgold, Blaze Star, Celeste, Columbia (Benton), Compact Stella, Early Star (Panaro 2), Frisco, Grace Star, Index, Lapins, New Star, Pacific Red, Royal Elaine, Royal Tioga, Royal Helen, Royal Edie, Sandor, Sandra Rose, Santina, Selah (Liberty bell), Skeena, Sofia (SPC 106), Sonata, Staccato (Summer Charm), Stardust, Starkimson, Stella, Sumesi, Sunburst, Sweet Aryana, Sweet Georgia, Sweet Saretta, Sweet Stephany, Sweet Valentine (Summac), Sweetheart, Symphony
References		Wünsch and Hormaza (2004b), Schuster (2012, 2017, Herrero et al. (2017), Quero-García et al. (2017)

Table 3.3 Incompatibility groups and *S*-genotype of 153 Japanese plum cultivars

I.G.	*S*-RNase genotype	Cultivars
I	*SaSb*	A-606, Angelo, Armstrong, Burmosa, Late Soldam, Mammoth, Red Beaut, Soldam, Sordum
II	*SbSc*	Blackamber, Black Beaut, Delbartazur, Early Sun, Flavor Granade, Fortune, Golden Plum, Golden Plumza, Green Sun, Gulfrose, Honey Red, Jupiter, Laroda, October Sun, Purple Queen, Sugared, Super Giant, TC Sun
III	*SbSf*	AU Amber, AU Road Side, Emarald, Frontier, Gran Colle, Verna Delicious
IV	*SbSh*	Betty Ann, Black Gem, Blue Knight, Eldorado, Freedom, Hiromi Red, Larry Ann, Queen Ann, Songria 10, Sundew, Yonemomo
VI	*SfSh*	Black Ruby, Kelsey Paulista, Kelsey, Mariposa
VII	*ScSh*	Angeleno, Gaia, Green Sun, Queen Rosa, Royal Diamond, Ruby Crunch, Ruby Queen, Sweet August
VIII	*SeSh*	African Delight, African Pride, Autumn Pride, Black Diamond, Black Gold, Black Late, Diamex, Earliqueen, John W., Ruby Star, Showtime, Souvenir
IX	*SfSg*	Golden Japan, Manchurian, Shiro, White Plum
X	*ShSk*	Elephant Heart, Explorer, Friar, Howard Sun, Golden Kiss, Redgold, Songold
XI	*ScSe*	AU Rosa, Autumn Giant, Black Splendor, Champion, Kesselman, Royal Garnet, Royal Zee, Roysum, Sweet Autumn, Sybarite
XII	*SbSe*	Black Jewell, Durado, Flavor King, Freya, Murietta, Pioneer, Saphire, Sparkly, Tomar
XIII	*SeSf*	Black Star, Morris, Primetime
XIV	*SaSc*	Crimson Glo, Sunkiss, White Queen
XV	*SgSh*	Bonnie, Ruby Sweet
XVI	*SfSk*	Kelsey, Weikeshum, Wickson
XVII	*SbSo*	Ambra, Olinda
XVIII	*SaSf*	Ozarkpremier, Terada
XIX	*SbSd*	Formosa, Harypickstone
XX	*SbSk*	Homeking Delicious
XXI	*SeSk*	Newyorker, Simon
XXII	*SaSe*	Dolly, Riou
XXIII	*ScSd*	Oishiwasesumomo
0 (cultivars with unique *S*-genotype)		Abundance*(*SaSk*), Botan (*SaSm*), Byron Gold (*ShSr*), Combination (*SgSl*), Gaia (*ScSh*), Joana Red (*SrSs*), Lantz (*SbSl*), Mitard (*SqSf*), October Red (*ShSp*), Red Heart (*ScSo*), Summer Queen (*ScSf*), Songria 15 (*SaSh*), Starkgold (SgSk), Superior (*SaSn*), Tecumseh (*SfSj*)
Self-compatible cultivars		African Rose, Beauty, Casselman, Honey Rosa, Karari, Laetitia, Late Santa Rosa, Methley, Nubiana, Pioneer, Red Rosa, Rio, Rubirosa, Santa Rosa, Simka, Souvenir, Zanzi Sun
References		Yamane et al. (1999), Beppu et al. (2002, 2003), Sapir et al. (2004), Halász et al. (2007a), Jun et al. (2007), Sapir et al. (2007), Zhang et al. (2007), Guerra et al. (2009, 2011, 2012, 2020) Guerra and Rodrigo (2015)

Table 3.4 Incompatibility groups and *S*-genotype of 160 apricot cultivars

I.G.	*S*-RNase genotype	Cultivars
I	S_1S_2	Castleton, Farmingdale, Giovanniello, Goldrich, Hargrand, Lambertin-1
II	S_8S_9	Perlecot, Pinkcot
III	S_2S_6	Avirine (Bergarouge), Iri Bitirgen, Moniqui
IV	S_2S_7	Artvin P.A., Ouardi, Priana
V	S_2S_8	Alyanak, Holly Cot, Sweet Cot, Ziraat Okulu
VI	S_2S_{19}	Dörtyol 4, Sebbiyiki
VIII	S_6S_9	Cataloglu, Cheyenne, Feria Cot, Ninja, Orangered, Ozal, Soganci, Stark Early Orange, Sunny Cot, Wonder Cot
XVIII	S_1S_3	Cooper Cot, Perfection
XIX	S_2S_3	Mayacot, Sun Glo
XX	S_2S_9	Goldstrike, Magic Cot
XXI	S_3S_8	Lilly Cot, Spring Blush
XXII	S_3S_9	Durobar (Almadulce), Flodea, Henderson, Kosmos, Tsunami
XXIII	S_7S_9	Goldbar
XXIV	S_1S_6	Primaya
0 (cultivars with unique *S*-genotype)		Castlebrite (S_2S_2), Cow-1 (S_1S_{31}), Cow-2 ($S_{20}S_{31}$), Estrella (S_1S_7), Harcot (S_1S_4), Harlayne (S_3S_{20}), Harmat ($S_{10}S_{11}$), Mariem (S_7S_{20}), Martinet (S_2S_2), Perla (S_2S_{20}), Portici (S_2S_{20}), Shalah (S_5S_{11}), Stella (S_6S_{20}), Veecot (S_2S_{20}), Velázquez (S_5S_{20})
Self-compatible cultivars		Alba, Aprix 20, Aprix 33, Aprix 9, Apriqueen, Bebecou, Beliana, Beliana, Berdejo, Bergecot, Bergeron, Budapest, Búlida, Callatis, Canino, Cebas Red, Charisma, Corbatoy, Cristalí, Currot, Delice Cot, Dorada, Dulcinea, Effect, Fantasme, Faralia, Farbaly, Farbela, Farclo, Fardao, Farfia, Farhial, Farius, Farlis, Fartoli, Flopria, Galta Vermella Valenciana, Galta Roja, Gandía, Ginesta, Gönci magyarkajszi, Grandir, Kalao, Kioto, Lady Cot, Lido, Lito, Lorna, Luizet, Manrí, Mauricio, Medflo, Mediabel, Mediva, Memphis, Milord, Mirlo Anaranjado, Mirlo Blanco, Mirlo Rojo, Mitger, Modesto, Murciana, Ninfa, Oscar, Palabras, Palau, Palsteyn, Patterson, Paviot, Peñaflor 02, Pepito del Rubio, Pisana, Pisana, Playa cot, Pricia, Primidi, Primorosa, Rakovszky, Rambo, Regibus, Rojo Carlet, Rouge Cot, Roxana, Rubista, Sam, Sandy cot, Sayeb, Sherpa, Sirena, Soledane, Sulmona, Swired, Tadeo, Tilton, Tirynthos, Tom Cot, Trevatt, Venus, Victor 1, Xirivello, Zaposdolye
References		Egea and Burgos (1996), Burgos et al. (1998), Alburquerque et al. (2002), Halász et al. (2005, 2007b, 2010) Vilanova et al. (2005), Donoso et al. (2009), Egea et al. (2010), Muñoz-Sanz et al. (2017), Herrera et al. (2018a, b), Ruiz et al. (2018)

1987), and after a variable period of mild temperatures, flower bud development reactivates, buds burst, and subsequent flowering proceeds rapidly (Campoy et al. 2011; Fadón et al. 2015).

Chilling requirements of particular cultivars may be estimated by statistical correlations between flowering dates and temperatures before flowering (Alonso et al.

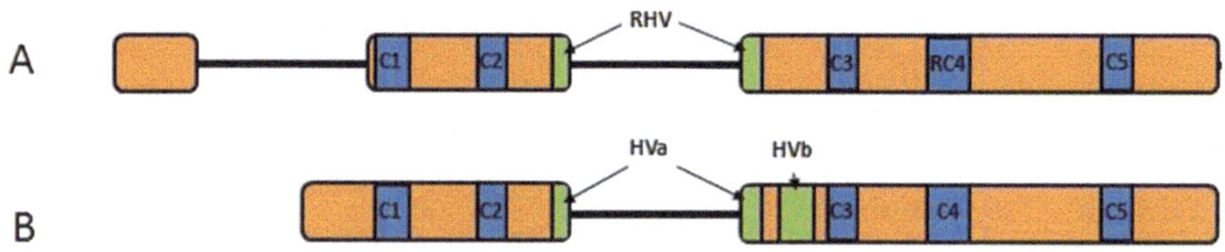

Fig. 3.6 Diagram of the *S-RNase* gene. Genomic sequence of the *S-RNase* gene showing the exons in square, including the five highly conserved regions in Rosaceae (**a**) and in Plantaginaceae and Solanaceae (**b**)

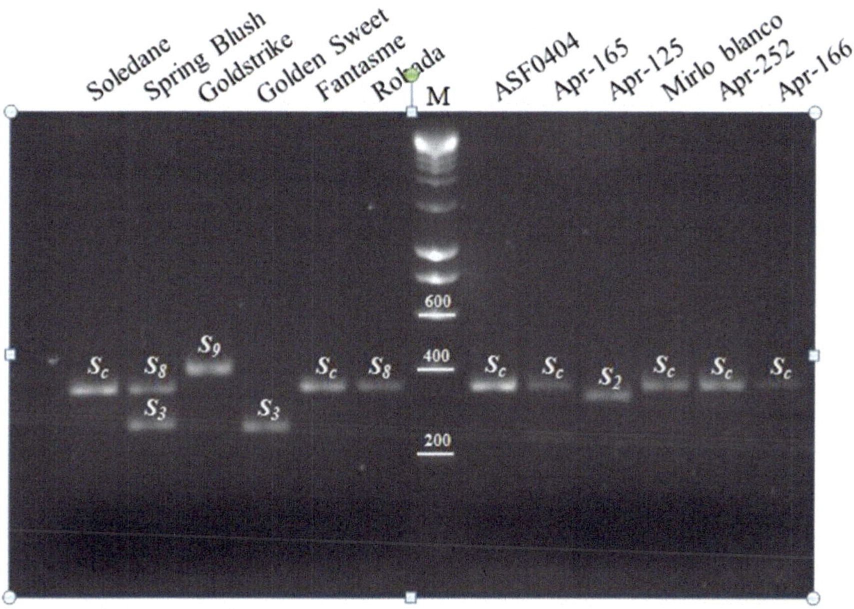

Fig. 3.7 PCR fragment amplification using primer pair primers SRc-(F/R) for the identification of *S*-alleles in 12 apricot cultivars. Size standard: 1 kb

2005; Luedeling et al. 2009; Luedeling and Gassner 2012) or by cutting shoots at different times during the winter and exposing them to warm temperatures to monitor if the flower buds have recovered their capacity to grow (Fadón and Rodrigo 2018). Once the date of chilling fulfillment is determined, the chilling temperatures previously accumulated are quantified following models such as "chilling hours" (Weinberger 1950), "Utah model" (chill units, Richardson et al. 1974), and "dynamic model" (chill portions, Fishman et al. 1987). These models have been widely used for the estimation of chilling requirements in *Prunus* spp. (Fadón et al. 2020). However, these approaches have some limitations since the methods for determining the time of endodormancy release and for calculating chilling requirements are not standardized and often result in high variations in the estimated values for particular cultivars depending on the location and the experimental conditions (Dennis

Table 3.5 Incompatibility groups and *S*-genotype of 51 European plum cultivars

S-RNase genotype	Cultivars
$S_2S_5S_6S_7S_{12}$	Mulata
$S_4S_6S_8S_9$	Negra
$S_2S_5S_7S_{10}S_8S_{12}$	HuevoChivato
$S_3S_5S_6S_9$	Verde
$S_4S_6S_{11}S_9$	Negra del país pequeña
$S_1S_5S_7S_9$	President
$S_1S_3S_5S_6S_{12}$	Ruth Gestteter
$S_1S_3S_2S_{11}S_9$	RC Dorada
$S_2S_8S_{10}S_6S_{12}$	Fraila
$S_1S_3S_6S_9$	Alcor-1
$S_1S_3S_5S_{11}S_9$	F-4 A-4
Self-compatible cultivars	Anna Späth, Belle de Louvain, Blufre, Cacanska Lepotica, California Blue, Cacanska Rodna, Czar, de Ente, Denniston Superb, Double Robe, Elena, Early Transparent, Early Mirabelle, German Prune, Giant, Gisborne's, Golden Transparent, Goliath, Guthrie's Late, Hanita, Harbella, Haroma, Herman, Jelica, Italian Prune, Jojo, Katinca, Martin, Monarch, Oullin's Golden Gage, Požegaca, Presenta, Reine Claude de Bavay, Reine Claude de Oullins, Ruth Gerstetter, Stanley, Tegera, Valjevka, Victoria, Wangenheims Frühzwetsche
References	Crane (1925), Cambra (1982), Nikolić and Milatović (2010), Neumüller (2011), Abdallah et al. (2019)

Table 3.6 Incompatibility groups and *S*-genotype of 58 sour cherry cultivars

S-RNase genotype	Cultivars
$S_1S_4S_{35}S_{36b}$	Crisana
$S_1S_{12}S_{13}S_{36b}$	Erdi Nagygyumolcsu
$S_1S_4S_{35}S_{36b}$	Pandy 38
$S_1S_4S_{35}S_{36b}$	Pandy 114
$S_1S_4S_{36a}S_{36b}$	Tarina
$S_9S_{13}S_{35}S_{36b2}$	Tschernokorka
$S_1S_{13'}S_{26}S_{35}$	Agat
$S_1S_{12}S_{13}S_{36b}$	Gubenska
$S_6S_9S_{13}S_{36b2}$	Griot Moskowskij
$S_9S_{13}S_{26}S_{36a}$	Vladimirskaja
$S_6S_{13}S_{14}S_{36a}$	Spanka
Self-compatible cultivars	Ametyst, Cigány 59, Diemitzer Amarelle, Dradem, Early Richmond, Englaise Timpurii, Erdi Botermo, Erdi Jubileum, Espera, Fanal, Galena, Ihruska, Karneol, Korund, Meteor, Montmorency, Morina, MSU III 18 (12), Nana, Oblačinska, Pitic de Iasi, Plodorodnaja Mitschurina, Rheinische Schattenmorelle, Safir, Surefire, Tamaris, Topas, Topaz, Újfehértói fürtős, Vowi, Vstryecha, Kelleriis 16, Stevnbaer, Achat, Bucovina, Carmine Jewel, Coraline, Kutahya, Gerema Ilva, Jachim, Jade, Lyubskaya, Nefris, Northstar, Sabina, Šumadinka, Tiki
References	Tsukamoto et al. (2008), Lisek et al. (2017), Schuster et al. (2017), Sebolt et al. (2017)

2003). However, the use of these models has been very useful for the selection of cultivars in particular areas, although in most *Prunus* species an increasing number of new cultivars of unknown requirements are being released (Fadón et al. 2020).

Low temperature stress resulting in death of flower organs is one of the main limiting factors affecting fruit production in stone fruits in some regions and seasons. Freeze injury may occur before bud acclimation in autumn and, especially, during deacclimation, bud burst, and flowering (Rodrigo 2000). High temperatures in the days before flowering can also cause floral malformations such as underdeveloped pistils (Rodrigo and Herrero 2002b).

When pollinizer cultivars are needed, an overlap in flowering time must occur between the main fruit producing cultivar and their pollinizers to allow the transfer of compatible pollen from the anthers to the stigma by pollinators. Insects play a main role in pollination for fruit production in most fruit tree crops, including stone fruits (Gallai et al. 2009). Among them, the major group of animal pollinators in commercial fruit orchards in temperate regions are bees (Kevan 1999), especially the European honeybee (*Apis mellifera* L. [Hymenoptera: Apidae]), but honeybee decline is a fact in different regions and environments (Goulson et al. 2015). Besides the effort needed to analyze and mitigate the causes behind honeybee decline, a complementary approach involves studying additional pollinating insects, such as solitary bees (Koh et al. 2018) or species of *Bombus* (Calzoni and Speranza 1998) that could diversify the availability of pollinators in order to optimize fruit production (Garibaldi et al. 2013; Kleijn et al. 2015). Different species of solitary bees have been shown to play a role in pollination of stone fruits, and a proper management of this diversity is necessary to deal with the diverse problems that affect honeybees, related to climate change, parasites, or diseases (Koh et al. 2018).

The decline of wild pollinating insects is making the introduction of honeybees or bumblebees in commercial orchards essential for ensuring fruit set in an increasing number of situations, not only in self-incompatible but also in self-compatible cultivars. If necessary, managed beehives should be placed in the orchard when the first flowers have already been opened, to prevent bees from establishing their foraging habits in other flowering species (Thompson 1996).

The optimum number of beehives in an orchard depends firstly on the size of the colonies. A good colony should show over 100 bees per minute entering the hive under adequate weather conditions (temperature above 18 °C without wind). A second important factor to determine the number of beehives is the number of trees per hectare and the flower density of the trees. In warm, sunny, and calm conditions, the observation of 25–35 bees working on a mature tree at full bloom can indicate enough pollinators in the orchard (Koumanov and Long 2017). Although ideally the beehives should be uniformly distributed in the orchard for optimal pollination efficiency, they may be placed in groups to facilitate handling, adjusting the distance between groups according to the needs (Thompson 1996).

3.7 Concluding Remarks and Perspectives

Adequate pollination during the short flower life span is essential to ensure fruit production in stone fruit crops. In many areas, climate change is negatively affecting different aspects of the reproductive biology of fruit trees jeopardizing fruit production, which is highlighting the impact of adequate pollination management in commercial orchards (Hedhly et al. 2009; Hedhly 2011). The reduction of chilling during winter is causing a delay for some cultivars to fulfill their chilling requirements or even not fulfill them. This often results in a delay in flowering and causes that inter-compatible cultivars with different chilling requirements do not coincide at blooming in those seasons of insufficient winter chilling (Luedeling et al. 2011; Luedeling 2012; Atkinson et al. 2013). Therefore, some pollinizer cultivars become inadequate and need to be replaced by others. Thus, knowing the pollination requirements of particular cultivars is increasingly necessary. Although self-(in)compatibility and inter-incompatibility between cultivars are known in many of the cultivars currently grown (see Sects. 3.4 and 3.5), especially in species such as sweet cherry, almond, or Japanese plum, the constant varietal renewal, with the introduction of a number of new cultivars each year, makes necessary to continue studying the pollination requirements in the self-incompatible species (Fig. 3.8).

Results obtained in the last decades have allowed establishing the pollination requirements of most commercial cultivars in the main stone fruit crops. For the coming years, it will be necessary to continue determining the self-(in)compatibility and the incompatibility relationships of the new releases. While the available technology allows a fast determination of the *S*-allele genotype of any new cultivar or offspring in the main stone fruit species, the complexity to establish self-(in)

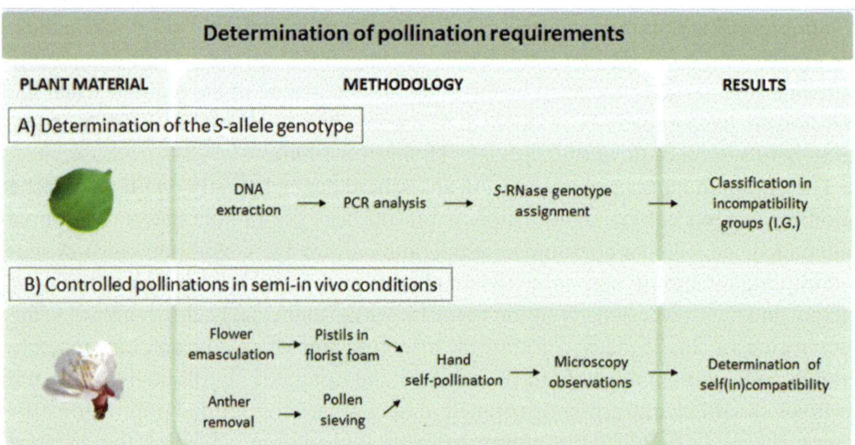

Fig. 3.8 Diagram of the experimental design to determine pollination requirements in stone fruit crops. Workflow of the *S*-allele identification by molecular approaches (**a**). Workflow of self-(in) compatibility determination by controlled pollinations in semi-in vivo conditions in the laboratory (**b**)

compatibility differs between species. Particular *S*-alleles associated with self-compatibility have been reported in almond (Kodad et al. 2010; Fernández i Martí et al. 2014), sour cherry (Yamane et al. 2003), and sweet cherry (Bošković and Tobutt 1996; Wünsch and Hormaza 2004a; Cachi and Wünsch 2014). In these species, the identification of *S*-alleles of each cultivar allows not only knowing its incompatibility relationships with other cultivars but also establishing its self-(in) compatibility. This situation has also been previously considered in apricot (Halász et al. 2005; Vilanova et al. 2006), although recent reports have shown self-compatible cultivars whose *S*-genotype does not include the *S*-allele previously associated with self-compatibility (Zuriaga et al. 2013; Muñoz-Sanz et al. 2017; Herrera et al. 2018a). In Japanese plum, self-compatibility was initially associated with the presence of particular *S*-alleles (Beppu et al. 2005, 2010; 2012a, b; Guerra et al. 2009). However, several cultivars carrying these *S*-alleles have been described as self-incompatible. Thus, in apricot, Japanese plum, European plum, and sour cherry, observing the pollen tube behavior under the microscope in self-pollinated pistils is still a recommended tool to assess self-(in)compatibility (Herrera et al. 2018a, 2020; Guerra et al. 2020).

On the other hand, there is a lack of information on the chilling requirements in most cultivars of stone fruit crops. In the next years, this character will be increasingly important to know the adaptability of cultivars to different geographical areas due to the changing conditions caused by climate change, which are causing insufficient yield in some cultivars previously considered as well adapted to particular regions (Campoy et al. 2011; Luedeling et al. 2011; Atkinson et al. 2013). The lack of estimations of chilling requirements is mainly due to the absence of a fast and reliable approach to quantify both chilling and heat requirements (Dennis 2003; Fadón et al. 2020) but also to the lack of a biological marker that allows knowing whether the chilling requirements have been fulfilled under orchard conditions (Fadón and Rodrigo 2018). Finding a biological marker by quantifying a physiological parameter or the expression of particular genes would clarify this situation, which in many cases is accentuated by the expansion of cultivars to new warmer areas looking for early ripening.

3.8 Conclusion

Under the generic name of stone fruit crops, different plant species of the *Prunus* genus in the Rosaceae that produce a drupe type of fruit are included, mainly almonds, apricots, cherries, peaches, and plums. As in most crops in which the main commercial interest is in the fruits, the reproductive phase is probably the main limiting factor for profitable production in stone fruit crops and, more specifically, the days following flower opening, the progamic phase that takes place from pollination to fertilization. The importance of appropriate pollination management approaches is increasing in recent years in these crops due to the release of new cultivars with unknown pollination requirements, the expansion of stone fruit crops to new cultivation areas, and the decrease in winter chilling in many regions caused by climate change. Consequently, a proper understanding of the pollination requirements of cultivars of the different stone fruits is urgently needed in order to optimize production. Although there is a high variability among stone fruit species in flower characteristics, chilling and heat requirements, pollination behavior, and flowering times, the main biological processes are conserved, and, consequently, a common approach to manage pollination problems can be followed.

Acknowledgments This research was funded by Ministerio de Ciencia, Innovación y Universidades—European Regional Development Fund, European Union (AGL2016–77267-R, PID2019-109566RB-I00, AGL2015–74071-JIN, and RYC-2017-21,909), Instituto Nacional de Investigación y Tecnología Agraria y Alimentaria (RFP2015–00015-00, RTA2017–00003-00), and Gobierno de Aragón—European Social Fund, European Union (Grupo Consolidado A12_17R).

References

Abdallah, D., Baraket, G., Perez, V., Ben Mustapha, S., Salhi-Hannachi, A., & Hormaza, J. I. (2019). Analysis of self-incompatibility and genetic diversity in diploid and hexaploid plum genotypes. *Frontiers in Plant Science, 10*, 1–14.

Abdallah, D., Baraket, G., Perez, V., Hannachi, A. S., Hormaza, J. I. (2020). Self-compatibility in peach [Prunus persica (L.) Batsch]: patterns of diversity surrounding the S-locus and analysis of SFB alleles. *Horticulture Research 7*, 170. https://doi.org/10.1038/s41438-020-00392-z.

Alburquerque, N., Egea, J., Pérez-Tornero, O., & Burgos, L. (2002). Genotyping apricot cultivars for self-(in)compatibility by means of RNases associated with *S* alleles. *Plant Breeding, 121*, 343–347.

Alonso, J. M., Anson, J. M., Espiau, M. T., & Socias i Company R. (2005). Determination of endodormancy break in almond flower buds by a correlation model using the average temperature of different day intervals and its application to the estimation of chill and heat requirements and blooming date. *Journal of the American Society for Horticultural Science, 130*, 308–318.

Anderson, M. A., Cornish, E. C., Mau, S. L., Williams, E. G., Hoggart, R., Atkinson, A., Bonig, I., Grego, B., Simpson, R., Roche, P. J., Haley, J. D., Penschow, J. D., Niall, H. D., Tregear, G. W., Coghlan, J. P., Crawford, R. J., & Clarke, A. E. (1986). Cloning of cDNA for a stylar glycoprotein associated with expression of self-incompatibility in *Nicotiana alata*. *Nature, 321*, 38–44. https://doi.org/10.1038/321038a0.

Atkinson, C. J., Brennan, R. M., & Jones, H. G. (2013). Declining chilling and its impact on temperate perennial crops. *Environmental and Experimental Botany, 91*, 48–62.

Badenes, M. L., & Byrne, D. H. (2012). *Fruit Breeding*. Berlin: Springer, ISBN 978-1-4419-0763-9.

Baggiolini, M. (1952). Les stades repérés des arbres fruitiers à noyau. *Revue Romande D'Agriculture et D'Aboriculture, 8*, 3–4.

Barckley, K. K., Uratsu, S. L., Gradziel, T. M., & Dandekar, A. M. (2006). Multidimensional analysis of *S*-alleles from cross-incompatible groups of California almond cultivars. *Journal of the American Society for Horticultural Science, 131*, 632–636.

Beppu, K., Yamane, H., Yaegaki, H., Yamaguchi, M., Kataoka, I., & Tao, R. (2002). Diversity of *S-RNase* genes and *S*-haplotypes in Japanese plum (*Prunus salicina* Lindl.). *The Journal of Horticultural Science and Biotechnology, 77*, 658–664.

Beppu, K., Takemoto, Y., Yamane, H., Yaegaki, H., Yamaguchi, M., Kataoka, I., & Tao, R. (2003). Determination of *S*-haplotypes of Japanese plum (*Prunus salicina*) cultivars by PCR and cross-pollination test. *The Journal of Horticultural Science and Biotechnology, 78*, 315–318.

Beppu, K., Komatsu, N., Yamane, H., Yaegaki, H., Yamaguchi, M., Tao, R., & Kataoka, I. (2005). *Se*-haplotype confers self-compatibility in Japanese plum (*Prunus salicina* Lindl.). *The Journal of Horticultural Science and Biotechnology, 80*, 760–764.

Beppu, K., Syogase, K., Yamane, H., Tao, R., & Kataoka, I. (2010). Inheritance of self-compatibility conferred by the *Se*-haplotype of Japanese plum and development of *Se-RNase* gene-specific PCR primers. *The Journal of Horticultural Science and Biotechnology, 85*, 215–218.

Beppu, K., Konishi, K., & Kataoka, I. (2012a). *S*-haplotypes and self-compatibility of the Japanese plum cultivar 'Karari'. *Acta Horticulturae, 929*, 261–266. https://doi.org/10.17660/ActaHortic.2012.929.38.

Beppu, K., Kumai, M., Yamane, H., Tao, R., & Kataoka, I. (2012b). Molecular and genetic analyses of the *S*-haplotype of the self-compatible Japanese plum (*Prunus salicina* Lindl.) 'Methley'. *The Journal of Horticultural Science and Biotechnology, 87*, 493–498.

Bošković, R., & Tobutt, K. R. (1996). Correlation of stylar ribonuclease zymograms with incompatibility alleles in sweet cherry. *Euphytica, 90*, 245–250.

Bošković, R., Tobutt, K. R., Batlle, I., & Duval, H. (1997). Correlation of ribonuclease zymograms and incompatibility genotypes in almond. *Euphytica, 97*, 167–176.

Bošković, R., Tobutt, K. R., Rovira, M., Romero, M. A., Batlle, I., Duval, H., & Dicenta, F. (1998). Inheritance of stylar ribonucleases in two almond progenies and their correlation with self-compatibility. Proceedings of the Second International Symposium on Pistachios and Almonds. *Acta Horticulturae, 470*, 118–122.

Bošković, R., Tobutt, K. R., Batlle, I., Duval, H., Martínez-Gómez, P., & Gradziel, T. M. (2003). Stylar ribonucleases in almond: correlation with and prediction of incompatibility genotypes. *Plant Breeding, 122*, 70–76.

Bošković, R. I., Tobutt, K. R., Ortega, E., Sutherland, B. G., & Godini, A. (2007). Self-(in)compatibility of the almonds *P. dulcis* and *P. webbii*: detection and cloning of 'wild-type *Sf*' and new self-compatibility alleles encoding inactive *S-RNases*. *Molecular Genetics and Genomics, 278*, 665–676. https://doi.org/10.1007/s00438-007-0283-4.

Boutard, A. (1999). *Détection et transmission allélique des ribonucléases stylaires impliquées dans l'incompatibilité pollinique chez différentes espèces de Prunus* (p. 56). Rennes: Ecole Nationale Superieure Agronomique de Rennes.

Bradbury, D. (1929). A comparative study of the developing and aborting fruits of *Prunus cerasus*. *American Journal of Botany, 16*, 525–542.

Broothaerts, W., Janssens, G. A., Proost, P., & Broekaert, W. F. (1995). cDNA cloning and molecular analysis of two self-incompatibility alleles from apple. *Plant Molecular Biology, 27*, 499–511. https://doi.org/10.1007/BF00019317.

Bukovac, M. J. (1963). Induction of parthenocarpic growth of apple fruits with gibberellins A3 and A4. *Botanical Gazette, 124*, 191–195. https://doi.org/10.1086/336191.

Burgos, L., Pérez-Tornero, O., Ballester, J., & Olmos, E. (1998). Detection and inheritance of stylar ribonucleases associated with incompatibility alleles in apricot. *Sexual Plant Reproduction, 11*, 153–158.

Byrne, D. H. (2005). Trends in stone fruit cultivar development. *HortTechnology, 15*, 494–500.

Cachi, A. M., & Wünsch, A. (2011). Characterization and mapping of non-*S* gametophytic self-compatibility in sweet cherry (*Prunus avium* L.). *Journal of Experimental Botany, 62*, 1847–1856. https://doi.org/10.1093/jxb/erq374.

Cachi, A. M., & Wünsch, A. (2014). S-genotyping of sweet cherry varieties from Spain and S-locus diversity in Europe. *Euphytica 197*, 229–236. https://doi.org/10.1007/s10681-014-1061-0.

Calzoni, G. L., & Speranza, A. (1998). Insect controlled pollination in Japanese plum (*Prunus salicina* Lindl.). *Scientia Horticulturae, 72*, 227–237. https://doi.org/10.1016/s0304-4238(97)00132-5.

Cambra, M. (1982). *La polinización de variedades de ciruelo europeo* (pp. 57–63). Zaragoza: Cuartas Jornadas Nacionales de Hortofruticultura.

Campoy, J. A., Ruiz, D., & Egea, J. (2011). Dormancy in temperate fruit trees in a global warming context: A review. *Scientia Horticulturae, 130*, 357–372.

Cerovic, R., & Micic, N. (1999). Functionality of embryo sacs as related to their viability and fertilization success in sour cherry. *Scientia Horticulturae, 79*, 227–235.

Certal, A. C., Almeida, R. B., Bošković, R., Oliveira, M. M., & Feijo, J. A. (2002). Structural and molecular analysis of self-incompatibility in almond (*Prunus dulcis*). *Sexual Plant Reproduction, 15*, 13–20.

Channuntapipat, C., Wirthensohn, M., Ramesh, S. A., Batlle, I., Arús, P., Sedgley, M., & Collins, G. (2003). Identification of incompatibility genotypes in almond (*Prunus dulcis* Mill.) using specific primers based on the introns of the *S*-alleles. *Plant Breeding, 122*, 164–168.

Crane, M. B. (1925). Self-sterility and cross-incompatibility in plums and cherries. *Journal of Genetics, 15*(3), 301–315. https://doi.org/10.1007/BF02983123.

del Pozo, J. C., & Estelle, M. (2000). F-box proteins and protein degradation: An emerging theme in cellular regulation. *Plant Molecular Biology, 44*, 123–128. https://doi.org/10.1023/A:1006413007456.

Dennis, F. G. (2003). Problems in standardizing methods for evaluating the chilling requirements for the breaking of dormancy in buds of woody plants. *HortScience, 38*, 347–350.

Diaz, D. H., Rasmussen, H. P., & Dennis, F. G. (1981). Scanning electron-microscope examination of flower bud differentiation in sour cherry. *Journal of the American Society for Horticultural Science, 106*, 513–515.

Donoso, J. M., Aros, D., Meneses, C., & Infante, R. (2009). Identification of *S*-alleles associated with self-incompatibility in apricots (*Prunus armeniaca* L.) using molecular markers. *Journal of Food Agriculture and Environment, 7*, 270–273.

Dorsey, M. J. (1919). A study of sterility in the plum. *Genetics, 4*, 417–488.

Egea, J., & Burgos, L. (1996). Detecting cross-incompatibility of three North American apricot cultivars and establishing the first incompatibility group in apricot. *Journal of the American Society for Horticultural Science, 121*, 1002–1005.

Egea, J., Burgos, L., García, J. E., & Egea, L. (1991). Stigma receptivity and style performance in several apricot cultivars. *Journal of Horticultural Science, 66*, 19–25.

Egea, J., Rubio, M., Campoy, J. A., Dicenta, F., Ortega, E., Nortes, M. D., Martínez-Gómez, P., Molina, A., Molina, A., & Ruiz, D. (2010). 'Mirlo Blanco', 'Mirlo Anaranjado', and 'Mirlo Rojo': Three new very early-season apricots for the fresh market. *HortScience, 45*, 1893–1894.

Entani, T., Iwano, M., Shiba, H., Che, F. S., Isogai, A., & Takayama, S. (2003). Comparative analysis of the self-incompatibility (*S*-) locus region of *Prunus mume*: Identification of a pollen-expressed *F-box* gene with allelic diversity. *Genes to Cells, 8*, 203–213.

Fadón, E., & Rodrigo, J. (2018). Unveiling winter dormancy through empirical experiments. *Environmental and Experimental Botany, 152*, 28–36.

Fadón, E., Herrero, M., & Rodrigo, J. (2015). Flower development in sweet cherry framed in the BBCH scale. *Scientia Horticulturae (Amsterdam), 192*, 141–147. https://doi.org/10.1016/j.scienta.2015.05.027.

Fadón, E., Herrero, M., & Rodrigo, J. (2018a). Dormant flower buds actively accumulate starch over winter. *Frontiers in Plant Science, 9*, 171.

Fadón, E., Rodrigo, J., & Herrero, M. (2018b). Is there a specific stage to rest? Flower bud development and winter dormancy in sweet cherry (*Prunus avium* L.). *Trees - Structure and Function, 32*, 1583–1594. https://doi.org/10.1007/s00468-018-1735-7.

Fadón, E., Herrera, S., Guerrero, B. I., Guerra, M. E., & Rodrigo, J. (2020). Chilling and forcing requirements of temperate stone fruit trees (*Prunus* sp.). *Agronomy, 10*, 409.

Fernández i Martí, A., Gradziel, T. M., & Socias i Company R. (2014). Methylation of the *Sf* locus in almond is associated with *S-RNase* loss of function. *Plant Molecular Biology, 86*, 681–689. https://doi.org/10.1007/s11103-014-0258-x.

Fishman, S., Erez, A., & Couvillon, G. A. (1987). The temperature dependence of dormancy breaking in plants: Mathematical analysis of a two-step model involving a cooperative transition. *Journal of Theoretical Biology, 124*, 473–483.

Gallai, N., Salles, J. M., Settele, J., & Vaissiere, B. E. (2009). Economic valuation of the vulnerability of world agriculture confronted with pollinator decline. *Ecological Economics, 68*, 810–821.

Garibaldi, L. A., Steffan-Dewenter, I., Winfree, R., Aizen, M. A., Bommarco, R., Cunningham, S. A., Kremen, C., Carvalheiro, L. G., Harder, L. D., Afik, O., Bartomeus, I., Benjamin, F., Boreux, V., Cariveau, D., Chacoff, N. P., Dudenhoffer, J. H., Freitas, B. M., Ghazoul, J., Greenleaf, S., Hipolito, J., Holzschuh, A., Howlett, B., Isaacs, R., Javorek, S. K., Kennedy, C. M., Krewenka, K. M., Krishnan, S., Mandelik, Y., Mayfield, M. M., Motzke, I., Munyuli,

T., Nault, B. A., Otieno, M., Petersen, J., Pisanty, G., Potts, S. G., Rader, R., Ricketts, T. H., Rundlof, M., Seymour, C. L., Schuepp, C., Szentgyorgyi, H., Taki, H., Tscharntke, T., Vergara, C. H., Viana, B. F., Wanger, T. C., Westphal, C., Williams, N., & Klein, A. M. (2013). Wild pollinators enhance fruit set of crops regardless of honey bee abundance. *Science, 339*, 1608–1611.

Gómez, E. M., Dicenta, F., Batlle, I., Romero, A., & Ortega, E. (2019). Cross-incompatibility in the cultivated almond (*Prunus dulcis*): Updating, revision and correction. *Scientia Horticulturae, 245*, 218–223.

Goulson, D., Nicholls, E., Botias, C., & Rotheray, E. L. (2015). Bee declines driven by combined stress from parasites, pesticides, and lack of flowers. *Science, 347*, 1435.

Griggs, W. H. (1953). *Pollination requirements of fruit and nuts*. Berkeley: College of Agriculture University of California.

Guerra, M. E., & Rodrigo, J. (2015). Japanese plum pollination: A review. *Scientia Horticulturae, 197*, 674–686.

Guerra, M. E., Rodrigo, J., López-Corrales, M., & Wünsch, A. (2009). *S-RNase* genotyping and incompatibility group assignment by PCR and pollination experiments in Japanese plum. *Plant Breeding, 128*, 304–311.

Guerra, M. E., Wünsch, A., López-Corrales, M., & Rodrigo, J. (2010). Flower emasculation as the cause for lack of fruit set in Japanese plum crosses. *Journal of the American Society for Horticultural Science, 135*, 556–562.

Guerra, M. E., López-Corrales, M., Wünsch, A., & Rodrigo, J. (2011). Lack of fruit set caused by ovule degeneration in Japanese Plum. *Journal of the American Society for Horticultural Science, 136*, 375–381.

Guerra, M. E., López-Corrales, M., & Wünsch, A. (2012). Improved *S* genotyping and new incompatibility groups in Japanese plum. *Euphytica, 186*, 445–452.

Guerra, M. E., Guerrero, B. I., Casadomet, C., & Rodrigo, J. (2020). Self-(in)compatibility, *S-RNase* allele identification, and selection of pollinizers in new Japanese plum-type cultivars. *Scientia Horticulturae (Amsterdam), 261*, 109022. https://doi.org/10.1016/j.scienta.2019.109022.

Guerrero-Prieto, V. M., Vasilakakis, M. D., & Lombard, P. B. (1985). Factors controlling fruit set of 'Napoleon' sweet cherry in western Oregon. *HortScience, 20*, 913–914.

Halász, J., Hegedűs, A., Hermán, R., Stefanovits-Bányai, É., & Pedryc, A. (2005). New self-incompatibility alleles in apricot (*Prunus armeniaca* L.) revealed by stylar ribonuclease assay and *S*-PCR analysis. *Euphytica, 145*, 57–66.

Halász, J., Hegedűs, A., Szabo, Z., Nyeki, J., & Pedryc, A. (2007a). DNA-based *S*-genotyping of Japanese plum and pluot cultivars to clarify incompatibility relationships. *HortScience, 42*, 46–50.

Halász, J., Pedryc, A., & Hegedűs, A. (2007b). Origin and dissemination of the pollen-part mutated SC haplotype which confers self-compatibility in apricot (*Prunus armeniaca*). *New Phytologist, 176*, 792–803. https://doi.org/10.1111/j.1469-8137.2007.02220.x.

Halász, J., Pedryc, A., Ercisli, S., Yilmaz, K. U., & Hegedűs, A. (2010). *S*-genotyping supports the genetic relationships between Turkish and Hungarian apricot germplasm. *Journal of the American Society for Horticultural Science, 135*, 410–417.

Harrold, T. J. (1935). Comparative study of the developing and aborting fruits of *Prunus persica*. *Botanical Gazette, 96*, 585–620.

Hedhly, A. (2011). Sensitivity of flowering plant gametophytes to temperature fluctuations. *Environmental and Experimental Botany, 74*, 9–16. https://doi.org/10.1016/j.envexpbot.2011.03.016.

Hedhly, A., Hormaza, J. I., & Herrero, M. (2005). The effect of temperature on pollen germination, pollen tube growth, and stigmatic receptivity in peach. *Plant Biology, 7*(5), 476–483.

Hedhly, A., Hormaza, J. I., & Herrero, M. (2007). Warm temperatures at bloom reduce fruit set in sweet cherry. *Journal of Applied Botany and Food Quality - Angewandte Botanik, 81*, 158–164.

Hedhly, A., Hormaza, J. I., & Herrero, M. (2009). Global warming and sexual plant reproduction. *Trends in Plant Science, 14*, 30–36.

Hegedűs, A., & Halász, J. (2006). Self-incompatibility in plums (*Prunus salicina* Lindl., *Prunus cerasifera* Ehrh. and *Prunus domestica* L.). A mini review. *International Journal of Horticultural Science, 12*, 137–140. https://doi.org/10.31421/IJHS/12/2/646.

Herrera, S., Lora, J., Hormaza, J. I., Herrero, M., & Rodrigo, J. (2018a). Optimizing production in the new generation of apricot cultivars: Self-incompatibility, *S-RNase* identification, and incompatibility group assignment. *Frontiers in Plant Science, 9*, 527. https://doi.org/10.3389/fpls.2018.00527.

Herrera, S., Rodrigo, J., Hormaza, J. I., & Lora, J. (2018b). Identification of self-incompatibility alleles by specific PCR analysis and *S-RNase* sequencing in apricot. *International Journal of Molecular Science, 19*, 3612. https://doi.org/10.3390/ijms19113612.

Herrera, S., Rodrigo, J., Hormaza, J. I., & Lora, J. (2020). Identification of self-incompatibility alleles by apecific PCR analysis and S-RNase sequencing in apricot. JoVE (Journal of Visualized Experiments), e60241. https://doi.org/10.3390/ijms19113612.

Herrero, M. (1992). From pollination to fertilization in fruit trees. *Journal of Plant Growth Regulation, 11*, 27–32.

Herrero, M. (2000). Changes in the ovary related to pollen tube guidance. *Annals of Botany, 85*, 79–85.

Herrero, M., & Arbeloa, A. (1989). Influence of the pistil on pollen tube kinetics in peach (*Prunus persica*). *American Journal of Botany, 76*, 1441–1447.

Herrero, M., Rodrigo, J., & Wünsch, A. (2017). Flowering, fruit set and development. In J. Quero-García, A. Iezzoni, A. Puławska, & G. Lang (Eds.), *Cherries: Botany, production and uses* (pp. 14–35). Boston: CAB International.

Hormaza, J. I., & Herrero, M. (1996). Dynamics of pollen tube growth under different competition regimes. *Sexual Plant Reproduction, 9*, 153–160.

Jia, H. J., He, F. J., Xiong, C. Z., Zhu, F. R., & Kamoto, G. (2008). Influences of cross pollination on pollen tube growth and fruit set in Zuili plums (*Prunus salicina*). *Journal of Integrative Plant Biology, 50*, 203–209. https://doi.org/10.1111/j.1744-7909.2007.00382.x.

Julian, C., Herrero, M., & Rodrigo, J. (2010). Flower bud differentiation and development in fruiting and non-fruiting shoots in relation to fruit set in apricot (*Prunus armeniaca* L.). *Trees - Structure and Function, 24*, 833–841. https://doi.org/10.1007/s00468-010-0453-6.

Jun, J., Kwon, J. H., & Chung, K. H. (2007). Identification of self-incompatibility genotypes in Japanese plum (*Prunus salicina* Lindl.) by polymerase chain reaction and cross-pollination tests. *Horticulture, Environment, and Biotechnology, 48*, 228–234.

Kao, T., & Huang, S. (1994). Gametophytic self-incompatibility: A mechanism for self/non-self discrimination during sexual reproduction. *Plant Physiology, 105*, 461–466. https://doi.org/10.1104/pp.105.2.461.

Kao, T. H., & McCubbin, A. G. (1996). How flowering plants discriminate between self and non-self pollen to prevent inbreeding. *Proceedings of the National Academy of Sciences of the United States of America, 93*, 12059–12065. https://doi.org/10.1073/pnas.93.22.12059.

Kester, D. E., & Gradziel, T. M. (1996). Almonds (*Prunus*). In J. N. Moore & J. Janick (Eds.), *Fruit breeding* (pp. 1–97). New York: Wiley.

Kester, D. E., & Griggs, W. H. (1959). Fruit setting in the almond: The pattern of flower and fruit drop. *Proceedings of the American Society for Horticultural Science, 74*, 214–219.

Kevan, P. G. (1999). Pollinators as bioindicators of the state of the environment: Species, activity and diversity. *Agriculture, Ecosystems and Environment, 74*, 373–393.

Kleijn, D., Winfree, R., Bartomeus, I., Carvalheiro, L. G., Henry, M., Isaacs, R., Klein, A. M., Kremen, C., M'Gonigle, L. K., Rader, R., Ricketts, T. H., Williams, N. M., Adamson, N. L., Ascher, J. S., Baldi, A., Batary, P., Benjamin, F., Biesmeijer, J. C., Blitzer, E. J., Bommarco, R., Brand, M. R., Bretagnolle, V., Button, L., Cariveau, D. P., Chifflet, R., Colville, J. F., Danforth, B. N., Elle, E., Garratt, M. P. D., Herzog, F., Holzschuh, A., Howlett, B. G., Jauker, F., Jha, S., Knop, E., Krewenka, K. M., Le Feon, V., Mandelik, Y., May, E. A., Park, M. G., Pisanty, G., Reemer, M., Riedinger, V., Rollin, O., Rundlof, M., Sardinas, H. S., Scheper, J., Sciligo, A. R., Smith, H. G., Steffan-Dewenter, I., Thorp, R., Tscharntke, T., Verhulst, J., Viana, B. F., Vaissiere, B. E., Veldtman, R., Westphal, C., & Potts, S. G. (2015). Delivery of crop pollination services is an insufficient argument for wild pollinator conservation. *Nature Communications, 6*, 7414.

Kodad, O., & Socias i Company R. (2009). Review and update of self-incompatibility alleles in almond. *Acta Horticulturae, 814*, 421–424.

Kodad, O., & Socias i Company R. (2013). Flower age and pollenizer could affect fruit set in late-blooming self-compatible almond cultivars under warm climatic conditions. *Scientia Horticulturae, 164*, 359–365.

Kodad, O., Alonso, J. M., Sánchez, A., Oliveira, M. M., & Socias i Company R. (2008a). Evaluation of genetic diversity of *S*-alleles in an almond germplasm collection. *The Journal of Horticultural Science and Biotechnology, 83*, 603–608.

Kodad, O., Sánchez, A., Saibo, N., Oliveira, M., & Socias i Company R. (2008b). Identification and characterization of new *S*-alleles associated with self-incompatibility in almond. *Plant Breeding, 127*, 632–638.

Kodad, O., Alonso, J. M., Fernández i Martí, A., Oliveira, M. M., & Socias i Company R. (2010). Molecular and physiological identification of new *S*-alleles associated with self-(in)compatibility in local Spanish almond cultivars. *Scientia Horticulturae, 123*, 308–311.

Koh, I., Lonsdorf, E. V., Artz, D. R., Pitts-Singer, T. L., & Ricketts, T. H. (2018). Ecology and economics of using native managed bees for almond pollination. *Journal of Economic Entomology, 111*(1), 16–25.

Koumanov, K. S., & Long, L. E. (2017). Site preparation and orchard infrastructure. In J. Quero-García, A. Iezzoni, A. Puławska, & G. Lang (Eds.), *Cherries: Botany, production and uses* (pp. 223–243). Boston: CAB International.

Lai, Z., Ma, W., Han, B., Liang, L., Zhang, Y., Hong, G., & Xue, Y. (2002). An *F-box* gene linked to the self-incompatibility (*S*) locus of *Antirrhinum* is expressed specifically in pollen and tapetum. *Plant Molecular Biology, 50*, 29–42. https://doi.org/10.1023/a:1016050018779.

Lamp, B. M., Connell, J. H., Duncan, R. A., Viveros, M., & Polito, S. V. (2001). Almond flower development: Floral initiation and organogenesis. *Journal of the American Society for Horticultural Science, 126*, 689–696.

Lang, G. A., Early, J. D., Martin, G. C., & Darnell, R. L. (1987). Endodormancy, paradormancy, and ecodormancy—Physiological terminology and classification for dormancy research. *HortScience, 22*, 371–377.

Lisek, A., Kucharska, D., Głowacka, A., & Rozpara, E. (2017). Identification of *S*-haplotypes of European cultivars of sour cherry. *The Journal of Horticultural Science and Biotechnology, 92*, 484–492.

Luedeling, E. (2012). Climate change impacts on winter chill for temperate fruit and nut production: A review. *Scientia Horticulturae, 144*, 218–229.

Luedeling, E., & Gassner, A. (2012). Partial least squares regression for analyzing walnut phenology in California. *Agricultural and Forest Meteorology, 158–159*, 43–52.

Luedeling, E., Zhang, M., McGranahan, G., & Leslie, C. (2009). Validation of winter chill models using historic records of walnut phenology. *Agricultural and Forest Meteorology, 149*, 1854–1864.

Luedeling, E., Girvetz, E. H., Semenov, M. A., & Brown, P. H. (2011). Climate change affects winter chill for temperate fruit and nut trees. *PLoS One, 6*, e20155.

Luu, D. T., Qin, X., Morse, D., & Cappadocia, M. (2000). *S-RNase* uptake by compatible pollen tubes in gametophytic self-incompatibility. *Nature, 407*, 649–651. https://doi.org/10.1038/35036623.

Marchese, A., Bošković, R., Martínez-García, P. J., & Tobutt, K. R. (2008). The origin of the self–compatible almond 'Supernova'. *Plant Breeding, 127*, 105–107.

Martínez-Gómez, P., Dandekar, A. M., Gradziel, T. M., López Valero, M., Batlle, I., Alonso, J. M., Ortega, E., Sánchez-Pérez, R., Dicenta, F., & Socias i Company R. (2003). Identification of self-incompatibility alleles in almond and related *Prunus* species using PCR. *Acta Horticulturae, 622*, 397–401.

Martínez-Tellez, J., & Crossa-Raynaud, P. (1982). Contribution à l'étude du processus de la fécondation chez trois espèces de *Prunus*: *P. persica* (L.) Batsch., *P. cerasifera* Ehrh., *P. mahaleb* L. grâce à l'utilisation de couples de variétés mâle-stériles et mâle-fertiles. *Agronomie, 2*, 333–340.

Matsumoto, D., & Tao, R. (2016). Recognition of a wide-range of *S-RNases* by *S* locus F-box like 2, a general-inhibitor candidate in the *Prunus*-specific *S-RNase*-based self-incompatibility system. *Plant Molecular Biology, 91*, 459–469. https://doi.org/10.1007/s11103-016-0479-2.

McClure, B. A., Haring, V., Ebert, P. R., Anderson, M. A., Simpson, R. J., Sakiyama, F., & Clarke, A. E. (1989). Style self-incompatibility gene products of *Nicotiana alata* are ribonucleases. *Nature, 342*, 955–957. https://doi.org/10.1038/342955a0.

McClure, B. A., Gray, J. E., Anderson, M. A., & Clarke, A. E. (1990). Self-incompatibility in *Nicotiana alata* involves degradation of pollen rRNA. *Nature, 347*, 757–760. https://doi.org/10.1038/347757a0.

Muñoz-Sanz, J. V., Zuriaga, E., López, I., Badenes, M. L., & Romero, C. (2017). Self-(in) compatibility in apricot germplasm is controlled by two major loci, S and M. *BMC Plant Biology, 17*, 82.

Neumüller, M. (2011). Fundamental and applied aspects of plum (*Prunus domestica*) breeding. *Fruit, Vegetable and Cereal Science and Biotechnology, 5*, 139–156.

Nikolić, D., & Milatović, D. (2010). Examining self-compatibility in plum (*Prunus domestica* L.) by fluorescence microscopy. *Genetika, 42*(2), 387–396. https://doi.org/10.2298/GENSR1002387N.

Ortega, E., Egea, J., & Dicenta, F. (2004). Effective pollination period in almond cultivars. *HortScience, 39*(1), 19–22.

Ortega, E., Sutherland, B. G., Dicenta, F., Bošković, R., & Tobutt, K. R. (2005). Determination of incompatibility genotypes in almond using first and second intron consensus primers: Detection of new *S* alleles and correction of reported *S* genotypes. *Plant Breeding, 124*, 188–196.

Ortega, E., Bošković, R. I., Sargent, D. J., & Tobutt, K. R. (2006). Analysis of *S-RNase* alleles of almond (*Prunus dulcis*): Characterization of new sequences, resolution of synonyms and evidence of intragenic recombination. *Molecular Genetics and Genomics, 276*, 413–426.

Pimienta, E., & Polito, V. S. (1983). Embryo sac development in almond [*Prunus dulcis* (Mill.) D. A. Webb] as affected by cross-, self- and non-pollination. *Annals of Botany, 51*, 469–479.

Qiao, H., Wang, H., Zhao, L., Zhou, J., Huang, J., Zhang, Y., & Xue, Y. (2004). The F-Box Protein AhSLF-S 2 physically interacts with *S-RNases* that may be inhibited by the Ubiquitin/26S proteasome pathway of protein degradation during compatible pollination in *Antirrhinum*. *Plant Cell, 16*, 582–595. https://doi.org/10.1105/tpc.017673.

Quero-García, J., Schuster, M., López-Ortega, G., & Charlot, G. (2017). Sweet cherry varieties and improvement. In J. Quero-García, A. Iezzoni, A. Puławska, & G. Lang (Eds.), *Cherries: Botany, production and uses* (pp. 60–94). Boston: CAB International.

Richardson, E. A., Seeley, S. D., & Walker, D. R. (1974). A model for estimating the completion of rest for 'Redhaven' and 'Elberta' peach trees. *HortScience, 9*, 331–332.

Rodrigo, J. (2000). Spring frosts in deciduous fruit trees: Morphological damage and flower hardiness. *Scientia Horticulturae, 85*, 155–173.

Rodrigo, J., & Herrero, M. (2002a). The onset of fruiting in apricot (*Prunus armeniaca* L.). *Journal of Applied Botany and Food Quality, 76*, 13–19.

Rodrigo, J., & Herrero, M. (2002b). Effects of pre-blossom temperatures on flower development and fruit set in apricot (*Prunus armeniaca* L.). *Scientia Horticulturae, 92*, 125–132.

Romero, C., Vilanova, S., Burgos, L., Martínez-Calvo, J., Vicente, M., Llácer, G., & Badenes, M. L. (2004). Analysis of the *S*-locus structure in *Prunus armeniaca* L. Identification of *S*-haplotype specific *S-RNase* and *F-box* genes. *Plant Molecular Biology, 56*, 145–157. https://doi.org/10.1007/s11103-004-2651-3.

Ruiz, D., Rubio, M., Martínez-Gómez, P., López-Alcolea, J., Dicenta, F., Ortega, E., Nortes, M. D., Molina, A., Molina, A., & Egea, J. (2018). 'Cebasred' and 'Primorosa' apricots: Two new self-compatible, Plum pox virus (Sharka)–resistant, and very early ripening cultivars for the fresh market. *HortScience, 53*(12), 1919–1921.

Sánchez-Pérez, R., Dicenta, F., & Martínez-Gómez, P. (2004). Identification of *S*-alleles in almond using multiplex PCR. *Euphytica, 138*(3), 263–269.

Sanzol, J., & Herrero, M. (2001). The "effective pollination period" in fruit trees. *Scientia Horticulturae (Amsterdam), 90*, 1–17.

Sapir, G., Stern, R. A., Eisikowitch, D., & Goldway, M. (2004). Cloning of four new Japanese plum *S*-alleles and determination of the compatibility between cultivars by PCR analysis. *The Journal of Horticultural Science and Biotechnology, 79*, 223–227.

Sapir, G., Stern, R. A., Goldway, M., & Shafir, S. (2007). *SFBs* of Japanese plum (*Prunus salicina*): cloning seven alleles and determining their linkage to the *S-RNase* gene. *HortScience, 42*, 1509–1512.

Sassa, H., Hirano, H., & Ikehashi, H. (1993). Identification and characterization of stylar glycoproteins associated with self-incompatibility genes of Japanese pear, *Pyrus serotina* Rehd. *Molecular and General Genetics, 241*, 17–25. https://doi.org/10.1007/bf00280196.

Sassa, H., Hirano, H., Nishio, T., & Koba, T. (1997). Style-specific self-compatible mutation caused by deletion of the *S-RNase* gene in Japanese pear (*Pyrus serotina*). *The Plant Journal, 12*, 223–227. https://doi.org/10.1046/j.1365-313X.1997.12010223.x.

Schuster, M. (2012). Incompatible (*S*-) genotypes of sweet cherry cultivars (*Prunus avium* L.). *Scientia Horticulturae, 148*, 59–73.

Schuster, M. (2017). Self-incompatibility (*S*) genotypes of cultivated sweet cherries—An overview. In *Open Agrar-Repositorium*. https://doi.org/10.5073/20171213-111734.

Schuster, M., Apostol, J., Iezzoni, A., Jensen, M., & Milatović, D. (2017). Sour cherry varieties and improvement. In J. Quero-García, A. Iezzoni, A. Puławska, & G. Lang (Eds.), *Cherries: Botany, production and uses* (pp. 95–116). Boston: CAB International.

Sebolt, A. M., Iezzoni, A. F., & Tsukamoto, T. (2017). *S*-genotyping of cultivars and breeding selections of sour cherry (*Prunus cerasus* L.) in the Michigan State University sour cherry breeding program. *Acta Horticulturae, 1161*, 31–40.

Sedgley, M., & Griffin, A. R. (1989). *Sexual reproduction of tree crops*. London: Academic Press.

Sijacic, P., Wang, X., Skirpan, A. L., Wang, Y., Dowd, P. E., McCubbin, A. G., Huang, S., & Kao, T. H. (2004). Identification of the pollen determinant of *S-RNase*-mediated self-incompatibility. *Nature, 429*, 302–305. https://doi.org/10.1038/nature025234.

Sonneveld, T., Robbins, T. P., Bošković, R., & Tobutt, K. R. (2001). Cloning of six cherry self-incompatibility alleles and development of allele-specific PCR detection. *Theoretical and Applied Genetics, 102*, 1046–1055. https://doi.org/10.1007/s001220000525.

Sterling, C. (1964). Comparative morphology of the carpel in the Rosaceae. I. Prunoideae: *Prunus*. *American Journal of Botany, 51*, 36–44.

Sutherland, B. G., Tobutt, K. R., & Robbins, T. P. (2004). Molecular genetics of self-incompatibility in plums. *Acta Horticulturae, 663*, 557–562. https://doi.org/10.17660/ActaHortic.2004.663.97.

Tao, R., & Iezzoni, A. F. (2010). The *S-RNase*-based gametophytic self-incompatibility system in *Prunus* exhibits distinct genetic and molecular features. *Scientia Horticulturae (Amsterdam), 124*, 423–433. https://doi.org/10.1016/J.SCIENTA.2010.01.025.

Tao, R., Yamane, H., Sassa, H., Mori, H., Gradziel, T. M., Dandekar, A. M., & Sugiura, A. (1997). Identification of stylar *RNases* associated with gametophytic self-incompatibility in almond (*Prunus dulcis*). *Plant and Cell Physiology, 38*, 304–311. https://doi.org/10.1093/oxfordjournals.pcp.a029167.

Tao, R., Yamane, H., Sugiura, A., Murayama, H., Sassa, H., & Mori, H. (1999). Molecular typing of *S*-alleles through identification, characterization and cDNA cloning for *S-RNases* in sweet cherry. *Journal of the American Society for Horticultural Science, 124*, 224–233.

Tao, R., Watari, A., Hanada, T., Habu, T., Yaegaki, H., Yamaguchi, M., & Yamane, H. (2006). Self-compatible peach (*Prunus persica*) has mutant versions of the *S* haplotypes found in self-incompatible *Prunus* species. *Plant Molecular Biology, 63*, 109–123. https://doi.org/10.1007/s11103-006-9076-0.

Thompson, M. (1996). Flowering, pollination and fruit set. In A. D. Webster & N. E. Looney (Eds.), *Cherries: Crop physiology, production and uses*. Wallingford: CAB International.

Thompson, M. M., & Liu, L. J. (1973). Temperature, fruit set, and embryo sac development in 'Italian' prune. *Journal of the American Society for Horticultural Science, 98*, 193–197.

Tsukamoto, T., Hauck, N. R., Tao, R., Jiang, N., & Iezzoni, A. F. (2006). Molecular characterization of three non-functional *S*-haplotypes in sour cherry (*Prunus cerasus*). *Plant Molecular Biology, 62*, 371–383. https://doi.org/10.1007/s11103-006-9026-x.

Tsukamoto, T., Potter, D., Tao, R., Vieira, C. P., Vieira, J., & Iezzoni, A. (2008). Genetic and molecular characterization of three novel *S*-haplotypes in sour cherry (*Prunus cerasus* L.). *Journal of Experimental Botany, 59*, 3169–3185.

Ushijima, K., Sassa, H., Dandekar, A. M., Gradziel, T. M., Tao, R., & Hirano, H. (2003). Structural and transcriptional analysis of the self-incompatibility locus of almond: Identification F-Box of a pollen-expressed gene with haplotype-specific polymorphism. *Plant Cell, 15*, 771–781. https://doi.org/10.1105/tpc.009290.

Ushijima, K., Yamane, H., Watari, A., Kakehi, E., Ikeda, K., Hauck, N. R., Iezzoni, A. F., & Tao, R. (2004). The *S* haplotype-specific F-box protein gene, *SFB*, is defective in self-compatible haplotypes of *Prunus avium* and *P. mume*. *The Plant Journal, 39*, 573–586. https://doi.org/10.1111/j.1365-313X.2004.02154.x.

Vilanova, S., Romero, C., Llácer, G., & Badenes, M. L. (2005). Identification of self- (in)compatibility alleles in apricot by PCR and sequence analysis. *Journal of the American Society for Horticultural Science, 130*(6), 893–898.

Vilanova, S., Badenes, M. L., Burgos, L., Martínez-Calvo, J., Llácer, G., & Romero, C. (2006). Self-compatibility of two apricot selections is associated with two pollen-part mutations of different nature. *Plant Physiology, 142*, 629–641. https://doi.org/10.1104/pp.106.083865.

Webster, A. D., & Goldwin, G. K. (1978). The use of hormone mixtures to increase the set of plum, *Prunus domestica* L. cv Victoria. *Journal of Horticultural Science, 53*, 123–129. https://doi.org/10.1080/00221589.1978.11514806.

Weinberger, J. H. (1950). Chilling requirements of peach varieties. *Proceedings of the American Society for Horticultural Science, 56*, 122–128.

Williams, R. R. (1970). Techniques used in fruit-set experiments. In R. R. Williams & D. Wilson (Eds.), *Towards regulated cropping* (pp. 57–61). London: Grower Books.

Wünsch, A., & Hormaza, J. I. (2004a). Genetic and molecular analysis in Cristobalina sweet cherry, a spontaneous self-compatible mutant. *Sexual Plant Reproduction, 17*, 203–210. https://doi.org/10.1007/s00497-004-0234-8.

Wünsch, A., & Hormaza, J. I. (2004b). *S*-allele identification by PCR analysis in sweet cherry cultivars. *Plant Breeding, 123*, 327–331.

Yamane, H., Tao, R., & Sugiura, A. (1999). Identification and cDNA cloning for *S-RNases* in self-incompatible Japanese plum (*Prunus salicina* Lindl. cv. Sordum). *Plant Biotechnology Journal, 16*, 389–396.

Yamane, H., Tao, R., Sugiura, A., Hauck, N. R., & Iezzoni, A. F. (2001). Identification and characterization of *S-RNases* in tetraploid sour cherry (*Prunus cerasus*). *Journal of the American Society for Horticultural Science, 126*, 661–667. https://doi.org/10.21273/JASHS.126.6.661.

Yamane, H., Ikeda, K., Hauck, N. R., Iezzoni, A. F., & Tao, R. (2003). Self-incompatibility (S) locus region of the mutated S*6*-haplotype of sour cherry (Prunus cerasus) contains a functional pollen S allele and a non-functional pistil S allele. *Journal of Experimental Botany, 54*(392), 2431–2437. https://doi.org/10.1093/jxb/erg271.

Yi, W. G., Law, S. E., McCoy, D., & Wetzstein, H. Y. (2006). Stigma development and receptivity in almond (*Prunus dulcis*). *Annals of Botany, 97*(1), 57–63.

Zhang, S. L., Huang, S. X., Kitashiba, H., & Nishio, T. (2007). Identification of *S*-haplotype-specific *F-box* gene in Japanese plum (*Prunus salicina* Lindl.). *Sexual Plant Reproduction, 20*, 1–8.

Zhang, L., Ferguson, L., & Whiting, M. D. (2018). Temperature effects on pistil viability and fruit set in sweet cherry. *Scientia Horticulturae, 241*, 8–17.

Zuriaga, E., Molina, L., Badenes, M. L., & Romero, C. (2012). Physical mapping of a pollen modifier locus controlling self-incompatibility in apricot and synteny analysis within the Rosaceae. *Plant Molecular Biology, 79*, 229–242. https://doi.org/10.1007/s11103-012-9908-z.

Zuriaga, E., Muñoz-Sanz, J. V., Molina, L., Gisbert, A. D., Badenes, M. L., & Romero, C. (2013). An *S*-locus independent pollen factor confers self-compatibility in "Katy" apricot. *PLoS One, 8*, e53947. https://doi.org/10.1371/journal.pone.0053947.

Canopy Management in Stone Fruits

4

Rifat Bhat, K. M. Bhat, Sharbat Hussain, M. Maqbool Mir, Umar Iqbal, and Mehvish Bashir

Abstract

Canopy management is one of the most important concerns of orchard management that growers stand facing each year as the fruit trees habitually produce additional fruit than necessary. Hence, to guarantee good size and fruit quality, canopy management is compulsory to lessen fruit-to-fruit competition, to allocate fruit enough opportunity to grow, to expose fruit to ample passable sunlight, and to improve the overall fruit quality. In order to lower the cost of production and increase the overall production and the quality of the fruit tree, canopy management, especially size control, has become a prerequisite. The design and shape of the canopy highly influences sunlight penetration and distribution with guaranteed financial profit to the orchardists. Thus, beforehand, manipulation of the height and management of the tree canopy are principal strategies which are supposed to be carried out in fruit crops for increased profit to the orchardists. In scores of fruit crops, the increase in the production along with well-appreciated fruit quality is acquired by managing canopies of squat heighted trees. Canopy management starts from selection of site, rootstock and scion combination correct spacing, training, pruning and the use of growth retardants also.

Keywords

Canopy management · Light interception · Training and pruning · Stone fruits

R. Bhat (✉) · K. M. Bhat · S. Hussain · M. M. Mir · U. Iqbal · M. Bashir
Division of Fruit Science, Sher-e-Kashmir University of Agricultural Sciences & Technology of Kashmir, Srinagar, Jammu and Kashmir, India

M. M. Mir et al. (eds.), *Production Technology of Stone Fruits*,
https://doi.org/10.1007/978-981-15-8920-1_4

4.1 Introduction

Management of the canopy of fruit trees is highly concerned with enhancing and maintaining the structure of the trees with respect to the shape and size for greater yield and enhanced quality. The chief objective in managing the canopy of a perennial tree is to receive the superlative sunlight and take advantage of the given area (land) and the environmental factors for an augmented yield. The key focus of canopy management is usually to decrease extreme canopy shading. The canopies of nearly all the fruit crops, both from temperate and tropical areas, get not as much of the 70% of radiated light, and rigorous management of the canopy is requisite for increasing light penetration and yield (Whiley et al. 2013). Managing the tree canopy, most importantly controlling the size, has turn into a prime concern for reducing the cost of production and enhancing the productivity and quality of the fruit crops. The design and shape of the canopy affect the interception of light with guaranteed higher economic income to the orchardists. Tree vigour, light, temperature and humidity are the key factors which play an important role in the production and quality of fruits. Therefore, in nutshell, canopy management can be better related to how efficient we control the tree vigour and take advantage of the available temperature and sunlight for increasing the overall production and the fruit quality to diminish the severe effects of climatic factors.

4.2 Canopy Management in Cherry

Cherry occupies an important status among temperate fruits all over the globe and is the season's pioneer fruit that reaches the market. The production of cherries is restricted to Kashmir, Himachal Pradesh and Uttar Pradesh in India. It is a delicious fruit that is rich in protein, sugars and minerals. The calorific value of cherry is higher than apple. Most of the present-day cherry cultivars are of European origin. Cherry orchard systems have gone through changes over several years which include transition from conventional systems of production customary to huge-sized trees with wider dimensions of spacing to high-density planting systems with comparatively shorter and smaller trees that are relatively spaced closely. Trees with seedling rootstock have been shaped in round to globular form and planted at a spacing of 6 × 6 m, and trees established on dwarfing rootstock are trained in a conical shape. Now trees on size-controlling rootstock produce crops at earlier stages of their life, with a consistency in higher yield, and produce excellent standard of the fruit and assure earlier gains on investment and increase profitability. Modern orchards with comparatively shorter and smaller trees and less spacing have also led to enhancement in the distribution of light and plant management methodologies regarding disease control, pruning, harvest, spraying and other practices.

However, canopy management starts from nursery conditions also. Nursery trees having a trunk girth of 3/8 in. mostly do not maintain lateral branches and can be cut back at a level of 45–60 cm from the base. Scaffolds or primary branches build up 15–25 cm below this cut, and these scaffolds need to be maintained at a lower height for ease in the operations like thinning, pruning and harvesting from the ground, or scaffold branches need to be kept at enough height so as to allow ease of orchard management practices like weed control, manuring and fertilisation and irrigation line maintenance. Nursery trees with diameter more than 1.27 cm might have already resulted in the growth of lateral branches, most of which need to be removed using pruning shears, retaining about three to five branches which are uniformly positioned in all the four directions. Cut back the remaining lateral branches 2–3 in. or closer to the trunk depending on their length and vigour to stimulate vigorous lateral growth. With most of the pruning left for the winter months when the trees have entered dormancy, trees should be developed in a way so as to result in much vegetative growth as possible during the first spring and summer. However, without much leaf surface reduction, young plantings can be pruned slightly in summer. Some growers select undesirable and vigorous branches to be removed as these may cause diversion of growth away from the scaffold branches. The unwanted branches ought to be pruned during the second and third summers, and excessive undesirable growth from the top portion of the trees needs to be pruned as well.

The various factors governing canopy management in cherry are as follows:

- Training.
- Pruning.
- Working with size-controlling rootstocks.
- Requirement of different scions.
- Use of growth regulator.

4.3 Training of Cherry

In this factor, the main objective is to develop nominal trunk which is permanent for different training systems practised under high density.

High density cherry orchards are trained like tall spindle axe (TSA), super spindle axe (SSA), central leader (CL), Kym Green bush (KGB), the upright fruiting offshoots (UFO), etc. (Figs. 4.1 and 4.2). Short-term fruiting structures are developed from this permanent wood, to which annual renewal pruning is done, i.e. a part/piece of this fruiting wood is cut, which is highly regarded as one of the factors to develop high standard fruit over a number of years; as the trunk matures, foliage and fruit buds are kept in a young age (Savini et al. 2007). These training systems (Fig. 4.3) hold numerous early expenditure costs, initial versus mature yield potentials, efficiencies for picking, etc. From an architectural point of view, the cherry tree is represented by the following characters. First, branches primarily pick up in an erect route (orthotropy) owing to the cultivar, a rearrangement of the branches with time as suited to their own mass. Second, blossoming appears laterally, i.e. inflorescences (umbels) are produced in the axil of the already formed foliage of every single shoot, either elongated or small (i.e. spurs), exit uncovered nodes in later years.

Both these characters depict the structural model of cherry, and in this model, two other characters specify cherry (Claverie and Lauri 2005), the first one being

Fig. 4.1 Super spindle axe (SSA), tall spindle axe (TSA)

Fig. 4.2 Kym Green bush (KGB), UFO system

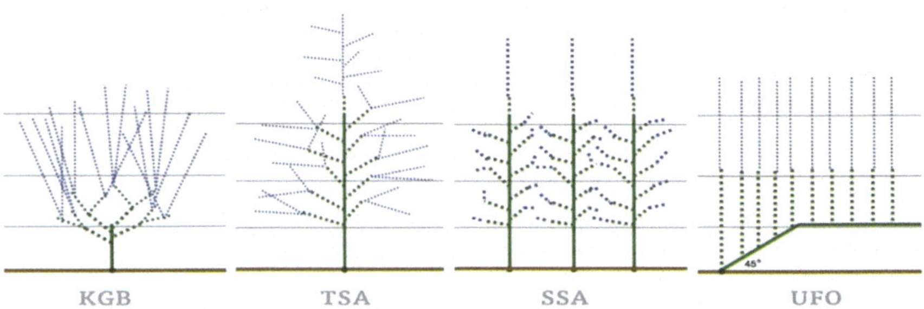

Fig. 4.3 All training systems of cherry

the effective dimorphism between the short shoot (consisting of ten already formed nodes covered in the over-wintering bud) (Rivals 1965) and the elongated shoot (which consists of a preformed portion succeeded by a new formed part). Owing to the age-based decline in length of annual shoot extension growth of long shoots, this dimorphism is generally not significant in old trees in comparison to young trees (Lauri 2005). Consequently, the amount of yield from spurs shows an increase with the age of the tree.

4.3.1 Pruning

A branch or tree that is pruned constantly tries to make not much of the total growth as compared to the one that is not pruned and left uncontrolled. Vegetative growth is generally generated by certain pruning cuts at the place of the cut and hence develops the impression of augmented growth which is always less than the sum of the portion removed and the growth it might have otherwise made. Since pruning tries to reduce the promising leaf surface for the succeeding year and stored reserves in the wood also are reduced, subsequently, its extent is calculated by the count of growing spots more precisely, instead of the mass of wood eliminated. Most of the pruning is carried out in the winter season. As each of the growing season comes to its end, the top portion of the tree and the roots are set stable and held in balance. While winter pruning prevents certain spots to grow, this shifts most of the resources for the retained buds. As a consequence, the invigoration partly recompenses for the wood separated, and until the pruning is conceded to extremes, the total loss is minimum. In a nutshell, light pruning has a profound influence on the spread of the growth within the tree than on net growth. Nevertheless, in case an enormously big amount of growing spots is removed, the net growth of the upcoming year will be considerably lowered, and the growth produced from the left behind emerging spots will be extremely high. Unlike dormant pruning, summer pruning is not invigorating (Fig. 4.4).

The initial growth of shoots, foliage and fruits is made at the cost of reserves which are stored in the woody tissues; in the late season, the leaves replenish these reserves. If active leaf surface is eliminated in midsummer, before shift of assimilates

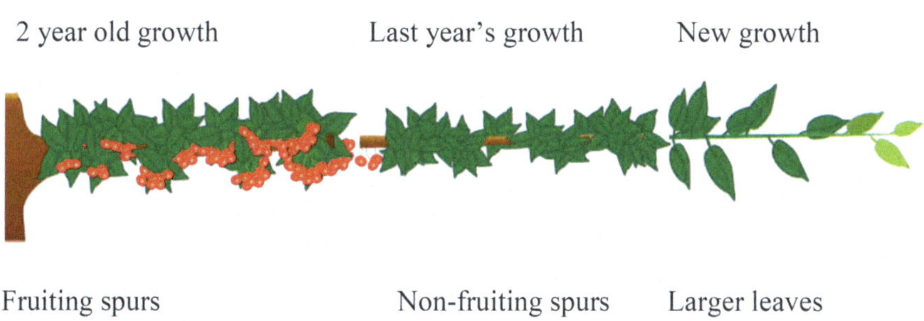

Fig. 4.4 Bearing habit of cherry

from the leaves back to the wood has started, the net effect is a depletion of tree reserves that is observed from the reduced vigour during the following season. This effect is directly proportional to the amount of leaf surface which is removed (Sansavini and Neri 2005). Pruning aids in tree size control; however, an inappropriate use of pruning techniques, such as the cutting of large branches or heavy winter pruning, has raised issues about the use of this technique, because they result in detrimental effects such as large-scale injuries which heal slowly. Highest yields were obtained by pre- and postharvest summer pruning, on young cv. 'Sweetheart' trees grafted on Mazzard seedlings, by removing either one-third or two-thirds of current season growth, in trees from fourth to seventh year of field growth (Webster and Schmidt 1996); however, preharvest treatment along with the removal of one-third of the vegetative portion leads to adverse effect on average fruit mass. The most severe pruning (two-thirds of current year's vegetation) has restricted production and allocation of carbohydrates and declined the number of spurs. The alternative strategies for controlling cherry tree size include partial branch breaking, root pruning or root restriction size (Webster 1998).

4.3.2 Growth Regulators

Paclobutrazol (PP.333 = Cultar) aids in size control, providing a compact scion tree. It was found that the endo-dormancy of tart cherry trees was altered by triazoles. This effect may also be as a result of time and method of treatment as these factors are linked to the seasonal stage of development as well as the degree of growth control achieved. Paclobutrazol has a profound effect on stone fruit tree management, though this growth regulator generally decreases shoot growth and lowers pruning costs yet depends on the training of the tree and the desired amount of growth removed.

4.3.3 Rootstock

The selection of rootstock has an important say in tree size control and cherry management. Clonal rootstocks play a vital role in tree physiology, so the choice of right rootstock is critical. Dwarfing rootstocks are essential for tree size control in high-density planting systems and provide better chances of easier rain cover against cracking. However, rootstocks that are dwarfing and which also lead to precocity and consistent cropping with high-quality fruits are yet to be developed for sweet cherry. A thorough evaluation was carried out in East Malling, UK, where a number of rootstocks were compared with 'Colt'. Several tested rootstocks, though not completely dwarfing, lead to high yield, precocity and yield efficiency and, with some scion-dependent differences, resulted in greater decline in tree size as compared to Colt. Only 'G 258' (*Prunus mugus* L.) showed signs for becoming a potential and promising rootstock, decreasing tree size of about 50% as compared to 'Colt' (De Salvador et al. 2005). However, G 258 showed difficulties in propagation and a certain degree of sensitivity to cold injury. 'Inmil' was found to be the most reducing rootstock as it showed 20–30% reduction in tree size as compared to 'F 12/1' (Salvador et al. 2006). 'Weiroot 158', 'Gisela 1', 'Gisela 4', 'Gisela 5' and 'AhrensburgA 473/10' are found to cause a tree reduction by 45–60%. 'Damil' has a potential to cause 25–50% tree size reduction. It was reported by Sansavini et al., in 2001 from the trials held in Bologna, Italy, that 'Inmil', 'Camil', 'Edabriz' and 'CAB 8F' were found to be the most dwarfing rootstocks, with a percentage vigour reduction of higher than 70% based on TCSA. Several rootstocks were tested in Oregon which showed interesting results for 'Damil', 'Gi 154–7', 'Gisela 5', 'Gisela 7' and 'Gisela 12' having a similar size. Also, a number of rootstocks were tested in Summerland, Canada. A certain degree of dwarfing potential was recorded for both 'GM 61/1' and 'GM9'. In experiments conducted in Spain, different rootstocks were grafted with scion 'Sunburst', and a comparison of their performances was held with those of 'Colt' and 'SL64'*15*. It was found that the most significant effect in controlling tree size by inducing a semidwarfing habit was by using two rootstocks, 'MaxMa 14' and 'MaxMa 97', while 'Damil' caused an excessive reduction of tree growth. Slender spindle and super spindle training systems showed a potential to be commercially successful by using dwarfing rootstock like 'Gisela 5', 'Weiroot 158' and 'Weiroot 72'. The NC140 cherry research trials held from 1987 to 2007 assessed new hybrid rootstocks that conferred previously unobserved levels of precocity (ability to begin fruiting at a young age) and varying degrees of tree vigour (or dwarfing). These resulted in the commercial release of the Gisela (Gi) rootstock series: Gi.3 (dwarfing, 30–45% of standard), Gi.5 (semidwarfing, 50–65% of standard) and Gi.6 and Gi.12 (semi-vigorous, 75–90% of standard).

For cherries, clonal rootstocks have been recently developed that showed significant signs for controlling tree size in commercial orchard plantings (Choi et al. 2002; Proebsting and Mills 1969). Rootstocks resultant from interspecific hybrids of *Prunus* in Belgium (the GM series), Gembloux and Giessen, Germany (the GI series) have produced 20% cherry trees or less than the size of trees on the standard rootstocks of Mahaleb or Mazzard. The most promising dwarfing cherry rootstocks produce trees between 20% and 50% of a standard size tree.

4.4 Canopy Management in Apricot

In apricot, fruiting behaviour and growth habit are soundly interrelated. The chilling requirement may be different which depends on the nature of bud (either floral or vegetative) and shoot (brindle, long brindle, sylleptic or spur) and, finally, might aid in modifying branch habit and fruiting behaviour (Guerriero and Viti 1997). The shoot is to be recognised by its particular growth rate succeeding the bud break in order to decide the method to manage pruning. So for this, its physiological behaviour must be constantly checked. The growth of sylleptic shoots is greatly possible when a critical level of the rate of growth exceeds (Zucconi 2003) which depicts that in certain cases apical dominance cannot restrict the lateral meristem growth and hence cause sylleptic (anticipated) shoots to originate instead of buds.

Mostly, the buds (once formed) switch to dormancy quickly and only shall show growth in the succeeding spring after a specific requirement of chilling is met and hence will originate proleptic shoots. Possibly, based on growing environmental conditions, sylleptic shoot emergence occurs many times along the shoot growth. Therefore, buds on sylleptic shoots which grow at different times in comparison to proleptic shoots might show a distinct fruiting performance and flowering time, which is more often confronted by the orchardists. As a result, the pruning time, either in the vegetative period (early summer pruning or late summer pruning) or in the winter prior to bud burst, has a sound effect in managing the fruiting of various cultivars. Also, knowing the pattern of the shoot growth is quite essential for effective pruning. Pruning in case of apricot must be regulated in extent and timing with respect to variety-environment interaction. Consequently, the particular physiology of the shoot and fruiting branch architecture of the varieties of apricot will ensure the variant mechanisms and systems for pruning. The general categorisation in northern Italy is given as: (a) with extremely vigorous and spreading behaviour and a capacity to fruit on spurs, brindles and sylleptic shoots; (b) with not much vigour and assurgent or semi-spreading behaviour, fruits are produced on spurs and vigorous shoots; and (c) with very vigorous, assurgent or mixed spreading behaviour and fruits on every type of shoot (Marini and Barden 1987; Neri 2003; Neri et al. 2010). Thus, in case of every apricot variety, it becomes necessary for determining the response (with respect to both the type and the number of lateral shoots) to heading back of branches and shoots in distinct times of the spring and summer months. Orchard management practices like fertilisation, irrigation, soil management and eventually forcing and protection strategies as well as pruning intensity play an efficient function in predicting the final outcome and the most ideal training system (Neri et al. 2011).

4.5 Training and Pruning

A very small portion of the fruit grows on 1-year-old lateral shoots. A major portion of fruit is produced laterally on spurs having a life span of 3 years. Apricot trees ought to be trained so as to ensure a strong structure which supports a huge crop load and for ease in the accommodation harvest tools and equipment. Apricots need to be pruned every year to maintain a consistency and uniformity in yields and good fruit size. Open centre is the most common training system. To prevent breakage from heavy crops, many trees take advantage from having a rope or wire strung around the scaffold limbs. Several new plantings are making use of high-density perpendicular 'V'-like arrangements. Some have training wires and some are freestanding. Since pruning wounds may get infected with the fungus *Eutypa lata*, pruning is carried out in late summer, early fall or late spring to prevent rains, which can cause disease spread. Smaller amounts of pruning may be practised in summer to enhance light penetrations. Larger pruning wounds may be treated using a fungicide to protect them from infection.

4.5.1 Thinning and Heading Back of Spring Shoot

The extent of heading back in developing shoots may be carried out between the following two extremes: short pruning (leaving just the portion of the base or half of the shoot with about three to five buds, as a spur) and long pruning (which includes reduction of the top portion of the shoot by pinching only some centimetres below the apex). Such pruning methodologies are generally employed in the spring season for rapidly emerging shoots. The pruning time can be either in early or late spring. The early pruning leads to the emergence of elongated sylleptic shoots. The late pruning produces no growth or the production of some short, sylleptic brindles along with an elevated differentiation potential of the flower. Shoot invigoration is greatly lowered, and the count of sylleptic shoots usually increases after spring heading back. Short head back pruning could not show much effect than long heading back in terms of flowering brindle induction. Plants show higher response to pruning in nutritious and watered soil. The late spring pruning produces a better effect if it is confined to the apical portion of the elongated shoot (long pruning). Delayed pruning in spring decreases the count of sylleptic shoots per one cut and lowers the flowering intensity also. Flowering intensity is found to be more when early spring pruning (May) is carried out in different varieties.

Shoot thinning is generally practised few weeks prior to harvest in order to enhance the quality of fruit in extremely invigorated trees. Generally, it is done to remove more crowded and mispositioned shoots. The conclusion is to attain adequate sunlight penetration and even distribution within the entire tree spread and conduction of low carbon to the water sprouts and suckers which otherwise becomes limited to be utilised for the production of fruits in the management of the branches later.

4.5.2 Summer Shoot Head Back Pruning and Thinning

Summer shoot head back focuses on promoting floral differentiation, but in case of apricot this can be achieved if there appears new growth of shoot, which can be produced by the application of irrigation after summer drought or by dense incisions such as late summer head back pruning. Shoot thinning in summer is practised with the target of enhancing the shoot quality which results from better light perception and distribution as well as carbon allocation. This generally lowers the necessity of winter pruning and may prove beneficial in regions where freeze damage might kill the blossoms, and hence winter pruning should not be done till the fruit sets and fruit are harvested successfully. Various pruning intensities regulate greater vigour of the vegetative growth when the shoot is accidently cut extremely short and promote the production of flowering brindles when it remains elongated. In northern Italian climate, pruning induces more vegetative growth in nourished and irrigated soils, hence leading to more shoot vigour. Eventually, sylleptic shoots blossom later than the remaining parts of the tree (advantageous in regions where late frost is usual) but produce fruits of smaller size, in certain varieties (Pirazzini 2004).

4.5.3 Pruning of Apricot in Several Training Systems

The exact training systems in apricot pertain specifically to each area of production. The two mostly adopted training systems in the northern Italy include free open vase (associated with different modifications, from delayed open vase to bush) in case of low-density planting system, hilly orchards and spindle in case of high-density planting systems in highly fertile levelled lands with less vigorous characterised rootstocks. The time and extent of pruning productively govern the framework of branches and the fruiting tendency of the cultivars. From such observations, it is evident that in case of apricot, summer pruning is a fundamental application in advanced orchards, but it ought to be acclimatised to local environmental conditions and genetic material. Hypothetically, physiology of the shoot is modelled with respect to rate of growth and may aid in the evaluation of the best time and the most efficacious intensity for the pruning of each new cultivar in the distinct systems of training specific to each growing region. We can infer that pruning in summer lowers vigour and stimulates greater flower formation. Early elongated shoot head back pruning is highly efficacious for very vigorous varieties and highly nourished soil conditions, while smaller shoot and branch heading back is advantageous for lean and spreading varieties. Varieties belonging to group A, for example, certain new varieties, are benefited from early summer pruning (early heavy shoot head back pruning) so as to stimulate the emergence of sylleptic shoots, and summer pruning (only with shoot thinning) to enhance shoots shows more lignifications. Group B shows better performance when winter pruned (thinning of shoots and head back pruning 2–3-year aged branches). Group C might be pruned during late summer or in ending winter (thinning of shoots, head back pruning of the branches), depending on agro-climatic conditions of local region. In order to restrict the formation of sylleptic shoots, which lead to the production of pygmy fruits only, it becomes effective to practise heading back in late summer rather than in winter. Generally, each pruning technique is to be evaluated on each variety prior to its adoption throughout commercial orchards because of the possible specific influence in varietal preferences of chilling requirements and particular differentiation physiology of the shoot and flower.

4.6 Plant Growth Regulators

4.6.1 Auxins and Gibberellins

In apricots, as a result of the application of three synthetic auxin compounds at the beginning of pit hardening, there was a significant increase in fruit size and total yield (Stern et al. 2007a). Anatomical studies showed that the principal effect was immediate stimulation of cell enlargement. The compounds 3,5,6-TPA, 2,4-DP and 2,4-D plus NAA neither showed any negative effect on fruit quality at harvest and after storage nor did they influence the return bloom in the following season. The synthetic auxin 2,4-DP-P resulted in greatest increase in growth rate with six times increased fruit size than the control. Nevertheless, these fruits were affected by severe internal browning after storage. Moreover, high concentrations resulted in fruit cracking, and hence concentrations ought to be carefully observed. The alternative strategy to hand thinning of stone crops (e.g. apricot) is the use of gibberellins (Southwick et al. 1995).

4.6.2 Gibberellin Synthesis Inhibitors

There are several compounds which obstruct gibberellin synthesis, thus inhibiting the endogenous gibberellin in the tree, and hence act as growth retardant. These compounds prove beneficial for controlling vigorously growing trees where vegetative growth can compete for nutrients with the fruit. Many cultivars of cherry, peach and plum show vigorous growth, producing large trees which are not easy to manage. This is specifically the case where the orchard is established on non-dwarfing rootstocks. The reduction in branch expansion allocates more nutrients to the fruit and often results in larger fruit size and higher yields per tree. The two anti-gibberellins that are generally applied as a soil drench or to the collar of the tree include paclobutrazol and uniconazole but are sometimes given as a foliar spray and are retained in the soil which can cause problems for future treatments. Prohexadione-Ca also causes interference with gibberellin synthesis and is usually applied as a foliar spray and does not show residual effect on tree growth. A study which included the comparison of the response of stone fruits to paclobutrazol application on the collar showed that the order of responsiveness to the compound was European plum > sweet cherry > apricot > peach, though the responses might be cultivar specific (Grochowska et al. 2004). Shoot growth was decreased to half and fruiting was increased. The same concentration did not influence shoot growth, fruiting or flowering when applied to mid-stem.

It was reported that paclobutrazol reduced shoot growth and tree size of cherry without showing adverse effects on yields or fruit quality (Looney and McKellar 1987; Webster et al. 1986). It too increased flower number and fruit set and resulted in increment of fruit size (Looney and McKellar 1987; Ogata et al. 1989). Soil drenching with the same chemical resulted in control of growth for a number of years, which was not advisable. In addition, if soil drenches are applied at too higher concentrations, then flowering and fruit set was found to be quite high, followed by fruit drop and poor yields.

4.7 Canopy Management in Plum

Plum trees are trained according to vigour and growth habitat of rootstock. Mostly open central system of training is recommended in which top of plant is headed back to 60 cm to regulate growth of lateral branches at the time of planting. First three to five scaffold branches are selected around the main stem during the first summer. The lowest branch ought to be 30 cm above the ground and others 15 cm apart in spiral manner. Only broad angled branches must be selected, and others should be removed. In the first dormant pruning, central leader is headed back and scaffold branches are allowed to grow. One-third growth of the scaffold branches is removed. The rest of the undesirable branches on the plants are also removed. In the second dormant pruning, two to three well-spaced secondary branches are selected on each primary scaffold, one-third to one-fourth portion of which is removed. The remaining weak and unwanted intercrossing shoots are also removed. In the third year thinning out and heading back of undesirable branches, diseased and interfering branches are carried out. Training is completed in the fourth year. During this year, only light pruning and heading back are practised (Lukic et al. 2012).

4.7.1 Pruning of Bearing Trees

In order to assure a stability and equilibrium between vegetative and reproductive growth, bearing trees are pruned. Pruning done during pre-bearing phase is usually light and corrective. Light heading back and removing of water sprouts and dead and diseased branches are carried out during this stage. Heavy heading back of branches must be avoided because it leads to the formation of elongated water sprouts. Fruit in plum is produced on spurs and 1-year-old shoots, and spur life is about 5–6 years (Fig. 4.5).

Pruning is quite important for the renewal of spur, and about 75–80% of new growth is removed in every season. For proper fruiting, plum requires about 25–30 cm of annual extension growth. About 25–30% of thinning along with one-third to one-half heading back of shoot is advised for Santa Rosa plum under Jammu

Fig. 4.5 Tree training with summer pruning in plum: (**a**) tree is left with unheaded central leader and slightly pruned side shoots; (**b**) in May, side shoots emerging at the top of the leader are removed, leaving only one for leader extension, and few lower shoots are twisted with clips; (**c**) treatments at top of the leader are repeated in May of the second year; (**d**) tree with fruiting potential in spring of the third year; (**e**) shoot bent with the help of a clip

and Kashmir. Plum has a potential to grow more upright than peaches and nectarines and shows more dense branching habit, hence requiring a specific pruning technique.

Shoot Twisting In this method, the primary focus is to bring the shoots to best possible position in initial stages of development. The stronger the twisting, the greater is the efficiency. Twisting controls vegetative growth and initiates cropping in the commencing years of spindle formation. It was reported by Gonda (2006) that the combined use of canopy management techniques along with twisting leads to early bearing of trees and reaching full productivity at an earlier date in many plum cultivars grown under high-density planting system.

Twisting in other stone fruit crops is not much effective (sour cherry) as in case of plums because it causes excessive shoot damage or dieback (sweet cherry).

4.8 Canopy Management in Peach and Almond

In peach industry, there has been increasing cost of the land, energy and salaries. So, to increase the productivity of the orchards and to reduce production cost, horticultural strategies have been developed in the last few years. Efficient and profitable land use has been achieved well through the method of tree spacing. Its main role is to confine the exploitation zone of the plant with respect to light, water and nutrients so that the highest total yield potential is attained from the smallest possible area (Rom and Blackburn 1998). Hence, high-density plantation system has considerably gained importance and acreage specifically to small land holdings during the previous few decades in the developed countries. Tree canopy management, particularly tree size control, has become a necessity for decreasing production cost and promoting fruit yield and quality. Canopy design and shape affect the light penetration and distribution with guaranteed greater financial returns to fruit growers. Tree vigour, light, temperature and humidity play an important role in the fruit production and fruit quality. Therefore, the essence of the canopy management is based on the fact on how effectively we control the tree vigour and utilise the available sunlight and temperature to enhance the productivity and quality to reduce the harsh effects of weather conditions.

Canopy in fruit tree pertains to its physical structure which includes the stem, branches, shoots and leaves. Canopy management deals with the development and maintenance of the structure of the fruit trees with respect to the size and shape for maximising productivity and quality so as to earn higher profit. The main focus in canopy management of a perennial tree is to make the efficient utilisation of land and the climatic aspects in order to elevate productivity in a three-dimensional approach. The basic aim of canopy management is generally to bring down unnecessary canopy shading. The canopies of majority of the fruit crops, both from temperate and tropical areas, receive not more than 70% of radiated light and thorough canopy management; hence, it is a requirement to promote light interception and productivity (Whiley et al. 2013).

4.8.1 Tools Used in Pruning

A number of tools for use in pruning are available in the market. Depending on the nature of pruning cut to be given, a pruning tool must be selected accordingly. Successful pruner must draw the knowledge of the pattern of tree growth and its response to different types of pruning cuts. It is also essential to draw the results of pruning. Short shoots, not more than 8 in. in length, have an enormous number of fruit buds but produce pygmy fruits, whereas shoots of 12–24 in. in length are highly productive. Elongated shoots which are branched produce less number of flower buds (Fig. 4.6).

Two types of buds usually appear on a peach tree. The bud at the end of the shoot is called the terminal bud that shows vegetative growth and develops a leafy shoot. The buds which are produced on current season's shoots at the bases of leaves during the summer are called as axillary buds that are either vegetative or floral buds. The reproductive buds of peach are known to be 'pure' or 'simple' as those comprise of only flower tissue. The flower bud in peach bears only one flower that can produce a single fruit. Every node (the spot where a leaf is fixed to the shoot) which occurs on a vegetative shoot may have zero to three buds. Terminal nodes generally consist of just single buds. Generally, buds which are small and pointed are vegetative in nature, while the plump and more hairy ones are reproductive in nature. Numerous nodes present on the basal two-third part of a shoot have about two to three buds which are organised in a parallel order. There can be any combination of the flower (F) and leaf (L) bud (FL, FF, FLF, FFF), yet often a leaf bud is bordered by flower buds (FLF). The count of flower or reproductive buds and their distribution on a shoot vary with vigour of the tree, cultivar and the light environment in which the development of the shoot takes place. Conclusions derived from a trial which was conducted in New Jersey specify that 'Jerseyglo'

Fig. 4.6 Growth habit of peach

and 'Springold' consisted of 20–23 floral buds per foot of the length of the shoot, while 'Harken' and 'Emery' had just 15 floral buds per foot of the length of the shoot. Shoots which show not more than 15 cm of growth normally comprise of more fruit buds per inch of growth. There is an increase in the total number of fruit buds per shoot as the growth of the shoot increases to about 2 ft. Moderately vigorous shoots represent higher node ratio with two flower buds. At most of the nodes, the leaf buds produce shoots laterally which might be productive in following years. Many axillary buds which are present on the vigorous shoots of current season (more than 60 cm in length) develop to be formed as secondary shoots. These shoots are never much productive as reproductive buds are produced on a less number of nodes which are found on secondary shoots. The ideal fruiting shoot is 30–60 cm in length and 3/16 to 1/4 in. wide at the base and does not have any secondary shoot. Orchard practices such as proper pruning, fertilisation, irrigation and fruit thinning must be carried out to ensure more shoot growth per year so as to develop adequate number of fruit buds for the subsequent season. Results from experiments conducted in Virginia reveal that fruit size and shoot length are related. Shorter shoots have the ability to develop small fruit as they have only fewer leaves to support fruit growth. Thus, shoots having 6 in. in length should be plucked out while pruning (Grossman and DeJong 1998). The energy required for supporting plant growth is derived from sunlight. Leaves capture sunlight, and the light energy is turned into chemical energy by photosynthesis. The chemical energy is stored in carbohydrate form, which is transferred to different parts of the tree and utilised for growth. It was observed from experiments carried out in Virginia that the threshold of light level in peach for development of flower bud is approximately 20% full sun (Fig. 4.4). Shoots must be subjected to sunlight in the days of June and early July for highest flower bud development. Greater levels of light in late July, August and September won't affect formation of flower bud. The rest of the studies revealed that shoots growing at the tree interior in 20% full sun developed just half as many flowers per foot of the length of the shoot as shoots growing at the margin of the tree canopy where there is a light interception of 70% full sun. Shoots growing in densely covered zones of the tree are inclined to death during the season and sometimes during the off season. While the centre of the peach tree is exposed through pruning for greater light interception, the peripheral shell of the leaves and shoots having 36–48 in. depth around the tree captures not less than 30% full sun. Yearly pruning enhances sunlight penetration within the entire tree canopy, which is essential for the growth of floral buds and fruit. In order to develop large, good coloured fruit having more sugar content, every region of the tree should get a minimum of 25% full sun (Kim et al. 1994).

4.9 Training

In peach, high-density orchards are based on the choice of a suitable training system rather than the selection of dwarfing rootstocks or variety which are yet unavailable. Various training systems like espalier, free spindle, open vase, palmette, fusetto,

Y-shaped, etc., are adopted for peach. The trees trained through these training systems show various patterns of sunlight interception and its distribution within the tree canopy. Light interception is the most important factor for the biomass production and its influence on the sugar and acid content of the fruit, flesh firmness and fruit colour. An orchard system involves a coherent form which includes two components, i.e. a training system and planting density. Nowadays, the orchard system assumes the interaction between spacing, applied training system and rootstock/cultivar combination. The suitable choice of the growing system is an important requirement for successful exploitation of agro-climatic conditions, which not only affects the yield and fruit quality but also the vegetative activity, orchard longevity, efficiency of mechanised as well as manual work and profitability of production. Peach orchards are traditionally planted at wider spacings as the trees tend to grow into large specimen. In peach tree, free-standing closed vase system with three erect leaders was introduced during the late 1970s for effective sunlight interception and distribution within the canopy and shoot development at the base on the periphery of the canopy (Rana et al. 1998).

A new peach tree training system named as *perpendicular-V system* has been adopted by some south-eastern peach growers which is relatively easy, requires low maintenance and can be used even in the home orchard. Trees trained by this system are generally planted closer within the row, but standard distance is maintained between the rows to allocate enough space for equipment movement through the orchard and light penetration within the canopy. Perpendicular-V tree training has been developed as a result of increasing land and labour costs. This intensive operation was carried out in the production of stone fruits as the comparable dwarfing rootstocks used in apple production have not been available to control extreme vigour of stone fruit trees. The perpendicular-V system is based much on cultural practices to bring down the excessive vigour of a simple tree structure. The perfect intensive orchard system results in the early production of higher yields rather than standard yields (during the initial 2–6 years) but can be undoubtedly managed in the succeeding years delivering yield analogous to that of the standard orchard system. The strategy to achieve this goal is to avoid the unnecessary vigour that would certainly compete with developing fruit wood. When trees are planted under high-density orchard system, a spacing of 6 ft is maintained within the row and 18–20 ft between rows (330–372 trees/acre). Row spacing can be reduced to 12–14 ft if path for the movement of equipment is not required, as in case of the home orchard setting (Caruso et al. 1993).

Just after planting, the trees must be headed back to a height of 20–30 in., and in case any lateral shoot is found, it must be removed, as in case of the open centre system. During spring when scaffold grows to 15–24 in. in length, two primary scaffolds are selected that are aligned in the same plane, perpendicular to the row. These branches must be directed 25–40 E from the vertical, assuming a *V shape* (Fig. 4.7). This scaffold selection initially during the first growing season shifts growth towards the two primary branches, since those branches won't compete with other branches of same size for water and nutrients. Moreover, the scaffolds are easier to spot, reach and remove if selected initially. It is also comparatively easy to

Fig. 4.7 Perpendicular-V system in peach and nectarine

cut a small portion of leaf area in cooler months than during midsummer. Avoid heading back or bench cutting of the selected scaffolds. Excessive heading cuts result in unwanted water sprout growth, which causes much shading and summer pruning cuts. This early tree training might require follow-up in midsummer to discard any unnecessary vigorous growth. In case water sprouts are found to form along the scaffolds or in the tree crotch, these can be simply removed by hand with a fast tearing motion if found early. By such early removal of the sprouts, the recurrence of numerous sprouts in the place of a single sprout is prevented. Avoid the removal of tender shoots that can fruit in next year. During the second leaf, in case the primary scaffolds were selected in the previous spring, the first dormant pruning with the risk of frost (early February) will be comparatively easy. Vigorous and dense non-fruiting wood should be removed, and only enough fruiting wood must be maintained to support 18–24 fruits per scaffold, depending on the variety, resulting in less fruit bud loss due to late spring frost. Accordingly, that much fruit shall be borne on 8–12 fruiting shoots per scaffold, with thinning two to three fruit per fruiting shoot. In succeeding seasons, that fruit load will be elevated to 50–75 per scaffold (Marshal et al. 2006). Summer pruning (early to mid-May) of the second leaf trees is again quite easy and quick. It requires water sprout removal and being careful with retaining the fruiting wood. Another pruning in July or August may prove quite beneficial for a vigorous orchard to remove water sprouts or vigorous upright shoots that may compete with the extension growth of following season's fruiting wood. Severe pruning late in the growing season must be avoided that may retard the tree growth or make the tree susceptible to early frost events prior to the stage where trees enter complete dormancy. Most orchards don't require second summer pruning.

The main target of this system is to arrest the growth of competing water sprouts and unwanted vigorous upright shoots quite early so as to prevent the shading out of other growth that tend to initiate flower buds. Unwanted vigorous shoots or sprouts are not potent to develop many flowers and, hence, are not regarded as good fruiting wood. Through keen maintenance and management of wood which has a potential of fruiting, premature loss of fruiting tissue in the lower portion of the canopy can be avoided, and that can be shaded out as the tree matures. If the dwarf stocks for peaches are not available, practising this simple training system that makes use of judicious pruning decisions can bring down the excessive vigorous growth and enhance the overall production of each tree or acre (Miller 1995).

4.9.1 Open Centre Method

Lower-headed trees are easier to prune, thin and harvest. High-headed trees carrying a crop are more susceptible to injury from strong winds. More trunk area is exposed to the sun in the winter, which results in more of what is known as 'southwest injury'. The bark is often severely damaged from the effects of alternating high temperatures in the day and freezing temperatures at night. High-headed trees are likely to sway in the wind causing openings to form between the trunk and the soil. Water accumulates in these openings in the fall and winter. Ice formation in these openings can seriously damage the trunk.

4.10 Selection and Pruning of Scaffold Limbs

Three to five main scaffold limbs can be used, with three or four being best. If more than three scaffolds are kept, they should be distributed around and up the central trunk and far enough apart so that they will not grow together later. They should be especially well spaced around the trunk. If scaffolds touch each other, a sharp angle is formed. The bark does not knit properly in the area, and canker is likely to develop. It is okay if only two suitable limbs are available because a side limb arising on one of these can serve as a third scaffold. The length to leave the scaffolds at the time the tree is planted will depend on their size and uniformity. If they are large and uniform in size, they can be left 10–12 in. in length. If they are rather slender and uneven in size, it would be better to cut them off to short stubs an inch or two long. New shoots will develop from the basal buds, and the best of these can be selected the next spring for main scaffold limb (DeSalvador and DeJong 1989) (Fig. 4.8).

Do not allow excess scaffolds to remain in the tree 2 or 3 years and then remove them. The scaffolds selected should have wide angles with the trunk at the point of attachment. This is very important. Scaffolds having sharp angles at the point split off the trunk easily. Also, bark and wood unite poorly in this narrow area, marking them very susceptible to injury from low temperatures and the entrance of canker-forming organisms. It is better to have two wide-angled scaffold branches than three which may grow together in the future. Pinching out the terminals of unwanted shoots in early summer will assist in proper early training. It is recommended that corrective pruning be done in the spring of the second and third years. It is too late

Fig. 4.8 Open centre system in peach and nectarine

to do corrective pruning in later years. Allow a few side branches coming off the original scaffolds within a foot or two of the trunk to remain so that the entire tree has up to a maximum of eight upright major scaffolds. Closer spaced trees will require fewer major scaffolds (Bellini et al. 2000).

4.10.1 Central-Leader Method

This method is a cone-shaped form, somewhat like a classic Christmas tree. Trees trained by the central-leader method should be somewhat larger than open centre trees during the first 3 or 4 years because less wood is removed at the time of planting. Typical spacing for a central leader ranges from (20′ × 15′) down to (15′ × 10′). Trees larger than 5/8″ caliper are less easy to work with because they may not have enough scaffolds in the right position or angle and may be more reluctant to put out new growth in the desired locations (Fig. 4.9). Smaller diameter trees without branches formed in the nursery are also very good to work.

The goal is to form a tree with a strong central axis (central leader) with a swirl of four to five branches in the region of 24–48″ from the ground with another weaker swirl in the 48–84″ range for wider spaced planting. For closer spacings,

Fig. 4.9 Central-leader method in peach

Y-system V-system T-system Palmette

Fig. 4.10 New and other training systems of peach

more scaffolds (6–8) with 12–18″ between branches are needed to encourage smaller limb diameters. Tree height should not exceed 12 ft. The lower swirl is generally considered permanent, and the upper swirl can be renewed by pruning. Prune between green tip and pink. Remove any branches that are dead. Remove any branches that are competing with the leader and any that have strong upright growth. If the leader is weak or has begun to lean with the wind, cut the central leader back to larger diameter wood with a branch into the wind. Thin out weak wood, small diameter wood or those pointing downwards. Reduce the length of the scaffolds to 1/2 to 2/3 of the length of the leader, with branches closer to the top of tree being shorter in length. During the summer, remove new vigorous growth that is competing with the leader and scaffolds, and begin to develop a second swirl of branches 2–3 ft above the first set (Caruso et al. 1998; Farina et al. 2005) (Fig. 4.10).

4.11 Pruning

Peach trees must not be pruned ahead of February; also immature trees must be always pruned next to old trees. A plenty of information from research and experience of the orchardists is known to reveal that pruning the peach trees during the starting of the season can lower their cold tolerance. Pruning must be avoided during some days of forecasted cold weather. In winter, pruning can decrease the cold hardiness of the tree for approximately 2 weeks. Trees pruned prior to harsh cold weather might result in poor flower bud survival, dieback of 1-year-old shoots and bruised and wounded bark on the trunk and majority of the branches. Even prior to blossoming, at the time of swollen flower buds and pink tissue at the bud tips, pruning can reduce the tolerance of the flower buds to frost. Pruning of peach trees during flowering or just after flowering is not feasible, yet it won't severely influence the tree or the growth of the fruit. It is advisable to prune slightly later instead of very early. Summer topping, through trimming the top portion or sides of the trees, is done by few orchardists to lower cost of pruning and control the size of the tree. Nevertheless, conclusions of research derived from summer topping done prior to the harvest depict that there is less economic benefit in comparison to winter pruning. Topping decreases costs of pruning, though this gain is counterbalanced by reduced size of the fruit. Also, penetration of sunlight, cold tolerance of the flower bud, shoot growth and fruit colour are not constantly modified via summer topping. In case a peach grower aims to lower the height of the tree, a costly procedure will be to cut the top of the tree just after harvest or in the ending winter. Mechanised topping must usually be succeeded by thorough pruning to discard stubs in the treetops which may develop heavy foliage and may cause shading of the tree centre. The use of 'collar cuts' is recommended. For a number of years, the application of 'flush cuts' is being suggested for pruning of the fruit trees to enhance speedy healing of the bruises and wounds. Latest research observations with peach and other tree species show that flush pruning cuts are more vulnerable to disease infection than cuts in which a part of the lateral branch is retained. The 'collar cut' method develops a pruning cut which retains an elevated collar of tissue at the branch junction. This method causes quick wound recovery and reduces chances of external dieback and disease infection (Chalmers et al. 1981).

4.12 Root Pruning

Root pruning has been beneficial in some of fruits but less effective with others. Root pruning can reduce resource uptake or lead to plant hormone imbalance which can severely influence shoot growth. It restricts water and nutrient uptake by controlling root systems of fruit trees. The timing of root pruning is a key factor, and root pruning in the dormant season or at full bloom was efficacious in decreasing shoot expansion than at June drop (Bargioni et al. 1983). Similar results were achieved with peach where root pruning in April was more efficacious for lowering shoot elongation than root pruning in June. Nevertheless, root pruning has not resulted in being consistent for shoot growth control in highly vigorous trees. Decreased fruit size and yield related to root pruning may restrict its usage as a cure where shoot development needs great corrective control. Spreading or bending in peach is aimed on inclined canopy systems (DeJong et al. 1994) rather than widening of individual branches to attain growth control.

4.13 Deficit Irrigation

Other factor which can aid in attaining a good canopy in peach is deficit irrigation. Irrigation at particular intervals of time in the growing season has proven to decrease vegetative growth without severely influencing yield in peach and pear (Chalmers et al. 1981). This method of irrigation, which limits shoot growth, is called as deficit irrigation. Dormant peach pruning can be decreased by one-third in case postharvest irrigation is restrained in areas with scarce or no summer rainfall. Holding back irrigation during the dry spell in summer hence can help to decrease vegetative growth in peach. Deficit irrigation might result in enhancement of the adaptation of the tree to dry conditions. Apple trees too show response to deficit irrigation, but such responses are not similar to peach. Rooting dimensions show interaction with available soil water so as to influence vegetative growth in peach trees. Decreased root volume did not show much effect on shoot growth in the case when deficit irrigation was provided with 30% replacement of water that had been applied (Proebsting 1989). Regulated deficit irrigation may prove efficacious to check shoot growth especially in dry areas, where stress is given early and rapidly, mostly in case of shallow soils.

4.14 Rootstocks in Peach

Rootstocks control tree size by directly influencing growth and indirectly enhancing crop load. Rootstocks which control size are most common in case of apple rather than in other temperate fruits. A huge number of size-controlling rootstocks are also available in stone fruit crops like peach, cherry, plum, etc., but dwarfing rootstocks in such species have not proven to be much successful than in apple. Rootstocks for reducing growth and size in peach have not been consistent and, mostly, unsuccessful. However, certain reports in the literature suggest size control up to 50% with *Prunus tomentosa* and *P. besseyi* rootstocks in peach (Rom 2000). Recently, clonal rootstocks have been developed for cherries which resulted in significant control of the tree size in commercial orchard plantings. Rootstocks that resulted from the interspecific hybrids of *Prunus* in Gembloux, Belgium (the GM series) and Giessen, Germany (the GI series) have formed cherry trees 20% or less than the size of trees on the standard rootstocks of Mahaleb or Mazzard. The highly successful and promising dwarfing cherry rootstocks produce trees between 20% and 50% of standard size.

4.15 Plant Growth Regulators

Controlling vegetative growth in the stone fruit crops has not been easier than for apple because of the nonavailability of appropriate size-controlling rootstocks or scions. Severe summer and winter pruning is generally needed to limit the tree within a given space and for the maintenance of good yield and quality of the fruit. Despite this, high-density plantings exceeding 8000 trees per hectare have continued to increase in case of several crop species like apricot (*P. armeniaca*) and plum (*P. domestica*). Hence, the need for methodologies to check vegetative growth has increased accordingly. Though growth retardants have been examined in the past, daminozide was the one and only compound which successfully suppressed the growth in young cherry (*P. avium*) trees. Recently, paclobutrazol and some triazole analogues have proven to be quite efficacious. The paclobutrazol application in stone fruit crops has imitated a similar pattern as in case of apple. Several procedures of application have been examined. Delay in flowering has been reported in other species and may be used as a measure of protection in the regions of potential freeze damage. In other cases, however, blossoming has been advanced by several days in apricot, peach (*P. persica*), cherry and plum. These differences appear to be associated with concentration of the dose and time of application; nevertheless, the cause for delay or advance in flowering is not well known. Generally, reduction in shoot growth is directly related with an increment in fruit set and flower bud density. In some cases, though, foliar sprays of paclobutrazol at the time of bloom have resulted in considerable reduction of fruit set. This depicts that paclobutrazol, or maybe other triazoles, could be utilised as chemical thinning agents which would become quite necessary because none is currently available for stone fruits.

There have been worthy experiments conducted on peaches using very less concentrations of paclobutrazol provided through drip irrigation. Triazoles applied over trunk bark dissolved in organic solvent carriers too have checked growth of old peach and cherry trees. Other peculiar process which has been observed in some of the cases is the reduction or loss of negative shoot geotropism in peach. As observed for some pear varieties, the 'weeping' growth characteristic noted in paclobutrazol-treated peaches may be linked to restriction of lignin synthesis or changes in shoot xylem integrity. There are controversial reports regarding the consequences of paclobutrazol use on the cold tolerance of stone fruit crops. Since gibberellic acid lowers cold tolerance of vegetative tissues in *Prunus*, it appears that probably the inhibitors of gibberellin formation might promote hardiness. No doubt, some reports have depicted that stone fruit trees may become more cold tolerant when applied with paclobutrazol, but others revealed the opposite. The effects on fruit quality have not been much significant but a negligible reduction of soluble solids. This might be as a consequence of an accelerated demand for photosynthates due to increase in fruit number. Moreover, fruit maturation might be highly uniform. The elevation in yield of stone fruit crops is quite consistent than that of pome fruits. The causes for this are yet to be understood but might be associated with the inhibition of GA formation in the fruit embryo. Pome fruits, which have more than one carpel and the probability of numerous seeds within a carpel, tend to produce greater GA and hence might be highly affected than stone fruits which usually consist of one seed. Certainly, there is a critical level beyond which any further increase in the number of fruit set will eventually set a reduction in the yield and fruit size. This differs with respect to the vigour, species, cultivar and rootstock. Nevertheless, a stone fruit tree will have greater fruit size and higher yields, for the most part, given the same number of fruit per trunk cross-sectional area. Many triazoles have been examined on stone fruit trees, though, currently, only paclobutrazol is recognised for application in peach, cherry, plum, nectarine, apricot and olive in 11 countries outside the USA.

In a number of physiological characteristics, mainly the tree characters, *almond* shows similarity with peach. Considering its bearing habit, it lies in between peach and domestic plums. The important target in almond growing is to get higher no. of fruits. Being a spur bearer, it is important to develop the tree in such a manner that a large no. of spurs are produced. Almond develops most of its fruit on spurs whose fruitfulness lasts for about 5 years. Pruning must be practised in a manner that one-fifth of the pruning wood is replaced each year. Hence, pruning should be carried out in such a way that new spur growth consistently replaces spurs that have lost the ability to produce fruit. Excessive water sprouts and suckers ought to be removed. Plants are headed back at a height of about 60–70 cm above the ground level just after 1–2 months of planting. Central modified leader system must be carried out by leaving three to four scaffold branches which develop stable and well-balanced trees. Strong scaffold limbs are attained by developing 45–60 crotch angles. In case of appearance of weak crotches, the branches must be tied by strong thread bound with pegs. A 1-year-old wood is pruned every year in December or January when trees have entered dormancy. Trees with not more than 10–12 years of age should produce 20–25 cm of annual growth, while older trees must make 15 cm of new shoot growth every year. In older trees that are less vigorous, growth can be enhanced by light severe pruning. The top is to be cut back to large lateral limbs, and the smaller and weaker wood is thinned out severely.

4.16 Conclusion

Managing the canopy, mainly controlling the tree size, has become a prime concern for decreasing the cost of production and enhancing the overall production and quality. The design and shape of the tree canopy affect the interception of sunlight and hence guarantee higher monitory income to orchardists. Establishing a better orchard canopy, the growers can attain greater yield and even high standard quality of the fruit. When there is a good light interception received by an orchard, well distribution of such light takes place within the entire tree canopy, and hence there appears a stability and equilibrium between vegetative growth and fruiting ('calm trees'). This can best be achieved with the maintenance of a narrow shape of the canopy and also by constant redevelopment and improvement of the limbs through pruning and also by establishing pendant-like fruiting branches. Prosperous orchardists try to maintain a stability and equilibrium between the vegetative growth and fruiting by constant renewal of the fruiting wood on the tree, providing medium levels of nitrogen, and by correct ways of crop load control. Orchardists should make efforts to acquire about 20-in.-long shoot growth of the leader during the initial year, about 30–40-in.-long leader shoot growth during the succeeding 2 years and again about 20 in. of leader shoot growth during the fourth year. If this is done in combination with light pruning and an early bearing rootstock, considerable yield will be achieved in the second to fourth years which will restrict vegetative growth in subsequent years, hence developing a 'calm' tree.

References

Bargioni, G., Loreti, F., & Pisani, P. L. (1983). Performance of peach and nectarine in a high density system in Italy. *HortScience, 18*, 143–146.

Bellini, E., Falqui, D., & Musso, O. (2000). Comparison between two training systems in peach protected culture in Sicily. *Acta Horticulturae, 513*, 427–433.

Caruso, T., Motisi, A., Marra, F. P., & Di Marco, L. (1993). Researches on planting densities for Y-shaped peach trees grown under greenhouses. *Acta Horticulturae, 349*, 85–88.

Caruso, T., Vaio, C. D., Inglese, P., & Pace, L. S. (1998). Crop load and fruit quality distribution within canopy of Spring Lady "peach trees trained to" central leader and "Y shape". *Acta Horticulturae, 465*, 621–628.

Chalmers, D. J., Mitchell, P. D., & VanHeek, L. (1981). Control of peach tree growth and productivity by regulated water supply, tree density and summer pruning. *Journal of the American Society for Horticultural Science, 106*, 307–312.

Choi, C., Wiersma, P. A., Toivonen, P., & Kappel, F. (2002). Fruit growth, firmness and cell wall hydrolytic enzyme activity during development of sweet cherry fruit treatedwithgibberellic acid. *Journal of. Horticultural Science and Biotechnology, 77*, 615–621.

Claverie, J., & Lauri, P. É. (2005). Extinction training of sweet cherries in France-appraisal after six years. *Acta Horticulturae, 667*, 367–371.

De Salvador, F. R., Di Tommaso, G., Piccioni, C., & Bonofiglio, P. (2005). Performance of new and standard cherry rootstocks in different soilsand climatic conditions. *Acta Horticultiurae, 667*, 191–200.

DeJong, T. M., Day, K. R., Doyle, J. F., & Johnson, R. S. (1994). The Kearney Agricultural Center perpendicular .V. (KAC-V) system for peaches and nectarines. *Horticultural Technology, 4*, 362–367.

DeSalvador, D. F. R., & DeJong, T. M. (1989). Observation of Sunlight interception and penetration into the canopies of peach trees in different planting densities and pruning configuration. *Acta Horticulturae, 254*, 341–347.

Farina, V., Bianco, R. C., & Inglese, P. (2005). Vertical distribution of crop load and fruit quality within vase-and Y shaped canopies of "Elegant Lady" peach. *HortScience, 40*(3), 587–591.

Gonda, I. (2006). The role of pruning in the intensification of plum production. *International Journal of Horticultural Science., 12*(3), 83–86.

Grochowska, M. J., Hodun, M., & Mika, A. (2004). Improving productivity of four fruitspecies by growth regulators applied once in ultra-low doses to the collar. *Journal of Horticultural Science and Biotechnology, 79*, 252–259.

Grossman, Y. L., & DeJong, T. M. (1998). Training and pruning system effects onvegetative growth potential, light interception, and cropping efficiency in peach trees. *Journal of the American Society for Horticultural Science, 123*, 1058–1064.

Guerriero, R., & Viti, R. (1997). Problemirelativiallabiologiafiorale e di fruttificazionedell'albicocco. *Italus Hortus, 4*, 29–36.

Kim, J. K., Kim, S. B., Kim, K. Y., Cho, M. D., Hong, J. S., & Kim, J. B. (1994). Effect of training system and planting density on tree growth, yield and fruit quality in peach orchard. *RDA - Journal of Agricultural Science and Technology, 36*(1), 460–464.

Lauri, P. E. (2005). Developments in high density cherries in France: Integration of tree architecture and manipulation. *Acta Horticulturae, 667*, 285–291.

Looney, N. E., & McKellar, J. E. (1987). Effect of foliar- and soil surface-applied paclobutrazol on vegetative growth and fruit quality of sweet cherries. *Journal of the American Society for Horticultural Science, 112*, 71–76.

Lukic, M., Mitrovic, M., Milosevic, N., Karaklajic, S. Z., Pesakovic, M., & Glisic, S. L. (2012). Biological properties of some plum cultivars grown under different training systems. *Acta Horiculturae, 968*, 227–232.

Marini, R. P., & Barden, J. A. (1987). Summer pruning of apple and peach trees. *Horticultural Reviews, 9*, 351–375.

Marshal, J., Lopez, G., Mata, M., & Girona, J. (2006). Branch removal and defruiting for the amelioration of water stress effects on fruit growth during Stage III of peach fruit development. *Scientia Horticulturae, 108*, 55–60.

Miller, S. S. (1995). Summer pruning affects fruit quality and light penetration in young peach trees. *HortScience, 22*, 390–393.

Neri, D. (2003). *Ipertestopotatura. CRPV, Cesena*, Italy.dellacolturadell'albicocco, Imola, Italy (pp. 35–44).

Neri, D., Giovannini, D., Massai, R., Divaio, C., Sansavini, S., Del Vechhio, G., Guarino, F., Mennone, C., Abeti, D., & Colombor, R. (2010). Efficienza produttiva egestionale degli impianti di pesco in un confronto Nord-Sud. *Italus Hortus, 17*(3), 46–62.

Neri, D., Masetanni, F., & Giorgi, I. V. (2011). *La potatura* (p. 370). Bologna: Edagricole.

Ogata, R., Saito, T., Araya, J., Nakagawara, I., & Kubo, T. (1989). Effect of paclobutrazol on vegetative growth and cropping of peach and cherry. *Acta Horticulturae, 239*, 297–299.

Pirazzini, P. (2004). Osservazioni sulla potatura di produzione di alcune varietà di albicocco. *Frutticoltura, 1*, 36–38.

Proebsting, E. L. (1989). The interaction between fruit size and yield in sweet cherry. *Fruit Variety Journal, 44*(3), 169–172.

Proebsting, E. L., & Mills, H. H. (1969). Ethephon increases cold hardiness in cherry. *Journal of the American Society for Horticultural Science, 101*, 31–33.

Rana, H. S., Awasthi, R. P., Sharma, R. M., & Jha, A. (1998). Effect of training system on canopy physiology, fruit yield and quality of peach. *Journal of Hill Research, 11*(1), 38–42.

Rival, P. (1965). Essaisur la croissance des arbresetsurleurssystèmes de floraison (application aux espècesfruitières). *Journal of Agricultural Tropical Botany Applied, 12*(12), 655–686.

Rom, C. R. (2000). Peach Rootstocks and Orchard Systems for the Arkansas-Oklahoma Region. In *Proceedings of the 19th Annual Oklahoma-Arkansas Horticulture Industries Show* (pp. 9–16).

Rom, C. R., & Blackburn, B. (1998). Preliminary observations of new peach training systems: The first 5 years of growth and production. In *Proceedings of the 17th Annual Horticulture Industries Show, Oklahoma-Arkansas Horticulture Industries* (pp. 113–117).

Salvador, A., Cuquerrelia, J., & Monterde, A. (2006). Effect of 1-methylcyclopropene on the post-harvest behaviour of apricot cv 'Canino'. *Acta Horticulturae, 717*, 591–594.

Sansavini, S., & Neri, D. (2005). Forme di allevamento e potatura (pp. 115–143). In C. Fideghelli, & S. Sansavini (Eds.), *Pesco* (p. 259). Bologna: Edagricole.

Sansavini, S., Lugli, S., Grandi, M., Gaddoni, M., & Correale, R. (2001). Impianto ad altadensita'diciliegiallevati a 'V': Confrontofraportinnestinanizzanti. *Rivista di Frutticoltura eOrtofloricoltura, 3*, 63–73.

Savini, G., Neri, D., Zucconi, F., & Mancini, G. (2007). Lateral shoot growth of apple, pear and cherry with selective disbudding on newly planted trees. *Acta Horticulturae, 732*, 587–592.

Southwick, S. M., Weis, K. G., Yeager, J. T., & Zhou, H. (1995). Controlling cropping in 'Loadel' cling peach using gibberellin: Effects on flower density, fruit distribution, fruit firmness, fruit thinning, and yield. *Journal of the American Society for Horticultural Science, 120*, 1087–1095.

Stern, R. A., Flaishman, M., & Ben-Arie, R. (2007a). The effect of synthetic auxins on fruit development, quality and final fruit size in 'Canino' apricot (*Prunus armeniaca* L.). *Journal of Horticultural Science and Biotechnology, 82*, 335–340.
Webster, A. D. (1998). Strategies for controlling the size of sweet cherry trees. *Acta Horticulturae, 468*, 229–240.
Webster, A. D., & Schmidt, H. (1996). Rootstocks for sweet and sour cherries. In A. D. Webster & N. E. Looney (Eds.), *Cherries: Crop physiology, productions and uses* (pp. 127–163). Oxon: CAB International.
Webster, A. D., Quinlan, J. D., & Richardson, P. J. (1986). The influence of paclobutrazol the growth and cropping of sweet cherry cultivars: I. The effect of annual soil treatments on the growth and cropping of cv Early Rivers. *Journal of Horticultural Science, 61*, 471–478.
Whiley, A. W., Wolstenholme, B. N., & Faber, B. A. (2013). Crop management. In B. Schaffer, B. N. Wolstenholme, & A. W. Whiley (Eds.), *The Avocado. Botany, production and uses* (pp. 342–379). Wallingford: CABI.
Zucconi, F. (2003). *Nuovetecniche per i fruttiferi* (p. 246). Bologna: Edagricole.

Rootstocks of Stone Fruit Crops

5

Amit Kumar, Jagdeesh Prasad Rathore, Umar Iqbal, Anil Sharma, Pawan K. Nagar, and Mohammad Maqbool Mir

Abstract

Among temperate fruits, stone fruits fall on the second position after pome fruits in case of area and production, but as they come early in the season (summer) and have high nutritive value in terms of high vitamins A and C, fibre and potassium and very low fat, calories and sugar, making them beneficial in weight management, they favour a great part of human diet. Area and production of stone fruits in the country has been limited due to many reasons, viz. nematode problems, disease and viruses, insect problem, edaphic reasons and incompatibility among rootstock and scion cultivars. Absences of suitable and compatible rootstocks restrict or limit the expansion in area and production of stone fruits. Still most of the stone fruits are grafted/budded on seedling rootstocks in the country which is also the main reason of low production as seedling rootstock suffers from most of the abovementioned problems. During last three to four decades, a great progress has been done in the development of clonal rootstock in *Prunus* species, and to date various clonal rootstocks have been developed in different countries of the world which offered a great potential for cultivation in respect to production and productivity.

A. Kumar (✉)
Division of Fruit Science, Faculty of Horticulture, SKUAST-Kashmir, Srinagar, Jammu and Kashmir, India

J. P. Rathore
Department of Horticulture, Rajasthan College of Agriculture, MPUAT-Udaipur, Udaipur, Rajasthan, India

U. Iqbal · M. M. Mir
Division of Fruit Science, Sher-e-Kashmir University of Agricultural Sciences & Technology of Kashmir, Srinagar, Jammu and Kashmir, India

A. Sharma
Department of Horticulture, Punjab Agricultural University, Ludhiana, Punjab, India

P. K. Nagar
Government of Gujarat, Gandhinagar, Gujarat, India

M. M. Mir et al. (eds.), *Production Technology of Stone Fruits*,
https://doi.org/10.1007/978-981-15-8920-1_5

Keywords

Rootstocks · Scion · Hybrid · Graft compatibility · Stone fruits

5.1 Introduction

Stone fruit is a generic term used to define fruits which includes peach, nectarine, plum, apricot, almond and cherry which are generally grown in temperate climatic conditions. The main feature of stone fruit is having fleshy layer, mesocarp, as edible pulp surrounding a relatively large, hard pit commonly known as 'stone' that shields and protects a seed. The commercial production of stone fruits is confined between the latitude of 30 and 40°N and S, although it is now grown almost all over the world. The major stone fruit-producing country is China accounting about 50% share of the total world production. In India, stone fruits are grown on a commercial scale in mid-hill Himalayan states, viz. Himachal Pradesh, Jammu and Kashmir, Uttarakhand, as well as in a limited scale in north-eastern states. These fruits are generally grown on soils having bulk density, parasitic nematodes, root rot problems, fungal pathogens or other soil and replant problems.

Rootstock plays an important role in deciding the success or failure of orcharding enterprise. Rootstocks are important in increasing productivity of the fruit plants by improving efficiency through improved plant survival techniques, controlling vigour of the plant and enhancing size and quality of fruit and ultimately yield. The productivity of stone fruits in India is considerably low, and one of the major reasons is the lack of locally suited clonal rootstocks. Currently, stone fruits are grown on seedling rootstocks of unknown origin and without any genetic background. Seedling stocks unlike clonal rootstocks lack uniformity in growth and development and are also prone to prevailing biotic and abiotic stresses. Rootstocks of clonal origin, on the other hand, influence the scion characteristics like uniformity to a greater extent besides having other inherent characteristics like tree size control, precocity, disease resistance and better adaptation.

5.2 Challenges with Stone Fruit Rootstock

Rootstocks are an essential component in modern fruit production because of their capability of adapting scion cultivars to diverse environmental conditions and cultural practices (Mestre et al. 2017). According to the requirement of the climate and growers, rootstock breeders always update their priorities.

In most of the fruit species, clonal rootstocks having low vigour are mostly preferred in breeding projects; however, under sustainable systems, semi-dwarf and moderately vigorous rootstocks are also considered. For high density stone fruit orchards, clonal rootstocks are the first preference having moderate to vigorous nature, precocious, compatible with most of the cultivars and suitable on heavy soils (Mika et al. 1998; Hrotkó 2005; Long et al. 2005).

Currently, stone fruits are grown on seedling rootstocks of unknown origin and without any genetic background. Apart from this, different problems are also associated with the different rootstocks of stone fruits such as soils having the nature of high bulk density, parasitic nematodes, fungal and viral pathogens, other edaphic related to soil and climate or replant disease problems, the major problem being incompatibility of the clonal rootstock with the most of scion cultivars. Major problems with the stone fruits are differentiated as follows.

5.2.1 Nematode Problem

1. Root-knot nematodes (*Meloidogyne* spp.)
2. Ring nematode (*Mesocriconema xenoplax*).
3. Lesion (*Pratylenchus* spp.)
4. Dagger (*Xiphinema* spp.)

5.2.2 Disease Problem

1. Fungi (*Armillaria*, *Phytophthora* and others).
2. Bacteria (crown gall, bacterial canker).
3. Mycoplasma-like organism (X-disease).
4. Viruses (TmRSV, PNRSV, prune dwarf).

5.2.3 Insect Problem

1. Borers (*Synanthedon* spp. and *Capnodis* spp.)
2. Root weevils (*Pachnaeus* spp.)

5.2.4 Edaphic Problem

1. Calcareous soils.
2. Salt tolerance.
3. Waterlogging.
4. Tolerance to drought.
5. Low fertility and low nutrition.
6. Cold hardiness.

5.2.5 Horticultural Problem

1. Vigour.
2. Bloom time.
3. Spring shock syndrome.
4. Low productivity.
5. Late bearing.
6. Low quality.
7. Incompatibility.

Besides seedling rootstocks, during the last two decades, clonal rootstocks for stone fruits gained popularity in the country, and new rootstocks for stone fruits were introduced through different institutes and commercial nurseries. These clonal rootstocks possess certain characteristics such as precocity, size controlling, very productive and resistant to insect pests and diseases. Due to these plus points, clonal rootstocks are preferred over seedling rootstocks for raising nursery plants of these fruits. Clonal rootstocks, especially dwarfing rootstocks, have been used extensively in other fruits, viz. apple, mango, pear, etc., due to which stone fruit industry also moves towards clonal rootstocks. Among different clonal rootstocks, few of hybrid rootstocks are originated from complex vegetatively propagated methods.

Dwarfing rootstocks is most economical, and during the past several decades, every effort has been made to develop dwarfing or semi-dwarfing rootstocks for stone fruits in different countries of the world using a number of species/varieties of *Prunus persica, P. insititia, P. davidiana, P. subcordata, P. inaritilna, P. tomentosa, P. pumila, P. besseyi, P. glandulosa, P. cerasifera and P. triloba* (Funt and Goulart 1981; Roberts and Westwood 1981; Rom 1983). Several important clonal rootstock developed which were widely used throughout the world having diverse characteristics in all the stone fruits are listed (Tables 5.1, 5.2, 5.3 and 5.4).

5.3 Plant Growth and Vigour

Growth of any plant can be judged on the basis of plant height, plant spread, plant volume and fresh and dry weight of different plant parts. Rootstock and scion cultivar greatly influence the tree vigour. In stone fruit, dwarfing rootstocks begin to make an impact on plum, peach and cherry production not only due to smaller size but also to be more precocious fruit bearing from a budded tree (Lang 2000; Reighard 2000).

Hardwood cuttings of Myrobalan plum were planted in two seasons (autumn and spring) with and without treating growth regulators, and it was reported that autumn planting treated with growth regulators was more superior than spring season planting (Garner and Hatcher 1955). Vigorous plants were produced by Myrobalan plum when several rootstocks were used as a rootstock for Japanese and European plums and apricots (Black 1959). Myrobalan plum rootstock produced larger plants than

Table 5.1 Rootstock for peach, nectarine and almond

Rootstock	Origin	Developed	Characteristics	References
Sirio	Italy	Seedling and open pollinated seedling of GF 557	Good resistance to limestone, peach × almond hybrid, flowers are pinkish colour, suitable for high-density planting	Loreti and Massai (1998)
PeMa	Hungary	*P. persica* × *P. amygdalus*	Good compatible with peach, almond and some apricot varieties	Nagy and Lantos (1998)
PeDa	Hungary	Open pollinated seedling of *P. persica* × *P. davidiana*	Compatible with peach and almond. It is more vigorous as compared to GF 677	Nagy and Lantos (1998)
GF 677	Bordeaux, France	*Prunus persica* × *P. amygdalus*	It is the most used rootstock in the European fruit growing in recent decades. It has an excellent affinity with varieties of peaches, nectarines, almonds and some varieties of plum trees. It adapts well to limestone soils and tolerates well ferric chlorosis. Moderate tolerance to drought and root asphyxia. It is one of the most sensitive rootstocks to nematodes. It confers a high vigour	Bernhard and Grasselly (1981)

(continued)

Table 5.1 (continued)

Rootstock	Origin	Developed	Characteristics	References
GF 305	France	Selected from seedling rootstock of peach at INRA	Uniform and very vigorous growth in nursery, susceptible to *Agrobacterium* and *Phytophthora.* Root-knot and root lesion nematodes and some viruses	Grasselly (1983)
Rubira	France	Selected from peach seedlings grown at INRA from a Californian seedlot in 1960	Uniform and vigorous growth, induces 15–20% lower vigour than GF 677 and is more precocious with good productivity, red foliage, uniform germination, slightly susceptible to powdery mildew and resistant to green peach aphid, sensitive to *Meloidogyne incognita* and *M. arenaria*	Loreti (1984)
Montelor	France	Seedling population of peach	Vigorous, sensitive to *Agrobacterium*, resistant to Fe and Mg deficiency, resistant to chlorosis	Webster (1997)
Higama	France	Selected from peach seedlings grown at INRA imported from Japan in 1960	Vigorous, resistant to Fe deficiency and replant disease, tolerant to *Meloidogyne javanica* and *M. incognita*, sensitive to lime-induced chlorosis and *Agrobacterium*	Grasselly (1983)

(continued)

Table 5.1 (continued)

Rootstock	Origin	Developed	Characteristics	References
PS A5	Department of Fruit Science and Crop Protection of Pisa University, Italy	Originated from peach seedling rootstock	Less vigorous (20–25%) than PS A6 rootstock and encourages uniform growth, precocity good seed germination, sensitive to poor drainage, resistant to *Verticillium* wilt, suitable for free spindle or delayed vase training system and is excellent for fertile soils	Scaramuzzi et al. (1976)
PS A6		Originated from peach seedling rootstock	Uniform and rapid growth, vigorous, susceptible to *Agrobacterium*, performs well in poor soils, good productivity and good seed germination	
PS B2		Originated from peach seedling rootstock	Less vigorous (10–15%) than PS A6 rootstock, good germination and yield efficiency, less sensitive to replant problem as it is resistant to *P. vulnus*, performs well in heavy soils	
Siberian C	Canada (Harrow Ontario)	Seedling population of peach	Uniform and cold resistant, some vigour control, induces good precocity and productivity, sensitive to drought, nematode and *Agrobacterium*	Okie (1998)
Harrow Blood	Canada (Harrow Ontario)	Seedling population of peach	Cold resistant, poor induction of precocity	Loreti et al. (1981)

(continued)

Table 5.1 (continued)

Rootstock	Origin	Developed	Characteristics	References
Rutgers Red leaf	USA (New Jersey)	Seedling population of peach	Small white-fleshed freestone fruit is of no value, but the tree is used as a source of hardy seedling rootstock that is readily identified by its red leaves in the nursery	Loreti et al. (1981)
Nemared	USA (California)	Redleaf selection from seedling population of peach	Produces seedlings with less lateral branches, more vigorous tree than Nemaguard, root-knot nematode resistant, but has increased susceptibility to bacterial canker	Ramming and Tanner (1983)
Controller 5	USA	*P. salicina* × *P. persica*	Dwarfing nature, compatible with peach and nectarine, no or little suckering, precocious, susceptible to root-knot, lesion and ring nematode, bacterial canker	Dejong et al. (2011)
Controller 9	USA	*P. salicina* × *P. persica*	Semi-dwarf, slightly less vigorous than Nemaguard, less water sprouts, compatible with peach and nectarine, no or little suckering, precocious, susceptible to root-knot and ring nematode and bacterial canker	Dejong et al. (2011)

(continued)

Table 5.1 (continued)

Rootstock	Origin	Developed	Characteristics	References
GF 305	France	Seedlings of peach population	Compatible with all peach and nectarine cultivars and has good growth and productivity, susceptible to *Agrobacterium*, *Phytophthora*, root-knot and root lesion nematodes and some viruses. Still used by virologists as a virus indicator in peach	Grasselly (1983)
Montclar	France	Seedlings of peach population	High seed production, uniform seedling growth and vigour in the nursery and increased vigour in scion cultivars	Grasselly (1983)
Adarcias	Spain	An almond-peach hybrid selected from open pollinated seedling population	Propagated by hardwood cuttings and micropropagation, suitable for calcareous and loam soils, induces lower vigour than 'Adafuel'. Reduces excessive tree growth, graft compatible with peach and nectarine cultivars, resistant to *C. beijerinckii* Oud. and *T. pruni-spinosae*	Moreno and Cambra (1994)

(continued)

Table 5.1 (continued)

Rootstock	Origin	Developed	Characteristics	References
Penta	Italy	Originated from open pollination of Imperial Epineuse	Easy propagation by hardwood cutting and also propagated by in vitro culture, vigorous, uniform and no suckers, suitable for waterlogging and anchorage is excellent. Excellent graft compatibility with peach and nectarines	Nicotra and Moser (1997)
Tetra	Italy	Originated from open pollination of Imperial Epineuse	Easy propagation by hardwood cutting and also propagated by in vitro culture. Uniform and less vigorous than Penta, suitable for heavy soil and waterlogging condition, very compatible with peach and nectarines	Nicotra and Moser (1997)
Kuban 86 (Krymsk 86)	Russia	Myrobalan plum × peach	Produces similar plants in size to Lovell. Its yield and yield efficiency were significantly lower than Lovell. Susceptible to nematodes but more tolerant to oak root fungus than Mariana 2624 and has less sensitivity to *Phytophthora* and less susceptibility to *Verticillium* than Lovell	

(continued)

Table 5.1 (continued)

Rootstock	Origin	Developed	Characteristics	References
Bailey	USA	Naturalized peach selection	Uniform and good vigour, cold hardiness, less tolerant to root lesion nematode, good survival in sandy soils, less vigour trees than Lovell, very productive, prone to root-knot nematodes, waterlogging and fungal root rot	Okie (1998)
Guardian	USA	*P. persica* × *P. davidiana*	Seed germination is low, resistance to root-knot nematode is low, higher tolerance to ring nematode, bacterial canker, susceptible to *Armillaria* root rot	Okie et al. (1994)
Adafuel	Spain	Seedling population obtained by open pollination of 'Marcona' almond cultivars	Propagates easily by hardwood cuttings, better rooting percentage than 'GF-677', extremely vigorous and suitable	Cambra (1990a)
Lovell	California	Seedling population of peach	Seed germination high, uniform seedling, compatible with peach and nectarine cultivars, no sucker production and prone to root-knot and root lesion nematodes, tolerant to ring nematodes and bacterial canker. Prone to waterlogging, crown gall, *Phytophthora* spp. and *Armillaria* spp.	Okie (1998)

(continued)

Table 5.1 (continued)

Rootstock	Origin	Developed	Characteristics	References
Densipac (Rootpac 20)	Spain	Clonal plum hybrid of *P. besseyi* × *P. cerasifera*	Propagated easily with tissue culture; semi-erect and compact form, a medium chilling requirement (600–800 h), low vigour similar to GF-655-2, advances fruit maturity with no suckering, tolerant to calcareous and wet soils, has moderate resistance to root-knot nematode, resistant to lesion nematode populations and to *Rosellinia necatrix* root fungus and has good compatibility with peach and nectarine varieties	Gasic and Preece (2014)
Greenpac	Spain	Cross of Felinem (almond × peach) × Cadaman (peach × *P. davidiana*)	Propagated easily with in vitro methods as compared to hardwood cuttings. Vigorous plants with green leaves, compatibility with peach and nectarine cultivars; well adapted to calcareous soils, resistant to lime-induced chlorosis, root-knot nematode and moderately resistant to *M. javanica*. Susceptible to crown gall and lesion nematodes	Pinochet (2009)

(continued)

Table 5.1 (continued)

Rootstock	Origin	Developed	Characteristics	References
Replantpac	Spain	Plum × almond rootstock	Most vigorous rootstock, though somewhat less productive than Lovell. It is very resistant to root-knot nematodes and slightly susceptible to lesion nematodes, and its survival rate in replanted soils has been high	Pinochet (2010)

peach rootstock (Westwood and Chaplin 1973) and was reported as most satisfactory rootstock for plum cultivars (Benjamin and Shoemaker 1978a).

Gall and Grasselly (1979) studying the behaviour of 12 almond varieties grafted on almond, peach and peach × almond rootstocks observed that peach seedling and peach × almond rootstocks were superior to almond seedling. Donno et al. (1976) reported that early trunk growth in almond cvs. Tuono and Filippo Ceo was faster on the sweet almond seedling rootstock Don Carlo than on bitter almond seedling rootstock, but the difference disappeared after 8–10 years. In peach, rootstocks also exhibited variable effects on scion growth and vigour. Myrobalan seed requires 100–200 days for ripening at 40–50 °F for raising seedling rootstocks (Benjamin and Shoemaker 1978b). Rooted hardwood cuttings of Myrobalan plum, Brompton plum and Marianna plum produced superior rootstocks for plum cultivars (Hudson et al. 1981). Kumar (1987) reported that the trees of Dhebar, IXL, Katha, Nonpareil and Merced almond cultivars were more vigorous and larger on wild peach than on Behmi and bitter almond rootstocks.

Bellini et al. (1993) observed reduced tree growth of May Crest peach and Maria Emila nectarine grafted on plum selections. Redhaven peach on GF 677 had greater root number and more roots with greater than 10 mm diameter (Lichev and Govedarov 1995). Micke et al. (1996) studied the effect of peach and almond rootstocks on the growth of Carmel and Nonpareil cultivars and observed that Lovell peach rootstock produced larger trees than the other rootstocks.

Rana et al. (1997) observed that peach cv. Sharbati budded on plum rootstock had the smallest trunk diameter as compared to plants produced from cuttings or on peach rootstock. 'Catherina' peach grafted on 'Montizo' and 'Monpol' rootstock produced similar TCSA as on Adesoto rootstock; however, both rootstocks were slightly more vigour than GF 655/2 rootstock (Felipe et al. 1997).

Sharma et al. (2004) recorded higher annual shoot growth and leaf area in Nonpareil almond cultivar than Merced cultivar when raised on wild peach than bitter almond rootstocks. Lanauskas (2006) evaluated two plum cultivars 'Stanley'

Table 5.2 Rootstocks for plum and apricot

Rootstock	Origin	Developed	Characters	References
Myrobalan seedling	USA	Seedlings are propagated from seed of Myrobalan plum seedling (*Prunus cerasifera*)	Each seedling is slightly different genetically, and seedlings show more variation than with rootstocks propagated from cuttings. The rootstock provides better anchorage than do other plum stocks and is less likely to lean and bow over; they produce few root suckers and only occasionally sucker from the crown, and they are more tolerant of boron and saline soil conditions than other rootstocks	Duval et al. (2004)
Myrobalan 29C	California	Selection from Myrobalan seedling from a population of *P. cerasifera*	Compatible with most cultivars of plum and apricot; tolerates different types of soil and climatic conditions; produces hardy, vigorous, long-lived, standard-sized tree; prone to suckering; and more tolerant to wet conditions	Andersen et al. (2006)
Marianna Plum	USA	Open pollinated seedling of Myrobalan plum (*P. cerasifera*) apparently a hybrid with *P. munsoniana*	Easily propagation through hardwood cuttings, moderately resistant to *Phytophthora* rot, susceptible to bacterial canker, resistant to root-knot nematodes, susceptible to lesion nematodes	Hartmann et al. (2007)
Marianna 2624	USA	Vigorous seedling selection of parent Marianna plum	Compatible with most cultivars of plum and apricot, semi-dwarf in nature, preferred in Northern California as it tolerates wet and heavy soils, adaptable in a wide range of soil and climatic conditions, resistant to oak root fungus, crown rot, crown gall and root-knot nematode but susceptible to bacterial canker, suckers badly, propagates through cuttings, has a shallow root system	Hartmann et al. (2007)
Pollizo	Spain	Selected from a local plum population	Adapted to the heavy and calcareous soil, graft compatibility with the stone fruit species is considered good in its area of origin	Felipe et al. (1994)

(continued)

Table 5.2 (continued)

Rootstock	Origin	Developed	Characters	References
Adara	Spain	Selected from an open pollinated population of Myrobalan	Good rooting ability, compatible with most of the sweet and sour cherry cultivars, few peaches and nectarines, Japanese plum and few apricots. Highly resistant to root asphyxia and associated diseases and has a higher yield efficiency, lower tendency of suckering	Moreno et al. (1995c)
Adesoto 101	Spain	Selected from open pollinated seedlings of 'Pollizo de Murcia'	Propagates easily by hardwood cuttings and in vitro and also adapts satisfactorily to calcareous, fine-textured and waterlogging soils. Resistant to *M. arenaria*, *M. incognita* and *M. javanica* and is moderately tolerant of *P. vulnus*	Moreno et al. (1995a)
Pixy	England	Seedling of St. Julien	Dwarfing rootstock for plum, ½ to 2/3 size of plants are produced, promotes precocity and regular cropping; reduction in fruit size was observed	Webster (1980)
Montizo	Spain	Clone selected from Pollizo plum rootstock	Compatible with most of the peach and nectarine and almond cultivars; vigour and productivity are similar to St. Julien rootstock, more productive than Monpol rootstock, good resistance to root-knot nematode, less suckering	Felipe et al. (1997)
Monpol	Spain	Clone selected from Pollizo plum rootstock	Compatible with most of the peach and nectarine and almond cultivars; vigour and productivity are similar to St. Julien rootstock, good resistance to root-knot nematode, less suckering	Felipe et al. (1997)
Ademir	Spain	Selected from open pollinated seedling populations of Myrobalan plum	Good rooting ability, low vigour, excellent plum graft compatibility and better graft compatibility with apricot cultivars. It is suitable for plum cultivars to avoid excessive vigour or to increase planting density. Well adapted to calcareous and compact soils, resistant to iron chlorosis and waterlogged conditions	Moreno et al. (1995b)

(continued)

Table 5.2 (continued)

Rootstock	Origin	Developed	Characters	References
Rival	Romania	Clonal plum rootstock derived from *Prunus insititia*	Medium vigour, good compatibility with most of plum cultivars, propagates through stem cuttings (79%) tolerance	Botu et al. (2007)
Krymsk 1	Russia	*Prunus tomentosa* and *Prunus cerasifera*	It is a promising rootstock for plums and apricots, reduces tree vigour and good fruit size, resistance to root-knot nematodes and some resistance to lesion nematode, good anchorage, few sucker production, also suitable for peach cultivars with high yield efficiency and enhanced fruit size. Not suitable for areas with high bacterial canker pressure	Maas et al. (2011)
St. Julien	UK	Seedling population of Plum	Most widely used rootstock for plums, semi-vigorous rootstock, compatible with all plum cultivars, come into bearing after 3–4 years, adapts in a wide range of soils having some tolerance of chalky soils	Hartmann et al. (2007)
Wavit	Austria	Seedling of a European plum variety called Wangenheim.	Good compatibility with plum cultivars even with apricot also, semi-vigorous tree, grow without support, somewhat similar to St. Julien, more precocious, bearing in third year, adapts a wide range of soils, tolerant to chalky soils	Southwick and Weis (1999)
Adaptabil	Romania	Western Sand Cherry (*Prunus besseyi)*	Compatible with plums (European) and apricots, vigour somewhat like St. Julien, more tolerant to sandy soils where mostly drought occurs	Dutu et al. (2001)
Torinel	France	Selected from plum seedling population	Compatible with most of the apricots as well as plums, somewhat similar to St. Julien, suitable for a wide range of soils	Audergon et al. (1991)

and ‘Kauno Vengrine’ grafted on four plum rootstocks, viz. seedlings of *Prunus cerasifera*, St. Julien A, St. Julien GF 655/2 and Marianna GF 8/1, and reported that plum trees grafted on St. Julien A and St. Julien GF-655/2 have reduction in trunk diameter of plant as compared to *Prunus cerasifera* seedlings.

Milosevic et al. (2011) grafted five apricot cultivars on Myrobalan rootstock and Blackthorn interstock to determine the effect of rootstock on vegetative growth and

Table 5.3 Rootstocks for cherry

Rootstock	Origin	Developed	Characteristics	References
Colt (Fb.2/58/21)	East Malling in the 1970s	Cross of *P. avium* × *P. pseudocerasus*	Semi-dwarfing rootstock, good compatibility with all types of cherry cultivars. It is precocious and early in cropping. Less winter hardy as compared to F12/1	Webster (1980)
Mazzard	USA	Selection of *Prunus avium*	Winter hardy rootstock, highly vigorous and having moderate productivity. Low precocity, less suckering habit and well adapted in soils ranging from sandy loam to clay loam, easily propagates from seeds. Susceptible to poorly drained or wet soils	Long and Kaiser (2010)
F 12/1	USA	Selection of Mazzard	Developed through vegetative method of propagation. More vigorous than 'Mazzard'. Resistant to bacterial canker but susceptible to crown gall caused by *Agrobacterium tumefaciens*, winter hardy	Long and Kaiser (2010)
Mahaleb	USA	Seedlings of *P. mahaleb*	Adapts different soil conditions even in drought, and root system is more frost resistant, produces slightly smaller tree than Mazzard and more disease resistant and has slightly better cold hardiness. Resists crown gall, bacterial canker and some nematodes	Long and Kaiser (2010)
Mahaleb 'CDR-1'	USA	A natural hybrid of *Prunus mahaleb*	Strong sprout capacity and branching ability, resistance ability to crown gall is superior. Sweet cherry grafted on it bears early having dwarfing effect. It has strong salt and alkaline tolerance ability.	Cai et al. (2013)

(continued)

Table 5.3 (continued)

Rootstock	Origin	Developed	Characteristics	References
Monrepos	Spain	Selected from an open pollinated population of Myrobalan	Good compatibility with most sweet cherry cultivars, suitable for calcareous and heavy soils, heavy producer with good rooting ability. Propagates with hardwood and in vitro methods, resistant to prune dwarf virus, necrotic ringspot virus, plum pox virus	Pina et al. (2011)
GiSelA 3	Germany	*P. cerasus Schattenmorelle* × *P. canescens*	Most dwarfing among Gisela series, recommended for high-density planting in deep fertile soils under super slender axe system. Provides high early yields, free from suckers, produce wide branch angles	Gruppe (1985), Long and Kaiser (2010)
GiSelA 5	Germany	*P. cerasus* × *P. canescens*	Most dwarfing rootstock, reduces vigour up to half of Mazzard seedlings, produces trees which have open and spreading nature with wide crotch angles, but branching may be sparse, not suitable for heavy soils and requires good drainage. Susceptible to replant problem	Gruppe (1985), Long and Kaiser (2010)
GiSelA 6	Germany	*P. cerasus* × *P. canescens*	Semi-dwarf rootstock that produces a tree about 80–90% of Mazzard, well suited for heavy soil types, precocious and tolerant to many cherry viruses, no suckering. Anchors well but may need support, especially in the first fruiting years due to early precocity. The new shoots are produced much easier in comparison to Gisela 5 which has gained popularity for this rootstock, shows good compatibility with different scions	Gruppe (1985); Long and Kaiser (2010)

(continued)

Table 5.3 (continued)

Rootstock	Origin	Developed	Characteristics	References
GiSelA 17	Germany	*P. canescens* × *P. avium*	Produce somewhat larger trees, less demanding than GiSela 5 and has performed well in replant sites, suitable and performed well with the more productive self-fertile cultivars	Gruppe (1985); Long and Kaiser (2010)
MaxMa	USA	Selection from an open pollinated 'Mahaleb' tree with *P. avium*	Precocious, semi-dwarfing, less suckering and resistance to iron-induced chlorosis caused by calcareous soils, produces larger trees than Mazzard, shows good scion compatibility, wider adaptation to soil types and environmental conditions	Long and Kaiser (2010)
Krymsk 5	Russia	*P. fruticosa* × *P. lannesiana*	It is a precocious cherry rootstock with vigour, similar to Gisela 6. Flower densities are lower than densities on Gisela. Virus-free budwood must be used due to the virus sensitivity of these rootstocks. Rootstock also produces some root suckers. Anchorage appears good	Long et al. (2014)
Krymsk 6	Russia	*P. cerasus* × (*P. cerasus* × *P. maackii*)	Produces plants 75–80% the size of Krymsk 5 or Gisela 12 and reduces tree size by 20–30% compared to Mazzard, has wider adaptability to both cold and hot climates, adaptable to heavier soils, well-anchored trees, low to moderate root suckering. Sensitive to *Prunus* ring spot and *Prunus* dwarf virus	Long et al. (2014)
Alkavo	USA	Selection from *P. avium*	Vigorous, often used for sweet cherry with espaliers system of training	Funk (1969)

(continued)

Table 5.3 (continued)

Rootstock	Origin	Developed	Characteristics	References
Charger	UK	–	Produces tree intermediate in size between Colt and F 12/1, vigorous, more productive, resistant to bacterial canker	Webster and Schmidt (1996)
Weiroot 10	Germany	Selected from wild cherry (*P. cerasus*) types	Propagated by softwood and semi-hardwood cuttings, vigorous, 20–30% smaller in comparison to F 12/1 and are good choice for poor soils	Treutter et al. (1993)
Weiroot 53	Germany	Selected from wild cherry (*P. cerasus*) types	Propagated by softwood and semi-hardwood cuttings, dwarfing nature 60–75% smaller	Treutter et al. (1993)
Weiroot 158	Germany	Selected from wild cherry (*P. cerasus*) types	Propagated by softwood and semi-hardwood cuttings, medium vigour, 50% smaller, suitable for fertile soils,	Treutter et al. (1993)
P-HL-A	Czech Republic	Selected from hybrid seedlings of 'Rtynska Ptacnine'	Propagates through in vitro culture, 70% reduction in tree size than F 12/1, encourages very early fruiting, suitable for high-density orchards and slender spindle systems, life cycle of orchards is 20–25 years, winter frost resistant and sensitive to cherry leaf spot	Paprstein et al. (2008)
P-HL-B	Czech Republic	Selected from hybrid seedlings of 'Rtynska Ptacnine'	Propagates through in vitro culture, 50% reduction in tree size than F 12/1, encourages very early fruiting, life cycle of orchards is 25–30 years, winter frost resistant and sensitive to cherry leaf spot	Paprstein et al. (2008)
P-HL-C	Czech Republic	Selected from hybrid seedlings of 'Rtynska Ptacnine'	Propagates through in vitro culture, 80% reduction in tree size than F 12/1, encourages very early fruiting, suitable for slender spindle systems, life cycle of orchards is 15 years, winter frost resistant and sensitive to cherry leaf spot	Paprstein et al. (2008)

(continued)

Table 5.3 (continued)

Rootstock	Origin	Developed	Characteristics	References
CAB-6 P		Selection of *P. cerasus*	Semi-dwarfing nature, strong root system, suckering ability, suitable for heavy textured soil, resistant to ground water, cold hardy, resistant to lime and drought, suitable for sweet and sour cherry	Faccioli et al. (1981)
Tabel Edabriz	France	Clone of *P. cerasus*	Graft compatibility is good, early bearer, dwarfing nature, 28–42% less vigorous than on Ma × Ma 14, require deep, fertile soils and irrigation, susceptible to chlorosis	Charlot et al. (2005)
Piku 1	Germany	*P. avium* × (*P. cyanescens* × *P. tomentosa*)	Moderate vigour, high productivity, adaptability, tolerant to prune dwarf virus and *Prunus* necrotic ringspot virus	Wolfram (1996)

observed that Myrobalan plum rootstock induces higher tree growth for apricot cultivars. Sitarek and Bartosiewicz (2011) grafted 'Morden 604' and 'Miodowa' apricot cultivars on seedling rootstocks 'Wangenheim Prune' and 'Erunosid', and Polish selection apricot genotypes A4 and M46 were compared with trees of the same cultivars on the standard *P. divaricata* rootstock.

Grzyb and Rozpara (2012) studying the plum cultivar 'Jolo' was grafted on Myrobalan plum seedling and Wangenheim Prune seedling rootstocks observed that Myrobalan Plum seedlings had larger trunk cross-sectional area than those grafted on Wangenheim Prune seedlings.

'Catherina' cv. of peach was grafted on seven plum rootstocks ('Adesoto', 'Monpol', 'Montizo', 'P. Soto-67-AD', 'PM-105-AD', 'GF 655/2' and a local plum rootstock Constanti 1) and found that most vigorous and highest cumulative yield was obtained in 'Constanti 1' rootstock, whereas higher yield efficiency was recorded in 'GF 655/2' and 'Montizo' rootstock. 'GF 655/2' was dwarfing or less vigorous in nature, whereas the most vigorous one was 'Constanti 1' and 'Monpol' rootstocks (Mestre et al. 2017).

5.4 Yield and Quality

Fruit quality attributes of any scion variety which is of great interest can be easily improved with rootstock. Reduced vigour and increased productivity are the positive traits from any combination of scion/rootstock; however, the negative effects of fruit size are as follows: often difficult to separate and if not managed properly leads

Table 5.4 Some hybrid rootstocks used for stone fruit species

Rootstock	Origin	Parentage	Characters	References
Nemaguard	USA	*P. persica* × *P. davidiana*	Produces uniform seedlings, resistant to both *M. incognita* and *M. javanica*, susceptible to ring nematode, provide strong well-anchored and high-yielding trees which are sensitive to wet conditions, adapted to well-drained, sandy soils. Susceptible to *Armillaria* root rot, *Phytophthora* rot and bacterial canker. Precocious and grows to 12–18 ft	Hartmann et al. (2007)
Florda guard	USA	*P. dulcis* × *P. persica*	Flordaguard is another peach seedling mainly compatible with peaches, nectarines, plums and apricots, causes early bud break due to low chill nature, root-knot nematode resistant	Sherman et al. (1991)
Barrier 1	USA	*P. persica* × *P. davidiana*	Vigour somewhat similar or greater than GF 677, possess extensive deep root system that ensures good anchorage. Resistant to root-knot nematodes and replant problem, induces higher productivity and larger fruit size than does GF 677	Moreno (2004)
Cadaman (Avimag)	Hungary/ France	*P. persica* × *P. davidiana*	Cadaman induces vigour similar to that of GF 677, but vigour tends to decrease 4 or 5 years after orchard establishment. Precocious, productive and larger fruit size in comparison to GF 677. Resistant to waterlogging, unlike peach-almond hybrids, tolerant to iron chlorosis and replant problem and nematodes	Edin and Garcin (1994)
GF 1869	France	*P. domestica* × *P. spinosa*	Semi-dwarf nature, compatible with most of the peach and plum cultivars, resistant to waterlogged conditions, more suckering, partly resistant to ring nematode and tolerant to *Phytophthora*	Salesses et al. (1988)

(continued)

Table 5.4 (continued)

Rootstock	Origin	Parentage	Characters	References
Julior 'Ferdor'	France	*P. insititia* × *P. domestica*	It is graft compatible with peach plum and apricot cultivars, and it is propagated by hardwood cuttings and micropropagation. It induces low to medium vigour although greater than GF 655/2 and is dwarfing rootstock for peach cultivars even in fertile soils, resistant to waterlogging, sensitive to calcareous soil above pH 8.2	Loreti and Massai (2002)
Ishtara (Ferciana)	France	Belsiana plum (*P. cerasifera* × *P. salicina*) × (*P. cerasifera* × *P. persica*)	It is a multiple species-compatible rootstock which induces medium vigour and can be used for peach, apricot and plum cultivars. It has good graft compatibility with peach and has shown good resistance to peach tree borer	Grasselly (1988), Renaud et al. (1988)
Myran (Yumer)	France	(*P. cerasifera* × *P. salicina*) × *P. persica*	Easily propagates through hardwood and semi-hardwood cuttings, induces medium to high vigour and has good graft compatibility, free from suckering, tolerant to waterlogging, tolerant to *A. mellea* and *Meloidogyne* spp. nematodes	Renaud et al. (1988)
Jaspi (Fereley)	France	*P. salicina* × *P. spinosa*	It is a multiple species-compatible rootstock and is more suitable for apricot and plum than peach as compatibility with peach is limited. It induces 20% less vigour than peach seedlings in Europe, very dwarfing nature (60% of standard) in the USA. It is adapted to difficult soil environments due to its tolerance to waterlogging, calcareous soil and replant sickness, but peach cultivars on it are extremely susceptible to bacterial blight.	Renaud and Salesses (1990)

(continued)

Table 5.4 (continued)

Rootstock	Origin	Parentage	Characters	References
Mr S. 2/5	Italy	Selected from open pollinated Myrobalan population. Probably a spontaneous *P. cerasifera* × *P. spinosa* hybrid	Easy propagation by layering, hardwood cuttings and micropropagation, good graft compatibility with peach and nectarine. Reduced scion vigour by 10–15%, tolerant to replant disease, less suckering nature, sensitive to rust, resistance to crown gall, calcareous soil and root waterlogging	Loreti et al. (1988)
Citation	USA	*P. salicina* × *P. persica*	Produces a tree 50–65% less of standard size in peaches and 75% less of standard size in apricots and plums. Strong well anchored, precocious and grows up to 12–16 ft, compatible with apricot and plum, induces early defoliation and dormancy in the nursery, tolerant of waterlogging and resistant to root-knot nematodes, susceptible to crown gall and oak root fungus. No sucker emergence and cold tolerant. The rootstock is not suitable for almond	Layne (1987)

to imbalanced crop load (Lang 2000). Influence on fruit maturity of potentially rootstock has been described in some fruit crops (Beckman and Cummins 1991; Beckman et al. 1992).

Rootstock and scion combinations have variable influence on tree productivity and fruit quality. Donno et al. (1972, 1976) observed that almond cv. Tuono had more yield on sweet almond than on bitter almond seedling rootstock; however, this difference disappeared after 8–10 years. Higher almond yields were recorded on peach and peach × almond hybrid than on almond rootstock (Gall and Grasselly 1979), on 2702, 2682, 2147, 8455 and 8475 rootstocks (Popak 1987), in Carmel and Nonpareil on peach than on almond rootstock (Micke et al. 1996) and Ferragues than Texas and Tuono grafted on almond × peach hybrid GF677 and peach 305 (Monastra 1976). Higher almond yields were also recorded in Ferragues grafted on almond, peach GF305 and peach × almond GF677 rootstock than in Tuono on almond rootstock (Barbera et al. 1994), Pizzuta d'Avala on Fasciunieddu Spammata (Alberghina, 1992) and Colorada and Clone Cebas on Garrigues rootstock (Egea and Burgos 1991).

Nonpareil almond cultivar on peach and bitter almond rootstocks produced higher yields than on Behmi rootstock (Kumar 1987; Das 1990). Highest fruit weight and size in Nonpareil, highest kernel percentage in IXL and Nonpareil, highest fat percentage in Katha almond on wild peach rootstock and highest per cent proteins in Afghanistan seedling on bitter almond rootstock have been reported by Kumar (1987).

'Stanley', 'Cacanska Lepotica' and 'Althan's Gage' cultivars of plum were grafted on rootstocks Myrobalan C 162/a seedling, Marianna 'GF 8–1', Myrobalan 'MY-BO-1', Myrobalan 'MY-KL-A', 'St. Julien GF 655/2' and *Prunus domestica* 'Feher besztercei', and it was found that 'Stanley' plum produced highest yield efficiency on vigorous Marianna GF 8–1; medium yield efficiency on Myrobalan C 162/a seedling, MY-BO-1, MY-KL-A and 'St. Julien GF 655/2'; and low yield efficiency on Feher besztercei (Hrotkó et al. 2001). No significant influence was observed for fruit weight when plum cultivars were grafted on different rootstocks. Maximum yield efficiency and fruit weight of 'Althan's Gage' plum was obtained when grafted on 'St. Julien' rootstock and minimum on 'MY-KL-A' rootstock.

Son and Kuden (2003) studied the influence of seedling and GF-31 rootstocks on the yield and quality of table apricot cvs. Tokalolu, Precoce De Tyrinthe, Joubert Foulon, Canino, Sakt 6, Beliana, Priana and Early Kishnevski and observed that fruit weight and yields was recorded on seedling rootstocks of apple for all cultivars; however, dates of full blooming and maturation were also found earlier on seedling rootstocks.

Sharma et al. (2004) recorded higher yield of green almonds in Nonpareil almond cultivar than Merced cultivar when raised on wild peach than bitter almond rootstocks. Lanauskas (2006) grafted two plum cultivars (Stanley and Kauno Vengrine) raised on four rootstocks (*Prunus cerasifera* seedlings, St. Julien A, St. Julien GF 655/2 and Marianna GF 8/1) and reported maximum fruit yield and yield efficiency was obtained on seedlings of *Prunus cerasifera* seedlings.

Wongtanet and Boonprakob (2010) studied the influence of nine peach rootstocks on growth of three scion cultivars and observed significant influence of rootstocks on scion height, branch weight and trunk size. Milosevic et al. (2011) reported that apricot cultivars grafted on Myrobalan plum produced better fruit weight and yield as compared to Blackthorn interstock.

Sitarek and Bartosiewicz (2011) grafted 'Morden 604' and 'Miodowa' apricot cv. on seedling rootstocks (Wangenheim Prune and Erunosid), Polish selection (A4 and M46) and *P. divaricata* rootstock and observed that Miodowa apricot registered maximum cumulative yields on *P. divaricata* and M46 rootstock and the lowest on Erunosid rootstock. 'Wangenheim Prune' rootstock significantly reduced mean fruit weight; however, soluble solid content was not affected by rootstocks.

Grzyb and Rozpara (2012) studying the plum cultivar 'Jolo' that was grafted on Myrobalan plum seedling and Wangenheim prune seedling rootstocks observed higher productivity index in Wangenheim prune seedling as compared to Myrobalan plum seedling, while Myrobalan plum seedling had higher soluble solid content compared to Wangenheim prune seedling.

Hajagos et al. (2012) studied combinations of five rootstocks ('GiSelA 5', 'GiSelA 6', 'Piku 1', 'PHL-C' and 'Weiroot 158') and two scions ('Regina' and 'Kordia') with regard to properties affecting consumer value, fruit appearance and flavour. They reported that rootstock effect was clearly identifiable in the development of fruit firmness, fruit weight and sugar and acid content, and based on these properties, 'PHL-C' was recommended for 'Kordia' scion.

Pisana cv. of apricot was grafted on Myrabolan 29C and apricot seedling rootstock to determine their influence on fruiting characteristics, and no significant influence was observed on flowering and fruiting; however, maximum fruit weight, total antioxidant capacity and total phenols were obtained with rootstock Myrobalan 29C (Bartolini et al. 2014).

'Catherina' cv. of peach was grafted on seven plum rootstocks ('Adesoto', 'Monpol', 'Montizo', 'P. Soto-67-AD', 'PM-105-AD', 'GF 655/2' and a local plum rootstock Constanti 1), and it was found that most vigorous and highest cumulative yield was obtained in 'Constanti 1' rootstock, whereas higher yield efficiency was recorded in 'GF 655/2' and 'Montizo' rootstock. 'GF 655/2' was dwarfing or less vigorous in nature, whereas the most vigorous one was 'Constanti 1' and 'Monpol' rootstocks (Mestre et al. 2017).

5.5 Nutrient Uptake

Rootstocks influence the nutrient content in the leaves of scion cultivar in different stone fruit crops. The effect of different *Prunus* rootstocks on the uptake of different nutrients by their respective studied scion cultivar is discussed here.

5.5.1 Nitrogen

Nitrogen is an essential component of protein, chlorophyll and protoplasm and stimulates vegetative and root growth in stone fruit plants. Rootstocks exert marked influence on the nitrogen status of scion cultivar. Almond trees on wild peach rootstock had higher leaf nitrogen content (Holeves et al. 1985; Das 1990; Upadhayay and Ananda 1991). Significantly higher leaf nitrogen content has been recorded in own-rooted Redhaven than in Redhaven peach on Bailey, in Loring on Siberian C than on Nemaguard (Couvillon 1982), in Italian prune on plum than on peach (Chaplin et al. 1972) and in sweet cherry cv. Bing on Mazzard than those raised on GI 195/1 and 196/4 rootstocks (Neilsen and Kappel 1996). However, Knowles et al. (1984) found only small and inconsistent difference in foliar nitrogen content of Loring and Redhaven peach cultivars on five seedling rootstocks. No significant difference in the nitrogen content of plum grown from suckers on Myrobalan (Dzamic et al. 1966) and seedling rootstocks (Vitanova 1982) could be recorded. Sharma et al. (2007) recorded higher leaf nitrogen content in almond leaves raised on wild peach than on bitter almond rootstock.

5.5.2 Phosphorus

Phosphorus is a component of energy compounds ADP and ATP and is important for growth, flowering, fruiting and seed formation in fruit plants. In almond, rootstocks did not show any significant differences in leaf phosphorous (Dass 1990; Upadhayay and Ananda 1991). However, on hybrids as well as commercial rootstocks, the leaves of compatible rootstock scion combinations had higher phosphorous concentrations than those on ungrafted plants used as control. Leaves of incompatible combinations had lower phosphorous concentration than the control (Mitasov et al. 1973). Stanley plum trees grown on Zhlta Dzhanka (*Prunus cerasifera*) rootstock had reduced leaf phosphorous content (Vitanova 1982). There was higher accumulation of phosphorous in scion leaves of plum on bitter almond and Behmi rootstocks (Sharma 1988) and in Bing sweet cherry on GM9 than on GM 61/1 rootstock (Neilsen and Kappel 1996). However, Hanson and Perry (1986), in an experiment with Montmorency cherries on seedling Mazzard and Mahaleb rootstock, found lower concentrations of leaf phosphorous on Mazzard than on Mahaleb rootstock. Sharma et al. (2007) recorded higher leaf phosphorous content in almond leaves raised on wild peach than on bitter almond rootstock.

5.5.3 Potassium

Potassium is an enzyme activator, regulates water relations and improves photosynthesis and development in fruit plants. Leaf potassium content is markedly influenced by different rootstocks. Leaf potassium content was lower in Carmel and Nonpareil almond grafted on almond than on Nemaguard and Lovell peach rootstocks (Micke et al. 1996), in almond on almond than on peach rootstock (Holeves et al. 1985) and in Bing Sweet cherry on GM9 than on GM61/1 rootstock (Neilsen and Kappel 1996). However, Das (1990) and Upadhayay and Ananda (1991) observed higher leaf potassium levels in almond plants on bitter almond rootstock. Sharma et al. (2007) recorded higher leaf potassium content in almond leaves raised on wild peach than on bitter almond rootstock.

5.5.4 Calcium

Calcium is a constituent of cell wall and is important in the formation of cell membrane. Rootstocks produce variable effects on leaf calcium contents of scion cultivar. Higher leaf calcium content was estimated in almond trees on wild peach than on bitter almond rootstock (Das 1990; Upadhayay and Ananda 1991) and on almond than on peach rootstock (Micke et al. 1996). Peach trees on Siberian C rootstock had reduced foliar calcium levels than on other rootstocks (Werner and Young 1987), whereas Car et al. (1995) found higher leaf calcium content of Maravilha peach on Hansen 536 than on Harrow Blood, Mrs. 2/5 or PSB2 rootstocks. Similarly, Vitanova (1982) reported lower calcium content of Stanley plum raised on M"tna

Boyaka Rakiinitsa (*Prunus domestica*) rootstock. However, Sharma (1988) observed higher accumulation of leaf calcium on Behmi and Myrobalan B rootstocks. Hanson and Perry (1986) found that the leaves of the Montmorency cherry on Mazzard contained higher calcium content than on Mahaleb. Rozpara et al. (1990) observed lower calcium content in the leaves of sweet cherry grafted on *Prunus mahaleb* as compared to those on *Prunus avium* rootstocks. Sharma et al. (2007) recorded higher leaf calcium content in almond leaves raised on wild peach than on bitter almond rootstock.

5.5.5 Magnesium

Magnesium regulates the processes of photosynthesis and carbohydrate metabolism and is also associated with protein synthesis. Magnesium content of the scion leaves was also influenced by the rootstock used. Higher leaf magnesium in almond grafted on wild peach than on bitter almond and Behmi rootstocks (Das 1990) and on peach than on almond rootstock (Holeves et al. 1985; Micke et al. 1996) has also been recorded. However, Upadhayay and Ananda (1991) observed that rootstock did not influence leaf magnesium content in the scion cultivars of almond. It has also been observed that there was higher accumulation of magnesium in the leaves of plum on Behmi and Myrobalan rootstock (Sharma 1988), Bing Sweet cherry on GM9 than on GM61/1 rootstock (Neilsen and Kappel 1996), Sweet cherry on *Prunus avium* than on *Prunus mahaleb* rootstock (Rozpara et al. 1990) and Montmorency cherry on Mahaleb than on Mazzard rootstock (Hanson and Perry 1986). Sharma et al. (2007) recorded higher leaf magnesium content in almond leaves raised on wild peach than on bitter almond rootstock.

5.6 Incompatibility

Lack of commercial rootstocks which are compatible with most of the *Prunus* cultivars is the main hurdle in widespread use of all the stone fruit species (Okie 1987; Zarrouk et al. 2006). For better performance, longevity, commercial success and adaption of any orchard, good scion/rootstock graft compatibility is required, and this graft compatibility is major concern in cherry, almond and apricot rather than in peach or plum (Lang and Ophardt 2000). Unhealthy trees, breaking of the graft union, immature death or later failure of the graft combination and unsuccessful in forming a strong union which lasts for long are the major drawbacks of graft incompatibility (Zarrouk et al. 2006).

Few of the clones have been preselected from plum and apricot varieties during the compatibility studies in the nursery which are compatible with most of the plum varieties, and such varieties were incompatible earlier with Myrobalan, Myrobalan B and Marianna rootstocks (Cambra and Cambra 1972, 1973). As compared to other commercial rootstocks, some of the selected clones were found to be compatible with most of the apricot varieties (Cambra 1979, 1990b). 'Bulida', 'Canino' and 'Moniqui' cultivars of apricot could not show any type of graft compatibility with most of the clonal rootstocks which were included under the non-congenial cultivars (Crossa-Raynaud and Audergon 1987).

Prunus cerasifera or Myrobalan spp. shows incompatibility symptoms when they were used as rootstocks for peach cultivars (Herrero 1951, 1955; Tabuenca 1960, 1962); however, in the past recent decades, few of the clones selected from the Myrobalan seedling population shows compatibility with most of the peach and nectarine cultivars (Tabuenca and Moreno 1988; Moreno et al. 1993; Moreno et al. 1994).

5.7 Resistance

Root-knot nematode is one of the major problems in stone fruit orchards worldwide which are plant parasitic nematodes (PPN) (Nyczepir and Esmenjaud 2008). Fruit tree crops like almond (*P. amygdalus* Batsch), cherry (*P. avium* L. and *P. cerasus* L.), peach (*P. persica*) and plum (*P. cerasifera, P. domestica* L. and *P. salicina*) are susceptible to PPN attack worldwide (Ye et al. 2009; Bosselut et al. 2011). The effects of PPN in crops are often underestimated, but in general, it is accepted that, on average, nematodes are annually reducing the global agricultural production by about 10–12% (Agrios 2005).

Resistance to soilborne pathogens (fungi and bacteria), phytoplasma and root-knot nematodes, tolerance to iron-induced chlorosis, salinity and root asphyxia are the main reasons due to which Myrobalan plum rootstock is suitable and compatible for plum and other stone fruits (Rowe and Catlin 1971; Dosba 1992; El-Motaium et al. 1994; Esmenjaud et al. 1994). A well-stated fundamental reason for the use of plum rootstocks is that they are more tolerant to waterlogging and heavy soils than other *Prunus* spp. (Bernhard and Grasselly 1959; Rowe and Catlin 1971; Salesses and Juste 1970; Bernhard et al. 1979). Graft compatibility with majority of the *Prunus* spp. and under calcareous soil conditions tolerance to iron-induced are the main reasons for the use of such rootstocks (Moreno et al. 1995a, b, c). Likewise, majority of the rootstocks is considered as precocious in nature and produces good-quality fruits (Bernhard and Grasselly 1959; Loreti and Massai 1990; Moreno et al. 1990).

Resistance of plants to root-knot nematode can be divided into three stages: pre-penetration, penetration and penetrated (Liu 2000). There is limited information about *Prunus*' resistance mechanisms to the root-knot nematode, especially at the pre-penetration and penetration stages. Root dissemination and root structural organization are factors that may affect nematode infection (Rodger et al. 2003; Lin et al. 1996; Yuan 2001). The studies on resistance against nematodes to completely inhibit or minimize the propagation of the nematode are preferred due to the lack of special application technique or equipment requirements, low costs and the environmentally friendly nature of the method (Cook and Evans 1987; Lopez-Perez et al. 2006).

Ozarslandan and Tanriver (2018) studied eight different rootstocks (Myrobalan 29-C, Mariana GF 8–1, Garnem, Cadaman, Patrones Arda, Patrones Toro, Myrobalan B and GF677) that were used for nematode resistance tests. Myrobalan 29-C, Mariana GF 8–1, Garnem, Cadaman, Patrones Arda and Patrones Toro rootstocks were considered resistant to *M. incognita*. The root nematodes in the GF 677 host are large and small gallings. So the exiles of the plant have stopped, and the plant is not growing. Choosing the wrong rootstock in the areas where the root nematode is located causes great economic loss.

Özbek et al. (2014) reported that in vitro-grown Myrobalan 29 C, Cadaman and Garnem clones were resistant to *M. incognita* and *M. javanica*. Marull et al. (1991) inoculated various *Prunus* rootstocks with *M. incognita* nematode species and determined that GN1, GN3 and GN9 rootstocks were resistant. Genotypes 9 and 31 were tolerant to *M. javanica*, while genotype 54 was classified as resistant to *M. incognita* (Kiziltan et al. 2016). There was no galling in Adesoto 101, Bruce, Ishtara, AC-952, Garnem and Cadamen rootstocks, and no nematode reproduction was recorded in the Adesoto 101, Adara, Myro-10, Constanti, AD 105 and Cadaman rootstocks, and GXN 17 exhibited no galling (Pinochet et al. 1999). Fluoride isolates resulted in galling in GF.557 rootstock. *M. arenaria*, *M. incognita* and *M. javanica* led to no galling in Nemaguard, Nemared, GXN 15 and GXN 22. The *Prunophora* rootstocks, P.2175, P.1079, P.2980 and Myro 29C, were not galled by any of the four isolates. Myrobalan (*Prunus cerasifera*) P.2175, P.1079 and P.2980 clones were highly resistant to root-knot nematodes (Lecouls et al. 1999). Myrobalan genotypes, (plum rootstocks), having resistance against *M. arenaria* (1 pop.), *M. incognita* (2 pop.) and *M. javanica* (1 pop) populations, determined that P. 1079 and P. 2175 genotypes exhibited resistance at low and high temperatures (Esmenjaud et al. 1996a).

Fernandez et al. (1993) reported that GF 677 was sensitive to root-knot nematodes, while Nemared, G × N no. 22, Marianna GF 8–1 and Myrobalan 29C persisted their resistance even at high temperatures. The almond, nectarine and peach rootstock GF 677 was sensitive to *M. arenaria*, *M. incognita* and *M. javanica*, while Nemared and Garnem (GXN 15, GXN 22) were resistant, and the plum rootstock Myrobalan 29C was also resistant (Esmenjaud et al. 1997). Ten peach rootstocks were tested against *M. javanica*, and it was found that they were resistant and GF 677 rootstock was sensitive (Pinochet 2009). *M. javanica* is very pathogenic to peaches in Italian populations and may cause severe crop losses in infected areas (Di-vito et al. 2005). The host suitability to RKN in Myrobalan plum material ranges from susceptible clones (P.2646, P.16.5 and P.2032) to highly resistant clones (P.2175, P.1079 and P.2980). Resistance to *M. arenaria* is monogenic and completely dominant in P.2175 (gene Ma1, heterozygous) and in P.1079 (gene Ma2, homozygous) (Esmenjaud et al. 1996b).

Under sandy soil conditions and soil having nematodes, peach or Myrobalan plum are more successful (Chandler 1978). Myrobalan plum seedlings have bacterial and canker disease resistance (Glenn 1962), and both peach and Myrobalan plum seedling rootstocks are recommended for plum even in those soils having root nematode problem.

In almond variety 'Nonpareil' grafted on 'Nemaguard' peach, Esparza et al. (2001) observed a decreased performance after 2 years of water stress applied during floral initiation, mainly explained by a decrease in the renewal of fruitwood, without affecting the weight of seeds by the deficit of last season. Varying productive responses to water deficit is also related with the rootstock used, in terms of firmness of fruit, soluble solid content and anthocyanin content. These variables can increase or decrease independently, improving or deteriorating various parameters of quality within a rootstock-scion combination, as has been observed in various combinations in peach (Besset et al. 2001).

5.8 Postharvest Management

The beneficial impacts of the rootstocks may also be extended beyond harvest. Some indicative examples are addressed below, although more research has to be carried out in the future for all related issues. Firstly, rootstock can improve the main quality indices of various harvested products, including the overall nutritive value and aromatic substances of horticultural products (Krumbein 2013; Legua et al. 2014). Rootstock may also enhance the levels of various particular health-promoting substances (Chavez-Mendoza et al. 2013; Cardenosa et al. 2015) and decrease those of health hazardous ones (e.g. organic pollutants) (Schwarz et al. 2010). Secondly, postharvest storage and shelf life of fruits are also rootstock-dependent (Ritenour et al., 2004). Thirdly, the effect of rootstock may also be critical in the processing of fruits and vegetables.

5.9 Replant Problem

Replant problems in stone fruit crop are of major concern and thus require a special mention. Replant disease is a syndrome expressed as a failure in tree establishment, suppressed growth and shortened productive life. One of the most common practices in fruit tree cultivation worldwide is repeated cropping. Soilborne pathogens (nematodes, fungi and bacteria) are essential components associated with replant disease. Other abiotic stress factors are also involved. However, it is difficult to determine the primary causal agent or the predominating factor for each replant situation. In most cases, the existence of a combination of factors whose damaging effects over the plant are accumulative is accepted. Thus, choosing a rootstock that would have multiple resistance or tolerance to several of these damage causing factors is a priority to assure success during the establishment of the tree and, afterwards, during its productive life (Calvet et al. 2000).

5.10 Conclusion

Rootstock decides the success or failure of any orcharding enterprise. Stone fruits in India are presently grown on seedling rootstocks of unknown origin which are heterozygous and prone to various biotic and abiotic stresses resulting in low productivity. However, clonal rootstocks are uniform, have capability in increasing the productivity and improve efficiency through improved tree survival, controlled tree vigour and increased fruit size, yield and quality. In the past few decades, a lot of the progress has been made in developing the clonal rootstock for stone fruits which are adaptable in a wide range of soils and compatible with most of the *Prunus* spp. Breeding programmes and modern genetic engineering technology have been employed in various parts of the world to screen and select rootstocks to overcome different production problems in stone fruits. The use of these rootstocks will improve production, productivity and quality of stone fruits.

References

Agrios, G. (2005). *Plant pathology* (5th ed.). Amsterdam: Elsevier Academic Press.

Alberghina, O. (1992). Productive ability of 31 almond cultivars on two rootstocks in an environment on the ionic coast of Sicily. *Rivista di Frutticoltura e di Ortofloricoltura, 54*(1), 69–74.

Andersen, R., Freer, J., & Robinson, T. (2006). Plum rootstock trial at Geneva: A progress report. *New York Fruit Quarterly, 14*(1), 27–28.

Audergon, J. M., Duquesne, J., Nicolas, J. C., & Audubert, A. (1991). A new selected plum rootstock for apricot varieties: Torinel(R). *Acta Horticultuare, 293*, 395–400.

Barbera, G., Mareo, L. D., La Mantia, T., & Schirra, M. (1994). Effect of rootstock on productive and qualitative response of two almond varieties. *Acta Horticulturae, 373*, 129–134.

Bartolini, S., Leccese, A., Lacona, C., Andreini, L., & Viti, R. (2014). Influence of rootstock on fruit entity, quality and antioxidant properties of fresh apricots cv. 'Pisana'. *New Zealand Journal of Crop and Horticulture, 42*(4), 265–274.

Beckman, T. G., & Cummins, J. N. (1991). Rootstocks for peaches. p. 974–975. In*:* J.N. Cummins, J.N. (ed.). Register of new fruit and nut varieties, Brooks and Olmo List 35. *HortScience, 26*, 951–986.

Beckman, T. G., Okie, W. R., & Myers, S. C. (1992). Rootstocks affect bloom date and fruit maturation of 'Redhaven' peach. *Journal of American Society for Horticultural Science, 117*, 377–379.

Bellini, E., Begnami, C., Nalali, S., & Moretu, G. (1993). Comparison of ten peach rootstocks on the Lazio marshes. In *21st Convegno Perchicolo*, Lugo, Revenna, Italy.

Benjamin, J. E. T., & Shoemaker, J. S. (1978a). Seedling rootstook. In J. E. T. Benjamin (Ed.), *Tree fruit production* (3rd ed., p. 385). Westport: AVI.

Benjamin, J. E. T., & Shoemaker, J. S. (1978b). Alter ripening of seed. In J. E. T. Benjamin (Ed.), *Tree fruit production* (3rd ed., p. 365). Westport: AVI.

Bernhard, R., & Grasselly, C. (1959). Les pruniers porte-greffes du pêcher. *Jour. Fruit et Maraich. d'Avignon*, 75–100.

Bernhard, R., & Grasselly, C. (1981). Les pechers amandiers. *L'Arboculture Fruitiere, 328*, 37–42.

Bernhard, R., Grasselly, C., & Salesses, G. (1979). Orientations des travaux de selection des porte-greffes du pêcher a la Station d'Arboriculture I.N.R.A. de Bordeaux. Eucarpia fruit Section (pp. 277–286).

Besset, J., Génard, M., Girard, T., Serra, V., & Bussi, C. (2001). Effect of water stress applied during the final stage of rapid growth on peach trees (cv. Big-Top). *Scientia Horticulturae, 91*, 289–303.

Black, J. D. F. (1959). Pointers from the plum and apricot rootstock trails. *Horticultural Abstracts, 30*, 3270–3270.

Bosselut, N., Ghelder, C. V., Claverie, M., Voisin, R., Onesto, J., & Rosso, M. (2011). Agrobacterium rhizogenes-mediated transformation of *Prunus* as an alternative for gene functional analysis in hairy-roots and composite plants. *Plant Cell Reports, 30*(7), 1313–1326.

Botu, I., Preda, S., Turcu, E., Achim, G., & Botu, M. (2007). 'Rival'—A new rootstock for plum. *Acta Horticulturae, 732*, 253–256.

Cai, Y. L., Feng, Y., Zhang, X., Wan, T., Gao, P., & Wang, Y. (2013). A new cheery dwarf rootstock variety—'Mahaleb CDR-1'. *Journal of Fruit Science, 30*(1), 177–178.

Calvet, C., Estaún, V., Camprubí, A., & Pinochet, J. Y. (2000). Enfermedades de replantación en los frutales. In E. Montesinos, P. Melgarejo, M. A. Cambra, & J. Y. Pinochet (Eds.), *Enfermedades de los frutales de pepita y hueso. Monografía de la Sociedad Española de Fitopatología No 3. 147 p* (pp. 107–109). Madrid: Ediciones Mundi-Prensa.

Cambra, R. (1979). Compatibilidad de variedades de albaricoquero con ciruelo Mirobolán (*Prunus cerasifera* Ehrh.) y Marianna (*P. cerasifera* Ehrh. X *P. munsoniana* Wight y Hedr.). *An Aula Dei, 14*, 371–375.

Cambra, R. (1990a). Adafuel, an almond x peach hybrid rootstock. *HortScience, 25*, 584.

Cambra, R. (1990b). El ciruelo Mirobolán 'Ademir' como patrón de ciruelo y albaricoquero. *Fruticultura Profesional, 30*, 22–25.

Cambra, R., & Cambra, M. (1972). Selección clonal de ciruelo mirobolán (*P. cerasifera* Ehrh). *Resumen de trabajos durante el período* 1950-1971 (p. 192).

Cambra, R., & Cambra, M. (1973). Selección clonal de ciruelo Mirobolán (*Prunus cerasifera* Ehrh.). Compatibilidad con variedades de ciruelo y albaricoquero. *An Aula Dei, 12*, 8–16.

Car, T., Inglese, P., Gwvannin, D., & Truce, E. (1995). Rootstock influence on dry matter and nutrient above ground content and partitioning in Maravilha peach trees. *Acta Horticulturae, 383*, 105–114.

Cardenosa, V., Barros, L., Barreira, J. C. M., Arenas, F., & Moreno-Rojas, J. M. (2015). Different citrus rootstocks present high dissimilarities in their antioxidant activity and vitamins content according to the ripening stage. *Journal of Plant Physiology, 174*, 124–130.

Chandler, W. H. (1978). *Rootstocks: Deciduous orchard. Thoroughly revised with 128 illustration* (3rd ed., p. 352). Philadelphia: Lea and Febige.

Chaplin, M. H., Westwood, M. N., & Roberts, A. N. (1972). Effects of rootstock on leaf element of Italian prune (*Prunus domestica* L.). *Journal of American Society for Horticultural Science, 97*, 641–644.

Charlot, G., Edin, M., Flochlay, F., Soing, P., & Boland, C. (2005). Tabel® Edabriz: A dwarf rootstock for intensive cherry orchards. *Acta Horticulturae, 667*, 217–222.

Chavez-Mendoza, C., Sanchez, E., Carvajal-Millan, E., Munoz-Marquez, E., & Guevara-Aguilar, A. (2013). Characterization of the nutraceutical quality and antioxidant activity in Bell pepper in response to grafting. *Molecules, 18*, 15689–15703.

Cook, R., & Evans, K. (1987). Resistance and tolerance. In B. R. Kerry & R. H. Brown (Eds.), *Principles and practice of nematode control in crop* (pp. 179–220). Samford: Academic Press.

Couvillon, G. A. (1982). Leaf elemental content comparisons of own-rooted peach cvs. to the same cvs. on several peach seedling rootstocks. *Journal of American Society for Horticultural Science, 107*(4), 555–558.

Crossa-Raynaud, P., & Audergon, J. M. (1987). Apricot rootstocks. In R. C. Rom & R. F. Carlson (Eds.), *Rootstocks for fruit crops* (pp. 295–320). New York: Wiley.

Das, B. (1990). *Influence of rootstock and scion cultivar on the uptake and removal of nutrients by almond trees*. M.Sc. Thesis submitted to Dr. Y.S.P. University of Horticulture and Forestry, Nauni, Solan, India.

Dejong, T. M., Johnson, R. S., & Day, K. R. (2011). Controller 5, Controller 9 and Hiawatha peach rootstocks: Their performance and physiology. *Acta Horticulturae, 903*, 221–228.

Di-vito, M., Simeone, A. M., & Catalano, F. (2005). Effect of the root-knot nematode, *Meloidogyne javanica*, on the growth of a peach (*Prunus persica*) rootstock in pots. *Nematology Mediterranean, 33*, 87–90.

Donno, G., Godini, A., & Ferrara, E. (1972). Comparison between the growth of sweet and bitter almond seedling rootstocks and their effect on growth, yield and nut: Kernel ratio in almond (preliminary contribution). *Annali della Facolta di Agraria, Universita di Bari, 25*, 513–518.

Donno, G., Godini, A., & Ferrara, E. (1976). Sweet and bitter almond seedling rootstocks and their influence on the development and yield of almonds. *Annali della Facolta di Agraria, Universita di Bari, 28*, 123–128.

Dosba, F. (1992). Espèces fruitières à noyau. Les maladies à mycoplasmes. Le point sur les recherches. *L'Arboriculture Fruitière, 454*, 24–28.

Dutu, I., Parnia, P., Viscol, I., Mazilu, C., & Ancu, S. (2001). Influence of two vegetative rootstocks on the fruit size and productivity of the Red Haven Peach variety. *Lucrari Stiintifice ale ICPP Pitesti Maracineni, 20*, 75–77. (In Romanian).

Duval, H., Signoret, V., & Joly, R. (2004). Use of Myrobalan species (Prunus cerasifera) as rootstocks for almond and peach. *Acta Horticulturae, 663*, 961–964.

Dzamic, H., Popovic, Z., & Pantelic, M. (1966). The nitrogen, phosphorus and potassium contents of plum leaves. *Zborn. Rad. Poljopriv. Fak. Beograd, 14*, 1–14.

Edin, M., & Garcin, A. (1994). Un nouveau portegreffe du pecher 'Cadaman' Avimag. *InfosCtifl N, 99*, 38–42.

Egea, C. J., & Burgos, O. L. (1991). Comparative study of some almond rootstocks. *ITEA Production Vegetal, 87*(1), 36–40.

El-Motaium, R., Hu, H., & Brown, P. H. (1994). The relative tolerance of six *Prunus* rootstocks to boron and salinity. *Journal of American Society for Horticultural Science, 119*(6), 1169–1175.

Esmenjaud, D., Minot, J. C., Voisin, R., Pinochet, J., & Salesses, G. (1994). Inter- and intraspecific resistance variability in myrobalan plum, peach, and peach-almond rootstock using 22 root-knot nematode populations. *Journal of American Society for Horticultural Science, 119*(1), 94–100.

Esmenjaud, D., Minot, J. C., & Voisin, R. (1996a). Effects of durable inoculum pressure and high temperature on root galling, nematode numbers and survival of Myrobalan plum genotypes (*Prunus cerasifera* Ehr.) highly resistant to *Meloidogyne* spp. *Fundamental and Applied Nematology, 19*(1), 85–90.

Esmenjaud, D., Salesses, G., Minot, J. C., Voisin, R., & Bonnet, A. (1996b). Inheritance of resistance to the root-knot nematode *Meloidogyne arenaria* in Myrobalan plum. *Theoretical and Applied Genetics, 92*, 873–879.

Esmenjaud, D., Minot, J. C., Voisin, R., Pinochet, J., Simard, M. H., & Salesses, G. (1997). Differential response to root-knot nematodes in *Prunus* species and correlative genetic implications. *Journal of Nematology, 29*(3), 370–380.

Esparza, G., DeJong, T., Weinbaum, S., & Klein, I. (2001). Effects of irrigation deprivation during the harvest period on yield determinants in mature almond trees. *Tree Physiology, 21*, 1073–1079.

Faccioli, F., Interieri, G., & Marangoni, B. (1981). Portinnesti nanizzanti del ciliego: le selezione CAB. Atti. Giorn. *Sulle Scelte Varietali in Frutticoltura, 19*, 125–128.

Felipe, A. J., Gomez-Aparisi, J., & Socias, R. (1994). 'Pollizo' plum: A rootstock for almond: Preliminary results. *Acta Horticultuare, 373*, 99–104.

Felipe, A., Carrera, M., Gomez-Aparisi, J., & Socias, R. (1997). 'Montizo' and 'Monpol', Two new plum rootstocks for peaches. *Acta Horticulturae, 451*, 273–278.

Fernandez, C., Pinochet, J., & Felipe, A. (1993). İnfluence of temperature on the expression of resistance in six *Prunus* rootstocks infected with *Meloidogyne incognita*. *Nematropica, 23*, 195–202.

Funk, T. (1969). Virusgetestete 'Kaukasische Vogelkirsche' alsneue Unterlage. *Obstbau, 9*, 140–142.

Funt, R. C., & Goulart, B. L. (1981). Performance of several peach cultivars on *Prunus tomentosa* and *Prunus besseyi* in Maryland. *Fruit Varieties Journal, 35*(1), 20–23.

Gall, H., & Grasselly, C. (1979). Trial on the behaviour of twelve almond varieties grafted on almond, peach and peach x almond rootstocks. *Station d'Arboriculture Fruitiere*, 215–227.
Garner, R. J., & Hatcher, E. S. J. (1955). The influence of source and growth substances on the behaviour of apple and plum cutting. *Horticultural Abstracts, 24*, 2444–2444.
Gasic, K., & Preece, J. E. (2014). Register of new fruit and nut cultivars. List 47. *HortScience, 49*, 413.
Glenn, E. M. (1962). Stem builder trail with plum. *Horticultural Abstracts, 32*, 297–297.
Grasselly, C. (1983). Nouvelles obtention INRA de pechers port-greffes: multiplies par semences. *L'Arboticulture Frutiere, 357*, 50–55.
Grasselly, C. (1988). New French stone fruit rootstocks. *Fruit Varieties Journal, 41*(2), 65–67.
Gruppe, W. (1985). An overview of the cherry rootstock breeding programme. *Acta Horticulturae, 169*, 189–198.
Grzyb, Z. S., & Rozpara, E. (2012). Effect of rootstock on the growth and yielding of 'Jolo' plum trees in an ecological orchard. *Acta Horticulturae, 968*, 133–136.
Hajagos, A., Spornberger, A., Modl, P., & Vegvari, G. (2012). The effect of rootstocks on the development of fruit quality parameters of some sweet cherry (*Prunus avium* L.) cultivars Regina and Kordia during the ripening process. *Agriculture and Environment, 4*, 59–70.
Hanson, E. J., & Perry, R. L. (1986). Comparison of nutrient uptake efficiency of Mazzard and Mahaleb cherry rootstock. In *Abstracts of Contributed Paper, Proceedings of 22nd International Horticultural Congress Davis*, USA. Abstract no. 68 (p. 117).
Hartmann, H. T., Kester, D. E., Davis, F. T., & Geneve, R. L. (2007). Propagation methods and rootstocks for fruit and nut species. In H. T. Hartmann & D. E. Kester (Eds.), *Plant propagation: Principal and practice.* (pp. 717–757). New York: Pearson Education.
Herrero, J. (1951). Studies of compatible and incompatible graft combinations with special reference to hardy fruit trees. *Journal of Horticulture Science, 26*, 186–237.
Herrero, J. (1955). Incompatibilidad entre patrón e injerto. II. Efecto de un intermediario en la incompatibilidad entre melocotonero y mirobolán. *An Aula Dei, 4*, 167–172.
Holeves, C. D., Stylianides, D. C., & Michaelides, Z. (1985). Nutrient element variability in the leaves of almond trees in relation to variety, rootstock and vegetative parts of the tree. *Options Mediterraneannes, 1*, 111–120.
Hrotkó, K. (2005). Developments in high density cherry production in Hungary. *Acta Horticulturae, 667*, 279–284.
Hrotkó, K., Magyar, L., Klenyán, T., & Simon, G. (2001). Effect of rootstocks on growth and yield efficiency of plum cultivars. *Acta Horticulturae, 577*, 105–110.
Hudson, T. H., Fofocker, W. J., & Kofrank, A. M. (1981). Hardwood cuttings. In H. T. Hartman, W. J. Flocker, & A. M. Korfranek (Eds.), *Plant science: Growth, development and utilization of cultivated plants* (p. 603). Englewood Cliffs: Prentice-Hall.
Kiziltan, et al. (2016). The determination of nematode tolerance and seedling performance of the selected almond (*Prunus amygdalus* L.) genotypes in Isparta Province. *Erwerbs-obstbau, 58*, 233–239.
Knowles, J. W., Dozier, W. A., Jr., Evans, C. E., Carlton, C. C., & McGuire, J. M. (1984). Peach rootstock influence on foliar and dormant stem nutrient content. *Journal of American Society for Horticultural Science, 109*, 440–444.
Krumbein, A. (2013). Grafting: A possibility to enhance health-promoting and flavour compounds in tomato fruits of shaded plants? *Scientia Horticulturae, 149*, 97–107.
Kumar, S. (1987). *Effect of rootstock and scion cultivars on foliar nutrient status in almonds*. M.Sc. thesis submitted to Dr. Y.S.P. University of Horticulture and Forestry, Nauni, Solan, India.
Lanauskas, J. (2006). Effect of rootstock on growth and yield of plum trees cv. 'Stanley' and 'Kauno Vengrine'. *Sosininkyste IR Darzininkyste, 25*(3), 243–249.
Lang, G. A. (2000). Precocious dwarfing and productive—How will new cherry rootstocks impact the sweet cherry industry. *HortTechnology, 10*, 719–725.
Lang, G. A., & Ophardt, D. (2000). Intensive crop regulation strategies in sweet cherries. *Acta Horticulturae, 514*, 227–234.

Layne, R. E. C. (1987). Peach rootstock. In R. C. Rom & R. F. Carlson (Eds.), *Rootstocks for fruit crops* (pp. 185–216). New York: Wiley.

Lecouls, A. C., Rubio-Cabetas, M. J., Minot, J. C., Voisin, R., Bonnet, A., Salesses, G., Dirlewanger, E., & Esmenjaud, D. (1999). RAPD and SCAR markers linked to the Ma1 root-knot nematode resistance gene in Myrobalan plum (*Prunus cerasifera* Ehr.). *Theoretical and Applied Genetics, 99*, 328–335.

Legua, P., Forner, J. B., Hernandez, F., & Forner-Giner, M.A. (2014). Total phenolics, organic acids, sugars and antioxidant activity of mandarin (*Citrus clementina* Hort. ex Tan.): Variation from rootstock. *Scientia Horticulturae, 174*, 60–64.

Lichev, V., & Govedarov, G. (1995). Investigation of root system of peach rootstock in a nursery. *Rasteniev dn Nauki, 32*(5), 264–267.

Lin, M. S., He, L. M., & Wen, L. (1996). Mechanism of morphological structure of sweet potato resistance to potato rot nematode (*Ditylenchus destructor*). *Scientia Agricultura Sinica, 29*, 8–12.

Liu, W. Z. (2000). *Plant parasitic nematodes*. Beijing: China Agriculture Press.

Long, L. E., & Kaiser, C. (2010). *Sweet cherry rootstocks*. Oregon: A Pacific Northwest extension publication, Oregon State University.

Long, L., Facteau, T., Nunez-Elisea, R., & Cahn, H. (2005). Developments in high density cherries in the USA. *Acta Horticulturae, 667*, 303–310.

Long, L. E., Brewer, L. J. & Kaiser, C. (2014). Cherry rootstocks for modern orchard. In *57th Annual IFTA Conference*, 2014 (pp. 24–28).

Lopez-Perez, J. A., Strange, M. L., Kaloshian, I., & Ploeg, A. T. (2006). Differential response of Mi gene resistant tomato rootstocks to root-knot nematodes (*Meloidogyne incognita*). *Crop Protection, 25*, 382–388.

Loreti, F. (1984). Attuali conoscenze su principali portinnesti degli alberi da frutta. *Frutticoltura, 9*, 9–60.

Loreti, F., & Massai, R. (1990). Los patrones del melocotonero y del almendro: situación actual, problemas y perspectivas. *ITEA, 9*, 73–116.

Loreti, F., & Massai, R. (1998). Sirio: new peach x almond hybrid rootstock for peach. *Acta Horticulturae, 465*, 229–236.

Loreti, F., & Massai, R. (2002). The high density peach planting system. Present status and perspectives. *Acta Horticulturae, 592*, 377–390.

Loreti, F., Guerriero, R., & Morini, S. (1981). Comparative trials on peach seedlings as rootstock for intensive nectarine orchards. *Acta Horticulturae, 114*, 218–220.

Loreti, F., Guerriero, R., & Massai, R. (1988). Una nova ed interessante selezione di sussino portimmesto: 'Mr.S.2/5'. In *Atti Convegno Nazionale l portinnesti delle piante da frutto*, Ferrara, Italy (pp. 45–50).

Maas, F. M., Balkhoven, J. M. T., Heijerman-Peppelman, G., & van-der -Steeg, P. A. H. (2011). Krymsk®1 (VVA-1), A dwarfing rootstock suitable for high density plum orchards in the Netherlands. *Acta Horticulturae, 903*, 547–554.

Marull, J., Pinochet, J., Verdejo, S., & Soler, A. (1991). Reaction of *Prunus* rootstocks to *Meloidogyne incognita* and *M. arenaria* in Spain. *Suppl. Journal of Nematology, 23*, 64–569.

Mestre, L., Reig, G., Betran, J. A., & Moreno, A. (2017). Influence of plum rootstocks on agronomic performance, leaf mineral nutrition and fruit quality of "Catherina" peach cultivar in heavy calcareous soil conditions. *Spanish Journal of Agricultural Research, 15*(1), 1–11.

Micke, W. C., Freeman, M. W., Beede, R. H., Kester, D. E., Yeager, J. T., Pelletreau, K. G., & Conelle, J. H. (1996). Almond trees grown on peach rootstock initially more productive. *California Agriculture, 50*(4), 29–31.

Mika, A., Buler, Z., & Chlebowska, D. (1998). The effect of training systems and planting density on growth and fruiting of plum trees grafted on two rootstocks. *Acta Horticulturae, 478*, 107–112.

Milosevic, T., Milosevic, N., & Glisica, I. (2011). Influence of stock on the early tree growth, yield and fruit quality traits of apricot (*Prunus armeniaca* L.). *Journal of Agricultural Sciences, 17*, 167–176.

Mitasov, I. M., Rikhter, A. A., Chemaarin, N. G., & Yaroslavtseva, I. P. (1973). Studies on the compatibility of almond with hybrid rootstocks. *Sel'Skokhozyaistvennaya Biologiya, 8*, 780–781.

Monastra, F. (1976). A comparative trial of eight almond cvs. grafted on various rootstocks. *Annali della Instituto Spermenlate perla Frutticoltura, 7*, 85–93.

Moreno, M. A. (2004). Breeding and selection of *Prunus* rootstocks at the Aula Dei Experimental Station, Zaragoza, Spain. *Acta Horticulturae, 658*(2), 519–528.

Moreno, M., & Cambra, R. (1994). Adarcias, an almond x peach hybrid rootstock. *HortScience, 29*, 925.

Moreno, M. A., Montañés, L., Sanz, M., & Tabuenca, M. C. (1990). Comportamiento y estado nutricional de la variedad de melocotonero Vesuvio sobre diversos patrones. In *En: III Symposium Nacional sobre la Nutrición de las Plantas* (pp. 137–142).

Moreno, M. A., Moing, A., Lansac, M., Gaudillere, J. P., & Salesses, G. (1993). Peach/myrobalan plum graft incompatibility in the nursery. *Journal of Horticultural Science, 68*, 705–714.

Moreno, M. A., Gaudillère, J. P., & Moing, A. (1994). Protein and amino acid content in compatible and incompatible peach/plum grafts. *Journal of Horticultural Science, 69*(6), 955–962.

Moreno, M. A., Tabuenca, M. C., & Cambra, R. (1995a). Adesoto 101, a plum rootstock for peaches and other stone fruits. *HortScience, 30*, 1314–1315.

Moreno, M. A., Tabuenca, M. C., & Cambra, R. (1995b). Ademir, A myrobalan rootstock for plums. *HortScience, 30*(7), 1475–1476.

Moreno, M. A., Tabuenca, M. C., & Cambra, R. (1995c). 'Adara', a plum rootstock for cherries and other stone fruit species. *HortScience, 30*, 1316–1317.

Nagy, P., & Lantos, A. (1998). Breeding stone fruit rootstock in Hungary. *Acta Horticulturae, 484*, 199–202.

Neilsen, G., & Kappel, F. (1996). Bing sweet cherry leaf nutrition as affected by rootstock. *HortScience, 31*(7), 1169–1172.

Nicotra, A., & Moser, L. (1997). Two new rootstock for peach and nectarines: Penta and Tetra. *Acta Horticulturae, 451*, 269–271.

Nyczepir, A., & Esmenjaud, D. (2008). Nematodes. In D. Layne & D. Bassi (Eds.), *The peach: Botany, production and uses* (pp. 505–535). Wallingford: CABI Publishing.

Okie, W. R. (1987). Plum rootstocks. In R. F. Carlson (Ed.), *Rootstocks for fruit crops* (pp. 321–360). New York: Wiley.

Okie, W. R. (1998). *Handbook of peach and nectarine varieties: Performance in South Eastern United States and index of names* (Agriculture Handbook No. 714). Washington, DC: US Government Printing Office.

Okie, W. R., Beckman, T. G., Nyezepir, A. P., Reighard, G. L., Newall, W. C., & Zehr, E. I. (1994). BY520-9, A peach rootstock for the South-Eastern United States that increase scion longevity. *HortScience, 29*, 705–706.

Ozarslandan, A., & Tanriver, E. (2018). Evaluation of some stone fruit rootstocks against resistance to rook knot nematode (*Meloidogyne incognita*). *International Journal of Agricultural and Natural Sciences, 1*(2), 137–141.

Özbek, B., Kayım, M., & Elekçioğlu, İ. H. (2014). Evaluation of resistance for root-knot nematodes (*Meloidogyne incognita*) and *Meloidogyne javanica* of some *Prunus* rootstocks grown *in-vitro* conditions. *Journal of Agricultural Faculty of Uludag University, 28*(2), 27–35.

Paprstein, F., Kloutvor, J., & Sedlak, J. (2008). P-HL dwarfing rootstocks for sweet cherries. *Acta Horticulturae, 795*, 299–302.

Pina, A., Errea, P., Wunsch, A., & Gella, R. (2011). 'Monrepos' a plum rootstock for cherries. *HortScience, 46*(2), 322–323.

Pinochet, J. (2009). 'Greenpac', A new peach hybrid rootstock adapted to Mediterranean conditions. *HortScience, 44*(5), 1456–1457.

Pinochet, J. (2010). 'Replantpac' (Rootpac® R), a Plum x almond hybrid rootstock for replant situations. *HortScience, 45*(2), 299–301.

Pinochet, J., Calvet, C., Hernandez-Dorrego, A., Bonet, A., Felipe, A., & Moreno, M. (1999). Resistance of peach and plum rootstocks from Spain, France and Italy to root knot nematode *Meloidogyne javanica*. *HortScience, 34*, 1259–1262.

Popak, N. G. (1987). Productivity of almonds on skeletal soils in relation to rootstocks. *Sbornik Nauchnykh Trudov Gosudarstvennogo Nikitskogo Botanicheskogo sada, 102*, 79–87.
Ramming, D. W., & Tanner, O. (1983). 'Nemared' peach rootstock. *HortScience, 18*, 376.
Rana, G. S., Daulta, B. S., & Rana, K. S. (1997). Effect of rootstocks and drip irrigation on trunk diameter of peach (*Prunus persica* Batsch.) under high density plantation. *Annals of Agricultural Research, 18*(2), 231–233.
Reighard, G. I. (2000). Peach rootstocks for the United States: Are foreign rootstocks the answer. *HortTechnology, 10*, 714–718.
Renaud, R., & Salesses, G. (1990). Prunier/pecher: deux nouveaux port-greffes. *Fruits et Legumes, 73*, 22–23.
Renaud, R., Bernhard, R., Grasselly, C., & Dosba, F. (1988). Diploid plum x peach hybrid rootstocks for stone fruit trees. *HortScience, 23*(1), 115–117.
Ritenour, M. A., Dou, H., Bowman, K. D., Boman, B. J., & Stover, E. (2004). Effect of rootstock on stem end rind breakdown and decay of fresh citrus. *HortTechnology, 14*, 315–319.
Roberts, A. N., & Westwood, M. N. (1981). Rootstock studies with peach and *Prunus subcordata* Benth. *Fruit Varieties Journal, 35*, 12–20.
Rodger, S., Bengough, A. G., & Griffiths, B. S. (2003). Does the presence of detached root border cells of *Zea mays* alter the activity of the pathogenic nematode *M. incognita*. *Phytopathology, 93*, 1111–1114.
Rom, R. C. (1983). The peach rootstock situation: An international perspective. *Fruit Varieties Journal, 37*, 3–14.
Rowe, R. N., & Catlin, P. B. (1971). Differential sensitivity to waterlogging and cyanogenesis by peach, apricot, and plum roots. *Journal of American Society for Horticultural Science, 96*, 305–308.
Rozpara, E., Grzyb, Z. S., & Olszewski, T. (1990). The mineral nutrient content in the leaves of two sweet cherry cvs. with interstem. *Acta Horticulturae, 274*, 405–411.
Salesses, G., & Juste, C. (1970). Recherches sur l'asphyxie radiculaire des arbres fruitières à noyau. I—Rôle éventuel de certaines substances présentes dans les racines du pêcher. *Prunus persica Ann Amélior Plantes, 20*, 87–103.
Salesses, G., Renaud, R., & Bonnet, A. (1988). Creation of plum rootstocks for peach and plum by interspecific hybridization. *Acta Horticulturae, 224*, 339–344.
Scaramuzzi, F., Loreti, F., & Guerriero, R. (1976). La selezione di nuovi portinnesti seguti dall'Istituto di Coltivazioni Arboree di Pisa. In *Atti incontro frutticolo SOI'l portinnesti degli alberi da frutto' SOI-ICA*, Pisa, Italy (pp. 15–26).
Schwarz, D., Rouphael, Y., Colla, G., & Venema, J. H. (2010). Grafting as a tool to improve tolerance of vegetables to abiotic stresses: Thermal stress, water stress and organic pollutants. *Scientia Horticulture, 127*, 162–171.
Sharma, R. P. (1988). *Reciprocal influence of rootstock and scion on growth and mineral composition of stone fruits*. Ph.D. thesis submitted to Dr Y.S.P. University of Horticulture and Forestry, Nauni, Solan, India.
Sharma, M. K., Joolka, N. K., & Kumar, S. (2004). Growth, water relations and productivity of almond as influenced by scion, rootstock and soil moisture. *Agriculture Science Digest, 24*(2), 115–117.
Sharma, M. K., Joolka, N. K., & Singh, S. R. (2007). Growth, yield and leaf nutrient status of almond as affected by scion, rootstock and soil moisture. *Environment & Ecology, 25*(1), 62–64.
Sherman, W. B., Lyrene, P. M., & Sharpe, R. H. (1991). Flordaguard: peach rootstock. *HortScience, 26*, 427–428.
Sitarek, M., & Bartosiewicz, B. (2011). Influence of a few seedling rootstocks on the growth, yield and fruit quality of apricot trees. *Journal of Fruit and Ornamental Plant Research, 19*(2), 81–86.
Son, L., & Kuden, A. (2003). Effects of seedling and GF-31 rootstocks on yield and fruit quality of some table apricot cultivars grown in Mersin. *Turkish Journal of Agriculture, 27*, 261–267.
Southwick, S. M., & Weis, K. G. (1999). Propagation and rootstocks for apricot production. *Acta Horticulturae, 488*, 403–410.

Tabuenca, M. C. (1960). Incompatibilidad entre patrón e injerto. IV. Comportamiento del melocotonero con distintos patrones clonales del género. *Prunus An Aula Dei, 6*, 173–180.

Tabuenca, M. C. (1962). Relaciones entre la composición química y el grado de compatibilidad en combinaciones de melocotonero y ciruelo. *An Aula Dei, 7*, 1–34.

Tabuenca, M. C., & Moreno, M. A. (1988). Incompatibilidad entre patrón e injerto. Comportamiento de un ciruelo como patrón de distintas especies frutales. *An Aula Dei, 19*, 251–263.

Treutter, D., Feucht, W., & Liebster, G. (1993). *40 Jahre Wissenschaftfür den Obstbau in Weihenstephan* (p. 170). Müchen: Obst- und Gartenbauverlag.

Upadhayay, S. K., & Ananda, S. A. (1991). Effect of rootstock and scion cultivars on the macronutrient content in almond leaves. *Indian Journal of Horticulture, 48*(4), 309–311.

Vitanova, I. (1982). Effect of rootstock on the nitrogen and mineral element contents in the leaves of the plum cv. Stanley. *Gradinarska i Lozarska Nauka, 19*(1), 34–40.

Webster, A. D. (1980). Pixy a new dwarfing rootstock for plums, *Prunus domestica* L. *Journal of Horticultural Science, 55*(4), 425–431.

Webster, A. D. (1997). A review of fruit tree rootstock research and development. *Acta Horticulturae, 451*, 53–73.

Webster, A. D., & Schmidt, H. (1996). Rootstocks for sweet and sour cherries. In A. D. Webster & N. E. Looney (Eds.), *Cherries: Crop physiology, productions and uses* (pp. 127–163). Oxon: CAB International.

Werner, D. J., & Young, E. (1987). Effect of 'Siberian C' rootstock, interstem and scion on foliar calcium content in peach. *Fruit Variety Journal, 41*, 140–141.

Westwood, M. N., & Chaplin, M. H. (1973). Effect of rootstock on growth, bloom, yield, maturity and fruit quality of plum (*Prunus domestica*). *Horticultural Abstracts, 44*, 2208–2208.

Wolfram, B. (1996). Advantages and problems of some selected cherry rootstocks in Dresden-Pilnitz. *Acta Horticulturae, 410*, 233–238.

Wongtanet, D., & Boonprakob, U. (2010). Effect of rootstocks on growth of peaches in the highland of Northern Thailand. *Acta Horticulture, 872*, 327–332.

Ye, H., Wang, W., Liu, G., Xhu, L., & Jia, K. (2009). Resistance mechanisms of *Prunus* rootstocks to root-knot nematode, *Meloidogyne incognita*. *Fruits, 64*, 295–303.

Yuan, F. (2001). *Resistant mechanisms to the race 3 of Soybean Cyst Nematode (Heterodera glycines).* Thesis, Shen Yang Agricultural University, Sheng Yang, China, pp. 30–49.

Zarrouk, O., Gogorcena, Y., Moreno, M. A., & Pinochet, J. (2006). Graft compatibility between peach cultivars and *Prunus* rootstocks. *American Society for Horticultural Science, 41*, 1389–1394.

Irrigation Management in Stone Fruits

6

Amit Kumar, Pramod Verma, and M. K. Sharma

Abstract

Commercial fruit production in wherever region depends on suitable ecological conditions, using fruitful cultivars, and contends with pests and diseases, regular pruning and especially fertilization and irrigation. Under the changing climate scenario, fruit production is facing severe water shortages at regular intervals. Optimum yield and proper growth of fruit plant can be achieved by providing adequate soil water availability to the root zone. Water requirements of any fruit plant can be fulfilled by providing accurate irrigation to them. When, how often and how much of water is required can be determined by proper irrigation scheduling. Soil characteristics, condition of weather, dormancy stage and flowering time of plant, efficiency of irrigation system, uniformity in application of water and status of evapotranspiration are required in developing proper irrigation schedule for any fruit crop. Among different stone fruits, requirement of cherry is higher followed by peach and plum. Although apricot and almond are resistant to drought and have some xeromorphic nature, they can tolerate water stress and loss of leaves, in dry and winter seasons, respectively.

Keywords

Irrigation · Scheduling · Methods · Quality · Water efficiency

A. Kumar (✉) · M. K. Sharma
Division of Fruit Science, Faculty of Horticulture, SKUAST-Kashmir, Srinagar, Jammu and Kashmir, India

P. Verma
Division of Fruit Science, College of Horticulture, Dr YSPUHF, Solan, Himachal Pradesh, India

M. M. Mir et al. (eds.), *Production Technology of Stone Fruits*,
https://doi.org/10.1007/978-981-15-8920-1_6

6.1 Introduction

Irrigation refers to the artificial application of water for harvesting the potential yield from the crop. Irrigation is important as water among all the substances is essential for plant life, is required in largest quantities and is the most abundant molecular species present in activity of growing plant cells. Irrigation in fruit crops plays a vital role in its cultivation by supplying adequate moisture, which is prerequisite for enhancing the growth and crop production. The requirement of water is also not the same for fruit crops belonging to all species and varieties. That is, there are some fruit crops whose requirement of water is less than other species or varieties whose requirements are higher, and also there are also others, whose necessity of water is still higher. Like any plant, in a fruit plant also, the amount of water that is needed varies not only in accordance with its age but also on the stage of growth, both vegetative and reproductive, which implies that requirement of water varies from season to season in a year. But this is somewhat difficult to be obtained from natural resource of water as more than 80% of the annual precipitation falls from July month to September month in most part of the country. Stone fruits like peach, apricot, plum, cherry, etc., are somewhat early in maturity. There is very less or no rain during fruit growth and development period of peach fruits, due to which water stress conditions develop and affect fruit size. Besides this, water resources are also limited, and available water must be utilized efficiently through efficient methods of irrigation. Hence, artificial irrigation in an orchard so as to ensure adequate and timely supply of water to the fruit plants growing there cannot be disregarded, if it is desired to be run to derive maximum benefit.

6.2 Irrigation Scheduling in Stone Fruits

Keeping in mind the increasing pressure of demand for water for various activities, irrigation management in crops assumes special significance (Saleth 1996; Vaidyanathan 1999), and demand management becomes the overall key strategy for managing scarce water resources (Molden et al. 2001). To achieve optimum performance of a crop, proper timing of irrigation is the most important. Hence, the timing and amount of irrigation to crops is referred to as irrigation scheduling. Irrigation scheduling plays a vital role in optimizing crop production with a given amount of water and avoiding the adverse effects of either over-irrigation or under-irrigation on soil environment. These objectives can be achieved by adopting proper scheduling based on scientific principles of soil-water-plant-atmosphere relationships. Approaches to irrigation scheduling vary depending on situations, e.g. (a) where adequate irrigation water is available on demand to secure potential yield and (b) where available supplies fall short of the full irrigation water requirement of crops over the entire command area. The following three aspects of irrigation management, (a) when to irrigate, (b) how much to irrigate and (c) how best to irrigate, are to be kept in mind while scheduling the water supply. Irrigation scheduling in stone fruits depends on the amount of available water capacity (mm of water, cm of soil) and intake rate (mm/h) with respect to type of soil (Table 6.1) and irrigation depth of fruit crops (Table 6.2).

Reduction in the fruit number and fruit size and increase in fruit drop were observed in stone fruits during the growing season under water stress conditions. Uniform distribution of rain round the year is required for successful cultivation of stone fruits, and during the dry spell period especially during the critical stages of the plant growth, supplementary irrigation is a must. Poor fruit setting, heavy fruit drop and poor-quality fruit with low production resulted under the water stress conditions. During flowering, after fruit set and even up to August during fruit maturity (if dry spell is more) in different stone fruits, water requirement is more as this is the most critical period. However, after fruit harvesting, when the manuring is done, i.e. from the month of November to December, orchards are to be irrigated. In summer season, if rains are not available, the fruit plants are irrigated at 7–10 days intervals, whereas after fruit set stage, the plants should be irrigated at weekly intervals. During maturity time of the fruits or before harvesting (fortnight), application of water markedly improves the fruit colour. Before the onset of dormancy, irrigation should be given at an interval of 3–4 weeks.

6.2.1 Peach

The young as well as mature peach trees should be irrigated regularly, as peach crops respond very well to irrigation and its requirement varies with the region, soil and even the type of the crop. Irrigation is mostly required during the dry and hot summer during which the depletion of the soil moisture occurs and rate of evapotranspiration occurs. This season usually coincides with the rapid vegetative and fruit growth (third fruit growth phase when fruit cells expand dramatically) in peaches culminating into a high demand of water. With the onset of spring, the trees put new growth flush, which in turn led to need of much greater amounts of water than in dormancy. However, irrigation may be given at about 8–10 days interval. For proper development of fruit size and quality, frequent irrigations should be given

Table 6.1 Ranges in available water capacity and intake rate for various soil textures

	Available water capacity (mm of water/cm of soil)		Intake rate (mm/h)	
Soil textures	Range	Average	Range	Average
Sands	0.5–0.8	0.65	12–20	16.0
Loamy sand	0.7–1.0	0.85	7–12	9.5
Sandy loam	0.9–1.2	1.05	7–12	9.5
Loam	1.3–1.7	1.50	7–12	9.5
Silty loam	1.4–1.7	1.55	4–7	5.5
Silty clay loam	1.5–2.0	1.75	4–7	5.5
Clay loam	1.5–1.8	1.65	4–7	5.5
Clay	1.5–1.7	1.60	2–5	3.5

Source: Tan and Layne (1990)

Table 6.2 Irrigation depth for various stone fruits

Fruit crops	Depth to irrigate (cm)
Peach	60
Cherries	60
Apricot	50

Source: Tan and Layne (1990)

during April with longer intervals (Daniell 1984). Less water is required after fruit harvest. Irrigation should be stopped during the winter when trees shed leaves and enter into dormancy. Excessive irrigation should be avoided as it results into rise of water table and thus causes injury to the peach roots and usually causes early yellowing and falling of peach leaves. Irrigation water applied to peach trees should be free from excess of alkali salts (Bal 1997).

More than 50% area of the plant canopy was covered when irrigation was applied through micro-sprinkler system; however, this efficiency of irrigation can be enhanced up to 80% with well-maintained and managed micro-sprinkler irrigation system (Haman et al. 2005).

6.2.2 Apricot

Apricot can tolerate dry atmosphere but also requires irrigation especially during critical periods of fruit growth and development. The most critical period for irrigation is being at pit hardening stage and thereafter till harvest. Lack of water at this stage interferes with fruit growth resulting in reduced fruit size. Hence, irrigation in apricot should be done at 8–10 days interval after fruit setting till harvesting which is usually from April to first week of June. Fruit quality was adversely affected with the excess application of water, which promotes nutritional imbalance and decreased fruit dry mass due to increase in vegetative growth (Herro and Guardia 1992). Under drip irrigation system, 7000 m^3/ha of water is needed in order to satisfy crop evapotranspiration (ETc) (Abrisqueta et al. 2001). However, in order to inspire saving of water without affecting the fruit yield and quality, regulated deficit irrigation (RDI) strategies should be adopted. RDI enforces water deficits during phonological stages when trees are less sensitive to water stress (noncritical periods). Perez-Pastor et al. (2007) reported RDI at 100% of ETc during the critical periods (second rapid fruit growth and early postharvest) and at 25% during the rest of the season enhances the fruit keeping quality and saves considerable amount of water.

6.2.3 Plum

Plum cultivation is mostly confined to rainfed conditions in India. Therefore, irrigation becomes a crucial method of meeting out the water requirement during critical period of fruit development, the most critical period being the April to June. Irrigation to young plum trees in early periods of their establishment is highly essential to overcome water stress. If available moisture becomes a limiting factor before the fruits are fully grown, then the percentage of bigger size fruits which are economical decreases drastically. The irrigation can be applied by furrow, portable pipe or basin flooding; drip irrigation is widely used because it saves and efficiently uses the water than flood or furrow irrigation. The irrigation requirement of plum depends on the soil type, topography and canopy arrangement of the trees (Chootummatat et al. 1990).

6.2.4 Cherry

Though cherries like and grow well in hot dry climate with free draining soil, most of the orchards need irrigation. The water requirement of cherries is more during spring and early summer. The irrigation should be applied when required; overwatering in cherries is not desirable as the plant is very sensitive to wet feet. The provision of drainage of excess rain water should be made. The average yield in sweet cherry increased when it irrigated before and during fruit ripening. Trees can be irrigated using under-tree sprinklers or drippers especially in the month before harvest as water deficit during this period may cause fruit splitting. Using drip irrigation can increase marketable fruit by 8–9% on average and reduce the irrigation water by 54.3% as compared to micro-sprinkler (Long et al. 2014). The incidence of fruit cracking may be higher in the tree with a dry spell followed by irrigation just before harvest (Sekse 1995). Evapotranspiration rates on average through the summer growing months may be in the order of 5–6 mm/day. Water of less than 500 mg/L of total dissolved salts is desirable. Water above 750 mg/L should be avoided or mixed with good-quality water before use (James 2011).

6.2.5 Almond

For proper growth and development, uniform and constant soil moisture is required. In the initial years of planting, when the plants are young and roots are developing, application of water is essential, and reduction in yield and fruit quality is observed under stress conditions during early growth of the plants which also leads to the susceptibility of insects and diseases. Immature defoliation and plant water stress was observed under irrigation deficiency (if happens for more than a month), especially in those orchards which are growing under high summer evaporative demand. Flowering, fruit set and yield were reduced under water deficiency especially during previous harvesting periods, as flower initiation for the subsequent year occurs during late August to early September (Goldhamer and Shackel 1990).

Before the commencement of stress conditions in plants, moisture should be replaced, where 12–19 L of water per week was required for young trees. Shallow root system of plants was produced with light and frequent watering. Irrigation in sandy soils may be given every 3–5 days, whereas in heavier soils watering will be given at the intervals of every 1.5–2 weeks.

Avoid irrigation nearby the trunk and lower branches as it promotes root and crown disease. Irrigation through drip system must be provided in the outer dimensions of the plant. Frequent and deep irrigation should be provided for 12–24 h at a depth of 1–2 m; however, this quantity must be reduced in late summer months and winter months.

Flowering in the month of February to March and fruit development from April to June are the critical stages for almond as this time mostly water shortages were noticed. For getting higher yield and quality nuts, irrigation must be given. Under Karewas land in Kashmir Valley, drip irrigation has been found very efficient method. Four drippers with a discharge of 4 L/h for 5 h at an interval of 4–5 days require about 1600 L per tree per season (Ahmed and Verma 2009).

6.3 Methods of Irrigation

Depending on the availability of water, topography, fruit species, age of the trees, intercrops, soil type and climate, different methods can be utilized to reach water to plant. However, the method adopted should be such that maximum water is available to active root zone to be effectively utilized by the plants. At farm level, irrigation may take place in two ways, i.e. by gravity (surface irrigation) and under pressure. There is another classification for applying irrigation method in stone fruits including surface irrigation, sub-surface irrigation, sprinkler irrigation and drip or trickle irrigation, and under each method, there are various systems such as under surface water could be given by flooding, ring basin, check basin and border strip and furrow methods of irrigation. The surface method of irrigation being conventional and old method of irrigation has various limitations that make them not suitable for fruit cultivation especially in stone fruits where water requirement during critical stages is more. The limitation includes lesser water saving due to lots of water losses which occurs during percolation, runoff and evaporation, accounts for lower water use efficiency (30–50%). Heavy water losses occurs during leaching, runoff, high weed infestation, higher soil erosion due to large stream sizes used for irrigation, accumulation of salts and ultimately adversely affecting the plant growth (Sivanappan 1994; Narayanamoorthy 1997).

In modern era, new method of irrigation, i.e. drip irrigation, is the most efficient method of irrigating fruit orchards, in which regular application of water is given directly on or below the soil surface near the root zone of plants. Application of water is given drop by drop at a very slow rate with the help of emitters or drippers at a specific point on or just below the soil surface. Irrigation through drip method includes plastic lateral pipes which are laid under the ground or along the ground or sometimes 10–15 cm above the ground level tied with the iron wires. Drip irrigation aims at increasing water use efficiency up to 70–90% as compared to conventional and sprinkler method of irrigation (Sivanappan 1998). Application of drip irrigation alone, water-soluble fertilizers through drip (fertigation) and use of mulch significantly had influenced growth, vigour, fruit set, fruit quality, yield and water use efficiency in stone fruit crops (Singh et al. 2002; Banyal et al. 2015).

6.4 Impact of Irrigation on Stone Fruits

6.4.1 Growth

Irrigation through modern method like drip irrigation has been found significantly in enhancing the growth and vigour of the plants (Sivanappan 1998). Daniell (1982) studied the effect of trickle irrigation on the growth and yield of 'Loring' peach trees. The supplemental water treatments of 0, 3.8, 7.6 and 11.4 L/h per tree were applied for five seasons to peach trees, beginning at planting in 1975 (the trees began to bear in 1977). On the basis of 70% pan evaporation method, the amount of water was given. The tree mantle volume was restricted by pruning only after the trees had filled their allotted space in 1978. The significant increase in trunk diameter was recorded with irrigation treatments up to full canopy of Loring peach trees, not afterwards. Rana and Daulta (1997) studied the combined effect of drip irrigation and rootstocks on peach trunk diameter under high-density plantation. They reported that the trunk diameter increased with increasing irrigation rate (1.5 L/day). Bryla et al. (2003) compared the effect of furrow, microjet, surface and sub-surface drip irrigation on vegetative growth of 'Crimson Lady' peach. Their study revealed that trees irrigated by surface and sub-surface drip method of irrigation attained significantly more height than those irrigated by other methods. For achieving equal vegetative growth, they suggested that through microjets more than twice amount of water is to be applied.

Malik et al. (1987) evaluated different irrigation methods on apricot and plum plants, viz. flood irrigation (water applied to maintain soil moisture at field capacity), water applied to 25% depletion of field capacity and water applied equivalent to evapotranspiration through drip method of irrigation, and reported that maximum trunk girth and shoot length was observed when water was applied through drip irrigation at 25% depletion of field capacity. Drip irrigation significantly increased the tree growth, leaf area and tree spread of plum (Treder et al. 1999). Reduced plant growth and complete leaf fall were reported as the irrigation requirement of almond after nut harvesting was 206 mm and the predawn leaf water potential dropped quickly after stopping water application which reached to −4.0 MPa before harvest (Goldhamer and Viveros 2000).

Under sub-surface drip irrigation conditions, the effect of regulated deficit irrigation on vegetative development of mature almond trees was determined (Romero et al. 2004). After 4 years of study, it was concluded that the vegetative (trunk and branch growth, canopy volume and pruning weight) development in trees irrigated at RDI at 100% ETc was significantly reduced which may be due to excess collective effect of water stress on growth process except at 20% ETc, i.e. kernel filling stage, and at 50% ETc, i.e. postharvest.

Kaya et al. (2010) reported mean higher values for trunk cross-sectional area (107.29 cm^2/tree) and crown volume (40.44 m^3/tree) when irrigation was applied to 'Salak' apricot trees for five growing seasons which was based on adjustment of coefficients of class A pan evaporation (1.50) and (1.25), respectively.

Verma and Chandel (2017) recorded maximum increase in tree height (14.68%), tree spread (12.41%), tree volume (21.83%), trunk girth (11.70%) and maximum annual shoot growth (98.30 cm) in peach trees irrigated with drip irrigation at 100% ETc treatment as compared to control (unirrigated trees). Similarly, Singh (2013) also observed increase in annual shoot growth, plant height, plant spread and volume of canopy in apricot growth parameters (annual shoot growth, tree height, spread and canopy volume) up to 32.0% high as compared to surface irrigation and 35.0% high under rainfed condition.

6.4.2 Fruit Set and Yield

Cropping parameters like fruit set and yield are likely to be influenced by the application of drip irrigation and have been found to be benefiting to a greater extent in comparison to conventional method of irrigations. Hutmacher et al. (1994) reported that in developing almond trees with the trickle irrigation, the nut retention on the basis of flower and fertile nut counts after flowering was enhanced by 10% with the increase in water application above 50% ETc in the semiarid San Joaquin Valley. Mean nut yield and mean kernel weights enhanced significantly with increasing water application from 50% to 150% ETc.

Lishchuk et al. (1988) carried out an experiment on peach to study the response of drip irrigation at 60–80% soil moisture capacity and reported that drip irrigation at 80% soil moisture capacity increased fruiting in peaches. Bloom density (52.2%) and fruit set (94.3%) were reduced in the following season when postharvest water deficiency occurred which also reduced fruit load up to 76.7% and kernel yield 73.6%. Postharvest irrigation deficiency affects more with respect to tree productivity than the preharvest irrigation deficiencies; however, the amount of water under postharvest deficiency was less than that which occurred with the more severe before harvest deficiencies (Goldhamer and Viveros 2000).

Furrow and microspray (weekly or biweekly), surface drip and sub-surface drip irrigation (daily) methods were used for 3 years to determine their effect on production and quality of fruits of mature 'Crimson Lady' peach trees (Bryla et al. 2005), and it was found that surface and sub-surface drip produced the bigger fruits (154.0 and 156.33 g) on average and the highest marketable yield (23.46 and 22.70 t/ha) as compared to other methods. Maximum moisture in soil nearby root zone was maintained with daily irrigations with drip methods whether it was applied above or below the ground level which also prevented cycles of water stress which was observed with furrow and microspray irrigation methods.

Ten-year-old Bulida cv. of apricot plants were subjected to three different drip irrigation treatments as (1) 100% irrigation of seasonal crop evapotranspiration (ETc), (2) irrigated at 50% with a continuous deficit irrigation and (3) irrigated at

100% of ETc as regulated deficit irrigation (RDI) during the critical periods of plant growth (stage III of fruit growth, 2 months after harvest and 25% and 40% of ETc during the noncritical periods in the 1–2 and 3–4 growing seasons, respectively) (Perez-Pastor et al. 2009). In the first two growing seasons, fruit yield and number of fruits per tree was decreased with longer and severe deficits of the regulated deficit irrigation treatment; however, in the third and fourth year of the growing season, fruit yield and number of fruit per tree is more or at par with that of control.

Six irrigation treatments based on adjustment coefficients of class A pan evaporation (S_1, 0.50; S_2, 0.75; S_3, 1.00; S_4, 1.25; and S_5, 1.50) were subjected to 'Salak' apricot trees in semiarid climatic conditions for five growing seasons to investigate their effects on vegetative growth, yield and quality (Kaya et al. 2010). However, the under sixth treatment (RDI), plants were irrigated by applying 100% of class A pan evaporation up to harvesting time only. Maximum yield (52.2 kg/tree), pulp hardness (2.25 kg) and TSS (13.1%) and minimum acidity (0.46%) were registered under the treatment S_5 (1.50).

Wang (2011) conducted multiyear experiment on peach trees to estimate deficit irrigation strategies for reduction in postharvest application of water and reported that with approximately 30–40% of water use during the season with deficit irrigation with furrows produced peach yield equal to full irrigation. Under severe deficit treatment in sub-surface drip irrigation method, small fruits were obtained.

Amount of irrigation water (four different pan coefficients as Kp 1:25, Kp 1:00, Kp 0:75 and Kp 0:50), irrigation interval (3 and 6 days) and water consumption were determined for getting maximum yield and quality fruits for 2 years in peaches cv. Redhaven which was irrigated by drip irrigation method in Aegean region of Turkey (Gunduz et al. 2011). Yield was significantly affected with the amount of irrigation, and maximum average yield (14.101 kg/ha) was obtained from Kp 1:00; however, average irrigation water amount was 482 mm, with a water consumption of 705 mm and a Kpc value of 0.785. Fruit weight varied from 203 to 253 g, length 6.3–6.6 cm, diameter 7.2–7.7 cm, soluble dry matter 10.8–14.5% and juice pH 4.14–4.37 under the studied irrigation treatment.

Alaoui et al. (2013) obtained higher yield (113.8 fruits/harvest) and quality fruits from peach plants with less irrigation water and adequate frequency (RDI 75% with 0.4 mm irrigation dose). However, adopting RDI at 50% with 0.4 mm irrigation dose in the early stages of fruit growth accelerate the fruit's maturity as compare to RDI at 75 % with 0.4 mm during fruit development.

Under Mediterranean climatic conditions, Vera et al. (2013) conducted an experiment for 5 consecutive years on the status of plant water and yield of early maturing peach trees (cv. Flordastar) with different deficit irrigation strategies. The treatments of deficit irrigation (DI) were continuous, regulated deficit irrigation (RDI), partial root zone drying and a soil water content base. Significant water saving was observed with deficit irrigation; however, reduction in yield was observed, clearing that deficit irrigation affects that number of fruits per tree more than that of fruit size.

Mechlia et al. (2012) studied the effect of deficit irrigation on yield of a 4-year-old late cultivar 'Carnival' of peach planted at 3 m × 6 m spacing for 5 consecutive years, and the experiment was laid out with four irrigation regimes. During phases I and II, i.e. early fruit growth, and phase III, i.e. late growth, and during all stages, water was reduced as compared to control, and 33% less irrigation water was restricted. For evaluation of applied restrictions, Penman-Monteith reference evapotranspiration (ETo) was used, and over the different growing stages, total water supply from precipitation (P) and irrigation (I) was watched. During the period of fruit set up to the end of phase III, the ratio between (I + P) and ETo was increased from 0.65 to 1.05 as compared to control. Depending on the year and treatment, 0–34% yield was reduced. Maximum water reduction was observed during the late growth stage, i.e. phase III. However, maximum yield was obtained during the continuous restriction of the three treatments.

With three irrigation intervals (7, 14 and 21 days) and five irrigation levels (0.0, 0.25, 0.50, 0.75 and 1.00), Bozkurt et al. (2015) conducted an experiment for 3 consecutive years on young apricot trees cv. Ninfa based on reduction coefficients (Kcp) of class A pan evaporation. Maximum average yield was obtained from T_{14} irrigation intervals (729.9 g/tree) and at I_{50} irrigation level (1031.3) with an increasing fruit size and number of fruits per tree. Under Mediterranean climate, young apricot trees were treated with 25% water deficit at a 14-day irrigation interval.

Khan et al. (2015) applied irrigation to almond cv. Shalimar with four different irrigation levels, viz. 0% ETc, 100% ETc, 75% ETc and 50% ETc, at four different crop phenological stages of growth and development, viz. fruit growth stage, kernel filling stage, preharvest stage and throughout the growth stage, and reported that highest nut yield (2.69 kg/tree) was recorded at 100% ETc during the first year of experimentation, which rose to 2.91 kg/tree during the subsequent year.

Durgac et al. (2017) carried out an experiment on young apricot (*Prunus armeniaca* L. cv Ninfa) plants with different irrigation intervals, i.e. 7 days as T_7, 14 days as T_{14} and 21 days as T_{21}, and water amount was calculated as reduction coefficients (Kcp) of class A pan evaporation (0.0 (I_0), 0.25 (I_{25}), 0.50 (I_{50}), 0.75 (I_{75}) and 1.00 (I_{100})), and it was observed that the interaction between irrigation intervals and the irrigation water amount treatments had significant effects on cumulative yield of the trees. The maximum cumulative yield/ha (859.5 kg) was obtained from I_{50} treatment; however, interaction between irrigation intervals and irrigation water amount depicts that $T_{14} \times I_{75}$ interaction had the highest cumulative yield (987.4 kg/ha). Verma and Chandel (2017) also recorded highest yield (16.21 kg/tree and 20.26 ton/ha) and proportion of three layer grade fruits (77.16%) in peach trees drip irrigated at 100% ETc.

6.4.3 Fruit Quality Parameters

Under different irrigation regimes (full irrigation at 100% ETc, deficit irrigation at 80% ETc, deficit irrigation at 40% ETc and drought period at 0% ETc) and different stages of fruit growth (stage I as fruit growth, stage II as kernel growth and stage III as preharvest period), Mousavi and Alimohamadi (2006) conducted an experiment on 'Mamaei' almond trees and observed a reduction in fruit size and fresh and dry weight of fruit during stage I under deficit irrigation and drought condition, whereas reduction in fruit fresh weight and fresh and kernel dry weight was observed during stage II. However, under irrigation treatments during stage III, i.e. preharvest period, no significant differences were observed for the measured parameters.

Ali et al. (2012) carried out an experiment on 'Marnaei' almond to determine the effects of deficit irrigation (T_1 at 100% ETc, full irrigation; T_2 at 80% ETc, deficit irrigation; T_3 at 40% ETc, deficit irrigation; and T_4 at 0% ETc, without irrigation) during phonological stages (stage I, fruit growth; stage II, kernel growth; and stage III, preharvest period) on fruit growth and development. Decrease in fruit size (length, width and diameter of fruit) and fresh and dry weight of fruit was observed during stage I with deficit irrigation and water stress; however, fruit drop percentage was increased. During stage II under deficit irrigation and water stress, fresh fruit weight and fresh and dry kernel weight were decreased. Yield, kernel dry weight and kernel percentage were decreased during all phonological stages of fruit growth under deficit irrigation.

Total dry matter and sugar content was improved with the increasing level of irrigation in late peach cv. 'Carnival' (Mechlia et al. 2012). Alcobendas et al. (2013) studied the effects of irrigation and fruit position on size, colour, firmness and sugar contents of fruits in a mid-late maturing peach cultivar. Trees were subjected to full irrigation (FI) and regulated deficit irrigation (RDI). Fruits from trees under RDI were firmer than those from FI trees but did not differ in weight and diameter. In contrast, fruits from RDI trees had more soluble solids, glucose, sorbitol and malic, citric and tartaric acids. The response of peach, plum and almond to water restrictions applied during slowdown periods of fruit growth was studied by Razouk et al. (2013). Water applied through drip irrigation at 50% ETc, 75% ETc and 100% ETc exerted a significant effect on fruit quality parameters in all three species. In case of peach and plum, higher fruit size and total soluble solids and lower acidity were recorded in 100% ETc, as compared to water application treatments of 50% ETc. For almond, kernel quality remained unaffected by water restriction at 75% ETc. However, the epidermal wrinkles of kernels were more embossed, in response to treatment irrigation at 50% ETc, which affected their appearance. Hussien et al. (2013) studied the effect of drip irrigation on some plum cultivars grown in clay loamy soil and concluded that fruit quality (weight, size, firmness, TSS and acidity) was significantly increased with drip irrigation method (2607.4 and 2553.6 m^3/fed./season), as compared to flood irrigation method (5180 and 5124 m^3/fed./season).

Durgac et al. (2017) applied irrigation at three different intervals, i.e. 7, 14 and 21 days, on young apricot (*Prunus armeniaca* L. cv Ninfa) plants for 2 consecutive years and reported that maximum fruit weight (24.4 g), fruit length (35.5 mm), soluble solids (12.0°B), fruit pH (3.61) and minimum acidity (0.89%) were obtained when irrigation was applied at 21-day interval; however, maximum fruit index (0.97), flesh firmness (3.88 kg/cm^2) and flesh/stone ratio (8.27) were registered under 14 days of irrigation intervals. Irrigation water amount significantly improved fruit weight, and maximum fruit weight (26.3 g) was obtained from I_{75}; however, highest flesh/seed ratio, TSS and pH values were obtained from I_{100} (8.55), I_{25} (12.4°brix) and I_{50} (3.80), respectively.

Verma and Chandel (2017) found maximum fruit length (6.12 cm) and diameter (5.56 cm), fruit weight (104.78), total soluble solids (12.20%) and titratable acidity (0.61%) in peach trees irrigated with drip irrigation at 100% ETc treatment as compared to other treatments.

Yield and water status of peach was evaluated using different irrigation systems for 2 consecutive years (2012–2013 and 2013–2014) influenced by irrigation scheduling (Zhang et al. 2017). Results showed that midday stem water potentials (ψ) for well-irrigated trees were maintained at a range of −0.5 to −1.2 MPa, while ψ of deficit-irrigated trees dropped to lower values. Fruit weight and number of fruit/tree were recorded maximum under well-watered trees; however, no statistically significant reduction in fruit size or quality was observed for trees irrigated with surface drip and microspray irrigation systems by deficit irrigation during the year of study.

6.4.4 Leaf and Fruit Nutrient and Leaf Chlorophyll Content

Studies on root system and nutritional status of peaches under drip and flood irrigation were conducted by Romo and Diaz (1985). They reported similar nutritional status of trees under both irrigation systems throughout the season. Seasonal differences for Mn were stable and low in drip irrigation but increased under flood irrigation. Significantly higher nitrogen and phosphorous content was obtained in the tissues of peach trees which was maintained at 70% of field capacity soil moisture as compared to 60% or 50% of field capacity (Dochev 1968). However, when drip irrigation was applied at 20 cm deep or channel irrigation at 60 cm deep, reduction in nitrogen, phosphorous and potassium content of leaf was obtained (Miculka 1983), whereas maximum magnesium content was obtained in peach plants when they were drip irrigated with low fertigation than other treatments. Rana et al. (2005) studied the combined effect of drip irrigation and rootstocks on N, P and K leaf content in peach cv. Flordasun under high-density plantation and reported that wider spacing and higher irrigation rate increased nitrogen, phosphorus and potassium content of leaves. The maximum nitrogen content was recorded in treatment combinations of 2.0 × 1.25 m and 3 L/day (2.55%) on peach rootstocks and minimum in 2.0 × 0.5 m and 1 L/day (1.58%) on plum rootstock. The maximum phosphorous content (0.253%) was recorded in the treatment combination of 2.0 × 1.25 m and 3 L/day in peach on peach rootstock. Treatment combination of 2.0 × 1.25 m and 3 L/day showed maximum K content (2.32%) in peach on peach rootstock. Similarly, the maximum leaf nitrogen (2.63%), phosphorous (0.15%), potassium (2.10%), calcium (2.24%), magnesium (0.38%), iron (204.94 ppm), manganese (52.03), zinc (19.94 ppm) and copper (9.69 ppm) content was observed in drip irrigation at 100% ETc in peach cv. Redhaven by Verma and Chandel (2017).

Decrease in leaf nitrogen and boron in fruit of Marnaei almond was observed for 3 consecutive years during stage II (i.e. kernel growth) under deficit irrigation and water stress (Ali et al. 2012). Khan et al. (2015) reported that almond cv. Shalimar leaves receiving the irrigation at 100% ETC level recorded maximum concentration of nitrogen (2.15%), phosphorous (0.19%) and potassium (1.59%).

Demirtas and Kirnak (2009) applied irrigation to a 4-year-old 'Hacihaliloglu' dried apricot cultivar using surface and mini-sprinkler irrigation methods with 15 days, 20 days and 25 days of irrigation intervals and reported that high leaf chlorophyll content was obtained in mini-sprinkler irrigation system, and with the increase in irrigation intervals, reduction in leaf chlorophyll content was observed. Maximum leaf chlorophyll-a content was obtained in surface (3.20 mg/mL) and mini-sprinkler (3.19 mg/mL) method of irrigation applied at 15 days of irrigation intervals. With 15 days of irrigation intervals and mini-sprinkler method, maximum leaf chlorophyll-b (1.81 mg/mL), total chlorophyll content (5.00 mg/mL) and leaf carotene (0.213 mg/mL) were observed.

6.5 Water Use Efficiency

Water use efficiency appeared to result both from reduced water losses in drip irrigation system and more efficient use by the plants. As compared to basin irrigation system, 68% water was saved under drip irrigation system (Sivanappan 1998). Vegetative and yield characteristics were examined of mature almond trees with regulated deficit irrigation under sub-surface drip irrigation conditions, and it was reported that SDI (sub-surface drip irrigation) treatments at 100% ETc significantly increased water use efficiency, except in the kernel filling stage at 20% ETc and postharvest at 75% ETc and 50% ETc.

Intrigliolo and Castel (2010) determined the combined effect of crop load and regulated deficit irrigation (RDI) on Japanese plum cv. 'Black Gold' grafted on 'Mariana GF81' for 3 consecutive years by applying regulated deficit irrigation during the second phase of fruit growth. Regulated deficit irrigation saved water up to 30% by increasing water use efficiency, and minimum effect on crop yield and fruit growth was recorded, thereby providing that plant water stress during the fruit growth period was low.

As compared to full water irrigation system, under deficit irrigation, consistently higher values were obtained for crop water stress index in peach tree experiment (Wang 2011).

Pan coefficients at Kp 1:00 (2:02 kg/m^3) registered higher water use efficiency among four different pan coefficients (Kp 1:25, Kp 1:00, Kp 0:75 and Kp 0:50) with two irrigation levels (3 and 6 days) and water consumption on Redhaven peaches when irrigated by drip irrigation in Aegean region of Turkey (Gunduz et al. 2011). Singh (2013) reported maximum water use efficiency, i.e. 0.41 t/ha, in apricot plantation under drip irrigation as compared to surface irrigation which registered 0.19 t/ha.

Bozkurt et al. (2015) observed maximum water use efficiency for T_7I_{50} (1.54 kg/ha/mm) closely followed by T_7I_{25} (1.71 kg/ha/mm) during a study conducted for 2 consecutive years, respectively, on young apricot trees; however, maximum irrigation water use efficiency was recorded in T_7I_{50} (2.30 kg/ha/mm) and T_7I_{25} (3.41 kg/

ha/mm) during both years, respectively. Maximum cumulative yield was produced by I_{25} irrigation level (860 g/tree) as compared to non-irrigated I_0 plot (126.9 g/tree) depicting the importance of irrigation level for young apricot trees. However, water use efficiency and irrigation water use efficiency were reduced with the increasing irrigation levels.

Maximum water use efficiency (0.60 kg/m^{-3}) was observed at 0% ETc level of irrigation followed by 50% ETc level (0.46 kg/m^{-3}) which was significantly and statistically at par with 75% ETc (0.40 kg/m^{-3}) (Khan et al. 2015).

Verma and Chandel (2017) recorded the highest water use efficiency in peach trees irrigated with drip irrigation at 80% ETc during the year 2015 and 2016, as compared to higher level of drip irrigation levels.

6.6 Fertilizer Application Through Drip Irrigation

Application of fertilizers through drip irrigation, known as fertigation, provides an additional input with water supply to enhance production and reduce the environmental pollution by improving fertilizer use efficiency, reducing fertilizer application and increasing profit on the fertilizer invested (Hagin et al. 2002). The efficiency of fertilizer enhances from 10% to 35% in fruit crops (Neilsen et al. 2001; Mussaddak 2007). The fertigation scheduling in stone fruits like peach includes application of water-soluble NPK fertilizers (20:10:10 and 19:19:19), along with 0:0:50 to supplement the left out K_2O content, and urea (46:0:0) to supplement the left out N content with respect to age of plant that should be applied in eight split doses at different eight fruit growth stages, i.e. half inch green leaf, pink bud, bloom, fruit set and 10, 20, 30 and 40 days after fruit set (Verma 2017).

6.7 Water Consumption and Other Parameters

Massai and Remorini (2000) carried out an experiment on eight 'Suncrest' peach plants grafted on GF-677 and Rubira (four plants per rootstock) for 3 consecutive years for estimation of water requirements during the fruit-bearing period. Among eight trees, four were irrigated with 25 L of water per day from June to September, and no irrigation was provided to the rest of four plants. Total leaf area and water availability in the soil affect the water consumption of the plants which was observed to be higher in the first (2000 g/m/day) year of the study as compared to second (1400 g/m/day) and third (1600 g/m/day) year of study.

Apricot plants of Bulida cultivar under drip irrigation method were subjected to water stress by withholding irrigation at five phonological periods, i.e. (a) flowering and fruit set period (T_1), (b) fruit growth stages (I and II) (T_2), (c) fruit growth stage III (T_3), (d) immediately after harvest for 1½ months (T_4) and (e) for 2 months during late postharvest (T_5) (Torrecillasa et al. 2000). Different treatments were compared with control where plants were irrigated throughout the year with an amount of water equivalent to 100% of the crop evapotranspiration (ETc). Comparing with the control volumetric soil water content, leaf water potential and leaf conductance showed maximum reduction under the T_4 and T_5 treatment.

Kaya et al. (2010) observed mean higher values for irrigation water (903 mm) and evapotranspiration (1132 mm) when irrigation was applied to 'Salak' apricot trees for five growing seasons based on adjustment coefficients of class A pan evaporation (1.50).

References

Abrisqueta, J. M., Ruiz, A., & Franco, J. A. (2001). Water balance of apricot trees (*Prunus armeniaca* L. cv. Bulida) under drip irrigation. *Agricultural Water Management, 50*, 211–227.

Ahmed, N., & Verma, M. K. (2009). *Scientific almond cultivation for higher returns* (pp. 8–10). New Delhi: M/s Royal Offset Printers.

Alaoui, S. M., Abouatallah, A., Salghi, R., Amahmid, Z., Bettouche, J., Zarrouk, A., & Hammouti, B. (2013). Impact assessment of deficit irrigation on yield and fruit quality in peach orchard. *Der Pharma Chemica, 5*(3), 236–243.

Alcobendas, R., Miras-Avalos, J. M., Alarcon, J. J., & Nicolas, E. (2013). Effects of irrigation and fruit position on size, colour, firmness and sugar contents of fruits in a mid-late maturing peach cultivar. *Scientia Horticulturae, 164*, 340–347.

Ali, M. R., Mousavi, A., Tatari, M., & Fattahi, A. (2012). Effects of deficit irrigation during different phonological stages of fruit growth and development on mineral elements and almond yield. *Iranian Journal of Water Research in Agriculture, 26*(2), 143–159.

Bal, J. S. (1997). Peach. In *Fruit growing* (pp. 363–374). New Delhi: Kalyani Publishers.

Banyal, S. K., Sharma, D., & Jarial, K. (2015). Effect of nitrogen fertigation on yield and fruit quality of low chilling peaches under sub-tropical conditions of Himachal Pradesh. *Indian Journal of Horticulture, 72*(4), 457–460.

Bozkurt, S., Odemis, B., & Durgac, C. (2015). Effects of deficit irrigation treatments on yield and plant growth of young apricot trees. *New Zealand Journal of Crop and Horticultural Science, 43*(2), 73–84.

Bryla, D. R., Trout, T. J., & Ayars, J. E. (2003). Growth and production of young Peach trees irrigated by furrow, microjet, surface drip or subsurface drip systems. *HortScience, 38*(6), 1112–1116.

Bryla, D. R., Dickson, E., Shenk, R., Johnson, R. S., Crisosto, C. H., & Trout, T. J. (2005). Influence of Irrigation method and scheduling on patterns of soil and tree water status and its relation to yield and fruit quality in peach. *HortScience, 40*(7), 2118–2124.

Chootummatat, V. D., Turner, W., & Cripps, J. E. (1990). Water use of plum trees (*Prunus salicina*) trained to four canopy arrangement. *Scientia Horticulturae, 43*, 255.

Daniell, J. W. (1982). Effect of trickle irrigation on growth and yield of Loring peach trees. *Journal of Horticultural Science, 57*, 393–399.

Daniell, J. W. (1984). Effect of glyphosate for weed control in eleven cultivars of peach trees. In *Proceedings of 37th Annual Meeting South Weed Science Society* (Abstract No. 126).

Demirtas, M. N., & Kirnak, H. (2009). Effects of different irrigation systems and intervals on physiological parameters in apricot. *Journal of Agricultural Sciences and Biotechnology, 19*(2), 79–83.

Dochev, D. (1968). A study on the irrigation of young peach trees. *Grdinarska I Lozarska Nauka, 5*(7), 3–16.

Durgac, C., Bozkurt, S., & Odemis, B. (2017). Different irrigation intervals and water amount studies in young apricot trees (cv. Ninfa). *Fresenius Environmental Bulletin, 26*, 1469–1476.

Goldhamer, D. A., & Shackel, K. (1990). Irrigation cut off and drought irrigation strategy effects on almond. In *Proceedings of 18th Almond Research Conference*, Fresno, CA (pp. 30–35).

Goldhamer, D. A., & Viveros, M. (2000). Effects of preharvest irrigation cutoff durations and postharvest water deprivation on almond tree performance. *Irrigation Science, 19*(3), 125–131.

Gunduz, M., Korkmaz, N., Asik, S., Unal, H. B., & Avci, M. (2011). Effects of various irrigation regimes on soil water balance, yield, and fruit quality of drip-irrigated peach trees. *Journal of Irrigation and Drainage Engineering, 137*(7), 426–434.

Hagin, J., Sneh, M., & Lowengart-Aycicegi, A. (2002). In A. E. Johnston (Ed.), *Fertigation – Fertilization through irrigation. IPI Research Topics No. 23*. Basel: International Potash Institute.

Haman, D. Z., Smajstrla, A. G., & Pitts, D. J. (2005). *Efficiencies of irrigation systems used in Florida nurseries. BUL312*. Gainesville: University of Florida Institute of Food and Agricultural Sciences.

Herro, A., & Guardia, J. (1992). *Conservacion de Frutos. Manual Tecnico*. Madrid: Ediciones Mundi-Prensa.

Hussien, S. M., Fathi, M. A., & Eid, T. A. (2013). Effect of shifting to drip irrigation on some plum cultivars grown in clay loamy soil. *Egypt Journal of Agricultural Research, 91*(1), 217–232.

Hutmacher, R. B., Nightingale, H. I., Rolston, D. E., Biggar, J. W., Dale, F., Vail, S. S., & Peters, D. (1994). Growth and yield responses of almond (*Prunus amygdalus*) to trickle irrigation. *Irrigation Science, 14*, 117–126.

Intrigliolo, D. S., & Castel, J. R. (2010). Response of plum trees to deficit irrigation under two crop levels: tree growth, yield and fruit quality. *Irrigation Science, 28*, 525–534.

James, P. (2011). Production aspects of sweet cherries. In *Australian Cherry Production Guide. TIAR* (pp. 37–41).

Kaya, S., Evren, S., Dasci, E., Adiguzel, M. C., & Yilmaz, H. (2010). Effects of different irrigation regimes on vegetative growth, fruit yield and quality of drip-irrigated apricot trees. *African Journal of Biotechnology, 9*(36), 5902–5907.

Khan, I. A., Wani, M. S., Mir, M. A., Rasool, K., & Simnani, S. A. (2015). Physiological and yield response of almond to different drip irrigation regimes under temperate conditions. *Indian Journal of Horticulture, 72*(2), 187–192.

Lishchuk, A. I., Semash, D. P., & Storchous, V. N. (1988). Transpiration intensity of apple and peach leaves with different irrigation methods. *Byulleten Gosudarstvenogo Nikitskogo Botanicheskogo Sada, 65*, 89–93.

Long, L. E., Yin, X., Huang, X. L., & Jaja, N. (2014). Response of sweet cherry water use and productivity and soil quality to alternate groundcover and irrigation systems. *Acta Horticulturae, 1020*, 331–338.

Malik, A. S., Kumar, A., Singh, J., Faroda, A. S., & Singh, J. (1987). Effect of methods of irrigation on grain yield, consumptive use, moisture extraction pattern and water use efficiency of raya and wheat. *Haryana Agriculture University Journal Research, 17*(4), 34–340.

Massai, R., & Remorini, D. (2000). Estimation of water requirements in a young peach orchard under irrigated and stressed conditions. *Acta Horticulturae, 537*, 77–86.

Mechlia, N. B., Ghrab, M., Zitouna, R., Mimoun, M. B., & Masmoudi, M. (2012). Cumulative effect over five years of deficit irrigation on peach yield and quality. *Acta Horticulturae, 592*, 301–307.

Miculka, B. (1983). Effect of positioned irrigation on nutrient concentration in peach leaves. *Sbornik Uvtiz Zahrgdnictvi, 10*(3), 185–194.

Molden, D. R., Sakthivadivel, R., & Habib, Z. (2001). *Basin-level use and productivity of water: Examples from South Asia, IWMI Research Report 49*. Colombo: International Water Management Institute (IWMI).

Mousavi, A., & Alimohamadi, R. (2006). Effects of deficit irrigation and drought during different phenological stages of fruit growth and development in almond production. *Acta Horticulturae, 726*, 489–494.

Mussaddak, J. (2007). Efficiency of nitrogen fertilizer for potato under fertigation utilizing a nitrogen tracer technique. *Communications in Soil Science and Plant Analysis, 38*, 2401–2422.

Narayanamoorthy, A. (1997). Drip Irrigation—A viable option for future irrigation development. *Productivity, 38*(3), 504–511.

Neilsen, D., Millard, P., Neilsen, G. H., & Hogue, E. J. (2001). Nitrogen uptake, efficiency of use and partitioning for growth in young apple trees. *Journal of the American Society for Horticultural Science, 126*, 144–150.

Perez-Pastor, A., Ruiz-Sanchez, M. C., Martinez, J. A., Nortes, P. A., Artes, F., & Domingo, R. (2007). Effect of deficit irrigation on apricot fruit quality at harvest and during storage. *Journal of Science Food and Agriculture, 87*, 2409–2415.

Perez-Pastor, A., Domingo, R., Torrecillas, A., & Ruiz-Sanchez, M. C. (2009). Response of apricot trees to deficit irrigation strategies. *Irrigation Science, 27*, 231–242.

Rana, G. S., & Daulta, B. S. (1997). Effect of different rootstocks, spacing and drip irrigation levels on plant height of peach (Prunus persica Batsch.) cv. Flordasun. *Crop Research, 14*(2), 293–296.

Rana, G. S., Sehrawat, S. K., Daulta, B. S., & Beniwal, B. S. (2005). Effect of drip irrigation and rootstock on N, P and K leaf content in peach under high density plantation. *Acta Horticulturae, 696*, 223–226.

Razouk, R., Lbijbijen, J., Kajji, A., & Mohammed, K. (2013). Response of peach, plum and almond to water restrictions applied during slowdown periods of fruit growth. *American Journal of Plant Sciences, 4*(3), 561–570.

Romero, P., Botia, P., & Garcia, F. (2004). Effects of regulated deficit irrigation under subsurface drip irrigation conditions on vegetative development and yield of mature almond trees. *Plant and Soil, 260*(1), 169–181.

Romo, R., & Diaz, D. H. (1985). Root system and nutritional status of peaches under drip or flood irrigation in warm climates. *Acta Horticlturae, 173*, 167–175.

Saleth, R. M. (1996). *Water institutions in India: Economics, law and policy*. New Delhi: Commonwealth Publishers.

Sekse, L. (1995). Cuticular fracturing in fruits of sweet cherry (*Prunus avium* L.) resulting from changing soil water content. *Journal of Horticultural Science, 70*(4), 631–635.

Singh, S. (2013). *Effect of drip irrigation and mulch on soil hydrothermal regimes, weed incidence, yield and quality of apricot cv. New Castle*. MSc Thesis. Department of Soil Sciences, Dr Y S Parmar UHF Nauni Solan, H.P. India.

Singh, R., Bhandari, A. R., & Thakur, B. C. (2002). Effect of drip irrigation regimes and plastic mulch on fruit growth and yield of apricot (*Prunus armeniaca*). *Indian Journal of Agricultural Sciences, 72*(6), 355–357.

Sivanappan, R. K. (1994). Prospects of micro-irrigation in India. *Irrigation and Drainage Systems, 8*, 49–58.

Sivanappan, R. K. (1998). Irrigation water management for sugarcane in VSI, pp II 100–125.

Tan, C. S., & Layne, R. E. C. (1990). *Irrigation scheduling for fruit crops* (Factsheet). Ministry of Agriculture and Food.

Torrecillasa, A., Domingob, R., Galegoc, R., & Ruiz-SaAncheza, M. C. (2000). Apricot tree response to withholding irrigation at different phenological periods. *Scientia Horticulturae, 85*, 201–215.

Treder, W., Grzyb, Z., & Rozpara, E. (1999). The influence of irrigation on growth and yield of plum trees cv. Valor grafted on Myrobalan and Wangenheim prune. *Acta Agrobotanica, 52*(1/2), 95–101.

Vaidyanathan, A. (1999). *Water resources management: Institutions and irrigation development in India*. New Delhi: Oxford University Press.

Vera, J., Abrisqueta, I., Abrisqueta, J. M., & Sanchez, R. (2013). Effect of deficit irrigation on early maturing peach tree performance. *Irrigation Science, 31*(4), 747–757.

Verma, P. (2017). *Studies on the effect of drip irrigation and fertigation in peach (Prunus persica (L.) Batsch.) cv. Redhaven*. Ph.D thesis, Department of Fruit Science, Dr Y S Parmar UHF Nauni Solan, HP, India.

Verma, P., & Chandel, J. S. (2017). Effect of different levels of drip and basin irrigation on growth, yield, fruit quality and leaf nutrient contents of peach cv. Redhaven. *The Bioscan, 2*(2), 1035–1039.

Wang, D. (2011). Deficit irrigation of peach trees to reduce water consumption. *Transactions on Ecology and the Environment, 145*, 497–505.

Zhang, H., Wang, D., & Gartung, J. L. (2017). Influence of irrigation scheduling using thermometry on peach tree water status and yield under different irrigation systems. *Agronomy, 7*, 1–15.

Physiological Disorders in Stone Fruits

7

A. Raouf Malik, R. H. S. Raja, and Rehana Javaid

Abstract

Abiotic changes in fruits which are associated with climatic or management practices are known as physiological disorders. These are commonly found in temperate fruits during their development on the tree and during storage. The disorders can be found in storage also without having any trace of them during the developmental stages of the fruit. Stone fruits which comprise peach, nectarines, plum, apricot and cherry are greatly affected by the occurrence of the physiological disorders which reduce their consumer acceptability and market value. Temperate fruit develops many physiological disorders during their developmental stages with all of them exhibiting particular symptoms. However, the processes involved in the development of these disorders have not been completely understood as of now. Some of the commonly observed physiological disorders in stone fruits are internal breakdown, gummosis, split and shattered pits, fruit discoloration, gel breakdown and pit burning in apricots, fruit cracking in cherry, etc. Proper identification, understanding the biology and correction measures like adaption of resistant varieties and proper cultural practices will help the farmers in reducing the losses due to these disorders.

Keywords

Stone fruits · Disorders · Physiology · Genetics · Management

A. R. Malik (✉) · R. H. S. Raja · R. Javaid
Division of Fruit Science, SKUAST-K, Srinagar, Jammu and Kashmir, India

M. M. Mir et al. (eds.), *Production Technology of Stone Fruits*,
https://doi.org/10.1007/978-981-15-8920-1_7

7.1 Introduction

Stone fruit can be defined as a fruit with a pit surrounded by a fleshy outer area. Botanically, stone fruit is a drupe with thin edible outer skin (exocarp), fleshy mesocarp and thick lignified ovary wall or endocarp that encloses a seed, commonly known as stone or pit (Crisosto et al. 2008; Crisosto and Day 2012). These are usually good sources of carbohydrates, organic acids, phenols and vitamins which make it attractable to consumers (Kader and Mitchell 1998). Stone fruits comprise climacteric fruits like peach, plum, nectarines and apricot which are harvested at the onset of maturity and non-climacteric fruit like cherry in which ripening occurs on trees before harvesting. They are perishable in nature and have very short shelf life and storage life and are also affected by many insects, pests, diseases and physiological disorders. Physiological disorder can be defined as any non-parasitic or inanimate disease or a malady of a plant. Physiological disorders usually breakdown the tissues. The breakdown takes place without being infected by any pathogen (which causes disease) or by any sort of mechanical injury. Most of the time, physiological disorders develop because of unfavourable environment conditions, improper management practices or nutritional deficiencies during the development of fruit on tree or after harvesting of fruits. The adverse environmental conditions include temperature, humidity, rain and sunlight. Improper management practices include faulty training, pruning, fertilization, irrigation and harvesting procedures. The unfavourable environment conditions, improper management practices or nutritional deficiencies during growth and development of fruits can change the microclimate endured by the plant or plant parts. Environment is one of the major influences on the severity of various physiological disorders in fruits crops (Ferguson et al. 1999). Therefore, any alteration in the metabolism of fruits that leads to cellular disorganization and cell death, which is caused mainly by the adverse environmental conditions or nutritional deficiencies, is referred to as physiological disorder. Physiological disorders are mostly irreversible, non-transmissible, clearly demarcated and affecting any developmental stage of plant and open the way for entry of pathogens inside the plant. The physiological disorders are commonly observed in deciduous fruit trees and affect discrete area of tissue. In some disorders, only the outer skin is affected, while in others, the flesh or core region is involved. There are a many physiological disorders in stone fruits. All of them have some peculiar symptoms. Susceptibility to a particular disorder is greatly affected by cultivar through permeability of cell wall, oxygen and carbon dioxide diffusivity by the skin, mineral composition and antioxidant status and is largely dependent upon environmental conditions (e.g. availability of light, rainwater contamination, humidity, carbon dioxide concentration in air, wind), water stress, salt content in plant cells (that lead to physiological drought), cultural practices and post-harvest effects (Singh et al. 2018).

In general, there are three types of physiological disorders based on the developmental stage they affect the fruits:

1. On the tree (internal breakdown in plums and others).

2. On the tree and during storage (wooliness and internal breakdown in peaches and nectarines).
3. During storage (brown core and low-oxygen injuries in peaches and nectarines).

The disorders which occur during storage (the third category) can be further divided into (1) those associated with senescence, (2) improper atmospheric conditions during the storage of fruits and (3) due to inappropriate temperature. The senescent disorders develop due to harvesting of fruits which have overmatured, nutritional imbalance like high-nitrogen, low-boron and low-calcium content or higher temperature during the development of fruits on tree. Physiological disorders developing due to inappropriate atmospheric manipulations, viz. lower oxygen concentration and higher carbon dioxide and nitrogen concentration, affect respiratory metabolic pathway, viz. glycolysis, tricarboxylic acid cycle, mitochondrial respiratory chain, fermentation and pathways involved in the synthesis of secondary metabolites, viz. phenolics, pigments, volatile compounds and ethylene. Succinate accumulation due to increased carbon dioxide concentration is usually toxic to the cells and has been thought to be responsible for carbon dioxide injury in fruits (Fernandiz et al. 2001). It may also be the result of its failure to not being able to maintain its energy balance or metabolic function, as such cannot be attributed to the accumulation of an injurious compound within fruit.

7.2 Chilling Injury

Low temperature reduces fruit metabolism and respiration rate as storing fruits at low temperature is very beneficial and enhances the shelf life of fruits. However, it must be noted that low-temperature storage does not suppress all the metabolic activities and leads to the production of certain by-products which may result into the cessation of normal cell functions; hence, the cell losses its structure and integrity. Chilling injury is the most common metabolic disturbance caused due to low-temperature storage (Crisosto and Day 2012). Chilling injury is commonly observed in case of stone fruits resulting from the exposure of susceptible tissues to a temperature which is less than 15 °C; however, the critical temperature for chilling injury to appear depends usually upon the stage of maturity at the time of harvest and fruit and cultivars' susceptibility to low temperature (Jing et al. 2018). Chilling injury should usually not be confused with the freezing injury as freezing injury results due to exposure of fruits to freezing temperature which results into the formation of ice crystals in their intercellular spaces and also inside the cells. Pitting which results from the collapse of the cells beneath its skin and browning of flesh tissues around their vascular bundles because of the decomposition of phenols by way of action of polyphenol oxidase enzyme released from vacuoles are some of the most common symptoms of chilling injury.

The mechanism involved in chilling injuries can be divided into primary events and secondary events. The primary events are usually reversible and instantaneous,

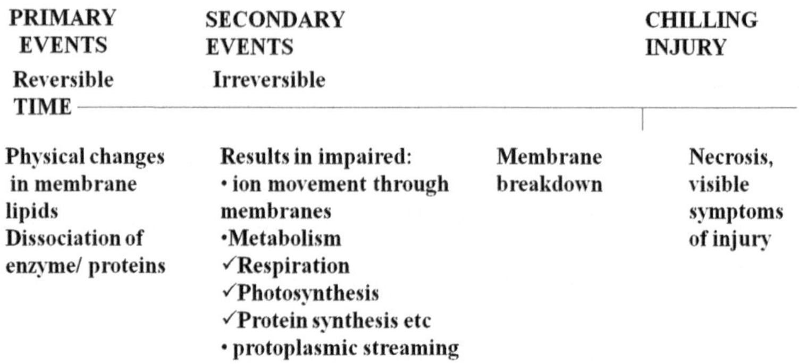

Fig. 7.1 Diagrammatic time sequence of events leading to chilling injury

while secondary events are irreversible which finally results into the death of cells. Primary events result from a change in the physical state of membrane lipids and dissociation of proteins and enzymes into subunits, while the secondary events result in impairment of the ion movement and other metabolic activities. Chilling injury in fruits can be controlled by determining the critical temperature for its development and not exposing the fruits below that temperature. The concept of chilling injury developed by Raison and Lyons (1986) is illustrated in Fig. 7.1.

7.3 Genetic Factors Involved in Chilling Injury

The stone fruits and their cultivars show different levels of susceptibility to chilling injury disorder, while some cultivars do not show any kind of symptoms or susceptibility. The early-season cultivars of nectarine and peach are less susceptible than their late-season counterparts (Crisosto et al. 1999a, b), while plum varieties do not show any seasonal pattern for their susceptibility towards chilling injury. However, overall susceptibility of nectarine cultivars to chilling injury is lesser than different cultivars of peach. Further, it can be seen that those cultivars of peach which have melting flesh show more susceptibility towards chilling injury than those cultivars which have firm and non-melting flesh (Brovelli et al. 1999; Crisosto et al. 1999a, b).

The cultivars with non-melting flesh have lesser endopolygalacturonase activity (Lester et al. 1996). The genetic locus for freestone fruits appears to contain a cluster of endopolygalacturonase genes (Callahan et al. 2004; Peace et al. 2005). Mealiness or woolliness can be considered as a phenomenon due to chilling injury which occurs in stone fruits particularly nectarine and peach cultivars stored at low temperatures for a longer period of time. This is caused due to altered cell wall metabolism during ripening which leads to a texture of the fruit which is drier and woolly. It has been found that endopolygalacturonase gene plays a quantitative role in controlling mealiness in stone fruit cultivars. The genes which influence mealiness susceptibility in stone fruit crops require a functional endo-PG gene for their expression.

The possible explanation for low-temperature injury encountered during cold storage is that a lower temperature for an extended period of time which is beyond a critical period can suppress the endopolygalacturonase gene activity. However, it does not suppress the pectin methylesterase activity in storage or even after that. This results in altered concentration of pectic compounds that can readily form gels resulting into a dry fruit flesh (Lill et al. 1989); this possibly consists of homogalacturonan coming from the middle lamella of cell wall (Jarvis et al. 2003; Vincken et al. 2003).

The juicy fruit mealiness has been found to be associated with a possible reduction in the activity of pectin methylesterase, in association with the reduction in the polygalacturonase enzyme activities. Low-temperature breakdown has been found to have changed isoprenoid metabolism as well as higher abscisic acid level, acetic ester level and acetic acid level. Altered cell wall enzyme activity during cold storage affects the metabolism of cell wall polysaccharides leading to the depolymerization of chelator-soluble polyuronides (Zhou et al. 2000). During chilling injury, chelator-soluble polyuronides are rapidly depolymerized from a very large to a small size during the final melting stage of ripening (Brummell et al. 2004).

7.4 Physiological Disorders and Their Management in Stone Fruits

- Splitting/Shaterring and Gumming of Stone Fruits
- This disorder occurs on dorsal and ventral side of large-sized fruits at pit-hardening stage where the downward movement of carbohydrates is blocked to the lower parts of plants and is accelerated with heavy rainfall. This occurs mostly as the gum fills the entire seed cavity and eventually leads to the abortion of seeds. The affected fruit losses its economic value and causes great loss to the farmers (Fig. 7.2).
- This malady can be controlled by avoiding long dry spells through frequent irrigations. It is observed that peach, plum, nectarines and cherry with early-season cultivars like 'June Gold' and 'Spring Gold' of peach are more susceptible than late-season cultivars because pit hardening and final swelling in fruits occur relatively at the same time (Khan and Bhat 2017). The heavy rains or the late spring frost which leads to crop losses during the critical developmental/growth stages, and cultural practices like girdling, excessive irrigation and heavy doses of nitro-

Fig. 7.2 Fruit splitting and gum disorder in peach fruits

gen near harvesting that promotes rapid fruit growth are the factors responsible for its incidences.

- Physiological Factors Involved in Split Pit

Split pit is a serious physiological disorder of stone fruits especially peach. It has been observed that fruits splitting during the early developmental phases mainly take place about 20 days after full bloom when the pit of the fruit is still soft. It may also take place after the pit hardens. Fruit splitting at this stage of the fruit is more damaging. During this stage of fruit splitting, the pit breaks along the suture due to pressure exerted by the flesh which keeps on expanding while still being attached with the pit. Split pit generally occurs due to the environmental factors (low temperature and frost damage during flowering and early fruit development, a late frost and heavy rains during critical growth period) or may occur due to different management practices followed by the grower to produce fruits of a larger size including increased watering, excessive thinning, fertilization close to harvest time which leads to increased swelling and higher growth of flesh. The rapid increase in flesh growth exerts pressure on the pit. During these stages, if the fruit swelled up while the flesh is still attached to the pit and before the bond between flesh and pit starts weakening, the pit is likely to get pulled apart. It may also lead to a fracture running in line with suture line and running down the side of the pit. This results into the split pit condition. The pit may break into pieces, which is known as pit shattering. The various characteristics of split pit fruits include their heavier weight than normal fruits. This is mainly due to mesocarp and exocarp which are usually heavier than normal weight of mesocarp and exocarp. Further, the moisture content in pits of those fruits which have split pit problem is usually lesser during pit-hardening stage of development when compared to their healthy counterparts. The concentrations of K, Ca, Mn, Zn and Fe and P, Mg, K, Mn, Zn and Cu in the pit and flesh of split pit fruits, respectively, is lesser as compared to healthy fruits when pit-hardening stage starts. However, at the end of pit-hardening stage, it is almost equivalent (Evert et al. 1988).

7.5 Genetic Factors Involved in Split Pit

Genetic factors influence the frequency with which split pit is formed in stone fruits. It has been observed that the varieties like June Gold and Spring Gold which are early maturing are more affected by shattered pit and split pit problems. The primary reason for being susceptible is that the final swell and pit-hardening stages of fruit development take place simultaneously almost during the same time. The cultivars which ripen late have these tow developmental phases with proper time gap which allows the bond between pit and surrounding flesh to loosen up before the flesh swells. The two genes in peach *PPERSHP* and *PPERFUL* that are homologues to the genes *SHATTERPROOF* and *FRUITFULL*, respectively, have been found to cause fruit splitting in case of *Arabidopsis thaliana*. These genes, viz. *PPERFUL* and *PPERSHP*, have been found to be expressed in peach throughout its fruit developmental stages starting from anthesis to the harvesting of fruits. Research indicates that the expression of gene *PPERFUL* has been found to be high at the time of fruit set; however, when pit-hardening stage starts, the gene *PPERSHP* expression decreases. It must be noted that the verities which are resistant have significantly lesser expression of gene *PPERSHP* than those varieties which are sensitive to split pit. However, at the time of final phases of fruit growth and development, the expression of gene *PPERFUL* increases significantly and has been found to be higher in the sensitive varieties than those varieties which have been found to be resistant. The genes *PPERFUL* and *PPERSHP* are involved in the formation of the separation layer, having a temporal expression pattern that allows earlier formation of the dehiscence zone in the varieties which are susceptible to split pit than those varieties which are resistant to the split pit (Eleni et al. 2007).

- Fruit Cracking in Cherry
- This is the most serious problem in sweet cherry as it reduces fruit quality and its market value. The malady is characterized by splitting of outer layer or exocarp due to absorption of rainwater from the surface. Sweet cheery varieties like Van, Lapins, Rainier, Sam and Sweetheart show lower incidence of cracking, while Bing cherry is highly susceptible to it. Cracking occurs as one of the three types: (a) stem end, (b) apical or nose end and (c) side or deep cracks (Figs. 7.3, 7.4, 7.5 and 7.6). The first two occurs at stem attachment point and towards apical end of fruits, respectively, while the side cracks appear along the cheeks and may extend deep up to the stone in any direction, 3 weeks prior to commercial harvesting, and are strongly controlled by season, i.e. incidence of rainfall (Measham 2011). A number of factors are responsible for the development of cherry fruit cracking; they are susceptibility of fruits to cracking (mostly hard-fleshed varieties are more prone to it than the soft-fleshed varieties), volume and distribution of rainfall at one time and soil moisture conditions. The occurrence of this disorder can be controlled by using resistant varieties and using of dry air sprays for removal of water from fruit surface and foliar sprays of $CaCl_2$ at 0.3% (Sharma 2006) at weekly interval before harvesting.

Fig. 7.3 Fruit cracking in cherry

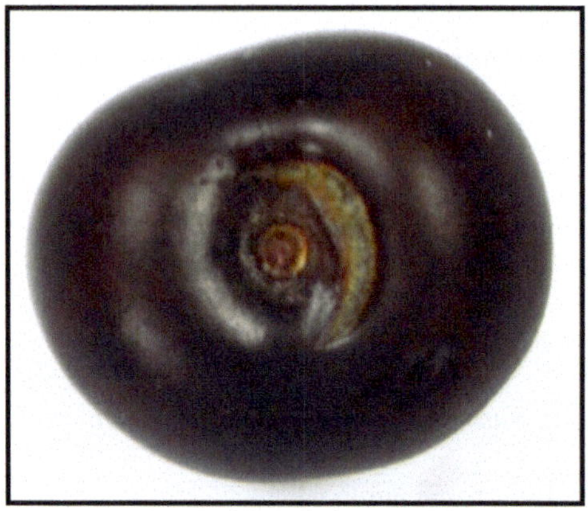

Fig. 7.4 Apical or nose end type of cherry cracking

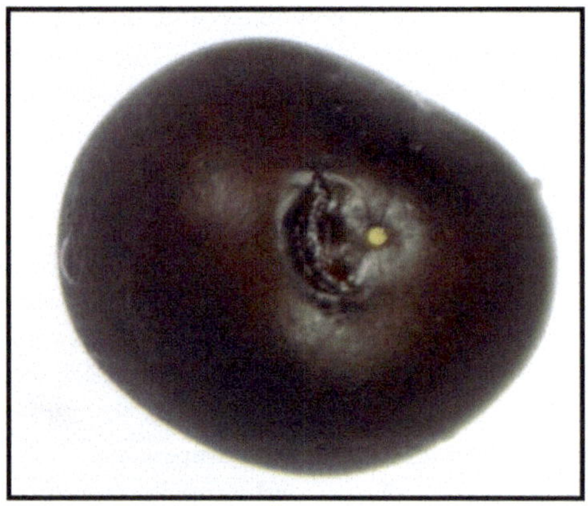

Fig. 7.5 Stem end type of cherry cracking

7.6 Physiology of Cracking

Fruit cracking which is one of the most important physiological disorders in cherry has been studied in detail. The main physiological aspects which have been studied include the anatomical aspects of the cells involved, the water uptake by different cells of the fruit, the osmotic potential of fruit, the different physical properties of the cell cuticle and the fruit growth dynamics. It has been found that fruits usually show cracking when a prolonged drought is followed by absorption of plenty of water (because of sudden rain or irrigation) by roots or directly by the fruits, both having a significant effect on fruit cracking. It has been found that fruits can develop cracking because of the absorbed water during fruit ripening which increased turgor pressure in fruit (Yamamoto et al. 1990; Measham et al. 2012). Further, water which is absorbed by fruits through their skin has also been found to be an important factor in development of fruit cracks (Christensen 1972). The main reason for the development of fruit cracks is the strengthening of xylem and phloem (because of the prolonged drought), thereby losing their ability to enlarge and divide. When water is

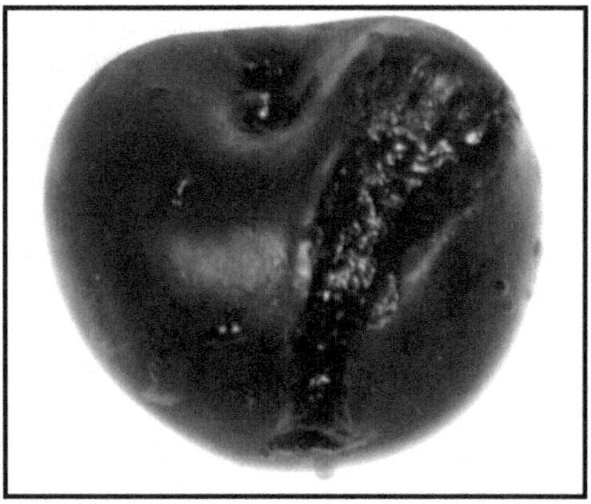

Fig. 7.6 Side or deep cracks in cherry

suddenly made available abundantly, the tissues which are meristematic resume their growth; however, phloem and xylem do not resume their division and enlargement process. As such, this creates a differential rate of growth between the growth patterns of strengthened tissues (xylem and phloem) and meristematic tissues, resulting in the rupturing of harder tissues, thereby causing fruit cracking (Sharma 2006). Many researchers have observed that the osmotic potential developed by the fruits and the cracking have a positive correlation between each other.

Fruit cracking is related to lower than normal osmotic potential which leads to the development of high internal pressure in the fruit causing tensile forces in the outer tissue to exceed the breaking point leading to fruit cracking (Sekse 1995; Christensen 1996). The anatomical studies of the fruits evidently show a correlation with fruit cracking especially the fruit size and fruit firmness. With increased water uptake, the fruit volume increases which more severely affects larger and firmer fruits (Simon 2006). Fruits susceptible to cracking are characterized by the absence of transition cells in between their fruit parenchyma and hypodermis. The hypodermal cells in fruits have been found to be thick walled, small, depressed and tangentially oriented. The cells of fruit parenchyma are isodiametric. It has been found that the hypodermal cell division ceases before the cell division of fruit parenchyma because of which there is a differential rate of growth between two parts and the outer fruit part is not able to follow the growth pattern of inner fruit part. The inability of the hypoderm to keep pace with the expansion of the fruit is due to the difference in cell wall composition and the consequent effect on wall extensibility, thereby forcing the fruit to crack (Sharma 2006). Therefore, it can be concluded that a combination of tensile forces acting on the fruit surface from inside the fruit and the loss of bearing structure of the fruit skin causes fruit to crack.

Stone fruits represent a double sigmoidal fruit growth pattern. The developmental period of fruit growth as such can be divided into three distinctive phases. First phase of development (Phase I) initiates with fruit set and lasts approximately 21 days. During this stage, active cell division occurs, determining mainly the number of cells per fruit at harvest. Fruit expansion growth leads to Phase II, and the endocarp lignifies. Finally, during Phase III, cell enlargement occurs followed by fruit ripening (Whiting 2001). The third and final phase of fruit growth (Phase III) is a very critical period for cracking which is linked to rapid expansion of fruit mesocarp. When the mesocarpic cells keep on enlarging, the expansion capacity of cuticle is not able to keep pace with the internal fruit pressure developed due to enlargement of mesocarpic cells during Phase III; as such, the incidence of cuticular fractures increases manifold due to which there is a large influx of water into the fruit (Knoche et al. 2004), thereby contributing to bulk of water diffusion (60%) into the fruit during their Phase III (Gibert et al. 2005). Further, at times, there is a prolonged period of high relative humidity; when the fruits are small, this can modify the cuticle composition, thereby losing its protective capacity. Therefore, water supply increase accompanied by a decreased loss of water loss from leaves due to saturated relative humidity promotes the cracking of fruit (Sharma 2006).

7.7 Genetic Factors Involved in Fruit Cracking

Fruit cracking is one of those physiological disorders which is under genetic control. The differential response of different varieties of the same crop clearly signifies the fact of it being controlled by genes. A classical quantitative trait locus (QTL) approach was implemented to study the cracking tolerance in sweet cherry at Institute of National Research in France (Garcia-Quero et al. 2014). Dirlewanger et al. (2004) developed a genetic map with simple sequence repeat (SSR) (microsatellite) markers. They developed it on the basis of full-sib hybrids which were developed by crossing Regina and Lapins. They concluded that Regina is one of those varieties which is most tolerant cultivar for cracking and is in commercial use. Further, it was concluded that Lapins has intermediate tolerance for fruit cracking. It has been found that cell wall is modified during ripening process and as such can be said to be linked with cracking of fruits. Fruit softening and ripening of fruits are due to combined action of various enzymes like pectinesterase (PE), expansin, *p*-galactosidase and polygalacturonase (PG) (Brummell and Harpster 2001). In cherry, during early stages of fruit development, *p*-galactosidase level is at its peak and decreases suddenly when the fruit starts to ripen, thereby providing flexibility to the cell wall of fruits which results in the emergence of microfractures that are found in most advanced stages of fruit growth (Knoche et al. 2001). Expansin proteins have recently been identified to be involved in breakdown of the bond in between cellulose xyloglucan molecules and microfibrils, thereby allowing expansion and relaxation of fruit cell wall under tensile stress of fruit softening as such helps in reducing the fruit surface cracking level (Kasai et al. 2008). Therefore, the analysis of expansion pattern of expansion proteins and other candidate genes may significantly help to understand different mechanisms involved in cracking of fruits.

- *Double Fruit or Twin Fruit/Cleft Suture:* It has been most of the times observed in different orchards with the extent of problem ranges from minor to major and is associated with previous years of high temperature and water stress during fruit bud differentiation stage (Fig. 7.7). The problem can be minimized by fre-

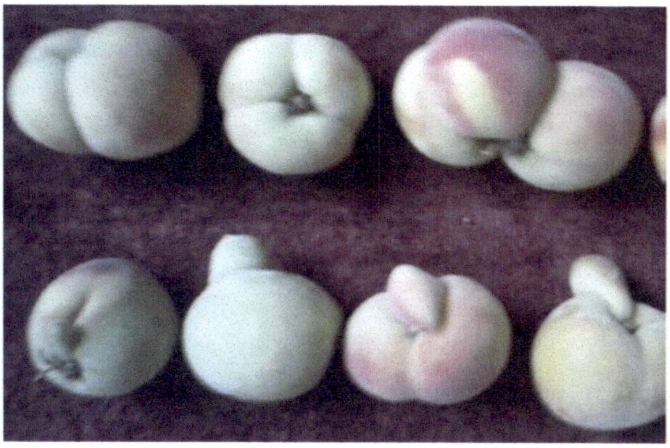

Fig. 7.7 Twin fruit in peach

quent irrigations on orchards during hot and dry summer months. When these double fruits develop in a way that one fruit remains small and the other large, it leads to deep and lengthened crease or fracture known as cleft or suture. In these junctures, there will be formation of gums which further reduces the appearance of fruits.

7.8 Physiology of Fruit Doubling

Double fruit or twin fruit formation can be considered as a morphological disorder. This disorder is formed due to a condition in the female flower part, viz. double pistil. The double pistil is formed because of an abnormal differentiation during the process of flower bud formation of pistil primordia (Guimond et al. 1998). Many researchers have found that if the temperature during mid-July to mid-August is higher, there will be higher fruit doubling during the preceding year (Beppu and Kataoka 2011). Studies have revealed that when there is higher temperature at the time of flower bud formation, it results in delaying of the progression of flower differentiation, which results into the double pistil formation. Further, it has been found that temperature which is higher than 30 °C is the most critical condition for double pistil formation. The duration for which fruits are exposed to higher temperature also influences the fruit doubling to a great extent and can be considered as a critical factor for it (Beppu et al. 2001).

Phytohormones have also been found to play a critical role in double-fruit formation. Ethylene increases the rate of double pistil formation while gibberellins decrease it. Ethephon delays the progress of flower differentiation in a way which is very similar to the way high temperature does it. Ethylene production is usually promoted in plant tissues due to an increased temperature condition. The peak ethylene production has been found to be at a temperature of 30–35 °C in many crop plants/species (Field 1985); as such, it can be said that ethylene can be a cause for double pistil formation induced by higher temperature conditions in most of the stone fruit crops.

Drought is another factor for fruit doubling; drought-like conditions during the floral initiation stage induce doubling in sweet cherry (Fukai 1995).

7.9 Genetic Factors Involved in Fruit Doubling

Fruit doubling is influenced by genetic factors as different varieties of stone fruits show differential response to this disorder. The frequency with which the doubling of fruit takes place has been found to be highly variable among cultivars of sweet cherry, such as 'Satonishiki', 'Bing', 'Chinook' and 'Karabodur'; these cultivars show about 30% or more of doubled fruits, while in cultivars such as 'Cherie', 'Jubilee' and 'Takasago', double-fruit formation seldom occurs (Roversi et al. 2008.). Also, a small percentage of fruits of cultivar Napoleon of sweet cherry show fruit doubles (Beppu and Kataoka 2011). From the above, it can be said that there is a possibility for breeding new cultivars which can show little occurrence of fruit doubling or double pistils.

The homeotic genes of C and B classes control pistil and stamen formation. These genes have been found to be involved in double-fruit formation. These genes generally together are responsible for the induction of stamens, while the genes of class C alone are responsible for the formation of pistils in stone fruits (Weigel and Meyerowitz 1994). Based on tissue-specific expression patterns and sequence similarity, PaPI and PaTM6 genes of sweet cherry have been found to be homologues of PISTILLATA (PI) and APETALLA3 (AP3) genes from *Arabidopsis thaliana*, respectively, which encoded MADS box and are expressed in stamens and petals of sweet cherry, respectively. The genes in class B PaPI and PaTM6 and genes of class C PaSHP and PaAG in the flower buds of sweet cherry are involved in pistil doubling (Beppu et al. 2015).

- Surface Discoloration/Inking/Black Staining
- It is one of the major physiological disorders in peaches and nectarines which affects the visual appearance of fruits and hence reduces its marketability (Cheng and Crisosto 1994). It is observed as black, blue, brown or some other coloured spots (Fig. 7.8) or stripes and is commonly noticed during harvesting, hauling (Crisosto et al. 1993), packing, storage and transportation (Phillips 1988) in fruit surface areas where there is accumulation of anthocyanins. This may be due to the exposure of physically damaged skin to heavy metals like iron or aluminium. Fruits treated with foliar application of nutrients, fungicides or insecticides containing theses heavy metals prior to harvesting in combination with abrasion damage are more prone to this disorder than the healthier fruits. The damage can be controlled by avoiding the application of heavy metal-containing nutrients, fungicides, etc. after fruit approaches maturity. Irrigation water and harvesting equipment should also be free from heavy metal contamination, and fruit should be kept dust-free.

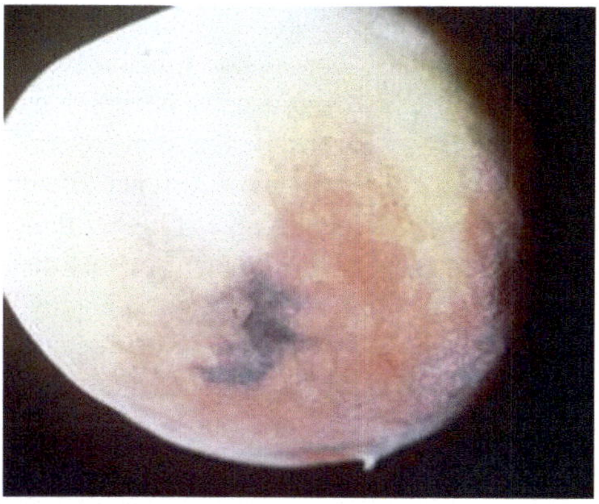

Fig. 7.8 Surface discolouration/black staining in peach

Fig. 7.9 Internal breakdown in plums

- *Internal Breakdown:* It is the most limiting factor in shipping of stone fruits and is characterized by flesh browning (Fig. 7.9), mealiness, black pit cavity, flesh bleeding, leatheriness, failure to ripen and development of off-flavours (Crisosto et al. 2008); even in some cultivars, flavours are lost before the evidence of symptoms (Crisosto and Labavitch 2002). Mealiness is characterized by dry soft texture of fruits and leathery fruits that are dry and firm (Ju et al. 2000) and is developed by the action of endopolygalacturonase (EndoPG) gene (Peace et al. 2006), while normal fruits are juicy, soft or firm. Mealiness in fruits develops due to formation of gel by combination of pectin substance with water, thereby reducing intercellular spaces (Brummell et al. 2004). Browning in fruits occurs due to mixing of polyphenol oxidase with phenolic compounds (Crisosto et al. 1999a, b). While bleeding results due to the spread of red pigment during cold storage (Peace et al. 2006). It appears after the removing of fruits from cold storage and placing at room temperature and is thus observed by consumers. Late-season cultivar is more susceptible to this disorder than early-season cultivar. This can be controlled by identified varieties which are resistant to internal breakdown and by storing the fruits below 0°C but above freezing point.

- Freezing Injury
- It appears as water-soaked translucent area in the flesh which turns brown on drying (Fig. 7.10). It usually occurs due to prolonged exposure of fruits that are accidently or purposely stored at freezing point. Fruit soluble solid (Table 7.1) content determines the occurrence of injury to the fruits. Generally, the fruits with low SS content freeze at higher temperature than the fruits with higher soluble solids. The development of this disorder can be controlled by storing the fruits at recommended temperature above freezing point.
- *Pit Burning in Apricot:* The disorder is marked by softening and browning of flesh around the pit that occur due to exposure of fruits to prolonged sunshine and higher temperature (above 38 °C) (Fig. 7.11). The disorder can be corrected through avoiding heat injuries by providing shading to the fruits (Singh et al. 2018).
- Buttons
- The fruits that fail to develop after the initial setting have been found to have dead embryos usually because of an incomplete fertilization process. The development of buttons is manifested due to unfavourable weather conditions like low chilling, spring frost or wet or cold weather during bloom (Khan and Bhat 2017).

Fig. 7.10 Freezing injury in peaches

Table 7.1 Relationship between soluble solid content (SSC) in stone fruits and freezing point as given by the University of California (Anonymous 2020)

	Safe freezing points	
SSC (%)	(0 °F)	(0 °C)
8.0	30.7	−0.7
10.0	30.3	−0.9
12.0	29.7	−1.3
14.0	29.4	−1.4
16.0	28.8	−1.8
18.0	28.5	−1.9

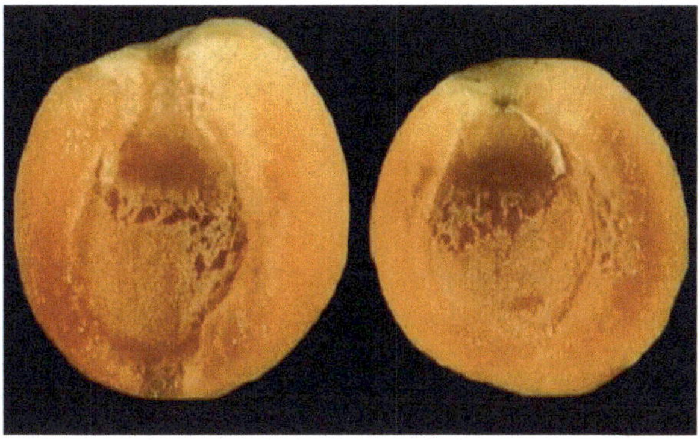

Fig. 7.11 Pit burning in apricot

The affected fruits are difficult to identify during thinning and provide place for insect pest to overwinter. The problem can be controlled by growing varieties which are well adapted to a specific area.

7.10 Physiology of Fruit Buttons

Fruit buttons are formed due to certain environmental conditions. Fruits usually have normal epidermis, mesocarp and endocarp tissues. The fruit buttons also have all these in a normal condition. As such, these tissues which are alive even though the fruitlets are injured have enough sink which allows the fruits to still persist on the trees. However, the embryo is dead inside which does not allow the normal fruit developmental processes in these fruitlets resulting in reduced size of fruits (Corrons 1922). After 30 days of full bloom which is green fruit thinning period typically for the production of fruits for commercial reasons, the fruit buttons usually are very similar to the healthy fruitlets in both size and appearance, thereby making it very difficult to differentiate them from fruitlets which are healthy during process of fruit thinning, which results in the retention of useless buttons by the growers, thereby reducing marketable fruit crop in stone fruits. In peaches, button fruits are produced due to frost damage during flowering which causes embryo abortion and hence forms buttons. It must be noted however that the buttons may also be formed due to embryo abortion because of a pollination failure in some cultivars and not only because of the frost (Chunxian et al. 2016).

7.11 Genetics of Fruit Buttons

Stone fruit cultivars show variation among their genotypes for frost damage as well as for incidence of fruit buttons. The fruit button condition has been found to be heritable; however, it is not expressed in all conditions and has been found to be heritable in certain conditions only. The button formation in peach has been found to occur when there is a frost condition which is cold enough to kill apparently the fruitlets' most vulnerable part which is the embryo. High button incidence has been found in many genetically related peach cultivars such as 'Goldprince' and 'Sunprince' varieties of peach. Both these varieties are descendants of 'Loring' cultivar which is highly vulnerable to low-temperature damage (Okie 1998). Further, the Loring cultivar is also prone to formation of fruit buttons, while as 'Flameprince', 'Julyprince' and 'Contender' varieties are tolerant to buttoning.

The genes associated with cold hardiness also control fruit button formation. Cold hardy cultivars of stone fruits do not form fruit buttons. Dehydrin gene, *ppdhn1*, isolated from peach and its expression are associated with cold hardiness and prevent buttoning of fruits in peach (Artlip et al. 1997). Development of new genotypes having resistance to freezing temperature will therefore help to minimize the fruit buttons.

- Gel Breakdown in Apricots
- The disorder actually develops in orchards (spreads from pit to flesh) but aggravates during cold storage (Ginsburg and Combrink 1972) and differs from internal breakdown in a way that mesocarp does not discolour initially but gives appearance of gelatinous mass (Dodd 1984). The incidence is more enhanced during storage if the harvesting is delayed and varies from year to year in the same cultivar (Taylor and De Kock 1991).
- Heat Spot or Kelsey Spot in Plums
- The disorder was first observed in South Africa in European pear due to high temperature and boron deficiency and deteriorates the fruit quality. The symptoms first appear as dark well-defined internal necrosis which develops cavities (Chamberlain et al. 1959). The problem can be controlled by foliar sprays of boron and by white washing of stem.
- Surface Pitting and Bruising in Cherry
- The small sunken areas on sweet cherry are pits, while large flattened areas are termed as bruises (Fig. 7.12). Pitting appears as coalition of epidermal cells due to mechanical compression which becomes apparent after several days of harvesting, while bruises are caused by picker during harvesting and appear much below the epidermis. These result in increase in cherry respiration and ethylene

Fig. 7.12 Surface pitting and bruising in cherry

Table 7.2 Control measures for various physiological disorders of stone fruits

S. no	Disorder	Control measures
1	Fruit cracking	• Spray of $CaCl_2$ at 0.3%. Four sprays at weekly intervals (foliar application) (Sharma 2006) • Rain cover protection 3 weeks before harvest (Meli 1982) • Use of cracking-tolerant rootstocks and varieties, e.g. 'Colt' and 'Regina' for sweet cherry, respectively (Hovland and Sekse 2003) • Post-harvest dip of fruits in borax solution of 0.25% (Powers and Bollen 1947) • Use of antitranspirants as Bioguard and Vapor Gard (Torres et al. 2009)
2	Fruit splitting	• Spray of $CaCl_2$ at 0.3%. Four sprays at weekly intervals (foliar application) (Sharma 2006) • One spray of paclobutrazol at 750 ppm or 750–1500 ppm after fruit set or soil application (Sharma 2006)
3	Chilling injury	• Use of cold hardy varieties, e.g. Elberta, Silver Gem, Goldcot and Goldstar in peach, nectarine, apricot and plum, respectively (Sharma 2006) • Use of cold hardy rootstocks, e.g. Gisela 5 and 6 and Lovell and Guardian in cherry and peach, respectively • Overhead irrigation, orchard heating, use of mulches/covers, use of wind machines and windbreaks • Two sprays of GA_3 at 100–200 ppm, before and during the occurrence of frost (Ju et al. 1999) • Use of bactericides as streptomycin or oxytetracycline 10 days before the expected frost (Sharma 2006) • Spray aminoethoxyvinylglycine (AVG), 2 weeks before harvest, and a one post-harvest fruit dip in AVG (Byers 1997)
4.	Split pit	• Foliar application of calcium, 3 weeks after full bloom (Saure 2005) • Spray of silicon (in the form of amorphous silica) foliar application after full bloom (Marchner 2002) • Avoid excessive thinning and irregular watering at pit-hardening stage. Avoid thinning until pits are mature and hard (Sharma 2006) • Avoid excessive nitrogen application close to harvest (Tani et al. 2007) • Use of resistant rootstocks and varieties
5.	Double fruits	• Provide adequate irrigation supply to the trees or avoid water stress • Apply N or GA_3 along with adequate irrigation during flower bud differentiation period (Taylor and Taylor 1998) • Avoid post-harvest water stress (Kader 2002) • Avoid the use of more sensitive varieties like Desert Red and Swelling in peach and Canino in apricot (Khan and Bhat 2017)
6.	Fruit buttons	• Use of resistant varieties as 'Flameprince', 'Julyprince' and 'contender' in peach • Proper enough spacing between the plants to avoid competition • Avoid the use of low-chill varieties

production which ultimately leads to the damage and decay of fruits. Large fruits with heavier weight are less susceptible than light fruits. The disorder can be controlled by foliar sprays of gibberellic acid, preventing fruit from falling on rough surface and proper picking and packaging of fruits (Singh et al. 2018) (Table 7.2).

7.12 Conclusion

Stone fruits are good source of vitamins, phenols and organic acids which strengthen the immune system of the human body. Due to its highly perishable nature, it has a very short shelf and storage life and is also affected by a number of insect pest diseases and physiological disorders which greatly reduce the market value of the fruit. One of the most important physiological disorders associated with stone fruits is chilling injury. Peach and nectarine cultivars which are late season and also have melting flesh are most vulnerable to the remedy than those cultivars which are early season and have a flesh which is non-melting type. Other economical important physiological disorders in stone fruits are fruit cracking, internal browning, surface discoloration, splitting, gummosis, freezing injury, etc. Their identification, understanding of their biology, identification of genes responsible for these disorders, correction measures like adaption of resistant varieties, and proper cultural practices will help in reducing the losses occurred by these problems.

References

Anonymous. (2020). *Physiological disorders*. Retrieved from http://kare.ucanr.edu/programs/Postharvest_for_ fruits_ and_nuts/Physiological_disorders/.

Artlip, T. S., Callahan, A. M., Bassett, C. L., & Wisniewski, M. E. (1997). Seasonal expression of a dehydrin gene in sibling deciduous and evergreen genotypes of peach (*Prunus persica* [L] Batsch). *Plant Molecular Biology, 33*, 61–70.

Beppu, K., Ikeda, T., & Kataoka, I. (2001). Effect of high temperature exposure time during flower bud formation on the occurrence of double pistils in 'Satonishiki' sweet cherry. *Scientia Horticulturae., 87*, 77–84.

Beppu, K., & Kataoka, I. (2011). Studies on pistil doubling and fruit set of sweet cherry in warm climate. *Journal of Japanese Society for Horticultural Science., 80*, 1–13.

Beppu, K., Sumida, H., & Kataoka, I. (2015). Cloning and characterisation of APETALA3-like and PISTILLATAlike B class MADS-box genes from sweet cherry. *Journal of Applied Horticulture*, 17(2): 87-91

Brovelli, E. A., Brecht, J. K., Sherman, W. B., & Sims, C. A. (1999). Anatomical and physiological responses of melting- and nonmelting-flesh peaches to postharvest chilling. *Journal of American Society for Horticultural Science, 123*, 668–674.

Brummell, D. A., Dal Cin, V., Lurie, S., Crisosto, C. H., & Labavitch, J. M. (2004). Cell wall metabolism during the development of mealiness in cold-stored peach fruit: 494 association of mealiness with arrested disassembly of cell wall pectins. *Journal of Experimental Botany, 55*, 2041–2052.

Brummell, D. A. & Harpster, M. H. (2001). Cell wall metabolism in fruit softening and quality and its manipulation in transgenic plants. *Plant Molecular Biology, 47*, 311–339

Byers, R. E. (1997). Peach and nectarine fruit softening following Aminoethyoxyvinylglycine sprays and dips. *Hortscience, 32*, 86–88.

Callahan, A. M., Scorza, R., Bassett, C., Nickerson, M., & Abeles, F. B. (2004). Deletions in an endopolygalacturonase gene cluster correlate with non-melting flesh texture in peach. *Functional Plant Biology, 31*, 159–168.

Chamberlain, E. E., Atkinson, J. D., & Hunter, J. A. (1959). Two diseases of plum causing distortion and internal necrosis of fruits. *New Zealand Journal of Agricultural Science, 2*, 174–183.

Cheng, G. W., & Crisosto, C. H. (1994). Development of dark skin discolouration of peach and nectarine fruits in response to exogeneous contaminations. *Journal of American Society of Horticultural Science, 11*, 529–533.

Christensen, J. V. (1972). Cracking in cherries III. Determination of cracking susceptibility. *Acta Agriculturae Scandinavica, 22*, 128–136.

Christensen, J. V. (1996). Rain-induced cracking of sweet cherries: Its causes and prevention. In A. D. Webster & N. E. Looney (Eds.), *Cherries: Crop physiology, production and uses* (pp. 297–330). London: CAB International.

Chunxian, C., William, R. O., & Thomas, G. B. (2016). Peach fruit set and buttoning after spring frost. *HortScience, 51*(7), 816–821.

Corrons, C. H. (1922). Plant breeding. A summary of results. *Proceeding of American Society of Horticultural Science, 19*, 108–115.

Crisosto, C., Johnson, R. S., Luza, J., & Day, K. (1993). Incidence of physical damage on peach and nectarine skin discoloration development: Anatomical studies. *Journal of American Horticultural Science, 118*, 796–800.

Crisosto, C. H., & Day, K. R. (2012). Stone fruits. In *Crop post-harvest: Science and technology* (pp. 212–225). Oxford: Black Well.

Crisosto, C. H., & Labavitch, J. M. (2002). Developing a quantitative method to evaluate peach (Prunus persica) flesh mealiness. *Postharvest Biology and Technology, 25*, 151–158.

Crisosto, C. H., Lurie, S., & Retamales, J. (2008). Stone fruits. In *Modified and controlled atmosphere* (pp. 287–316). Retrieved from www.researchgate.net/publication/21120508.

Crisosto, C. H., Mitchell, F. G., & Ju, Z. (1999a). Susceptibility of chilling injury of peach, nectarine and plum cultivars grown in California. *Hortscience, 34*, 264–266.

Crisosto, C. H., Mitchell, F. G., & Ju, Z. (1999b). Susceptibility to chilling injury of peach, nectarine, and plum cultivars grown in California. *HortScience, 34*, 1116–1118.

Dirlewanger, E., Graziano, E., Joobeur, T., Francesc, G. C., Cosson, P., Howad, W., & Arus, P. (2004). Comparative mapping and marker assisted selection in Rosaceae fruit crops. *Proceedings of National Academy of Science USA, 101*(26), 9891–9896.

Dodd, M. C. (1984). Internal breakdown in plums. *Deciduous Fruit Grower, 34*, 255–256.

Eleni, T., Alexios, N., Polidoro, S., & Athanasios, S. (2007). Characterization and expression analysis of *FRUITFULL*- and *SHATTERPROOF*-like genes from peach (*Prunus persica*) and their role in split-pit formation. *Tree Physiology, 27*, 649–659.

Evert, D. R., Gallies, T. P., & Mullinix, B. G. (1988). Effects of split pit on elemental concentrations of peach fruit during pit hardening. *Scientia Horticulture, 34*(2), 55–65.

Ferguson, I., Volz, R., & Woolf, A. (1999). Pre-harvest factors affecting physiological disorders of fruits. *Post-Harvest Biology and Technology, 15*, 255–262.

Fernandiz, T. J. P., Nock, J. P., & Watkins, C. B. (2001). Superficial scald, carbon dioxide injury and changes of fermentation products and organic acids in 'Cortland' and 'Law Rome apple' fruit after high carbon dioxide stress treatment. *Journal of American Society of Horticultural Science, 126*, 235–241.

Field, R. J. (1985). The effect of temperature on ethylene production by plant tissues. In J. A. Roberts & G. A. Tucker (Eds.), *Ethylene and plant development* (pp. 47–69). Sevenoaks: Butterworths.

Fukai, N. (1995). Cherry physiological disorder. In N. Fukai (Ed.), *Cultivation technology in sweet cherry and pear (in Japanese)* (pp. 115–117). Tokyo: Yokendo.

Garcia-Quero, J., Fodor, A., Reigneir, A., Capdeville, G., Joly, J., Tauzin, Y., Foullhaux, L., & Dirlewanger, E. (2014). QTL detection of important agronomic traits for sweet cherry breeding. *Acta Horticulture, 1020*, 57–64.

Gibert, C., Lescorruet, F., Génard, M., Vercambre, G., & Perez, A. (2005). Modelling the effect of fruit growth on surface conductance to water vapour diffusion. *Annals of Botany, 95*, 673–683.

Ginsburg, L., & Combrink, J. C. (1972). Cold storage of apricots. *Dried Fruit, 4*, 19–23.

Guimond, C. M., Andrews, P. K., & Lang, G. A. (1998). Scanning electron microscopy of floral initiation in sweet cherry. *Journal of American Society for Horticultural Science, 123*, 509–512.

Hovland, K. L., & Sekse, L. (2003). The development of cuticular fractures in fruits of sweet cherries (*Prunus avium* L.) can vary with cultivar and rootstock. *Journal of American Pomological Society, 57*, 58–62.

Jarvis, M. C., Briggs, S. P., & Knox, J. P. (2003). Intercellular adhesion and cell separation in plants. *Plant, Cell and Environment, 26*, 977–989.

Jing, Y., Ma, X., Jin, P., & Zhu, X. (2018). Effect of harvest maturity on chilling injury and storage quality of apricots. *Journal of Food Quality, 18*, 1–8.
Ju, Z., Duan, Y., & Ju, Z. (1999). Combinations of GA3 and AVG delay fruit maturation, increase fruit size and improve storage life of 'Feicheng' peaches. *Journal of Horticultural Science and Biotechnology, 74*, 579–583.
Ju, Z., Duan, Y., & Ju, Z. (2000). Leatheriness and mealiness of peaches in relation to fruit maturity and storage temperature. *Journal of Horticultural Science and Biotechnology, 75*, 86–91.
Kader, A. (2002). *Postharvest technology of horticultural crops*. Publication 3311. University of California Agricultural and Natural Resources.
Kader, A. A., & Mitchell, F. G. (1998). Postharvest physiology. In J. H. LaRue & R. S. Johnson (Eds.), *Peaches, plums, nectarines: Growing and handling for fresh market. University of California DANR Publication 3331* (pp. 154–164). Davis: University of California, Davis.
Kasai, S., Hayama, H., Kashimura, Y., Kudo, S., & Osanai, Y. (2008). Relationship between fruit cracking and expression of expansin gene MdEXPA3 in Fuji apples (Malus domestica Borkh.). *Scientia Horticulturae, 116*, 194–198.
Khan, F. A., & Bhat, S. A. (2017). *Physiological disorders and their management in stone fruits*. Retrieved form http/researchgatepublication/313611588.
Knoche, M., Beyer, M., Peschel, S., Hinz, M., & Bukovoc, M. G. (2001). Studies on water transport through the sweet cherry fruit surface conductance of the cuticle in relation to fruit development. *Planta, 213*, 927–936.
Knoche, M., Beyer, M., Peschel, S., Operlakov, B., & Bukovoc, M. G. (2004). Changes in strain and deposition of cuticle in sweet cherry fruit. *Journal of Plant Physiology, 120*, 667–677.
Lester, D. R., Sherman, W. B., & Atwell, B. J. (1996). Endopolygalacturonase and the melting flesh (M) locus in peach. *Journal of American Society for Hortcultural Science, 121*, 231–235.
Lill, R. E., Donoghue, E. M., & King, G. A. (1989). Postharvest physiology of peaches and nectarines. *Horticultural Review, 11*, 413–452.
Marchner, H. (2002). *Mineral nutrition of higher plants*. San Diego: Academic Press.
Measham, P. (2011). *Rain induced fruit cracking in sweet cherry* (p. 18). Thesis, School of Agricultural Science, University of Tasmania.
Measham, P. F., Garcie, A. J., Bound, S., & Wilson, S. J. (2012). Crop load manipulation and fruit cracking in sweet cherry (Prunus avium. L.). *Advances in Horticultural Sciences, 26*(1), 25–31.
Meli, T. (1982). Abdecken von Susskirschen-heckenmitPolyathylenfolien: I. Voraus-setzungen fur das Abdecken. II. Technik des Abdeckens. III. Betriebs- und arbeit- swirtschaftlicheAspekte (Covering sweet cherry hedges with polyethylene. I. Prerequisites of covering. II. Techniques of covering. III. Plantation and labour economy aspects). *Scweizerische Zeitschrift fur Obstund Weinbau, 110*(25–27), 761–768, 778–794, 812–821.
Okie, W. R. (1998). *Handbook of peach and nectarine varieties: Performance in the southeastern United States and index of names* (p. 714). Washington, DC: U.S. Department of Agriculture, Agricultural Research Service.
Peace, C. P., Crisosto, C. H., Garner, D. T., Dandekar, A. M., Gradziel, T. M., & Bliss, F. A. (2006). Genetic control of internal breakdown in peach. *Acta Horticulture, 713*, 489–496.
Peace, C. P., Crisosto, C. H., & Gradziel, T. M. (2005). Endopolygalacturonase: A candidate gene for freestone and melting flesh in peach. *Journal of Molecular Breeding., 15*, 420–427.
Phillips, D. J. (1988). Reduction of transit injury-associated black discoloration of fresh peaches with EDTA treatments. *Plant Disease, 72*, 118–120.
Powers, W. L., & Bollen, W. B. (1947). Control of cracking of fruit by rain. *Science, 105*, 334–335.
Raison, J. K., & Lyons, J. M. (1986). Chilling injury: A plea for uniform terminology. *Plant, Cell & Environment, 9*, 685–686.
Roversi, A., Monteforte, A., Panelli, D., & Folini, L. (2008). Observations on the occurrence of sweet cherry double-fruits in Italy and Slovenia. *Acta Horticulturae., 795*, 849–854.
Saure, M. C. (2005). Calcium translocation to fleshy fruit: Its mechanism and endogenous control. *Scientia Horticulturae., 105*, 65–89. https://doi.org/10.1016/j.scienta.2004.10.003.
Sekse, L. (1995). Fruit cracking in sweet cherries (*Prunus avium* L.). some physiological aspects - a mini review. *Scientia Horticulturae, 63*, 135–141.

Sharma, R. R. (2006). Fruit cracking: Causes and control. In *Fruit production: Problems and solutions* (pp. 73–86). Lucknow: International Book Distributing Co.

Simon, G. (2006). Review on rain induced fruit cracking of sweet cherries (*Prunus avium* L.), its causes and the possibilities of prevention. *International Journal of Horticultural Science., 12*, 27–35.

Singh, N., Singh, G., Thakur, K. K, Kumar, S., & Sharma, S. S. (2018). *Causes and Remedies of physiological disorders in Stone fruits crops*. Retrieved from www.researchgate.net/publication/328937571.

Tani, E., Polidoros, A. N., & Tsaftaris, A. S. (2007). Characterization and expression analysis of FRUITFULL-and SHATTER-PROOF-like genes from peach (*Prunus persica*) and their role in split-pit formation. *Tree Physiology, 27*, 649–659.

Taylor, B. H., & Taylor, D. G. (1998). Flower bud thinning and winter survival of 'Redhaven' and 'Cresthaven' peach in response to GA_3 sprays. *Journal of the American Society for Horticultural Science., 123*(4), 500–508.

Taylor, M. A., & De Kock, V. A. (1991). Effect of harvest maturity and storage regimes on the storage quality of Peeka apricots. *Deciduous Fruit Grower, 41*, 139–143.

Torres, C. A., Yuri, A., Bastias, R., & Lepe, V. (2009). Use of lipophilic coatings to reduce sweet cherry (*Prunus avium* L.) rain cracking. In: *Proceedings of the 6th international cherry symposium*, Renaca.

Vincken, J. P., Schols, H. A., Oomen, R. J., McCann, M. C., Ulvskov, P., Voragen, A. G., & Visser, R. G. (2003). If homogalacturonan were a side chain of rhamnogalacturonan. I. Implications for cell wall architecture. *Plant Physiology, 132*, 1781–1789.

Weigel, D., & Meyerowitz, E. M. (1994). The ABCs of floral homeotic genes. *Cell, 78*, 203–209.

Whiting, M. D. (2001). *Whole canopy source-sink relations and fruit quality in 'Bing' sweet cherry trees on a dwarfing, precocious rootstock*. Unpublished PhD dissertation, Washington State University, Pullman.

Yamamoto, T., Kudo, M., & Watanabe, S. (1990). Fruit cracking and characteristics on fruit thickening in Satonishiki cherry. *Journal of Japanese Society for Horticultural Science., 59*, 325–332.

Zhou, H. W., Ben-Arie, R., & Lurie, S. (2000). Pectin esterase, polygalacturonase and gel formation in peach pectin fractions. *Phytochemistry, 55*, 191–195.

Orchard Factors Affecting Postharvest Quality of Stone Fruits

8

Kalpana Choudhary, Nirmal Kumar Meena, and Uma Prajapati

Abstract

Quality is the major factor which is perceived by the consumer. In stone fruits, quality criteria can be judged by appearance, colour, size, fragrance and their internal composition such as sugars, phenolics, specific gravity, total solids and other biochemical constitutes. Stone fruits are considered as a rich basket of phytonutrients. There are numerous factors which determine quality of the fruits. Orchard factors like orchard soil, microclimate, quality of irrigation water, application of nutrients, rootstock, canopy management, pollination and use of hormones are more prominent which influence overall quality and acceptance by the consumer. The reason being they regulate different catabolic and anabolic processes which lead to further improvement in quality characteristics. Climatic conditions affect from its cultivation to harvest. High temperature leads to injury, whereas frost also caused damage to cells. Fruit size and colour development in temperate stone fruits are highly variable and dependent upon temperature. Mineral nutrition determines various levels of photosynthates, sugars, acids and other compounds. Deficiency of these leads to various physiological disorders. Suitable rootstocks control accumulation of nutrients and water from the soil and also influence vigour of the tree. Some salt-resistant rootstock improves fruit quality in mango, peach and ber. Similarly, training and pruning are major aspect in temperate stone fruits. Hormones play an important role in governing various metabolic processes and enzymes. They create resistance against various diseases and stresses. Therefore, combination of these orchard factors not only

K. Choudhary · N. K. Meena (✉)
Department of Fruit Science, College of Horticulture and Forestry, (Agriculture University, Kota), Jhalawar, India

U. Prajapati
Division of Food Science and Postharvest Technology, ICAR-Indian Agricultural Research Institute, New Delhi, India

M. M. Mir et al. (eds.), *Production Technology of Stone Fruits*,
https://doi.org/10.1007/978-981-15-8920-1_8

influences the quality but also alters the shelf life. This chapter will be discussing about various orchard factors which influence quality and food value of stone fruits.

Keywords

Stone fruits · Minerals · Quality · Hormones · Firmness

8.1 Introduction

Fruits play an important role in the economy of a country and sustainable development of the people. Fruits not only are pleasing to the olfactory sense but also protect from several diseases owing to their nutraceutical values. During the supply chain management, maintaining the optimum quality and shelf life is a major challenge. The quality of fruits creates a dynamic composite of the physicochemical properties concerning horticultural commodities and consumer view (Kyriacou and Rouphael 2018). Basically, it is made up of numerous elements of the product such as intrinsic features that include key external attributes like shape, size, colour, texture, sweetness, acidity, flavour, aroma and nutritional value, and it is free from defects, likewise extrinsic traits of production and distribution systems (including the chemicals used during production), type of packaging with their recycling capacity and sustainability of production and distribution (in relation to energy utilization), which influence consumer acceptability and decision. Quality cannot be enhanced after harvest; it can only be retained. As the consumer demand is increasing with changing trends, therefore, a great attention must be emphasized on quality of produce. Most of the temperate stone fruits belong to genus *Prunus* which consists of exocarp, mesocarp and hardened endocarp with seed kernel (Pande Kamal 2017; Hazra et al. 2012). Peach, plum, nectarine, cherry, apricot, almond, walnut, etc. are temperate stone fruits, whereas mango, ber, coconut, lasoda, etc. are tropical and subtropical stone fruits. Due to their high-perishable nature, a large amount of produce goes to waste. As far as quality is concerned, preharvest factors or orchard factors play determining role (Meena and Asrey 2018a, b). Most of stone fruits are climacteric in nature (except cherry and ber) which tend to deteriorate rapidly after harvesting. Orchard factors starting from orchard climate; soil condition; genotypes; rootstocks; interstocks; canopy management; thinning; girdling; use of plant growth promoters, fertilizer and nutrient management; assisted pollination; age of tree and rootstock; etc. help in shaping the final quality (Asrey et al. 2018; Asrey et al. 2013; Meena and Asrey 2018a, b; Baghel et al. 2019). Heavy stress in the form of either excess or deficiency is detrimental to plant growth and development. Heavy nitrogenous fertilizer triggers vegetative growth and leads to susceptibility towards diseases with inferior quality of fruits (Daane et al. 1995; Michailides et al. 1992).

In the climate-resilient scenario, cumulative effects of stresses are also reported as detrimental. Under such scenario, use of good horticultural practices may be effective to enhance inherent quality and shelf life. Use of heat and drought-resistant varieties and rootstocks, interventions in canopy management, use of novel phytohormones and use of eco-friendly and cost-effective protection measures could be fruitful in betterment of quality and sustainability of stone fruit production. This chapter reviews several orchard factors which greatly influence stone fruit quality and shelf life.

8.2 Effect of Climatic Factors

Ecological factors like light, temperature, relative humidity, CO_2 and rainfall are major direct and indirect factors which affect the quality of stone fruits. These factors directly affect photosynthesis of plant which provides energy to the plant. Many environmental factors (such as temperature, light and atmospheric composition) actively regulate major physiological and biochemical processes such as photosynthesis, respiration, transpiration, transportation of solutes and anabolic and catabolic metabolism from fruit development to ripening which affects fruit quality (Yahia 2019). Various environmental factors (like temperature, solar radiation, photoperiod, precipitation and soil profile) affect growing environment which results in a wide variation in quality of harvested peach fruits (Lopresti et al. 2014).

8.3 Effect of Quality and Quantity of Water

The quality of water in fruit production refers to its suitability for irrigation, and the availability of quality water is crucial for cultivation of fruit crops. Recently, good-quality water is becoming a limiting factor for agricultural production due to competitiveness of other sectors (Asrey et al. 2018). Irrigation with quality water may affect yield, colour, appearance and the composition of stone fruits. The quality of pear fruits in terms of sugar content and acidity was improved when irrigated with saline water through subsurface drip irrigation (Oron et al. 2002). Water stress favours the synthesis of volatile antioxidants in fruits. It was found that water deficit during stage II significantly improves the quality of nectarine in terms of total sugars, sucrose and sorbitol and decreases total acids and malic acid, while in peach, water stress during stage II, advanced fruit ripening, improves important sensory attributes, such as sweetness and flavour intensity (Thakur and Singh 2013; Valverdu et al. 2012). Alcobendas et al. (2013) found in peach that fruits from regulated deficit irrigation trees had more soluble solids, glucose, sorbitol and malic, citric and tartaric acids. The effect of deficit irrigation in apricot was found to be beneficial over full irrigation; it slightly increases total soluble solid (TSS) and fruit firmness at harvest as well as during cold storage (Perez-pastor et al. 2007). Similarly, it was observed by Maatallah et al. (2015) that deficit irrigation (at 50% ETc) during fruit growth stage increases soluble solids, fruit firmness, total phenolics and flavonoids, whereas deficit irrigation during stage II of fruit growth makes fruits more sweeter as compared to fruits from the trees with full irrigation (Intrigliolo et al. 2013) in plum.

8.4 Effect of Mineral Nutrition (Manure/Fertilizers)

Mineral nutrition is the major preharvest factor that influences the quality of stone fruits. Fruit quality problems are related to fruit mineral deficiencies (Aktas and Ates 1998; Spectrum Analytic Inc 2006). The balanced use of nutrients is very important since absence/deficiency as well as the toxicity of these nutrients affects fruit quality. The deficiency of one element can negatively affect plant development and thus inhibit optimum uptake, utilization or distribution of other elements. The absence or deficiency of any plant nutrient can affect different plant processes such as inhibiting the uptake and distribution of other nutrients (Johnson and Uriu 1989) that cause many pre- and postharvest physiological maladies. However, it mainly depends upon soil type and fruit crop. It is well known that amount, formulations, rate and time of application of different fertilizers affect the quality parameters of stone fruit crops. The role of calcium in stone fruit crops is well recognized, maintenance of cell membrane integrity (Yamaguchi et al. 1986). Preharvest spray of Ca-based formulations helps in maintaining high firmness, thickness of cell wall and lower rate of degrading enzymes like PME and PG. Use of calcium chloride as preharvest treatments helps in reduction of losses in weight, activity of enzymes in peach and postharvest decay in cherry (Barwal and Kumar 2014; Vangdal et al. 2006). The role of calcium in stone fruit crops is well recognized, maintenance of cell membrane integrity (Yamaguchi et al. 1986). In mango, lesser use of micronutrients Zn and B is responsible for the reduction in yield and fruit quality in mango orchards (Wahdan et al. 2011), whereas the yield and quality increase in response to combined application of Zn and B which may be due to improvement in sugar concentrations, vitamins and other physiological features in this crop (Saleh and Abd El-Moneim 2003). However, higher doses of N are thought to negatively influence fruit quality (Souza et al. 2013). It was reported in a study that the firmness of apricots was reduced with increased rate of nitrogen from the recommended dose for the crop (Rettke et al. 2006). It was observed that amounts of nutrients significantly influence the quality in terms of increase in TSS, sucrose and malic acid content, whereas glucose and fructose decrease (during fruit maturation) in peach (Fajt and Veberic 2002). Application of nutrients with irrigation water also affects the stone fruits' quality in terms of TSS, acidity and total sugar content in cherry, nectarine and peach (Ahmad et al. 2010; Singh et al. 2015; Verma et al. 2017).

8.5 Rootstocks/Interstock

It is well understood that selection of appropriate rootstocks is associated with the success of any fruit orchard. These are the essential component of fruit production due to their ability of adapting a specific scion to varied environmental conditions. Rootstocks have potential to modify the architecture of trees for efficient utilization of resources. Rootstock can ameliorate soil with enhancement in water and nutrient use efficiency leading to conservation. Rootstocks play an important role in improving quality of stone fruit crops. Rootstocks influence the quality of stone fruits such as peach (Reig et al. 2017), plum (Rato et al. 2008), cherries (Ersoy et al. 2010), apricot (Milošević et al. 2015) and ber (Prasad and Bankar 2006).

The Rainha Claudia Verde cultivar of plum produced the largest fruit size and fruits with higher calcium level when grafted on GF8-1 rootstock, whereas the fruits from GF10-2 rootstock exhibited the highest firmness of pulp (Rato et al. 2008). When peach cultivar Rich May was grafted on Penta rootstock, it yielded larger-size fruits with better skin colour and higher total polyphenol content and antioxidant capacity (Gullo et al. 2014). López-Ortega et al. (2016) reported that Piku 3 rootstock provides the highest-quality fruits in terms of fruit size, fruit firmness and soluble solids in sweet cherry cultivar Newstar. Similarly, the fruits of 'Jolico' cherry planted on rootstock *Cornus amomum* seedlings showed an increased content of anthocyanin and higher content of polyphenolic compounds and total anthocyanin (Ochmian et al. 2019). The hexaploid rootstocks of peach on 'Pollizo' plums, Adesoto 101 and PM 150 AD, can increase sweetness (including higher sugar and organic acid content) in the Big Top nectarine; likewise, diploid plum-based hybrids (PADAC 04-01, PADAC 99-05 and Rootpac R) increases sugar and beneficial antioxidants such as total phenols, flavonoids and vitamin C (Forcada et al. 2019). In apricot, the highest total soluble solid content was detected in cultivar Buda, and highest total acid content was reported in cultivar 'Novosadska Kasnocvetna' when grafted on interstock *Prunus spinosa* L. (Miodragović et al. 2019). The highest sugar content (sucrose) was observed as 24.178 mg 100 g^{-1} in Hacıhaliloğlu cultivar of apricot when it was grafted on Pixy rootstock, whereas in the same variety grafted on Tokaloğlu rootstock, an increment was recorded in phenolic compounds (219.440 mg 100 g^{-1} chlorogenic acid), organic acid (1030.730 mg 100 g^{-1} malic acid) and antioxidant capacity (Gündoğdu 2019). Vazquez-Luna et al. (2011) reported higher firmness as well as 3-carene levels and main flavonoid content when Manila mango was grafted on Criollo rootstock. The fruit quality of ber cultivar Seb in terms of size, weight and TSS was found better when budded on *Ziziphus rotundifolia* and *Z. mauritiana* rootstocks (Prasad and Bankar 2006). Effect of rootstocks on quality and shelf life on various fruits is depicted in Table 8.1.

8.6 Effect of Canopy Management

Canopy in fruit tree refers to physical composition of the tree which is composed of the stem, branches, shoots and leaves (Vandan et al. 2017). It is the foremost driver of energy from sunlight (source) to the sink organs particularly the fruits. Hence, transformation of sunlight energy into fruits is its major task in stone fruit trees. Canopy management of the fruit trees deals with the maintenance of proper structure in relation to the shape and size for maximum productivity and quality produce. To make the best use of land and the climatic factors to increase productivity in a three-dimensional approach is the basic concept in canopy management of a perennial tree (Sharma 2016). The major emphasis of canopy management is usually to provide proper sunlight and aeration to the plant by reducing excessive canopy shading, because it is not always wise to allow a plant

Table 8.1 Quality of stone fruits as affected by different rootstocks/interstock

Crop	Cultivar	Rootstock	Inferences	References
Mango	Manila	Criollo	Higher firmness as well as 3-carene levels and main flavonoid content	Vazquez-Luna et al. (2011)
Ber	Seb	*Ziziphus rotundifolia* and *Z. mauritiana*	Better size, weight and TSS	Prasad and Bankar (2006)
Peach	Rich May	Penta	Larger fruit with better skin colour, greater total polyphenol content and antioxidant capacity	Gullo et al. (2014)
Plum	Rainha Claudia Verde	GF8-1, GF10-2	Largest fruit size, higher calcium fruit level and highest firmness	Rato et al. (2008)
Nectarine	Big Top nectarines	Adesoto 101 and PM 150 AD	Higher TSS	Forcada et al. (2019)
Apricot	Buda, Novosadska Kasnocvetna	Interstock *Prunus spinosa* L.	Highest total soluble solids and highest total acid content	Miodragović et al. (2019)
Apricot	Hacıhaliloğlu cultivar	Pixy	The highest sugar content (sucrose)	Gündoğdu (2019)
Cherry	Jolico	*Cornus amomum* seedlings	Increased content of anthocyanin, a higher content of polyphenolic compounds and total anthocyanin	Ochmian et al. (2019)
	Newstar	Piku 3	Highest-quality fruits in terms of fruit size, fruit firmness and soluble solids	López-Ortega et al. (2016)

to grow naturally, since unwanted portions may develop at the disbursement of those which are essential. Training and pruning are the major tools of canopy management in stone fruit crops.

Pruning involves the judicious removal of vegetative growth (sometimes fruit thinning also) and plays an important role in canopy management of fruit trees. Proper pruning keeps the plant in such shape and condition as to yield fruits of desired quantity and quality. An unpruned tree is habitually very large and reduces light penetration inside the canopy (Asrey et al. 2013).

It was found in peach that pruning, leaf removal and other canopy management practices during the later stages of fruit growth, SSC and size of fruit can be effectively improved if there is better light interception (He et al. 2008). Pruning intensities or severity of pruning also affects the quality of stone fruits. Gopu et al. (2014) observed that maximum total soluble solids (TSS), total sugars and non-reducing sugars of mango fruits were recorded in total removal of past season's growth, whereas highest total carotenoid content was recorded in moderate pruning and reducing sugars in heavy pruning. Highest TSS was reported in the severely pruned trees, while TSS/acid ratio was higher in the lightly pruned mango trees (Singh et al.

2010). Better performance of fruits from pruned trees in respect to total carotenoid content might be due to elevation in plant photosynthesis and the congenial microclimate created through pruning (Asrey et al. 2013).

It was proved that quality parameters of stone fruits are significantly affected by different training systems (Iannini et al. 2001; Lu et al. 2003). Lewallen (2000) observed in 'Norman' peach trained to an open-vase form that soluble solid content was significantly reduced in artificially shaded branches of trees. Sobierajski et al. (2019) showed that the fruiting wall training system could be an alternative in peach to Brazilian's growers for producing higher fruit quality in terms of fruit weight, pulp firmness and titratable acidity (Table 8.2). It was found that the Y-shaped Tatura trellis tree training system is better than other traditional tree training systems and could be potentially utilized for higher-quality commercial nectarine cultivation (Lal et al. 2017). Roussos et al. (2011) reported that the quality of apricot fruit in relation to carbohydrate concentration and sweetness index total phenol concentration improved with thinning. It enhances fruit weight with a subsequent decrease in fruit firmness, while antioxidant capacity of the pulp was not influenced by thinning. Fruit firmness is mainly affected by sun exposure and mineral nutrition; thus, canopy management to increase the penetration of sunlight typically leads to firmer fruits (Yahia 2019).

8.7 Effect of Pollination and Pollinizers

Pollination is an important tool for quality fruit production. It is transfer of viable pollen grains from mature anther to receptive stigma. Pollination in fruit trees can take place by their own pollen grain called as self-pollination or can take place by pollen grains of other varieties or species called as cross-pollination. Common

Table 8.2 Effect of training and pruning on quality parameters of fruits

Crop	Training/pruning	Outcomes	References
Nectarine	Fruiting wall training system	Higher fruit quality in terms of fruit weight, pulp firmness and titratable acidity	Sobierajski et al. (2019)
Peach cv. Ruipan 5	Slanting central leader system	Good fruit quality	Lu et al. (2003)
'Norman' peach	Open-vase form	Soluble solid content was significantly reduced in artificially shaded branches of trees	Lewallen (2000)
Mango	Severely pruning	TSS and reducing sugars	Gopu et al. (2014), Singh et al. (2010)
Apricot	Thinning	Carbohydrate concentration, sweetness index total phenol concentration improved	Roussos et al. (2011)
Peach	Pruning, leaf removal	SSC and size of fruit can be effectively improved	He et al. (2008)

example of self-pollinated fruits are apricots, nectarines, peaches and sour cherries, while cross-pollinated crops are apples, pears, plums, strawberry and sweet cherries. Cross-pollination can take place by many ways, viz. by insect, bird, wind, water, animals and hand pollination.

Cross-pollination by insects in almond fruits was found to have better-quality kernels in terms of fat and vitamin E composition in comparison to self-pollinated fruits. On the other hand, self-pollinated fruits produce lower proportion of oleic acid and linolenic acid which is less desirable from health and commercial point of view (Brittain et al. 2014). Apart from mode of pollination, pollen source also affects quality of fruit and fruit set percentage. Studies conducted on Japanese plum showed that pollen sources significantly affect fruit characteristics such as fruit length, diameter, fresh weight, dry matter content and firmness. Biochemical quality parameters like pH, total acid, ascorbic acid and total phenol content were not affected by pollen source. Out of six pollinizer varieties viz. 'Goje', 'Sabz', 'Simka', 'Shablon', 'Methley', 'Myrobalan' and 'Black Star', Simka showed best results when used as pollen source for 'Black Star' cultivar of Japanese plum (Barzamini and Ghazvini 2017).

8.8 Effect of Hormones

Plant hormones are widely known as plant growth regulators and are involved in fruit development and ripening process. There are many plant growth regulators which are involved in production of quality stone fruits. Plant hormones are categorized in to two basic groups, viz. plant growth promoters (auxin, cytokinin and gibberellin) and plant growth inhibitors (abscisic acid and ethylene). Stone fruits like cherry, apricot, nectarines, peaches and plums show diverse ripening behaviour, viz. non-climacteric and climacteric. Plant hormones play diverse role in managing quality parameters of stone fruits.

Auxin is considered as the main regulator for fruit development in both climacteric and non-climacteric fruits (Lurie 2010a, b). Small fruit size is one of the major limiting factors in marketing of stone fruits which happens because of excessive fruit set. Synthetic auxin was found to be effective in fruit growth when applied to the second stage of fruit development. Auxin plays an important role in increasing the cell size, thus improving the fruit size even without thinning in stone fruits. Application of synthetic auxin on 'Canino' apricot at the start of pit hardening stimulated cell enlargement in the mesocarp and improved keeping quality of fruits. Apart from positive effect of auxins, its higher concentration caused cracking and internal browning of fruits (Stern et al. 2007a). Similar studies in cherry fruits were done where synthetic auxins, viz. 2,3,6-TPA, 2,4-DP and 2,4-D, were applied with NAA and were found to be effective in increasing total yield and fruit quality in terms of increased fruit size and enhanced colour. Hence, auxin was also found effective in advancing harvest by a few days (Stern et al. 2007b). Application of auxin at pit-hardening stage in Japanese plum increased fruit size and yield (Stern et al. 2007c).

Another important growth promoter applied to stone fruits is gibberellin. Application of gibberellin shows varied response in different stone fruits. It helps in fruit thinning, reduces differentiation of flower buds, influences fruit quality, delays fruit harvesting and improves storability of fruits. GA3 was found effective in maintaining firmness, decreasing softening, delaying maturity (Choi et al. 2002) and reducing fruit cracking (Usenik et al. 2005) of cherry fruits. Other studies on cherry fruits also showed positive effect of GA3 on improving sensory quality, higher firmness, reducing bruises and delayed pitting during fruit storage (Lurie 2010a, b). GA3 was not able to maintain antioxidant content, and phenol content was found similar to untreated fruits (Erogul and Sen 2015). One hundred mg L^{-1} of GA3 was applied on harvested peaches and stored at 2 °C for 14 days. Treatment was found effective in reducing the susceptibility of the fruit to mechanical damage and reducing ethylene emission and respiration rate which reflects a delayed ripening process (Martinez-Romero et al. 2000). GA3 effects were also analyzed in 'Patterson' apricot fruits where trees were sprayed with GA3 at 100 ppm and were found effective in advancing maturity as indicated by colour. Fruit soluble solids and fruit weight were increased in fruits treated with GA3 at 50 and 100 ppm, while fruit firmness was reduced in both the treatments (Southwick et al. 1995). Other than gibberellin, its biosynthesis inhibitors like paclobutrazol and uniconazole also play a role in increasing fruiting and reducing shoot growth to half in stone fruits (Lurie 2010a, b).

Cytokinin is another plant growth regulator which promotes cell division and acts as antisenscent hormone. Besides auxin and gibberellin, cytokinin is also known to induce fruit set in several fruit crops (Matsuo et al. 2012). The level of endogenous cytokinin is directly correlated with fruit growth by stimulation of cell division. External application of cytokinin leads to formation of parthenocarpic fruits (Kumar et al. 2014). Effect of cytokinin was studied on 'Bing' sweet cherry fruits where it was found to increase fruit firmness and fruit soluble solid concentration but delayed exocarp colouration (Zhang and Whiting 2011).

Ethylene is the gaseous plant hormone associated with fruit ripening and senescence. It is produced in large quantity by climacteric fruits. Ethylene is involved in many catabolic and anabolic changes that occur during ripening such as degradation of chlorophyll, softening, respiration and production of volatiles. Its inhibition or removal from atmosphere can slow down the ripening process. Non-climacteric fruits produce ethylene but in very small quantity (Lurie 2010a, b). Exogenous application of ethylene has been utilized to greater extent so far. In apricot, ethylene treatment after fruit harvest has shown to enhance yellow colour development (Brecht et al. 1982). Propylene treatments to apricots have been shown to induce ethylene production and carotenoid accumulation in apricots (Kita et al. 2007). Effect of ethylene on peach fruit woolliness was analyzed where it was found that ethylene is not recommended for its control as it encourages fruit decay and a significant loss of pulp firmness (Girardi et al. 2005). In contrary to this, woolliness in nectarine showed almost opposite response on the application of ethylene. In other fruit crops, ethylene is the reason behind incipient damage, but in nectarines, it has shown positive effects (Zhou et al. 2001). Exogenous effect of ethylene at 0 °C storage temperature was not found to affect storage disorder and quality parameters in 'Red Rosa' plum, but it affected fruit-softening process, hence reducing fruit firmness. In contrast to this, 1-MCP (ethylene inhibitor) was found to maintain fruit firmness and can be commercially utilized for extending shelf life of plum fruits (Dong et al. 2001).

Abscisic acid (ABA) is considered as another ripening control factor after ethylene. It is found in very low amount in unripe fruits and increases as fruit ripens; therefore, it is believed that ABA plays an important role in regulating the rate of fruit ripening. ABA level increases before maturation and decreases towards harvest which showed that it may be involved in ripening process of sweet cherry fruits (Kondo and Gemma 1993). Endogenous ABA was involved in the beginning of ripening process in sweet cheery, and application of exogenous ABA stimulated ethylene production by regulating ACO gene expression and hence accelerated fruit ripening (Ren et al. 2011). In 'Hakuho' peach, application of ABA (500 mg/L) on fruit 90 days after full bloom (DAFB) have enhanced sugar accumulation due to stimulated activity of sorbitol oxidase (SOX) enzyme (Kobashi et al. 1999). Hence from these studies, it was found that ABA not only plays a role in stress mechanism but also is involved in fruit ripening and quality improvement of stone fruits.

There are some other growth regulators such as salicylic acid, brassinosteroids, polyamines and jasmonic acid which were found to induce defence reactions in fruits that help in preventing fungal infection and inducing resistance to low-temperature injury. Polyamine (putrescine, 1 mM) treatment on peach fruit has been found to increase postharvest shelf life by inducing resistance against different diseases and chilling injury. It was also found to reduce respiration rate and maintain firmness and acidity of peach fruits under storage condition (Bal 2013). Effect of salicylic acid and putrescine was analyzed on 'Santa Rosa' plum fruit storability and quality attributes. Exogenous application of putrescine and salicylic acid was found effective in maintaining titratable acidity, ascorbic acid, total phenolics and antioxidant activity during storage at 4 °C. It was also able to reduce weight loss and fruit softening, thus delaying the ripening process and extending the shelf life of plum with its acceptable fruit quality (Davarynejad et al. 2015). Exogenous application of brassinosteroids (BR) at different level has shown to influence both yield and quality characteristic (titratable acidity, anthocyanin and phenol content) of sweet cherry fruits. BRs have also triggered the active cell division process which was associated with increased synthetic anthocyanins (Roghabadi and Pakkish 2014). Application of 0.2 mmol/L methyl jasmonate and 2 mmol/L salicylic acid on apricot fruit under cold storage has shown significant reduction in weight loss, softening and juice pH. It maintained soluble solid content and acidity over the whole storage period. Hence, both methyl jasmonate and salicylic acid treatments were considered best for prolonging apricot storability and maintain quality attribute under cold storage (Ezzat et al. 2017).

8.9 Conclusion

Stone fruits are important commodities in horticultural crops and highly perishable in nature. In spite of huge production, only little amount has export potential. People are focusing more and more on quantity aspect only, whereas quality concept remains neglected. In the changing scenario, both intrinsic and extrinsic qualities must be explored and emphasized. Most of the orchard factors like soil status, rootstocks, pollination, training and pruning are critical which ensure the quality characteristics. The production and supplying of high-quality fruits require more attention to all aspects of the cultivation from field to market, but orchard factors are needed to be well addressed during production. Judicious uses of resources, proper canopy managements and use of recent interventions in orchards could be beneficial to enhance quality and shelf life.

References

Ahmad, M. F., Samanta, A., & Jabeen, A. (2010). Response of sweet cherry (*Prunus avium*) to fertigation of nitrogen, phosphorus and potassium under kerawa land of Kashmir valley. *Indian Journal of Agricultural Sciences, 80*(6), 512–516.

Aktas, M., & Ates, M. (1998). *Bitkilerde beslenme bozuklukları, nedenleri ve tanınmaları*. Ankara: Engin Yayınevi.

Alcobendas, R., Mirás-Avalos, J. M., Alarcón, J. J., & Nicolás, E. (2013). Effects of irrigation and fruit position on size, colour, firmness and sugar contents of fruits in a mid-late maturing peach cultivar. *Scientia Horticulturae, 164*, 340–347.

Asrey, R., Kumar, S., & Meena, N. K. (2018). Influence of water quality on postharvest fruit and vegetable quality. In *Preharvest modulation of postharvest fruit and vegetable quality* (pp. 169–187). London: Academic Press.

Asrey, R., Patel, V. B., Barman, K., & Pal, R. K. (2013). Pruning affects fruit yield and postharvest quality in mango (*Mangifera indica* L.) cv. Amrapali. *Fruits, 68*(5), 367–380.

Baghel, M., Nagaraja, A., Srivastav, M., Meena, N. K., Kumar, M. S., Kumar, A., & Sharma, R. R. (2019). Pleiotropic influences of brassinosteroids on fruit crops: a review. *Plant Growth Regulation, 87*(2), 375–388.

Bal, E. (2013). Effects of exogenous polyamine and ultrasound treatment to improve peach storability. *Chilean Journal of Agricultural Research, 73*(4), 435–440.

Barwal, V. S., & Kumar, J. (2014). Effect of pre-harvest calcium sprays and harvesting time on quality and shelf-life of nectarines. *Advances in Applied Research, 6*(1), 53–56.

Barzamini, S., & Ghazvini, R. F. (2017). Pollinizer influence on fruit quality traits in Japanese plum (*Prunus salicina* Lindl.). *International Journal of Horticultural Science and Technology, 4*(2), 229–237.

Brecht, J. K., Kader, A. A., Heintz, C. M., & Norona, R. C. (1982). Controlled atmosphere and ethylene effects on the quality of California canning apricots and clingstone peaches. *Journal of Food Science, 47*, 432–436.

Brittain, C., Kremen, C., Garber, A., & Klein, A. M. (2014). Pollination and plant resources change the nutritional quality of almonds for human health. *PLoS One, 9*(2), e90082.

Choi, C., Wiersma, P. A., Toivonen, P., & Kappel, F. (2002). Fruit growth, firmness and cell wall hydrolytic enzyme activity during development of sweet cherry fruit treated with gibberellic acid. *The Journal of Horticultural Science and Biotechnology, 77*, 615–621.

Daane, K. M., Johnson, R. S., Michailides, T. J., Crisosto, C. H., Dlott, J. W., Ramirez, H. T., & Morgan, D. P. (1995). Nitrogen fertilization affects nectarine fruit yield, storage qualities, and susceptibility to brown rot and insect damage. *California Agriculture, 49*(4), 13–18.

Davarynejad, G. H., Zarei, M., Nasrabadi, M. E., & Ardakani, E. (2015). Effects of salicylic acid and putrescine on storability, quality attributes and antioxidant activity of plum cv. 'Santa Rosa'. *Journal of Food and Science Technology, 52*(4), 2053–2062.

Dong, L., Zhou, H. W., Sonego, L., Lers, A., & Lurie, S. A. (2001). Ripening of 'red Rosa' plums: Effect of ethylene and 1-methylcyclopropene. *Australian Journal of Plant Physiology, 28*, 1039–1045.

Erogul, D., & Sen, F. (2015). Effects of gibberellic acid treatments on fruit thinning and fruit quality in Japanese plum (*Prunus salicina* Lindl.). *Scientia Horticulturae, 186*, 137–142.

Ersoy, N., Kalyoncu, I. H., Aydin, M., & Yilmaz, M. (2010). Effects of some humidity and IBA hormone dose applications on rooting of M9 apple clonal rootstock softwood top cuttings. *African Journal of Biotechnology, 9*(17), 2510–2514.

Ezzat, A., Ammar, A., Szabo, Z., Nyéki, J., & Imre, J. H. (2017). Postharvest treatments with methyl jasmonate and salicylic acid for maintaining physico-chemical characteristics and sensory quality properties of apricot fruit during cold storage and shelf-life. *Polish Journal of Food and Nutrition Science, 67*(2), 159–166.

Fajt, N., & Veberic, R. (2002). The influence of fertirrigation on the development of fruits and the contents of sugars and organic acids in 'Redhaven' peach. *Acta Horticulturae, 592*, 323–329.

Forcada, C. F., Reig, G., Giménez, R., Mignard, P., Mestre, L., & Moreno, M. Á. (2019). Sugars and organic acids profile and antioxidant compounds of nectarine fruits influenced by different rootstocks. *Scientia Horticulturae, 248*, 145–153.

Girardi, C. L., Corrent, A. R., Lucchetta, L., Zanuzo, M. R., da Costa, T. S., Brackmann, A., Twyman, R. M., Nora, F. R., Nora, L., Silva, J. A., & Rombaldi, C. V. (2005). Effect of ethylene, intermittent warming and controlled atmosphere on postharvest quality and the occurrence of woolliness in peach (*Prunus persica* cv. Chiripa) during cold storage. *Postharvest Biology and Technology, 38*(1), 25–33.

Gopu, B., Balamohan, T. N., Soman, P., & Jeyakumar, P. (2014). Canopy management in mango (*Mangifera indica* L.) cv. Alphonso with reference to flowering, yield and quality characters under ultra high density planting. *Journal of Applied Horticulture, 16*(1), 50–53.

Gullo, G., Motisi, A., Zappia, R., Dattola, A., Diamanti, J., & Mezzetti, B. (2014). Rootstock and fruit canopy position affect peach [*Prunus persica* (L.) Batsch] (cv. Rich May) plant productivity and fruit sensorial and nutritional quality. *Food Chemistry, 153*, 234–242.

Gündoğdu, M. (2019). Effect of rootstocks on phytochemical properties of apricot fruit. *Turkish Journal of Agriculture and Forestry, 43*(1), 1–10.

Hazra, P., Ghosh, S. K., Maity, T. K., Pandit, M. K., & Som, M. G. (2012). *Glossary of horticulture*. New Delhi: Kalyani Publishers.

He, F.-L., Wang, F., Wei, Q.-P., Wang, X.-W., & Zhang, Q. (2008). Relationships between the distribution of relative canopy light intensity and the peach yield and quality. *Agricultural Science in China, 7*, 297–302.

Iannini, C., Cirillo, C., Basile, B., & Forlani, M. (2001, July). Estimation of nectarine yield efficiency and light interception by the canopy in different training systems. *In V International Peach Symposium, 592*, 357–365.

Intrigliolo, D. S., Ballester, C., & Castel, J. R. (2013). Carry-over effects of deficit irrigation applied over seven seasons in a developing Japanese plum orchard. *Agricultural Water Management, 128*, 13–18.

Johnson, R. S., & Uriu, K. (1989). Mineral nutrition. In J. La Rue & S. Johnson (Eds.), *Peaches, plums and nectarines, growing and handling for fresh market* (p. 69). Oakland: University of California, Division of Agriculture and Natural Resources.

Kita, M., Kato, M., Ban, Y., Honda, C., Yaegaki, H., Ikoma, Y., & Moriguchi, T. (2007). Carotenoid accumulation in Japanese apricot (*Prunus mume* Siebold & Zucc.): Molecular analysis of carotenogenic gene expression and ethylene regulation. *Journal of Agricultural and Food Chemistry, 55*, 3414–3420.

Kobashi, K., Gemma, H., & Iwahori, S. (1999). Sugar accumulation in peach fruit as affected by Abscisic acid (ABA) treatment in relation to some sugar metabolizing enzymes. *Journal of Japanese Society for Horticultural Science, 68*(3), 465–470.

Kondo, S., & Gemma, H. (1993). Relationship between Abscisic acid (ABA) content and maturation of the sweet cherry. *Journal of Japanese Society for Horticultural Science, 62*, 63–68.

Kumar, R., Khurana, A., & Sharma, A. K. (2014). Role of plant hormones and their interplay in development and ripening of fleshy fruits. *Journal of Experimental Botany, 65*(16), 4561–4575.

Kyriacou, M. C., & Rouphael, Y. (2018). Towards a new definition of quality for fresh fruits and vegetables. *Scientia Horticulturae, 234*, 463–469.

Lal, S., Sharma, O. C., & Singh, D. B. (2017). Effect of tree architecture on fruit quality and yield attributes of nectarine (*Prunus persica var. nectarina*) cv. 'Fantasia' under temperate condition. *Indian Journal of Agricultural Science, 87*(8), 24–28.

Lewallen, K. S. (2000). *Effects of light availability and canopy position on peach fruit quality*. Masters Thesis, Virginia Polytechnic Institute and State University, Blacksburg.

López-Ortega, G., García-Montiel, F., Bayo-Canha, A., Frutos-Ruiz, C., & Frutos-Tomás, D. (2016). Rootstock effects on the growth, yield and fruit quality of sweet cherry cv. 'Newstar' in the growing conditions of the Region of Murcia. *Scientia Horticulturae, 198*, 326–335.

Lopresti, J., Goodwin, I., McGlasson, B., Holford, P., & Golding, J. (2014). Variability in size and soluble solids concentration in peaches and nectarines. *Horticultural Reviews, 42*, 253–299.

Lu, R. Q., Wang, X. W., Guo, B. L., Liu, J., & Wei, Q. P. (2003). Effects of light distribution on yield and fruit quality of ruipan 5 flat peach variety trained in slanting central leader system. *Journal of Fruit Science, 20*(6), 509–511.

Lurie, S. (2010a). Plant growth regulators for improving postharvest stone fruit quality. In G. Costa (Ed.), *Proceedings of XIth IS on Plant Bioregulators in Fruit Production, Acta Horticuturae* 884, ISHS2010 (pp. 189–198).

Lurie, S. (2010b). Plant growth regulators for improving postharvest stone fruit quality. *Acta Horticulturae, 884*, 189–197. https://doi.org/10.17660/ActaHortic.2010.884.21.

Maatallah, S., Guizani, M., Hjlaoui, H., Boughattas, N. E. H., Lopez-Lauri, F., & Ennajeh, M. (2015). Improvement of fruit quality by moderate water deficit in three plum cultivars (*Prunus salicina* L.) cultivated in a semi-arid region. *Fruits, 70*, 325–332.

Martinez-Romero, D., Valero, D., Serrano, M., Burlo, F., Carbonell, A., Burgos, L., & Riquelme, F. (2000). Exogenous polyamines and gibberellic acid effects on peach (*prunus persica* l.) storability improvement. *JFS: Food Chemistry and Toxicology, 65*(2), 288–294.

Matsuo, S., Kikuchi, K., Fukuda, M., Honda, I., & Imanishi, S. (2012). Roles and regulation of cytokinins in tomato fruit development. *Journal of Experimental Botany, 63*, 5569–5579.

Meena, N. K., & Asrey, R. (2018a). Tree age affects physico-chemical, functional quality and storability of Amrapali mango (*Mangifera indica* L.) fruits. *Journal of the Science of Food and Agriculture, 98*(9), 3255–3262.

Meena, N. K., & Asrey, R. (2018b). Tree age affects postharvest attributes and mineral content in Amrapali mango (Mangifera indica) fruits. *Horticultural Plant Journal, 4*(2), 55–61.

Michailides, T. J., Ramirez, H. T., Morgan, D. P., Crisosto, C. H., & Johnson, R. S. (1992). Nitrogen fertilization affects fruit quality and susceptibility of nectarine fruits to brown rot. In *1992 Research Reports for California Peaches and Nectarines, California Tree Fruit Agreement*, Sacramento, CA, 24p.

Milošević, T., Milošević, N., & Glišić, I. (2015). Apricot vegetative growth, tree mortality, productivity, fruit quality and leaf nutrient composition as affected by Myrobalan rootstock and blackthorn inter-stem. *Erwerbs-Obstbau, 57*(2), 77–91.

Miodragović, M., Magazin, N., Keserović, Z., Milić, B., Popović, B., Blagojević, B., & Kalajdžić, J. (2019). The early performance and fruit properties of apricot cultivars grafted on Prunus spinosa L. interstock. *Scientia Horticulturae, 250*, 199–206.

Ochmian, I., Oszmiański, J., Lachowicz, S., & Krupa-Małkiewicz, M. (2019). Rootstock effect on physico-chemical properties and content of bioactive compounds of four cultivars Cornelian cherry fruits. *Scientia Horticulturae, 256*, 108588.

Oron, G., DeMalach, Y., Gillerman, L., David, I., & Lurie, S. (2002). SW—Soil and water: Effect of water salinity and irrigation technology on yield and quality of pears. *Biosystems Engineering, 81*(2), 237–247.

Pande Kamal, K. (2017). Advances in pre and post harvest treatments for extending shelf life of stone fruits: A Review. *International Journal of Agriculture Sciences, ISSN*, 0975–3710.

Perez-Pastor, A., Ruíz-Sánchez, M. C., Martínez, J. A., Nortes, P. A., Artes, F., & Domingo, R. (2007). Effect of deficit irrigation on apricot fruit quality at harvest and during storage. *Journal of the Science of Food and Agriculture, 87*, 2409–2415.

Prasad, R. N., & Bankar, G. J. (2006). Effects of different rootstocks on growth, yield and quality of ber cv. Seb under rainfed conditions of arid region. *Annals of Arid Zone, 45*(1), 49.

Rato, A. E., Agulheiro, A. C., Barroso, J. M., & Riquelme, F. (2008). Soil and rootstock influence on fruit quality of plums (*Prunus domestica* L.). *Scientia Horticulturae, 118*(3), 218–222.

Reig, G., Alegre, S., Cantín, C. M., Gatius, F., Puy, J., & Iglesias, I. (2017). Tree ripening and post-harvest firmness loss of eleven commercial nectarine cultivars under Mediterranean conditions. *Scientia Horticulturae, 219*, 335–343.

Ren, J., Chen, P., Dai, S. J., Li, P., Li, Q., Ji, K., Wang, Y. P., & Leng, P. (2011). Role of abscisic acid and ethylene in sweet cherry fruit maturation: Molecular aspects. *New Zealand Journal of Crop and Horticultural Science, 39*(3), 161–174.

Rettke, M. A., Pitt, T. R., Maier, N. A., & Jones, J. A. (2006). Quality of fresh and dried fruit of apricot (cv. Moorpark) in response to soil-applied nitrogen. *Australian Journal of Experimental Agriculture, 46*(1), 123–129.

Roghabadi, M. A., & Pakkish, Z. (2014). Role of Brassinosteroid on yield, fruit quality and postharvest storage of 'Tak Danehe Mashhad' sweet cherry (*Prunus avium* L.). *Agricultural Communications, 2*(4), 49–56.

Roussos, P. A., Sefferou, V., Denaxa, N. K., Tsantili, E., & Stathis, V. (2011). Apricot (*Prunus armeniaca* L.) fruit quality attributes and phytochemicals under different crop load. *Scientia Horticulturae, 129*(3), 472–478.

Saleh, M. M., & Abd El-Moneim, E. A. (2003). Improving the productivity of Fagri Kalan mango trees grown under sandy soil conditions using potassium, boron and sucrose as foliar spray. *Annals of Agricultural Science, 48*, 747–756.

Sharma, Y. (2016). *Studies on canopy management in peach (Prunus persica (L.) Batsch)*. Doctoral dissertation, Punjab Agricultural University, Ludhiana.

Singh, D., Sharma, S. D., & Kumar, P. (2015). Nitrogen fertigation for nectarines (*Prunus persica var. nucipersica*): Lateral and vertical nutrient acquisition and cropping behaviour in rainfed agro-ecosystem. *Indian Journal of Agricultural Science, 85*(11), 1440–1447.

Singh, S. K., Sharma, R. R., Srivastav, M., & Patel, V. B. (2010). Influence of pruning intensity on leaf nutrient composition in some mango cultivars planted under high density. *Indian Journal of Horticultural, 67*(1), 16–20.

Sobierajski, G. D. R., Silva, T. S., Hernandes, J. L., Júnior, P., & José, M. (2019). Y-shaped and fruiting wall peach orchard training system in subtropical Brazil. *Bragantia, 78*, 229–235.

Southwick, S. M., Yeager, J. T., & Zhou, H. (1995). Flowering and fruiting in 'Patterson' apricot (*Prunus armeniaca*) in response to postharvest application of gibberellic acid. *Scientia Horticulturae, 60*, 267–277.

Souza, F., Argenta, L. C., Nava, G., Ernani, P. R., & Amarante, C. V. T. (2013). Quality of Fuji' apples affected by nitrogen and potassium fertilization in two soil types. *The Revista Brasileira de Fruticultura, 35*(1), 305–315.

Spectrum Analytic Inc. (2006). *Fertilizing apples*. Retrieved April 20, 2015, from http://www.spectrumanalytic.com/support/library/pdf/fertilizingg_apple_trees.pdf.

Stern, R. A., Flaishman, M., Applebaum, S., & Ben-Arie, R. (2007b). Effect of synthetic auxins on fruit development of 'Bing' cherry. *Scientia Horticulturae, 114*, 275–280.

Stern, R. A., Flaishman, M., & Ben-Arie, R. (2007a). The effect of synthetic auxins on fruit development, quality and final fruit size in 'Canino' apricot (*Prunus armeniaca* L.). *Journal of Horticultural Science and Biotechnology, 82*(2), 335–340.

Stern, R. A., Flaishman, M., & Ben-Arie, R. (2007c). Effect of synthetic auxins on fruit size of five cultivars of Japanese plum (*Prunus salicina* Lindl.). *Scientia Horticulturae, 112*, 304–309.

Thakur, A., & Singh, Z. (2013). Deficit irrigation in nectarine: Fruit quality, return bloom and incidence of double fruits. *European Journal of Horticultural Science, 78*, 67–75.

Usenik, V., Kastelec, D., & Štampar, F. (2005). Physicochemical changes of sweet cherry fruits related to application of gibberellic acid. *Food Chemistry, 90*(4), 663–671.

Valverdu, X., Girona, J., Echeverria, G., Marsal, J., Behboudian, M. H., & Lopez, G. (2012). Sensory quality and consumer acceptance of 'Tardibelle' peach are improved by deficit irrigation applied during stage II of fruit development. *Horticultural Science, 47*, 656–659.

Vandan, S. P., Solanki, S., & Lamo, K. (2017). Canopy Management: Way to Develop Fruit Tree Architecture. Biomolecules Reports.

Vangdal, E., Hovland, K. L., Børve, J., Sekse, L., & Slimestad, R. (2006). Foliar application of calcium reduces postharvest decay in sweet cherry fruit by various mechanisms. In *XXVII international horticultural congress-ihc2006: international symposium on the role of postharvest technology in the 768* (pp. 143–148).

Vazquez-Luna, A., Rivera-Cabrera, F., Perez-Flores, L. J., & Diaz-Sobac, R. (2011). Effect of rootstock on mango fruit susceptibility to infestation by *Anastrepha obliqua*. *Journal of Economic Entomology, 104*(6).

Verma, P., Chandel, J. S., Sharma, N. C., & Thakur, Y. (2017). Effect of fertigation on growth, yield, fruit quality and fertilizer use efficiency of peach. *Journal of Hill Agriculture, 8*(2), 181–186.

Wahdan, M. T., Habib, S. E., Bassal, M. A., & Qaoud, E. M. (2011). Effect of some chemicals on growth, fruiting, yield and fruit quality of mango "Succary Abiad". *Journal of American Science, 7*(2), 651–658.

Yahia, E. M. (Ed.). (2019). *Postharvest technology of perishable horticultural commodities* (pp. 99–128). Cambridge: Woodhead Publishing.

Yamaguchi, T., Hara, T., & Sonoda, Y. (1986). Distribution of calcium and boron in the pectin fraction of tomato leaf cell wall. *Plant & Cell Physiology, 27*, 729–732.

Zhang, C., & Whiting, M. D. (2011). Improving 'Bing' sweet cherry fruit quality with plant growth regulators. *Scientia Horticulturae, 127*(3), 341–346.

Zhou, H. W., Dong, L., Ben-Arie, R., & Lurie, S. (2001). The role of ethylene in the prevention of chilling injury in nectarines. *Journal of Plant Physiology, 158*, 55–61.

Nutritional Composition of Stone Fruits

9

Nirmal Kumar Meena, Kalpana Choudhary, Narender Negi, Vijay Singh Meena, and Vaishali Gupta

Abstract

Stone fruits are emerging in the market in response to increased consumer desire for health-promoting foods. These fruits have an important role in mitigating nutritionally related diseases because of their high level of nutraceutical properties. They fulfill our nutritional requirements and enriching our healthy diet. These fruits are an abundant source of carbohydrates, lipids, organic acids, vitamins, minerals, carotenoids, phenolic, anthocyanins, and other secondary metabolites that enhance the defense-related systems in the body and help in curing different chronic diseases. This inherent potential is to be explored. The inherent level of nutrients may be affected by a number of preharvest factors including genotype, rootstock, canopy management, agronomic practices, and postharvest factors. There is an urgent need to adopt some good agricultural practices to maximize the quality and proportion. An integrated approach could reserve the quality of the fruits and their efficient use. This chapter presents the nutritional composition in different stone fruits.

Keywords

Stone fruits · Nutrients · Peach · Minerals · Phenols · Carotenoids · Quality

N. K. Meena (✉) · K. Choudhary
Department of Fruit Science, College of Horticulture and Forestry, (Agriculture University, Kota), Jhalawar, India

N. Negi · V. S. Meena
Indian Council of Agricultural Research-NBPGR, New Delhi, India

V. Gupta
Department of Post Harvest Technology, College of Horticulture and Forestry, (Agriculture University, Kota), Jhalawar, Rajasthan, India

M. M. Mir et al. (eds.), *Production Technology of Stone Fruits*,
https://doi.org/10.1007/978-981-15-8920-1_9

9.1 Introduction

Fruits are an important source of phytonutrients, which are an integral part of our diet. Fruits provide many nutrients that are needed to cure many diseases and malnutrition. These phytonutrients are the fuel for our body. Stone fruits are a major category with significant contributions in delivering nutrients. Stone fruits include peaches, nectarines, plums, cherries, apricots, almonds, mangos, and jujube, which contain an endocarp and a single hard stone form of seed. In the past few years, production and marketing of stone fruits has increased tremendously, especially in Asian countries. Peaches and plums are two major temperate stone fruits that occupy 18,000 and 22,000 hectares (ha) area with a total production of 107,000 and 76,000 metric tonnes, respectively, in 2016–2017 (Table 9.1). Apricots and almonds are included as fruits (*Prunus*) belonging to family Rosaceae. Fruit is consumed in both fresh and processed forms. Dried apricots are very popular. The apricot kernel is rich in various nutrients, vitamins, and other health-benefiting compounds. Some of the varieties are grown especially for dried products. The almond occupies a distinct position among stone fruits because it has a multifaceted compositional contour and significant functional potential. The mango is one of the choicest fruits in the world. The tree produces fruit from April to August. The mesocarp is the edible portion, having an abundant amount of pulp. A large number of varieties is available in India. The Totapuri cultivar is rich in pulp whereas Chousa and Dashehari are rich in juice. Raw green mango also consumed in various forms. Mango fruits are rich in carbohydrates, fibers, pectin, and other starchy materials. Ber (*Ziziphus mauritiana* L.) is also one of the arid stone fruits. Its several local species are consumed traditionally. Commercially, *Z. jujube* is cultivated, whereas *Z. rotundifolia* and *Z. nummularia* are found in the wild. These species produce small sweet climacteric fruits, which are harvested by the local people and consumed fresh as well in the processed form. *Ziziphus* produces fruits from December to March depending upon weather conditions.

These fruits are rich in various bioactive compounds that keep calories under control and likely decrease the chances of developing many chronic diseases, such as cancer, heart disease, and diabetes. Besides being an energy source, these fruits are rich in vitamins, minerals such as potassium, calcium, and iron, and antioxidants, and also provide fiber with many other health-promoting compounds. The

Table 9.1 Area and production of some selected stone fruits in India

Sample	Name of fruits	Area [hectares (ha) × 1000]	Production (MT × 1000)
1.	Mango	2263	19,687
2.	Ber	49	481
3.	Peach	18	107
4.	Plum	22	76
5.	Almond	12	8
6.	Walnut	92	228

Source: Horticulture statistics at a glance, 2017

content of nutrients may vary from fruit to fruit and variety to variety. Most of these are colored fruits and enriched with anthocyanins and carotenoids, which act as antioxidants. Many phenols, carotenes, and xanthophylls help in fighting against different chronic diseases (Holst and Williamson 2008). The plum is the most common stone fruit after the peach, consumed fresh as well as in the dried form, called prunes. Plums contain low carbohydrates and high nutrients. They are high in potassium content and therefore preferred by patient suffering from arterial hypertension (Lucas et al. 2004). Plum fruits are rich in vitamins C and A and in anthocyanins and phenols. The autumn varieties have a greater amount of antioxidants and anthocyanins than the summer varieties (Arion et al. 2014). Similarly, among tropical and subtropical fruits, mango (the "king of fruits") is rich in vitamin A, antioxidants, ascorbic acid, phenolics, minerals, and total carotenoids (Meena and Asrey 2018a). A wide range of factors associated with the level of nutrients and their retention from harvesting to consumer. Composition and level of nutrients varies with cultivars and agronomic practices, rootstocks, pre- and postharvest application of hormones, pruning, and age of trees (Meena and Asrey 2018a, b; Baghel et al. 2019; Asrey et al. 2013).

9.2 Carbohydrates

Carbohydrates are the most abundant constituent in fruits, responsible for structural formation of tissue by acting as an energy reservoir. Carbohydrates are the largest constituent in fruits after water (50–80% of total dry weight). Carbohydrates are the polysaccharides, the major source of energy in foods. Carbohydrates yield 4 kcal/g energy. Sugar is the chief constituent of carbohydrates in fruits. It is present in the fruits as fructose, glucose, sucrose, arabinose, and sorbitol. However, sorbitol is found in lesser quantities as compared to other sugars, although some amounts have been detected in plum and cherry. Stone fruits contain an appreciable amount of carbohydrates in the form of starch, sugars, and pectin. The concentration varies from fruit to fruit and also among varieties.

The peach fruit is a rich source of carbohydrates. According to the USDA one small peach weighing around 130 g contains 12 g of carbohydrates (Anonymous 2019). The fruits contain three main sugars with sucrose (54–75%) predominant, followed by glucose (4–11%) and fructose (9–21%) (Elsadr and Sherif 2016). Fruit pulp varies in sugar percent according to variety. Saidani et al. (2017) reported that sugar content in the cultivar Calanda Tardio ranges from 10.8 to 15.7 g/100 g, and the Big Top nectarine contains about 8.64–11.5 g/100 g fresh weight (FW).

In the plum, carbohydrates are found in the form of sucrose, glucose, fructose, and some amount of sorbitol (Forni et al. 1992; Wills et al. 1983). However, the value varies among cultivars. Sucrose sugar has been recorded from 1000 to 6270 mg/100 g FW in 'Santa Rosa' and 'Sel D8 Dark,' respectively (Forni et al. 1992). Total sugar content in plum fruits varies from 9.63 to 29.47% in the 23 cultivars evaluated (Sahamishirazi et al. 2017). Progressive increase in total sugar content with the ripening makes fruit sweeter and could be considered as criteria for maturity (Butac et al. 2012). Butac et al. (2012) reported that total sugar in plum fruits should be more than 16% for fresh consumption. Fresh plums contained about 5.07 g glucose, 3.07 g fructose, and 1.57 g sucrose; these quantities were enhanced in dried form with values recorded as 25.46 g glucose, 12.45 g fructose, and 0.15 g sucrose.

Carbohydrates are the most abundant energy source present in cherries (Pacifico et al. 2014; Bastos et al. 2015). Sweet cherry fruits exhibit more carbohydrates, 12.2 and 17 g/100 g per edible portion, whereas sour cherry fruits have on average 12.2 g/100 g edible portion (USDA ARS 2016). Glucose is the predominant sugar in both sweet and sour cherries, ranging from 6.0 to 10.0 g/100 g fruit weight, depending upon the genotype as well as environmental conditions. Fructose is the second most abundant sugar present in cherries, ranging from 5.0 to 7.6 g/100 g FW in sweet cherry and from 3.5 to 4.9 g/100 g FW in sour cherry (Papp et al. 2010; Ballistreri et al. 2013). Sweet cherry also exhibited some amount of sorbitol, ranging from 0.9 to 26.7 mg/100 g FW (Usenik et al. 2008; Ballistreri et al. 2013).

Apricot is rich in carbohydrates, varying among varieties and from place to place. The fresh kernel of apricot contains about 11–13% carbohydrates and can provide 50 kcal energy per 100 g (Lichou et al. 2003; USDA-ARS 2005; Leccese et al. 2007). Sucrose is the chief carbohydrate source, ranging from 3.9 g/100 g FW in 'Chuan Zhi Hong' to 8.8 g/100 g FW in the 'Hargrand' cultivar (Aubert and Chanforan 2007; Schmitzer et al. 2011). Akin et al. (2008) found an average sugar content up to 90 mg/100 g dry weight basis. Sorbitol is the least abundant carbohydrate and rarely detected in only a few cultivars.

Dry matter content of fresh almond kernel is relatively high, 97–98% (Fatima et al. 2018). The major sugars found in developing almond kernels were sucrose, glucose, fructose, mannose, and arabinose, although at harvest, sucrose was the major sugar component [1.6–2.6 g/100 g dry weight (DW)]. In the kernel, glucose and fructose were found in very small amounts whereas mannose and arabinose were no longer detected (Egea et al. 2009). The nutritional composition of Indian almond (*Prunus amygdalus*) as studied by Agunbiade and Olanlokun (2006) revealed that carbohydrate content in almond kernel is about 54.87%, whereas it is 25.47% in tropical almond (*Terminalia catappa*) (Akpakpan and Akpabio 2012).

Yerlikaya et al. (2012) reported that carbohydrate was the chief constituent of walnut, after oil, which constitutes about 9.05–18.92%. Similarly, Ozkan and Koyuncu (2005) carried out chemical analysis of ten walnut genotypes and found the carbohydrate value to be about 8.05–13.23%. Pereira et al. (2008) estimated carbohydrate value in different six Portugal cultivars to range from 3.75 to 6.10%.

Mango has a total carbohydrate range from 27.33 g/199 g in the Amrapali cultivar to 4.49 g/100 g in Guti cultivar (Ara et al. 2014). Fruits of ripe mango contain a range of sugars: the flesh of a ripe mango fruit contains about 15% of total sugars (Maldonado-Celis et al. 2019). It was observed that during the preclimacteric phase the predominant sugar is fructose whereas sucrose is the major component during the ripening of fruits (Bernardes-Silva et al. 2008; Krishnamurthy et al. 1971; Liu et al. 2013). Rumainum et al. (2018) reported that among all sugars, sucrose content was highest, from 19 to 55.4 g/100 g DW. In unripe mango cultivars, starch is the chief carbohydrate component, which is hydrolyzed during ripening and converted to simple sugars. It has been observed that ripening triggers the ratio of sucrose, fructose, and glucose, and their quantities increase with ripening and after a certain period the levels start decreasing. Pectin is also abundantly found in unripe mango but slowly begins to degrade with the onset of ripening by hydrolysis of pectin enzymes.

Carbohydrate content is found to be variable in jujube fruits. Fresh fruits contain less sugar than the dried fruits. The average sugar content of dried *Zizyphus jujube* is 50.3–86.9 g/kg. Soluble sugars content in ber continuously increases throughout growth and development of fruit; the highest increase is seen between 40–48 and 80–88 days after petal fall. The soluble sugars content in ripe jujube fruit was recorded up to 121.58 mg/g (Lu et al. 2012). However, it was recorded as about 85–145 mg/g in Indian jujube fruits by Teotia et al. (1974). It was found as low as 58–79 mg/g in 'Mallacy' and 'Bambawi' (*Zizyphus spina-christi* L. Willd.) jujube fruits (Abbas et al. 1994) and as high as 179 mg/g in 'Zaytoni' jujube fruits (Abbas and Fandi 2002). Reducing sugars were increased from 40 to 72 days after petal fall and then decreased until maturity and ripening (Lu et al. 2012). However, in some Indian jujube cultivars ('Zaytoni,' 'Umran,' 'Sanaur,' and 'Kaithli'), the reverse trend was reported (Abbas and Fandi 2002). Reducing sugars tended to accumulate over most of the growth period in jujube fruit cultivars 'Bambawi' and 'Mallacy' (Abbas et al. 1994). Reducing and nonreducing sugars increased up to maturity (Bal et al. 1979; Jawanda and Bal 1980). Total sugars increased gradually up to a certain point in growth and then increased rapidly (Pandey et al. 1990; Kadam et al. 1993). In 'Umran' ber, sucrose and fructose continued to increase whereas glucose decreased slightly as ripening advanced. The stages of harvest had a significant effect on total sugars (Kudachikar et al. 2001). Delaying harvesting of fruits to later maturity stages resulted in higher sugar levels after ripening (Bal and Chauhan 1981; Bal 1986). Studying four Indian jujube cultivars ('Gaolangyihao,' 'Xinshiji,' 'Mizao,' 'Miandianchangguo') in China, Ling et al. (2008) reported that the soluble sugars consisted mainly of sucrose, glucose, and fructose. The rate of increase in sucrose accumulation was highest during the mid- to late growth period to ripening in all cultivars studied; however, the rate of sucrose accumulation in 'Gaolangyihao' was faster than that of the other three cultivars. The cultivar difference was found in glucose and fructose content of total sugars, and fructose was significantly higher than glucose in 'Mizao' and 'Miandianchangguo' fruit (Ling et al. 2008).

9.3 Lipids and Fat

Lipids in fruits and vegetables are also an important source of energy, found in the form of triglycerides. They are an important constituent of cell membranes and the natural wax present over the fruit and vegetable surfaces. A natural lipid layer can be seen over the surface of plums. The biochemistry of lipids is largely determined by the types of fatty acids present. As fatty acids are required in the human body to regulate and absorb fat-soluble vitamins, they also regulate some other metabolic activities in the human system. Stone fruits are a good source of fatty acids. Almond, walnut, and apricot contain a significant amount of fat and fatty acids. They are rich sources of good cholesterol and omega-3 fatty acids.

Walnuts contain about 65% fat by weight. As other nuts, most of the energy in walnuts comes from fat. The kernel is rich in oil content, making it an energy-dense and high-calorie food. However, even though walnuts are rich in fat and calories, studies indicate that they do not increase the risk of obesity when replacing other foods in the diet. Walnut kernels are enriched with oil content that contains health-promoting compounds. Prasad (1994) reported that walnut contains approximately 60% oil. Polyunsaturated fatty acids (PUFA) are essential dietary fatty acids that are present abundantly in walnut (Amaral et al. 2003). Walnut is also rich in both omega-6 and omega-3 fatty acids, which are safer for health reasons. Yerlikaya et al. (2012) found oil as a chief constituent ranges from 61.32 to 69.35%. The major constituents of the oil are triacylglycerols: free fatty acids, diacylglycerols, monoacylglycerols, sterols, sterol esters, and phosphatides are all only present in minor quantities (Prasad 1994). The major fatty acids found in walnut oil are oleic, linoleic, and linolenic. The most abundant is an omega-6 fatty acid called linoleic acid. Walnuts also contain a relatively high percentage of a healthy omega-3 fat called alpha-linolenic acid (ALA) that comprises about 8–14% of the total fat content (Ruggeri et al. 1996). In fact, walnuts are the only nut that contains significant amounts of ALA, which is considered to be especially beneficial for heart health. It also helps reduce inflammation and improve the composition of blood fats. ALA is also a precursor for the long-chain omega-3 fatty acids EPA and DHA, which have been linked with numerous health benefits (Ruggeri et al. 1996).

Almonds are a rich source of fat and lipids composed of monounsaturated and polyunsaturated fatty acids (USDA 2010; Venkatachalam and Sathe 2006). Yada et al. (2011) observed 44–61% fat in almond. Several reports on fatty acid profile of almond show the kernel oil contains oleic and linoleic fatty acids that contribute about 90% of the total lipid. Lipid contributes to the typical flavor and aromatic profile of nuts. Askin et al. (2007) examined 26 genotypes and found lipid content to vary from 25.29 to 60.77%; kernel weight is correlated with different levels of fatty acids. Momchilova and Nikolova-Damyanova (2007) reported oil content through hexane extract in mature almond reached $42.3 \pm 0.8\%$; the oil was composed of triacylglycerols (98.2%), sterols (0.6%), diacylglycerols (0.6%), and polar lipids (0.3%). Major fatty acid composition was found by Sathe et al. (2008) to be in the range of 5.15–6.65% (palmitic), 59.52–73.80% (oleic acid), and 19.49–33.29% (linoleic acid). They concluded that oleic acid and linoleic acids were the two major fatty acids present in almond oil.

The apricot kernel is a good source of fat and lipids. The oil obtained from apricot seeds has a pale yellow color; it is tasteless and is used in cosmetics, confectionaries, and many industrial applications (Dixit et al. 2010; Bachheti et al. 2012). Apricot kernel oil contains significant amounts of both saturated and unsaturated fatty acids. The major fatty acids are palmitic, stearic, myristic, oleic, palmitoleic, linoleic, and linolenic, which provide several health benefits. An apricot seed is rich in oil, recorded as 37% in Poland. Stryjecka et al. (2019) estimated oil composition in various cultivars to be between 32.2 and 44.2%, which may vary with cultivar and location. It has been reported that higher oil content (44.2%) was found in seeds of the 'Somo' cultivar whereas the least was reported in 'Harcot' (Stryjecka et al. 2019). The most common fatty acids in apricot kernel oil are oleic, varying between 62.07and 70.06%, and linoleic, from 20.5 to 27.76% (Gupta et al. 2012). Beyer and Melton (1990) also reported similar findings for fatty acid composition: 69.0% oleic acid and 26.0% linoleic acid in apricot seed oil. Sharlatifar et al. (2017) reported different fatty acids ranges present in kernel seed oil: apricot kernel oil contains 60.01–70.56% oleic acid, 19.74–23.52% linoleic acid, 2.35–5.97% palmitic acid, 0.8–1.5% stearic acid, and 0.2–0.9% palmitoleic acid. Seed kernel oil contains a high amount of monounsaturated fatty acids such as oleic acid and a low concentration of saturated fatty acids, which is favorable for human health.

Sweet and sour cherries have a low fat content, less than 1.0 g/100 g edible portion, mostly saturated fat; thus, cherries are a cholesterol-free fruit (Ferretti et al. 2010; McCune et al. 2011; Pacifico et al. 2014).

Lipids are also an important component but present in small quantities in mango pulp and peel. However, the mango stone contains significant amounts of lipids, chiefly palmitic, stearic, oleic, and linoleic acids. Several authors have reported that the fat and lipid content in mango cultivars ranges from 0.75 to 1.7% in the peel and 0.8 to 1.36% in the pulp in Mullgoa, Totapuri, Benishan, Sundari, and Neelam cultivars (Pathak and Sarada 1974; Selvaraj et al. 1989). Triglycerides are the major component in fatty acids. The ratio of these fatty acids is responsible for flavor and aroma. Established a relationship between these acids and the flavor chemistry of mango as these are considered as precursors of lactones. Bandyopadhyay and Gholap (1973) developed an index for aroma and flavor when the ratio of palmitic acid to palmitoleic acid is more than 1; a ratio of less than 1 generates mild and strong aromas.

Generally, the fat content in ber ranges from 0.2 to 0.4 g/kg. Ber fruits contain 0.07% fat content (Morton 1987). Li et al. (2007a, b) investigated the proximate composition of five cultivars: 'Jinsixiaozao,' 'Yazao,' 'Jianzao,' 'Junzao,' and 'Sanbianhong'; significant variations were recorded for lipids (0.37–1.02%). In some cultivars of Spain triglycerides of medium chain fatty acids were present in abundant quantity. On a total saponifiable oil basis, 12:0, 10:0, 18:2n6, 16:1n7, 16:0, and 18:1n9 were identified as major fatty acids. On a DW basis, fruits contain 1.33 ± 0.17 g/100 g saponifiable oil.

9.4 Protein

Protein is composed of individual amino acids, some of which are essential in the diet. The protein content of cherries varies according to genotype; for sweet cherries it ranges between 0.8 g and 1.4 g/100 g edible portion, whereas for sour cherries it is less than 1.0 g/100 g edible portion (Serradilla et al. 2017; Ferretti et al. 2010).

Proteins are present in abundant quantity in walnuts. The value of protein content of walnuts is reported to range from 13.6 to 18.1 g crude protein/100 g dry matter

(DM) (Savage 2000). Walnuts contain a relatively low content of lysine and high levels of arginine (Ruggeri et al. 1996). On an average, most Turkish cultivars contain 12–24% protein. Yerlikaya et al. (2012) found protein content in various Turkish cultivars to range from 10.58 to 18.19%, whereas 3.59–22.99% was reported in 35 different genotypes by Şahin and Akbaş (2001), and 5.17–19.24% in ten genotypes was reported by Ozkan and Koyuncu (2005). In Portugal, Pereira et al. (2008) reported that the protein content of six walnut cultivars ranged from 14.38 to 18.03%. The high levels of arginine in walnut have already been identified as a positive feature because arginine can be converted into nitric oxide, a potent vasodilator, which can inhibit platelet adhesion and aggregation (Sabaté et al. 1993). Savage et al. (2001) showed that the ratio of lysine to arginine for 12 different cultivars grown in New Zealand to be 0.24, which is much lower than other common proteins (Lavedrine et al. 1999). A low ratio of lysine to arginine in a protein has been identified to reduce the development of atherosclerosis in laboratory animals (Kritchevsky 1988).

Protein concentration was found variable and probably low in Poland cultivars, New Zealand cultivars (20.6%), and Turkey cultivars (15.7–17.1%) as compared to Indian apricot cultivars (31.18%) (Beyer and Melton 1990; Gezer et al. 2011; Bachheti et al. 2012).

Mango seed contains about 10.06% crude protein, and mango pulp contains about 17 types of amino acids. It was reported that the seed of mango had maximum glutamate, about 13.00 g/100 g protein), whereas methionine was found to be least in quantity (Fowomola 2010).

Ber fruits as well as leaves are good source of proteins. The dried fruit contain a higher amount of protein than fresh pulp. The ber stone also contains a significant amount of proteins and is an important source of feed for animals in arid parts of India. It has been reported that jujube fruit is rich in protein content, 0.8% protein (Morton 1987). Pareek (2001) recorded protein content in jujube pulp at about 2.9%. The dried jujube fruit provides 3.3–4.4 g/kg protein, which is higher than in the fresh fruit. Ber fruit contains 3.3–4.0 g/kg protein. Chinese jujube also contains 18 types of amino acids, including eight essential amino acids (Anonymous 1989). In various studies it was found that protein content in commercial California-grown almond cultivars varies between 16 and 23 g/100 g (Ahrens et al. 2005; Venkatachalam and Sathe 2006). Aslantas et al. (2001) found protein content ranging from 19 to 24 g/100 g (dry weight basis, dwb). Askin et al. (2007) reported a wider range of protein content (16–31%) in 26 native genotypes of almond selected from Turkey.

9.5 Vitamins

Vitamins are also an integral part of nutrient value. Vitamins A, B complex, C, D, E, and K are common in fruits. However, vitamin D is not reported in fruits and vegetables. Among these vitamins, A, D, E, and K are fat soluble and the others are water soluble. Vitamin A deficiency is more common as its deficiency promotes night blindness. Vitamin B is also a part of flavoproteins. Its deficiency leads to beriberi, pellagra, and other diseases. Almonds and cherries are good sources of riboflavin and niacin. Vitamin E is known for its antioxidant properties. Alpha-tocopherol is the most active and dominant form of vitamin E; almonds and oil-rich fruits are good sources. Vitamin K is essential for blood coagulation, but its deficiency is not as common. Vitamin C is highly sensitive to loss; therefore, fruits and vegetables need protection from direct exposure to light. Dehydroascorbic acid is the first oxidized product of ascorbic acid. Stone fruits are rich in vitamins, especially vitamin A and ascorbic acid. Of higher carotenoids, β-carotene imparts vitamin A-like activities and others act as antioxidants. Ascorbic acid (AA) is a water-soluble vitamin that is considered to be an antioxidant vitamin.

Peaches and nectarines are important source of ascorbic acid (vitamin C) and carotenoids (Elsadr and Sherif 2016). Depending on the cultivar, vitamin C ranges from 1 to 14 mg AA in 100 g FW (Elsadr and Sherif 2016). Peach peel contains higher vitamin C content than the flesh. Genetic background, flesh type, and climatic conditions also affect ascorbic acid content in fruits.

The yellow-colored nectarines, 'Red Jim' (peel, 13.00 mg/100 g) and 'August Red' (flesh, 5.80 mg/100 g), contain the most ascorbic acid in the peel and flesh whereas the white-fleshed nectarine cultivar 'Arctic Snow' (peel, 20.00 mg/100 g; flesh, 12.20 mg/100 g) is richest in this vitamin. There are no differences in vitamin C content between peaches and nectarines, although it is higher in white-than in yellow-fleshed fruits (Bassi et al. 2016).

Plum fruits also contain a range of vitamins but ascorbic acid is the major contributory vitamin. The content of ascorbic acid ranges from 4 to 6 mg per 100 g (Wills et al. 1983). All other vitamins were found in very small ranges. Plum contains ascorbic acid (9.5 mg); vitamin A (17 μg), tocopherol (0.26 mg), niacin (0.4 mg), and vitamin B complex are present to a limited extent.

Cherry fruits are a good source of ascorbic acid (vitamin C), from 6 to 10 mg/100 g fruit weight, although some Estonian cultivars contain up to 27 mg/100 g FW (Serradilla et al. 2017). Moreover, the cultivar '470' contains 27 mg/100 g FW (Serrano et al. 2005). In contrast, the cultivars 'Van,' 'Noir De Guben,' and '0900 Ziraat' contain less than 4 mg/100 g FW (Vursavuş et al. 2006). Sour cherries showed higher vitamin A (1283 IU/100 g) and β-carotene (770 μg/100 g) content than sweet cherries (64 IU vitamin A, 38 μg/100 g β-carotene) (McCune et al. 2011). Vitamin E amounts of 0.07–0.2 mg tocopherol equivalent/100 g of edible portion were found in cherries (Serradilla et al. 2017). Contents of the other water-soluble vitamins in cherries are much lower. Contents of niacin and pantothenic acid vary between 0.15 and 0.40 mg/100 g edible portion, whereas thiamine, riboflavin, and pyridoxine content ranged from 0.02 to 0.05 mg/100 g of edible portion. Folate and biotin exist in microgram concentrations (Serradilla et al. 2017).

Walnuts are a rich source of several vitamins and minerals, including vitamin E. As compared to other nuts, walnuts contain high levels of a special form of vitamin E called gamma-tocopherol (γ-tocopherol). Vitamin E has high antioxidant activity, with an important role against oxidation of fats in lipid membranes. Lavedrine et al. (1999) presented some data on the vitamin E content of walnuts grown in France and the USA; they identified γ-tocopherol as the main tocopherol in walnut oil. The tocopherol content of New Zealand walnuts ranges from 290 to 435 mg/g oil. Walnut is also an abundant source of vitamin B complex (thiamine, riboflavin, niacin, pantothenic acid, folate, B_6) (Sen 2013). It has been reported that walnut contains almost all types of vitamins including vitamin A (20 IU), thiamine (0.341 mg), riboflavin (0.15 mg), niacin (1.125 mg), B_5 (0.570 mg), B_6 (0.537 mg), and folate (98 μg. The kernel also contains vitamin C (1.3 mg) and vitamin K (2.7 μg) (USDA-ARS 2010).

The apricot kernel is rich in vitamins such as ascorbic acid, β-carotene, and thiamine but low in vitamin E (Lee and Kader 2000; Chauhan et al. 2001). Akin et al. (2008) reported that vitamin C content in apricot varieties could reach 100 mg/100 g DW.

Mango contains a range of vitamins, especially vitamin A and ascorbic acid. It is suggested that vitamin A-deficient persons should consume more mango to fulfill the need for vitamin A. Several authors reported different ranges of these vitamins. Vitamin A in the fruit varies from 1000 to 6000 IU (Matheyambath et al. 2016). Ascorbic acid content varies from 9.79 to 186 mg/100 g of mango pulp (Vazquez-Salinas and Lakshminarayana 1985; Manthey and Perkins-Veazie 2009; Wongmetha and Ke 2012; Matheyambath et al. 2016). Vitamin C and E content was estimated in dry peel to be 188–392 and 205–509 μg/g, respectively. It is assumed that in mango fruit vitamins A, C, and E constitute 25%, 76%, and 9% of the Dietary Reference Intake (DRI) in a 165-g serving (Fowomola 2010). Vitamin B_6 (pyridoxine, 11% DRI), vitamin K (9% DRI), and other B vitamins are also present in significant amounts. 'Ataulfo,' a Mexican cultivar, contains vitamin C 125.4 mg/100 g pulp on average. Vitamin B complex and tocopherol are reported in minor quantities. Vitamin D has not been reported in any mango cultivars.

Zizyphus jujube fruits are very rich in vitamin C, thiamine, and riboflavin (Kuliev and Guseinova 1974). Chinese jujube is rich in ascorbic acid, from 192 to 359 mg/100 g among the cultivars studied. Consumption of one ber fruit in a day would meet the dietary requirements for vitamin C and vitamin B complex of an adult man, as recommended by WHO. The vitamin A value was found to be 38 μg RE (retinol equivalent)/100 g on a fresh weight basis (Guil-Guerrero et al. 2004). It is also known to have high vitamin P (bioflavonoid) content. The ascorbic acid content of ber fruits is initially low and continues to increase until the fruit reaches physiological maturity (Abbas 1997). The increase in ascorbic acid as ripening advanced was observed in ber fruit to reach peak value, 559 mg/100 g, on the 15th day of storage. Bal et al. (1995) also noted the increase in vitamin C content as maturity advanced in 'Umran' ber fruits. The highest content of ascorbic acid was observed at 56 days after petal fall, and it then continuously decreased until maturity (Lu et al. 2012). A comparatively lesser amount of ascorbic acid (250–600 mg/g FW) was reported by Wu et al. (2010) whereas it was as high as 721 mg/g FW at 88 days after petal fall (Lu et al. 2012). Ber fruit is also rich in vitamins other than ascorbic acid and contains about 0.02 mg/100 g carotene, and vitamin B_1 (0.020–0.038 mg/100 g), vitamin B_2 (0.7–0.9 mg/100 g), and vitamin B_3 (Morton 1987; Pareek and Dhaka 2008). Ascorbic acid content was reported to be

136–363 mg/100 g in fresh pulp (Ciressa et al. 1984; Pareek 2001) Almond kernels are also recognized as a good source of riboflavin (vitamin B_2) (Chen et al. 2006).

9.6 Minerals

Mineral content is sometimes expressed in general terms as ash content, the inorganic residue remaining after incineration of the plant tissue.

Plum contains 157 mg potassium, 16 mg phosphorus, 6 mg calcium, and 7 mg magnesium. Wills, (1983) reported niacin ranging from 0.3 to 0.9 mg/100 g FW in Narrabeen and Santa Rosa, respectively. Similarly, vitamin B_2 was present at 0.02 mg/100 g in Narrabeen plum; it was higher in Greengage (0.05 mg/100 g FW). Plum fruits are rich in minerals and contain abundant amounts of potassium, sodium, calcium, and magnesium, plus some minor minerals such as boron and iron (Bhutani and Joshi 1995; Kunachowicz et al. 2005; Yagmur and Taskin 2011). Milosevic and Milosevic (2012) estimated total ash content in Serbian plum cultivars relative to minerals in plum was 4.54%. It was complemented with nitrogen (0.78%), phosphorus (0.06%), potassium (1.45%), calcium (0.07%), magnesium (0.16%), iron (19.37 μg/g), manganese (10.21 μg/g), copper (3.21 μg/g), zinc (19.29 μg/g), and boron (22.83 μg/g) on a dry weight basis. Almond kernels contain about 3 g ash/100 g (FW); the limited data reported in the literature for ash content include an average of 3.05 g/100 g for a mixture of three almond cultivars grown in Spain, 3.4 g/100 g in Lebanese-grown almonds (Cowan et al. 1963), 2.3–3.7 g/100 g in various almond cultivars grown in Italy (Barbera et al. 1994; Ruggeri et al. 1998), and 3.8 g/100 g in selected Turkish almond types (Aslantas et al. 2001). The ash contents of various commercial California-grown cultivars have been reported to range from 2.6 to 4.6 g/100 g (Ahrens et al. 2005; Hall et al. 1958; Sathe 1992). In almond skins, ash contents of 3.4% (Hall et al. 1958) and 4.0% (Saura-Calixto et al. 1983) have been measured.

Walnuts are also a very good source of several minerals, including potassium, calcium, phosphorus, magnesium, and sodium. Total mineral content ranges from 1.7 to 2.0%, with significant amounts of potassium (390–700 mg/100 g), phosphorus (310–510 mg/100 g), magnesium (90–140 mg/100 g), and in small amounts, sodium (1–15 mg/100 g) (Souci et al. 1994; Lavedrine et al. 2000; Savage 2001; Akca et al. 2005). About 1% of our body is made up of phosphorus, a mineral that is mainly present in bones.

Mango pulp as well as mango seed contains plenty of minerals such as calcium, potassium, phosphorus, magnesium, boron, iron, and zinc. Meena and Asrey (2018a, b) reported that Amrapali mango contains about 914–1184 mg/kg potassium, 110–148 mg/kg Ca, and 110–139 mg/kg magnesium. However, copper, manganese, and sodium were detected in very low concentration. Fowomola (2010) estimated mango seed mineral composition, finding a variation among minerals: the seed of mango contained sodium (21.0 mg/100 g), potassium (22.3 mg/100 g), calcium (111.3 mg/100 g), magnesium (94.8 mg/100 g), iron (11.9 mg/100 g), zinc,

(1.1 mg/100 g), and copper (0.1 mg/100 g), which has good potential to supply minerals.

Cherry fruits are considered as a good source of dietary potassium; about 260 mg/100 g of edible portion is present in sweet cherry (McCune et al. 2011), and in sour cherry it is 200.0 mg/100 g edible portion. Cherries contain other minerals in low concentrations. In sweet cherry, calcium ranged from 13.0 to 20.0 mg, magnesium ranged from 8.0 to 13.0 mg, phosphorus varied between 15.0 and 18.0 mg, and sodium between 1.0 and 8.0 mg per 100 g edible portion (USDA ARS 2016). In sour cherry, calcium ranged from 9.0 to 14.0 mg, magnesium was 7.0–10.0 mg, and phosphorus was 9.0–20.0 mg per 100 g edible portion (Mitić et al. 2012; USDA ARS 2016).

Among the minerals K, P, Ca, and Mn contributed the major portion, although Fe, Na, Zn, and Cu were also found in good amounts in ber. Jujube fruits contain high amounts of minerals such as potassium, phosphorus, manganese, and calcium. Some trace amounts of sodium, zinc, iron, and copper are also reported. San et al. 2009 studied mineral composition of leaves and fruits of four promising genotypes of Turkey and reported that 93% of total mineral composition was composed of nitrogen, potassium, and calcium. Nitrogen content varied from 170 to 506.67 mg/100 g DW basis. They also recorded potassium ranging from 314.67 to 420 mg, calcium 79.33 to 121.33 mg, and magnesium from 15.77 to 20.87 mg/100 g. The jujube fruit contains a wide range of minerals such as iron (0.76–1.8%) and 0.03% each of calcium and phosphorus (Pareek 2013). Ascorbic acid content ranges from 65.8 to 76 mg/100 g in the fresh fruit (Pareek and Dhaka 2008; Pareek et al. 2009).

9.7 Fiber

Dietary fibers are a group of carbohydrates, heterogeneous substances composed of cellulose, hemicellulose, pectins, lignins, gums, and polysaccharides (Asp 1996). Dietary fibers can be generally classified into two main categories, insoluble and soluble, depending on their solubility in water. Fiber is mentioned as one of the eight possible positive constituents of nuts (Hu et al. 1998).

According to various studies, the crude fiber (CF) values of peach fruit were found to be 1.2 and between 1.9 and 3.9 g/100 g of fruit weight, respectively, using the CF and total dietary fiber (TDF) methods (Elsadr and Sherif 2016). Cherry fruits are a moderate source of dietary fiber, from 1.3 to 2.1 g/100 g edible portion (McCune et al. 2011). Walnuts may protect against coronary heart disease through a number of mechanisms (Fraser 1994). The TDF content of 12 different cultivars of walnuts from New Zealand ranged from 3.1 to 5.2 g/100 g DM (Savage 2000). Lintas and Cappelloni (1992) found 0.78 g soluble dietary fiber (15.2 g TDF) and 14.4 g insoluble dietary fiber per 100 g of blended almond samples.

Apricot contains an appreciable amount of fiber, from 1.5 to 2.4 g, which provides roughage, stimulates gastric mobility, and prevents constipation (Ali et al. 2011; Hacıseferoğulları et al. 2007; Akin et al. 2008; Tamura et al. 2011). Fresh plum fruit contains about 1.4–1.7 g dietary fiber and dried prunes contain about 7.1 g TDF.

Ajila et al. (2007) estimated TDF in mango peel was 45–78%, and higher values were detected in the ripe peel of the fruits. Edible mango peel has potential to offer dietary fiber (Ajila and Rao 2008). Fiber content in mango seeds is about 2.40% on a dry weight basis, although it is higher in the peel. Total dietary fiber content was 40.6–72.5% (Ajila and Rao 2013); their complete estimated analysis of dietary fiber reported soluble dietary fiber content in mango peel powder was 12.8–23.0% and insoluble dietary fiber content was 27.8–49.5%. Ahmed and Mohamed (2015) characterized some Sudan-based mango cultivars and found fiber in pulp from 7 to 45%. Othman and Mbogo (2009) determined low crude fiber content as 0.85 g/100 g FW in two local cultivars of Tanzania, Dodo and Viringe.

Ber fruit contains about 1.3 g/100 g fiber (Morton 1987). Li et al. (2007a, b) estimated the composition of major jujube cultivars and reported that jujube fruits contained around 0.57–2.79% soluble fiber and 5.24–7.18% insoluble fiber.

9.8 Antioxidants, Phenols, and Secondary Metabolites

Antioxidants are also abundantly found in mango. Total carotenoids, ascorbic acids, and phenols are interrelated and contribute to imparting antioxidant activities in combined form. Velioglu et al. (1998) stated that natural antioxidants include phenols, nitrogenous compounds, and carotenoids. Phenolics possess a wide range of pharmaceutical properties. Carotenoids are lipid-soluble pigments that produce yellow, orange, and red colors: these are well known as terpenoids and having provitamin A activity. Carotene and xanthophylls are two broad classifications of carotenoids; some carotenoids also help in protection against photooxidation. Anthocyanins are another important group of secondary metabolites, which have antioxidant activity and are responsible for purple, black, and blue color in fruits.

Total antioxidants were reported to be 20 times higher in peel than flesh in *Prunus domestica* (Usenik et al. 2013). (Piga et al. 2003) reported that neochlorogenic acid is the predominant phenolic component in plums, followed by chlorogenic acid and rutin. In the peel, procyanidin B4, followed by procyanidin B1, was the major phenolic (Tomás-Barberán et al. 2001). The total phenolic content was reported as highest [3200 mg gallic acid equivalents (GAE) kg^{-1} FW] in the local traditional cultivar 'Sugar' (Pega et al. 2003). Sahamishirazi et al. (2017) evaluated 23 cultivars for phenolic and total antioxidants: the Cacaks Spaete cultivar had a good amount of total phenols and this content was stable throughout the year. They also reported that the 'Hohenheim breed 4894' cultivar contained high phenolic, anthocyanin, and sugars during the overall period. Total phenolics differed in cultivars, recorded as 38.45–841.50 mg GAE 100 g^{-1} FW. Deep dark colored varieties contain more anthocyanin as compared to yellow varieties. The plum cultivar Black Amber contains a very high amount of anthocyanin, 10 mg/100 g, as compare to other varieties (Fanning et al. 2014). It is well established that the chief anthocyanin-contributing compounds are cyanidin-3-glucoside and cyanidin-3-rutinoside in Japanese plum. Plum cultivars also contain carotenoids ranging from 0.09 to 1.9 mg/100 g FW. β-Carotene is a major contributory (>90%), and small amounts of β-cryptoxanthin are also reported (Gil et al. 2002).

Peaches and nectarines are a rich source of bioactive compounds such as phenolic compounds, ascorbic acid, carotenoids, and anthocyanin. The antioxidant activity of red-fleshed peaches is higher as compare to light-fleshed peaches. It varies from 2074 to 13,505 mg Trolox/g of fruit weight in red-flesh peaches and between 437 and 1128 mg Trolox/g of fruit weight in lighter flesh peaches (Elsadr and Sherif 2016). Saidani et al. (2017) reported that total phenolic content ranged from 88.9 to 277 mg GAE/100 g fruit weight, flavonoids varied from 39.3 to 245 mg catechin equivalents (CE)/100 g fruit weight, and relative antioxidant capacity (RAC) ranged from 133 to 401 mg Trolox equivalents (TE)/100 g FW, whereas the total anthocyanin values ranged from 0.55 mg C3GE/100 g FW in Calanda Tardio to 17.6 mg cyanidin-3-glucoside relative (C3GE)/100 g FW in Big Top. Anthocyanins are one of the primary components responsible the flesh and skin colors of peach fruit. There are two mainly found anthocyanin pigments in peaches and nectarines are cyanidin-3-glucoside and cyanidin-3-rutinoside. Total anthocyanins content in peeled peach and nectarine varieties greatly vary among genotypes (0.1–26.7 mg of cyanidin-3-glucoside equivalents per kg).

The bioactive compounds present in sweet cherries are polyphenolics, in particular, anthocyanins, flavonols, and hydroxyl cinnamic acids, which highly influence the antioxidant capacity of these fruits (Gonçalves et al. 2019). Sweet cherries are rich in phenolic acids (Tomás-Barberán et al. 2013). These phenols include derivatives of the hydroxyl cinnamic acids, which are the main type of phenols reported in sweet cherries: a flavan-3-ol, and a flavonol (Mozetič et al. 2006). Blando et al. (2004) reported the total anthocyanin content of sour cherries was between 27.8 and 80.4 mg/100 g FW. However, total anthocyanin content differed according to sour cherry cultivar (Simunic et al. 2005). Alrgei et al. (2016) studied many 'Oblačinska' clones, showing substantial levels of total anthocyanin content, more than 100.0 mg cyanidin 3-*O*-glucoside/100 g FW. Total phenolic content ranged from 5.18 to 8.53 mg/g.

Walnuts contain a complex mixture of bioactive plant compounds. The walnut kernel is a rich source of flavonoids, sterols, pectic substances, phenolic acids, and related polyphenols (Crews et al. 2005). The main benefits of walnut kernels include lowering of cholesterol, increasing the ratio of high-density lipoprotein cholesterol to total cholesterol, reducing inflammation, and improving arterial function (Martinez et al. 2010). The kernels are exceptionally rich in antioxidants, which are concentrated in the brown skin. Ellagic acid is found in high amounts in walnuts, along with other related compounds such as ellagitannins. Ellagic acid may reduce the risk of heart disease and cancer. Catechin is a flavonoid antioxidant that may have various health benefits, including promoting heart health. Phytic acid, or phytate, is a beneficial antioxidant; although it can reduce the absorption of iron and zinc from the same meal, this effect is only of concern for those following imbalanced diets.

Apricot contains abundant amounts of the antioxidants required to cure many potential diseases. A range of phenolic compounds such as catechin, epicatechin, rutin, chlorogenic acid, and neochlorogenic acid are present in apricot fruits (Roussos et al. 2011; Radi et al. 2003; Ruiz et al. 2005a; Schmitzer et al. 2011). Drogoudi et al. (2008) reported that total phenolic content was highest in the Robada cultivar with an average value of 560 mg GAE/100 g. Estimation of AOX value through FRAP assay showed higher values in 'Super Gold' (12.3 mmol Fe kg^{-1} FW, whereas the least value (0.5 mmol Fe kg^{-1} FW) was found in 'Tonda di Costigliole' (Contessa et al. 2013). β-Carotene is predominant among carotenoids, followed by β-cryptoxanthin. The amount of total carotenoids varies from 1512 to 16,500 μg/100 g of edible portion (Ruiz et al. 2005a, b). Phenolic compounds in different varietals range from 4233.70 to 8180.49 mg GAE/100 g DW, and β-carotene was 5.74–48.69 mg/100 g DW, which could impart higher antioxidative value (Akin et al. 2008). Polyphenolics were in the range of 1.22 mM GAE/l in the Somo cultivar but minimum at 0.85 mM GAE/l in the Goldrich cultivar (Stryjecka et al. 2019). In a study of bioactive compounds in oils by Popa et al. (2011), apricot seed oil contained about 0.88–1.30 mM GAE/l polyphenolic compounds.

Almond and other nuts contain good amounts of polyphenols and flavonoids, which are conjugated with sugars and polyols via alpha-glycosidic or ester bonds (Croft 1998). It was reported that the almond kernel possesses polyphenolics and antioxidant action (Chen and Blumberg 2008). Milbury et al. (2006) determined phenolics and flavonoids distribution in almonds, finding that total phenolics ranged from 127 mg GAE/100 g FW (in Fritz cultivar) to 241 mg GAE/100 g FW (in Padre Cultivar). They also observed the values of individual flavonoids that were predominant in almond cultivars, including some major flavonoids such as isorhamnetin-3-*O*-rutinoside and isorhamnetin-3-*O*-glucoside (in combination) (16.81 mg/100 g), catechin (1.93 mg/100 g), kaempferol-3-*O*-rutinoside (1.17 mg/100 g), epicatechin (0.85 mg/100 g), quercetin-3-*O*-galactoside (0.83 mg/100 g), and isorhamnetin-3-*O*-galactoside (0.50 mg/100 g) of almonds by fresh weight. Reported phenolics content in almond up to 18 mg GAE/100 g. Kornsteiner et al. (2006) estimated total phenols by the acetone solvent extraction method to be about 239 mg GAE/100 g in chopped almonds. It was also found that major polyphenolics were present in seed skin as compared to the kernel, only 47 mg GAE/100 g in kernel (Kornsteiner et al. 2006; Milbury et al. 2006).

Asrey et al. (2013) reported higher amounts of AOX 424.48 μmol Trolox in unripe mature fruits as compared to 536.64 μmol Trolox in ripe mature prunes. They also reported that antioxidant capacity ranges from 23 to 68 mg (Trolox) equivalents/100 g. Meena and Asrey (2018b) showed a decreasing pattern of antioxidant-contributing factors (except carotenoids) with the age of the tree. It is well established that mango peel contains higher total polyphenols. These polyphenols counteract free radicals in several disease mechanisms. Main polyphenols include quercetin, kaempferol, gallic acid, caffeic acid, catechins, tannins, xanthone, and mangiferin (Singh et al. 2004). Carotene, lutein, and α-carotene are antioxidative compounds. Rumainum et al. (2018) evaluated six cultivars for their antioxidant, phenol, and carotenoid values and found that the 'Thongdum' cultivar had higher carotenoid content (15 mg/100 g DW), and total flavonoids and total phenolics were from 33 to 67 mg/100 g and 8 to 22 mg/100 g, respectively. Amrapali is rich in carotenoid content (Meena and Asrey 2018a; Asrey et al. 2013): they reported total carotenoids as 27.20–37.80 mg/kg FW and total phenols as 438.17 μg GAE/g to 410.64 μg GAE/g. Ascorbic acid ranged from 568.40 mg/kg FW to 444.20 mg/kg FW and AOX from 2.72 to 2.14 μmol Trolox in the Amrapali cultivar.

Jujube fruits contain a significant amount of phenolic compounds. On an average, ber fruit contains about 5.18–8.53 mg/g total phenolic content (Pareek 2013). The major polyphenolics in ber are *p*-hydroxybenzoic, caffeic, ferulic, and *p*-coumaric acids, at concentrations of 366, 31, 20, and 19 mg/kg dry mass, respectively. Another minor polyphenol is vanillic acid, at about 2.5 mg/kg (Pareek 2013). Li et al. (2007a, b) recorded total antioxidant value in different cultivars ranging from 342 FRAP μmol/g in the Sanbianhong cultivar to 1173 FRAP μmol/g in the Jinsixiaozao cultivar.

The reduction in phenolics compounds during ripening could be the result of hydrolysis into sugars, acids, or other compounds or be caused by their transformation from a soluble into an insoluble form (Singh et al. 1981). The total phenols decrease with advanced maturity (Bal et al. 1995). Both on-tree as well as stored 'Umran' ber fruits showed decrease in total phenols as ripening advanced (Sharma 1996), and tannins also decreased (Kadam et al. 1993). Total phenolics were decreased in Chinese ber after 3 days at 20°C and again increased on 15 days of storage (Kader et al. 1982). Total phenols increased from 40 to 48 days after petal fall, decreased from 48 to 56 days after petal fall, again increased between 56 and 64 days after petal fall, and decreased steadily after 64 days to maturity (Lu et al. 2012). Carotenes in Spanish cultivars of ber fruit vary from 4.12 to 5.98 mg/100 g on a DW basis. Carotenoids content increased from 40 to 56 days after petal fall, staying more or less unchanged until 72 days after petal fall. At the later fruit development stage, the carotenoid content rose again (Lu et al. 2012).

9.9 Conclusion

Stone fruits are good sources of polyphenols, anthocyanins, minerals, and vitamins. Because they are a rich source of bioactive compounds, these fruits help reduce the chances of many chronic diseases such as cancer, heart attack, and diabetes. However, the composition and concentration of bioactive compounds varies and depends on genetic makeup, growing conditions, and agronomic practices.

References

Abbas, M. F. (1997). Jujube. In S. K. Mitra (Ed.), *Postharvest physiology and storage of tropical and sub-tropical fruits* (pp. 405–415). London: CAB International.

Abbas, M. F., Al-Niami, J. H., & Al-Sareh, E. A. (1994). The effect of ethephon on the ripening of fruits of jujube. *Journal of Horticultural Science, 69*(3), 465–466.

Abbas, M. F., & Fandi, B. S. (2002). Respiration rate, ethylene production and biochemical changes during fruit development and maturation of jujube (*Zizyphus mauritiana* Lamk). *Journal of the Science of Food and Agriculture, 82*, 1472–1476.

Agunbiade, S. O., & Olanlokun, J. O. (2006). Evaluation of some nutritional characteristics of Indian almond (*Prunus amygdalus*) nut. *Pakistan Journal of Nutrition, 5*(4), 316–318.

Ahmed, T. H. M., & Mohamed, Z. M. A. (2015). Diversity of Mango (Mangifera indica L.) cultivars in Shendi area: Morphological fruit characterization. *International Journal of Research in Agricultural Sciences, 2*(4), 2348–3997.

Ahrens, S., Venkatachalam, M., Mistry, A. M., Lapsley, K., & Sathe, S. K. (2005). Almond (*Prunus dulcis* L.) protein quality. *Plant Foods for Human Nutrition, 60*(3), 123–128.

Ajila, C. M., & Prasada Rao, U. J. (2008). Protection against hydrogen peroxide induced oxidative damage in rat erythrocytes by *Mangifera indica* L. peel extract. *Food and Chemical Toxicology, 46*(1), 303–309.

Ajila, C. M., & Rao, U. P. (2008). Protection against hydrogen peroxide induced oxidative damage in rat erythrocytes by Mangifera indica L. peel extract. *Food and Chemical Toxicology, 46*(1), 303–309.

Ajila, C. M., & Rao, U. P. (2013). Mango peel dietary fibre: Composition and associated bound phenolics. *Journal of Functional Foods, 5*(1), 444–450.

Ajila, C. M., Bhat, S. G., & Rao, U. J. S. P. (2007). Valuable components of raw and ripe peels from two Indian mango varieties. *Food Chemistry, 102*(4), 1006–1011.

Ajila, C. M., Leelavathi, K. U. J. S., & Rao, U. P. (2008). Improvement of dietary fiber content and antioxidant properties in soft dough biscuits with the incorporation of mango peel powder. *Journal of Cereal Science, 48*(2), 319–326.

Akca, Y., Sutyemez, M., Ozgen, M., Tuzen, M., & Mendil, D. (2005). Determination of chemical properties of walnut (*Juglans regia* L.) cultivars grown in Turkey. *Asian Journal of Chemistry, 17*(1), 548–552.

Akin, E. B., Karabulut, I., & Topcu, A. (2008). Some compositional properties of main Malatya apricot (*Prunus armeniaca* L.) varieties. *Food Chemistry, 107*, 939–948.

Akpakpan, A. E., & Akpabio, U. D. (2012). Evaluation of proximate composition, mineral element and anti-nutrient in almond (*Terminalia catappa*) seeds. *Research Journal of Applied Sciences, 7*(9–12), 489–493.

Ali, S., Masud, T., & Abbasi, K. S. (2011). Physicochemical characteristics of apricot (*Prunus armeniaca* L.) grown in northern areas of Pakistan. *Scientia Horticulturae, 130*(2), 386–392.

Alrgei, H. O., Dabić, D., Natić, M., Rakonjac, V., Milojković-Opsenica, D., Tešić, Ž., & FotirićAkšić, M. (2016). Chemical profile of major taste- and health-related compounds of Oblačinska sour cherry. *Journal of the Science of Food and Agriculture, 96*, 1241–1251.

Amaral, J. S., Casal, S., Pereira, J. A., Seabra, R. M., & Oliveira, B. P. P. (2003). Determination of sterol and fatty acid compositions, oxidative stability, and nutritional value of six walnut (*Juglans regia* L.) cultivars grown in Portugal. *Journal of Agricultural and Food Chemistry, 51*(26), 7698–7702.

Anonymous. (2019). *Peach, raw*. Food Data Central. Washington, DC: U.S. Department of Food & Agriculture.

Anonymous. (1989). *Recommended dietary allowances* (10th edn). Food Nutrition Board, National Research Council. Washington, DC: National Academies Press.

Askin, M. A., Balta, M. F., Tekintas, F. E., Kazankaya, A., & Balta, F. (2007). Fatty acid composition affected by kernel weight in almond [*Prunus dulcis* (Mill.) D.A. Webb] genetic resources. *Journal of Food Composition and Analysis, 30*, 7–12.

Ara, R., Motalab, M., Uddin, M. N., Fakhruddin, A. N. M., & Saha, B. K. (2014). Nutritional evaluation of different mango varieties available in Bangladesh. *International Food Research Journal, 21*(6), 2169–2174.

Arion, C. M., Tabart, J., Kevers, C., Niculaua, M., Filimon, R., Beceanu, D., & Dommes, J. (2014, Mar 1). Antioxidant potential of different plum cultivars during storage. *Food Chemistry, 146*, 485–491.

Aslantas, R., Guleryuz, M., & Turan, M. (2001). Some chemical contents of selected almond (Prunus amygdalus Batsch) types. *Cahiers Options Méditerranéennes, 56*, 347–350.

Asp, N. G. (1996). Dietary carbohydrates: Classification by chemistry and physiology. *Food Chemistry, 57*(1), 9–14.

Asrey, R., Patel, V. B., Barman, K., & Pal, R. K. (2013). Pruning affects fruit yield and postharvest quality in mango (Mangifera indica L.) cv. Amrapali. *Fruits, 68*(5), 367–380.

Aubert, C., & Chanforan, C. (2007). Postharvest changes in physicochemical properties and volatile constituents of apricot (Prunus armeniaca L.). Characterization of 28 cultivars. *Journal of Agricultural and Food Chemistry, 55*(8), 3074–3082.

Bachheti, R. K., Rai, I., Joshi, A., & Rana, V. (2012). Physicochemical study of seed oil of *Prunus armeniaca* L. grown in Garhwal region (India) and its comparison with some conventional food oils. *International Food Research Journal, 19*, 577–581.

Baghel, M., Nagaraja, A., Srivastav, M., Meena, N. K., Kumar, M. S., Kumar, A., & Sharma, R. R. (2019). Pleiotropic influences of brassinosteroids on fruit crops: A review. *Plant Growth Regulation, 87*(2), 375–388.

Bal, J. S. (1986). Pattern of fruit development in Indian jujube. *Hortscience, 21*, 841.

Bal, J. S., & Chauhan, G. S. (1981). Effect of ethephon on ripening of ber. *Punjab Hort. J., 21*, 188–191.

Bal, J. S., Jawanda, J. S., & Singh, S. N. (1979). Developmental physiology of ber var. Umran. Changes in amino-acids and sugars (sucrose, glucose and fructose) at different stages of fruit ripening. *Indian Food Packer, 33*, 33–35.

Bal, J. S., Kahlon, P. S., Jawanda, J. S., & Randhawa, J. S. (1995). Studies on artificial ripening of ber fruits (*Zizyphus mauritiana* Lamk.). *Phala Samaskaran, 1*, 325–333.

Ballistreri, G., Continella, A., Gentile, A., Amenta, M., Fabroni, S., & Rapisarda, P. (2013). Fruit quality and bioactive compounds relevant to human health of sweet cherry (*Prunus avium* L.) cultivars grown in Italy. *Food Chemistry, 140*(4), 630–638.

Bandyopadhyay, C., & Gholap, A. S. (1973). Relationship of aroma and flavour characteristics of mango (Mangifera indica L.) to fatty acid composition. *Journal of the Science of Food and Agriculture, 24*(12), 1497–1503.

Barbera, G., Di Marco, L., & Schirra, M. (1994). Effects of rootstock on productive and qualitative response of two almond varieties. *Acta Horticulturae, 373*, 129–134.

Bassi, D., Mignani, I., Spinardi, A., & Tura, D. (2016). Peach (Prunus persica (L.) Batsch). In *Nutritional composition of fruit cultivars* (pp. 535–571). Academic Press.

Bastos, R., Coelho, E., & Coimbra, M. A. (2015). Modifications of Saccharomyces pastorianus cell wall polysaccharides with brewing process. *Carbohydrate Polymers, 124*, 322–330.

Beyer, R., & Melton, L. D. (1990). Composition of New Zealand apricot kernels. *New Zealand Journal of Crop and Horticultural Science, 18*, 39–42.

Bhutani, V. P., & Joshi, V. K. (1995). Plum. In D. K. Salunkhe & S. S. Kadam (Eds.), *Fruit science and technology, production, composition, storage and processing* (pp. 203–241). New York: Dekker.

Blando, F., Gerardi, C., & Nicoletti, I. (2004). Sour cherry (*Prunus cerasus* L.) anthocyanins as 612 ingredients for functional foods. *Journal of Biomedicine & Biotechnology, 5*, 253–258.

Butac, M., Militaru, M., & Budan, S. (2012). Criteria for harvesting plums at the optimum maturity for Romanian varieties of the fresh fruit market. *Fruit Growing Research, 28*, 1–6.

Caliskan, O., Bayazit, S., & Sumbul, A. (2012). Fruit quality and phytochemical attributes of some apricot (Prunus armeniaca L.) cultivars as affected by genotypes and seasons. *Notulae Botanicae Horti Agrobotanici Cluj-Napoca, 40*(2), 284–294.

Calixto, F. S., Bauza, M., Martinez de Toda, F., & Argamenteria, A. (1981). Amino acids, sugars, and inorganic elements in the sweet almond (Prunus amygdalus). *Journal of Agricultural and Food Chemistry, 29*(3), 509–511.

Chauhan, S. K., Tyagi, S. M., & Singh, D. (2001). Pectinolytic liquefaction of apricot, plum, and mango pulps for juice extraction. *International Journal of Food Properties, 4*(1), 103–109.

Chen, C. Y. O., & Blumberg, J. B. (2008). In vitro activity of almond skin polyphenols for scavenging free radicals and inducing quinone reductase. *Journal of Agricultural and Food Chemistry, 56*(12), 4427–4434.

Chen, C. Y., Lapsley, K., & Blumberg, J. (2006). A nutrition and health perspective on almonds. *Journal of the Science of Food and Agriculture, 86*(14), 2245–2250.

Ciressa, V., Rosu, F., & Ursache, C. (1984). Ziziphus jujuba Mill., a promising new fruit. *Horticultura, 27*(28), 103–104.

Cowan, J. W., Sabry, Z. I., Rinnu, F. J., & Campbell, J. A. (1963). Evaluation of protein in Middle Eastern diets: I. Almond (Prunus amygdalus). *The Journal of Nutrition, 81*(3), 235–240.

Crews, C., Hough, P., Godward, J., Brereton, P., Lees, M., & Guiet, S. (2005). Study of the main constituents of some authentic walnut oils. *Journal of Agricultural and Food Chemistry, 53*, 4853–4860.

Croft, K. D. (1998). The chemistry and biological effects of flavonoids and phenolic acids. *Annals of the New York Academy of Sciences, 854*, 435–442.

Dar, M. S., Oak, P., Chidley, H., Deshpande, A., Giri, A., & Gupta, V. (2016). Nutrient and flavor content of mango (Mangifera indica L.) cultivars: An appurtenance to the list of staple foods. In *Nutritional composition of fruit cultivars* (pp. 445–467). Academic Press.

Deshpande, A. B., Anamika, K., Jha, V., Chidley, H. G., Oak, P. S., Kadoo, N. Y., Pujari, K. H., Giri, A. P., & Gupta, V. S. (2017). Transcriptional transitions in Alphonso mango (Mangifera indica L.) during fruit development and ripening explain its distinct aroma and shelf life characteristics. *Scientific Reports, 7*(1), 1–19.

Dixit, A. K., Sharma, P. C., Nanda, S. K., & Aleksha Kudos, S. K. (2010). Impact of processing technology in hilly region: a study on extraction of apricot kernel oil. *Agricultural Economics Research Review, 23*, 405–410.

Drogoudi, P. D., Vemmos, S., Pantelidis, G., Petri, E., Tzoutzoukou, C., & Karayiannis, I. (2008). Physical characters and antioxidant, sugar, and mineral nutrient contents in fruit from 29 apricot (*Prunus armeniaca* L.) cultivars and hybrids. *Journal of Agricultural and Food Chemistry, 56*, 10754–10760.

Egea, G., González-Real, M. M., Baille, A., Nortes, P. A., Sánchez-Bel, P., & Domingo, R. (2009). The effects of contrasted deficit irrigation strategies on the fruit growth and kernel quality of mature almond trees. *Agricultural Water Management, 96*(11), 1605–1614.

Elsadr, H. T., & Sherif, S. (2016). Peaches and nectarines. *Book: The Encyclopedia of Food and Health, 122*(1), 270–276. https://doi.org/10.1016/B978-0-12-384947-2.00527-4.

Fanning, K. J., Topp, B., Russell, D., Stanley, R., & Netzel, M. (2014). Japanese plums (Prunus salicina Lindl.) and phytochemicals–breeding, horticultural practice, postharvest storage, processing and bioactivity. *Journal of the Science of Food and Agriculture, 94*(11), 2137–2147.

Fatima, T., Bashir, O., Gani, G., Bhat, T., & Jan, N. (2018). Nutritional and health benefits of apricots. *International Journal of Unani and Integrative Medicine, 2*(2), 05–09.

Ferretti, G., Bacchetti, T., Belleggia, A., & Neri, D. (2010). Cherry antioxidants: From farm to table. *Molecules, 15*(10), 6993–7005.

Forni, E., Erba, M. L., Maestrelli, A., & Polesello, A. (1992). Sorbitol and free sugar contents in plums. *Food Chemistry, 44*(4), 269–275.

Fowomola, M. A. (2010). Some nutrients and antinutrients contents of mango (Magnifera indica) seed. *African Journal of Food Science, 4*(8), 472–476.

Fraser, G. E. (1994). Diet and coronary heart diseases: beyond dietary fats and low density lipoprotein cholesterol. *The American Journal of Clinical Nutrition, 59*, 1117–11235.

Gezer, I., Haciseferoğullari, H., Ozcan, M. M., Arslan, D., Asma, B. M., & Ünver, A. (2011). Physicochemical properties of apricot (*Prunus armeniaca* L.) kernels. *Southwestern Journal of Horticulture, Biology and Environment, 2*, 1–13.

Gil, M. I., Tomas-Barberan, F. A., Hess-Pierce, B., & Kader, A. A. (2002). Antioxidant capacities, phenolic compounds, carotenoids, and vitamin C contents of nectarine, peach, and plum cultivars from California. *Journal of Agricultural and Food Chemistry, 50*, 4976–4982.

Gonçalves, A. C., Bento, C., Silva, B., Simões, M., & Silva, L. R. (2019). Nutrients, bioactive compounds and bioactivity: The health benefits of sweet cherries (Prunus avium L.). *Current Nutrition & Food Science, 15*(3), 208–227.

Guil-Guerrero, J. L., Diaz Delgado, A., Matallana Gonzalez, M. C., & TorijaIsasa, M. E. (2004). Fatty acids and carotenes in some ber (*Zizyphus jujube* Mill.) varieties. *Plant Foods for Human Nutrition, 59*, 23–27.

Gupta, A., Sharma, P. C., Tilakratne, B. M. K. S., & Verma, A. K. (2012). Studies on physico-chemical characteristics and fatty acid composition of wild apricot (*Prunus armeniaca* Linn.) kernel oil. *Indian Journal of Natural Products and Resources, 3*, 366–370.

Hacıseferoğulları, H., Gezer, İ., Özcan, M. M., & MuratAsma, B. (2007). Post-harvest chemical and physical–mechanical properties of some apricot varieties cultivated in Turkey. *Journal of Food Engineering, 79*, 364–373.

Hall, A. P., Moore, J. G., Gunning, B., & Cook, B. B. (1958). Nutrients in nuts, the nutritive value of fresh and roasted California grown nonpareil almonds. *Journal of Agricultural and Food Chemistry, 6*(5), 377–382.

Holst, B., & Williamson, G. (2008). Nutrients and phytochemicals: from bioavailability to bioefficacy beyond antioxidants. *Current Opinion in Biotechnology, 19*(2), 73–82.

Hu, F. B., Stampfer, M. J., Manson, J. E., Rim, E. B., Colditz, G. A., & Rosner, B. A. (1998). Frequent nut consumption and risk of coronary heart disease in women: prospective cohort study. *British Medical Journal, 317*, 1341–1345.

Jawanda, J. S., & Bal, J. S. (1980). A comparative study on growth and development of ZG-2 and Kaithli cultivars of ber. *Punjab Horticultural Journal, 20*, 41–46.

Kadam, S. S., Kotecha, P. M., & Adsule, R. N. (1993). Changes in physicochemical characteristics and enzyme activities during ripening of ber (*Ziziphus mauritiana* Lamk.). *Indian Food Packer, 48*, 5–10.

Kader, A. A., Li, Y., & Chordas, A. (1982). Postharvest respiration, ethylene production, and compositional changes of Chinese jujube fruits. *Horticultural Science, 17*, 678–679.

Kornsteiner, M., Wagner, K.-H., & Elmadfa, I. (2006). Tocopherols and total phenolics in 10 different nut types. *Food Chemistry, 98*, 381–387.

Krishnamurthy, S., Patwardhan, M. V., & Subramanyam, H. (1971). Biochemical changes during ripening of the mango fruit. *Phytochemistry, 10*, 2577–2581.

Kritchevsky, D. (1988). Dietary fibre. *Annual Review of Nutrition, 8*, 301–328.

Kudachikar, V. B., Ramana, K. V. R., & Eipeson, W. E. (2001). Pre and post-harvest factors influencing the shelf life on ber (Ziziphus mauritiana Lamk.): A review. *Indian Food Packer, 55*, 81–90.

Kuliev, A. A., & Guseinova, N. K. (1974). The content of vitamin C B1,B2 and E in some fruits. *Referativnyi Zhurnel, 2*, 69–73.

Kunachowicz, H., Nadolna, I., Iwanow, K., & Przygoda, B. (2005). *Wartość odżywcza wybranych produktów spożywczych i typowych potraw* [Nutritional value of selected food products and typical dishes]. PZWL Warszawa [in Polish].

Lavedrine, F., Zmirou, D., Ravel, A., Balducci, F., & Alary, J. (1999). Blood cholesterol and walnut consumption: a cross-sectional survey in France. *Preventive Medicine, 28*, 333–339.

Lavedrine, F., Ravel, A., Villet, A., Ducros, V., & Alary, J. (2000). Mineral composition of two walnut cultivars originating in France and California. *Food Chemistry, 68*, 347–351.

Leccese, A., Bartolini, S., & Viti, R. (2007). Total antioxidant capacity and phenolics content in apricot fruits. *International Journal of Fruit Science, 7*(2), 3–16.

Lee, S. K., & Kader, A. A. (2000). Pre-harvestand post-harvest factors influencing vitamin C content of horticultural crops. *Postharvest Biology and Technology, 20*, 207–220.

Li, J. W., Fan, L. P., Ding, S. D., & Ding, X. L. (2007a). Nutritional composition of five cultivars of Chinese jujube. *Food Chemistry, 103*(2), 454–460.

Li, M., Yang, G. L., Min, S., Gao, X. Y., Wang, Y., & Li, M. R. (2007b). Extract process of cyclic adenosine on phosphate (cAMP) in *Ziziphus jujube*. *Journal of Chinese Medicinal Materials, 30*, 1143–1145.

Lichou, J., Jay, M., Vaysse, P., & Lespinasse, N. (2003). Recognising apricot varieties. In *Reconnaître les variétés d'abricots 2003* (pp. 90, pp. ref. 6).

Ling, L., X. Jiang-hui, S. Guang-ming, H. YanBiao, Z. Xiao-ping and W. Chang-bin. 2008. Sugar accumulation in fruit of Zizyphus mauritiana Lam during its development. Acta Agriculturae Universitatis Jiangxiensis. CNKI:SUN:JXND.0.2008-06-018.

Lintas, C., & Cappelloni, M. (1992). Dietary fiber content of Italian fruit and nuts. *Journal of Food Composition and Analysis, 5*(2), 146–151.

Liu, F. X., Fu, S. F., Bi, X. F., Chen, F., Liao, X. J., Hu, X. S., & Wu, J. H. (2013). Physico-chemical and antioxidant properties of four mango (Mangifera indica L.) cultivars in China. *Food Chemistry, 138*(1), 396–405.

Lu, H., Lou, H., Zheng, H., Hu, Y., & Li, I. (2012). Nondestructive evaluation of quality changes and the optimum time for harvesting during jujube (*Zizyphus jujuba* Mill. cv. Changhong) fruits development. *Food and Bioprocess Technology, 5*(6), 2586–2595.

Lucas, E. A., Mocanu, V., Smith, B. J., Soung, D. Y., & Daggy, B. P. (2004). Daily consumption of dried plum by postmenopausal women does not cause undesirable changes in bowel function. *Energy (kcal), 259*, 239.

Maldonado-Celis, M. E., Yahia, E. M., Bedoya, R., Landázuri, P., Loango, N., Aguillón, J., Restrepo, B., & Ospina, J. C. G. (2019). Chemical composition of mango (Mangifera indica L.) fruit: Nutritional and phytochemical compounds. *Frontiers in Plant Science, 10*.

Manthey, J. A., & Perkins-Veazie, P. (2009). Influences of harvest date and location on the levels of β-carotene, ascorbic acid, total phenols, the in vitro antioxidant capacity, and phenolic profiles of five commercial varieties of mango (Mangifera indica L.). *Journal of Agricultural and Food Chemistry, 57*(22), 10825–10830.

Matheyambath, A. C., Subramanian, J., & Paliyath, G. (2016). Mangoes reference module in food science. In B. C. Paul & F. F. Told (Eds.), *Encyclopedia of food and health* (pp. 641–645). Switzerland: Elsevier. https://doi.org/10.1016/B978-0-12-384947-2.00442-6.

Martinez, M. L., Labuckas, D. O., Lamargu, A. L., & Maestri, D. M. (2010). Walnuts (*Juglans regia* L.): genetic resources, chemistry, by-products. *Journal of the Science of Food and Agriculture, 90*, 1959–1967.

McCune, L. M., Kubota, C., Stendell-Hollis, N. R., & Thomson, C. A. (2011). Cherries and health: a review. *Critical Reviews in Food Science and Nutrition, 51*, 1–12.

Meena, N. K., & Asrey, R. (2018a). Tree age affects physicochemical, functional quality and storability of Amrapali mango (Mangifera indica L.) fruits. *Journal of the Science of Food and Agriculture, 98*(9), 3255–3262.

Meena, N. K., & Asrey, R. (2018b). Tree age affects postharvest attributes and mineral content in Amrapali mango (Mangifera indica) fruits. *Horticultural Plant Journal, 4*(2), 55–61.

Milbury, P. E., Chen, C. Y., Dolnikowski, G. G., & Blumberg, J. B. (2006). Determination of flavonoids and phenolics and their distribution in almonds. *Journal of Agricultural and Food Chemistry, 54*(14), 5027–5033.

Milosevic, T., & Milosevic, N. (2012). Factors influencing mineral composition of plum fruits. *Journal of Elementology, 17*(3), 453–464. https://doi.org/10.5601/jelem.2012.17.3.08.

Mitić, M. N., Obradović, M. V., Kostić, D. A., Micić, R. J., & Pecev, E. T. (2012). Polyphenol content and antioxidant activity of sour cherries from Serbia. *Chemical Industry and Chemical Engineering Quarterly, 18*, 53–62.

Momchilova, S., & Nikolova-Damyanova, B. (2007). Quantitative TLC and gas chromatography determination of the lipid composition of raw and microwave roasted walnuts, hazelnuts, and almonds. *Journal of Liquid Chromatography and Related Technologies, 30*, 2267–2285.

Morton, J. (1987). *Indian Jujube*. Retrieved December 20, 2011, from http://www.hort.purdue.edu/newcrop/morton/Indian jujube.html.

Mozetič, B., Simčič, M., & Trebše, P. (2006). Anthocyanins and hydroxycinnamic acids of Lambert compact cherries (*Prunus avium* L.) after cold storage and 1-methylcyclopropene treatment. *Food Chemistry, 97*(2), 302–309.

Othman, O. C., & Mbogo, G. P. (2009). Physico-chemical characteristics of storage-ripened mango (Mangifera indica L.) fruits varieties of eastern Tanzania. *Tanzania Journal of Science, 35*(1), 57–65.

Ozkan, G., & Koyuncu, M. A. (2005). Physical and chemical composition of some walnut (*Juglans regia* L.) genotypes grown in Turkey. *Grasas y Aceites, 56*(2), 141–146.
Pacifico, S., Di Maro, A., Petriccione, M., Galasso, S., Piccolella, S., Di Giuseppe, A. M., & Monaco, P. (2014). Chemical composition, nutritional value and antioxidant properties of autochthonous *Prunus avium* cultivars from Campania Region. *Food Research International, 64*, 188–199.
Pandey, R. C., Pathak, R. A., & Pathak, R. K. (1990). Physicochemical changes associated with growth and development of fruits in ber (*Ziziphus mauritiana* Lamk.). *Indian Journal of Horticulture, 47*, 266–270.
Papp, N., Szilvássy, B., Abrankó, L., Szabó, T., Pfeiffer, P., Szabó, Z., Nyéki, J., Ercisli, S., Stefanovits-Bányai, É., & Hegedűs, A. (2010). Main quality attributes and antioxidants in Hungarian sour cherries: identification of genotypes with enhanced functional properties. *International Journal of Food Science and Technology, 45*(2), 395–402.
Pareek, S. (2013). Nutritional composition of jujube fruit. *Emirates Journal of Food and Agriculture*, 463–470.
Pareek, O. P. (2001). *Ber*. Southampton: International Centre for Underutilized Crops, University of Southampton.
Pareek, S., & Dhaka, R. S. (2008). Association analysis for quality attributes in ber. *Indian Journal of Arid Horticulture, 3*, 77–80.
Pareek, S., Kitinoja, L., Kaushik, R. A., & Paliwal, R. (2009). Postharvest physiology and storage of ber. *Stewart Postharvest Review, 5*(5), 1–10.
Pathak, S. R., & Sarada, R. (1974). Lipids of mango (Mangifera indica). *Current Science*.
Pereira, J. A., Oliveira, I., Sousa, A., Ferreira, I. C. F. R., Bento, A., & Estevinho, L. (2008). Bioactive properties and chemical composition of six walnut (*Juglans regia* L.) cultivars. *Food and Chemical Toxicology, 46*, 2103–2111.
Piga, A., Del Caro, A., & Corda, G. (2003). From plums to prunes: Influence of drying parameters on polyphe nols and antioxidant activity. *Journal of Agricultural and Food Chemistry, 51*, 3675–3681.
Popa, V. M., Bele, C., Poiana, M. A., Dumbrava, D., Raba, D. N., & Jianu, C. (2011). Evaluation of bioactive compounds and of antioxidant properties in some oils obtained from food industry by-products. *Romanian Biotechnology Letters, 16*, 6234–6241.
Prasad, R. B. N. (1994). Walnut and pecans. In *Encysclo-paedia of food science, food technology and nutrition*.
Radi, M., Mahrouz, M., Jaouad, A., & Amiot, M. (2003). Influence of mineral fertilization (NPK) on the quality of apricot fruit (cv. Canino) The effect of the mode of nitrogen supply. *Agronomie, 23*, 737–745.
Roussos, P. A., Sefferou, V., Denaxa, N.-K., Tsantili, E., & Stathis, V. (2011). Apricot (*Prunus armeniaca* L.) fruit quality attributes and phytochemicals under different crop load. *Scientia Horticulturae, 129*, 472–478.
Ruggeri, S., Cappelloni, M., Gambelli, L., Nicoli, S., & Carnovale, E. (1996). Chemical composition and nutritive value of nuts grown in Italy. *Italian Journal of Food Science, 3*, 243–252.
Ruggeri, S., Cappelloni, M., Gambelli, L., Nicoli, S., & Carnovale, E. (1998). Chemical composition and nutritive value of nuts grown in Italy. *Italian Journal of Food Science, 10*(3), 243–252.
Ruiz, D., Egea, J., Gil, M. I., & Tomás-Barberán, F. A. (2005a). Characterization and quantitation of phenolic compounds in new apricot (Prunus armeniaca L.) varieties. *Journal of Agricultural and Food Chemistry, 53*(24), 9544–9552.
Ruiz, D., Egea, J., Tomás-Barberán, F. A., & Gil, M. I. (2005b). Carotenoids from new apricot (Prunus armeniaca L.) varieties and their relationship with flesh and skin color. *Journal of Agricultural and Food Chemistry, 53*(16), 6368–6374. https://doi.org/10.1021/jf0480703.
Rumainum, I. M., Worarad, K., Srilaong, V., & Yamane, K. (2018). Fruit quality and antioxidant capacity of six Thai mango cultivars. *Agriculture and Natural Resources, 52*(2), 208–214.
Sabaté, J., Fraser, G. E., Burke, K., Knutsen, S. F., Bennett, H., & Linstead, K. D. (1993). Effects of walnuts on serum lipid levels and blood pressure in normal men. *The New England Journal of Medicine, 329*, 603–660.

Şahin, İ., & Akbaş, H. (2001). Researches on the technological properties of walnuts gathered from different types and areas (in Turkish). In: *First National Walnut Symposium of Turkey*, 2001, Tokat, Turkey (pp. 104–114).
Sahamishirazi, S., Moehring, J., Claupein, W., & Graeff-Hoenninger, S. (2017). Quality assessment of 178 cultivars of plum regarding phenolic, anthocyanin and sugar content. *Food Chemistry, 214*, 694–701.
Saidani, F., Giménez, R., Aubert, C., Chalot, G., Betrán, J. A., & Gogorcena, Y. (2017). Phenolic, sugar and acid profiles and the antioxidant composition in the peel and pulp of peach fruits. *Journal of Food Composition and Analysis, 62*, 126–133.
San, B., Yildirim, A. N., Polat, M., & Yildirim, F. (2009). Mineral composition of leaves and fruits of some promising Jujube (Zizyphus jujuba miller) genotypes. *Asian Journal of Chemistry, 21*(4), 2898.
Sathe, S. K. (1992). Solubilization, electrophoretic characterization and in vitro digestibility of almond (Prunus amygdalus) proteins 1, 2. *Journal of Food Biochemistry, 16*(4), 249–264.
Sathe, S. K., Seeram, N. P., Kshirsagar, H. H., Heber, D., & Lapsley, K. A. (2008). Fatty acid composition of California grown almonds. *Journal of Food Science, 73*(9), C607–C614.
Saura-Calixto, F., Cañellas, J., & Soler, L. (1983). Dietary fibre and components of the nitrogen-free extract of almond kernels. *Journal of the Science of Food and Agriculture, 34*(12), 1419–1422.
Savage, G. P. (2001). Chemical composition of walnuts (Juglans regia L.) grown in New Zealand. *Plant Foods for Human Nutrition, 56*(1), 75–82.
Savage, G. P., McNeil, D., & Sterberg, K. (2001). Oxidative stability of walnuts during long term storage. 4th International Walnut Symposium. *Acta Horticulturae, 544*, 591–597.
Savage, G. P. (2000). Chemical composition of walnuts (*Juglands regina* L.) grown in New Zealand. *Plant Foods for Human Nutrition, 56*, 75–82.
Schmitzer, V., Slatnar, A., Mikulic-Petkovsek, M., Veberic, R., Krska, B., & Stampar, F. (2011). Comparative study of primary and secondary metabolites in apricot (*Prunus armeniaca* L.) cultivars. *Journal of the Science of Food and Agriculture, 91*, 860–866.
Selvaraj, Y., Kumar, R., & Pal, D. K. (1989). Changes in sugars, organic acids, amino acids, lipid constituents and aroma characteristics of ripening mango (Mangifera indica L) fruit. *Journal of Food Science and Technology, 26*(6), 308–313.
Sen, S. M. (2013). *Walnut diet: eating walnut living healthy (in Turkish)* (p. 216S). Turkey: Alper Ankara.
Serradilla, M. J., Akšic, M. F., Manganaris, G. A., Ercisli, S., González-Gómez, D., & Valero, D. (2017). 17 Fruit chemistry, nutritional benefits and social aspects of cherries. In *Cherries: Botany, production and uses* (p. 420). Wallingford: CABI.
Serrano, M., Guillén, F., Martínez-Romero, D., Castillo, S., & Valero, D. (2005). Chemical constituents and antioxidant activity of sweet cherry at different ripening stages. *Journal of Agricultural and Food Chemistry, 53*(7), 2741–2745.
Sharlatifar, N., Pourfard, I. M., Khaniki, G. J., Nabizadeh, R., Akbarzadeh, A., & Nejad, A. S. M. (2017). Mineral composition, physicochemical properties and fatty acids profile of *Prunus armeniaca* apricot seed oil. *Asian Journal of Chemistry, 29*, 2011–2015.
Sharma, R. K. (1996). *Physiological and biochemical studies in ber during ripening on tree and in storage*. Ph.D. Thesis, Choudhary Charan Singh Hisar Agriculture University, Hisar, Harayana.
Silva, A. P. F. B., Nascimento, J. R. O. D., Lajolo, F. M., & Cordenunsi, B. R. (2008). Starch mobilization and sucrose accumulation in the pulp of Keitt mangoes during postharvest ripening. *Journal of Food Biochemistry, 32*(3), 384–395.
Simunic, V., Kovac, S., Gaso-Sokac, D., Pfannhauser, W., & Murkovic, M. (2005). Determination of anthocyanins in four Croatian cultivars of sour cherries (*Prunus cerasus*). *European Food Research and Technology, 220*, 575–578.
Singh, B. P., Singh, S. P., & Chauhan, K. S. (1981). Certain chemical changes and rate of respiration in different cultivars of ber during ripening. *Haryana Agricultural University Journal of Research, 11*, 60–64.

Singh, U. P., Singh, D. P., Singh, M., Maurya, S., Srivastava, J. S., Singh, R. B., & Singh, S. P. (2004). Characterization of phenolic compounds in some Indian mango cultivars. *International Journal of Food Sciences and Nutrition, 55*(2), 163–169.

Souci, S. W., Fachmann, W., & Kraut, H. (1994). *Food composition and nutrition tables*. Stuttgart: Medpharm, CRC Press.

Stryjecka, M., Kiełtyka-Dadasiewicz, A., Michalak, M., Rachoń, L., & Głowacka, A. (2019). Chemical composition and antioxidant properties of oils from the seeds of five apricot (Prunus armeniaca L.) cultivars. *Journal of Oleo Science*, ess19121.

Tamura, M., Ohnishi, Y., Kotani, T., & Gato, N. (2011). Effects of new dietary fiber from Japanese Apricot (Prunus mume Sieb. et Zucc.) on gut function and intestinal microflora in adult mice. *International Journal of Molecular Sciences, 12*(4), 2088–2099.

Teotia, S. S., Dubey, P. S., Awasthi, R. K., & Upadhyay, N. P. (1974). Studies on physic-chemical characteristics of some important ber varieties (*Ziziphus mauritiana* Lamk.). *Progressive Horticulture, 5*(4), 81–88.

Tomás-Barberán, F. A., Gil, M. I., Cremin, P., Waterhouse, A. L., Hess-Pierce, B., & Kader, A. A. (2001). HPLC–DAD–ESIMS analysis of phenolic compounds in nectarines, peaches, and plums. *Journal of Agricultural and Food Chemistry, 49*, 4748–4760.

Tomás-Barberán, F. A., Ruiz, D., Valero, D., Rivera, D., Obón, C., Sánchez-Roca, C., & Gil, M. (2013). Health benefits from pomegranates and stone fruit, including plums, peaches, apricots and cherries. In M. Skinner & D. Hunter (Eds.), *Bioactives in fruit: Health benefits and functional foods* (1st ed., pp. 125–167). Chichester: Wiley.

USDA-ARS. (2005). U.S. Department of Agriculture. USDA National Nutrient Database for Standard Reference.

USDA-ARS. (2010). U.S. Department of Agriculture, Agricultural Research Service, 2010. USDA National Nutrient Database for Standard Reference. Nutrient Data Laboratory Home Page. http://www.ars.usda.gov/ba/bhnrc/ndl

USDA-ARS. (2016). *National nutrient database for standard reference, release 26*. Nutrient Data Laboratory, US Department of Agriculture, Agricultural Research Service. Available at https://ndb.nal.usda.gov/

USDA (US Department of Agriculture), Agricultural Research Service. (2010). *USDA National Nutrient Database for Standard Reference, release 23*. Retrieved October 4, 2010, from Nutrient Data Laboratory Home Page: http://www.ars.usda.gov/ba/bhnrc/ndl.

Usenik, V., Fabčič, J., & Štampar, F. (2008). Sugars, organic acids, phenolic composition and antioxidant activity of sweet cherry (*Prunus avium* L.). *Food Chemistry, 107*(1), 185–192.

Usenik, V., Stampar, F., & Kastelec, D. (2013). Phytochemicals in fruits of two Prunus domestica L. plum cultivars during ripening. *Journal of the Science of Food and Agriculture, 93*(3), 681–692.

Vazquez-Salinas, C., & Lakshminarayana, S. (1985). Compositional changes in mango fruit during ripening at different storage temperatures. *Journal of Food Science, 50*(6), 1646–1648.

Velioglu, Y. S., Mazza, G., Gao, L., & Oomah, B. D. (1998). Antioxidant activity and total phenolics in selected fruits, vegetables, and grain products. *Journal of Agricultural and Food Chemistry, 46*(10), 4113–4117.

Venkatachalam, M., & Sathe, S. K. (2006). Chemical composition of selected edible nut seeds. *Journal of Agricultural and Food Chemistry, 54*, 4705–4714.

Vursavuş, K., Kelebek, H., & Selli, S. (2006). A study on some chemical and physico-mechanic properties of three sweet cherry varieties (Prunus avium L.) in Turkey. *Journal of Food Engineering, 74*(4), 568–575.

Wills, R. B. H., Scriven, F. M., & Greenfield, H. (1983). Nutrient composition of stone fruit (*Prunus* spp.) cultivars: apricot, cherry, nectarine, peach and plum. *Journal of the Science of Food and Agriculture, 34*, 1383–1389.

Wongmetha, O., & Ke, L. S. (2012). The quality maintenance and extending storage life of mango fruit after postharvest treatments. *World Academic of Science Engineering and Technology, 6*(9), 844–849.

Wu, H., Wang, D., Shi, J., Xue, S., & Gao, M. (2010). Effect of the complex of zinc (II) and cerium (IV) with chitosan on the preservation quality and degradation of organophosphorus pesticides in Chinese jujube (Zizyphus jujuba Mill. cv. Dongzao). *Journal of Agricultural and Food Chemistry, 58*(9), 5757–5762.

Yada, S., Lapsley, K., & Huang, G. (2011). A review of composition studies of cultivated almonds: macronutrients and micronutrients. *Journal of Food Composition and Analysis, 24*(4–5), 469–480.

Yagmur, C., & Taskin, M. (2011). Study on changes in mineral content of plum (*Prunus domestica*) and strawberry (*Fragaria × ananassa*) during canning. *Indian Journal of Agricultural Sciences, 81*(8), 723–728.

Yerlikaya, C., Yucel, S., Erturk, U., & Korukluoğlu, M. (2012). Proximate composition, minerals and fatty acid composition of *Juglans regia* L. genotypes and cultivars grown in Turkey. *Brazilian Archives of Biology and Technology, 55*(5), 677–683.

Chemical Treatments for Shelf Life Enhancement of Stone Fruits

10

Satyabrata Pradhan, Ipsita Panigrahi, Sunil Kumar, and Naveen Kumar Maurya

Abstract

Stone fruits, including mango, peach, plum, and apricot, are known for their luscious taste and higher nutritional and processing value, creating a year around market demand. However, these fruits are harvested in a particular time period in a year. Market glut during the peak season is a common hindrance for a good economic return of the cultivators. However, the situation gets aggravated due to the perishable nature of these stone fruits which limits the longer duration storage and transportation under ambient temperature. For longer storage and transportation, these fruits require a cold storage accompanied with modified atmospheric storage (MAS) or controlled atmospheric storage (CAS) facility and cold chain. However, cold storage and cold chain may not provide a viable solution to some stone fruits like mango and peach due to their vulnerability to cold-induced or chilling injury (CI). Pre- or postharvest treatment with chemicals like 1-methylcyclopropene (1-MCP), calcium chloride, methyl jasmonate (MeJA), salicylic acid (SA), oxalic acid, and melatonin at very low concentration can be a feasible option for these problems. Apart from enhancing shelf life and storability, these chemical treatments can also maintain nutritional attributes. In some stone fruits, chemical treatments followed by cold temperature storage enhance longer duration storability. Various chemical treatments and their underlying mechanism to enhance the shelf life of stone fruits will be discussed in this chapter.

S. Pradhan · S. Kumar (✉) · N. K. Maurya
Division of Fruits and Horticultural Technology, ICAR-Indian Agricultural Research Institute, New Delhi, India

I. Panigrahi
Division of Vegetable Science, ICAR-Indian Agricultural Research Institute, New Delhi, India

M. M. Mir et al. (eds.), *Production Technology of Stone Fruits*,
https://doi.org/10.1007/978-981-15-8920-1_10

Keywords

Stone fruits · 1-Methylcyclopropene · Methyl jasmonate · Putrescine · Ethylene · Respiration rate

10.1 Introduction

Most stone fruits, like mango, peach, and plum, cannot be stored for longer duration at room temperature owing to their highly perishable nature. However, storage at cold or low temperature may induce chilling injury (CI) in stone fruits like mango and peaches. Apart from some tropical stone fruits like mango, most of the commercial stone fruits are temperate in nature. Being temperate in nature, these fruits are available only for a shorter duration of time period in the year. Along with this, these fruits can be cultivated in confined geographical areas (temperate climatic condition) owing to their chilling requirement. However, due to their higher nutritional value accompanied with their luscious taste and processing values, these fetch a great year around market demand. The similar problems also persist in mango which is available for a particular time period during a year, leading to market glut during the peak harvesting season (Singh et al. 2017). Therefore, stone fruits are needed to be stored and transported to long distance. However, most of the stone fruits have lower shelf life, and fruit decay is one of the major problems associated with postharvest storage of fruits for fresh marketing and also for the purpose of processing (Benichou et al. 2018; Eckert and Ogawa 1988). Ripening in climacteric stone fruits, like mangoes, peaches, apricots, and plums, is stimulated by the plant growth regulator ethylene (Lurie and Pesis 1992; Martinez-Romero et al. 2003). These fruits are highly perishable in nature during storage at ambient temperature because of their rapid ripening nature and are also highly vulnerable to postharvest pathogens (Barkai-Golan and Follett 2017). The delicate skin of non-climacteric sweet cherry makes it susceptible to mechanical injury and bruising during handling and rapid moisture loss during storage which shortens its shelf life. The rapid ripening and softening of these fruits makes a shorter shelf life which renders the longer time storability and marketability.

During storage, stone fruits need a rapid cold storage with modified atmospheric storage (MAS) or controlled atmospheric storage (CAS) facility and cold chain for transport (Malakou and Nanos 2005). In present days, low-temperature storage is one of the mostly used methods to enhance postharvest life and maintain the quality of most of the stone fruits. However, these may be a costlier affair and require higher initial investment. Fruits like mangoes and peaches also develop chilling injury (CI) symptoms during postharvest cold. Chemical treatments of stone fruits, either pre- or postharvest, with 1-methylcyclopropene (1-MCP), calcium chloride, salicylic acid (SA), methyl jasmonate (MeJA), melatonin, oxalic acid, etc., offers a feasible option. These not only enhance the shelf life but also maintain the nutritional attributes. During storage and transportation, these chemical treatments at lower concentrations are used to maintain fruit quality by reducing discoloration and softening, delaying ethylene production, and decreasing the respiration rate. Chemical treatments accompanied with low-temperature storage facility are also followed to get better storability. This chapter highlights various chemical treatments used to enhance the shelf life of stone fruits.

10.2 Chemical Treatments for Shelf Life Enhancement of Stone Fruits

10.2.1 1-Methylcyclopropene

Ethylene-induced plant senescence can be inhibited by various cyclic olefinic compounds, including 1-methylcyclopropene (1-MCP), 3,3-dimethyl-cyclopropene, and 2,5-norbornadiene. These compounds can effectively inhibit the ethylene action at very lower concentration (*n* to μL/L) (Sisler et al. 1990; Sisler and Serek 1999). In plant cells, the irreversible binding of 1-MCP to ethylene receptor blocks them from binding to ethylene, which reduces autocatalytic ethylene production and hinders the process of ripening in many climacteric fruits (Blankenship and Dole 2003; Thompson and Bishop 2016). 1-MCP has been used to enhance the postharvest shelf life and increase the marketability of fresh stone fruits. Various physiological and biochemical mechanisms underlying the effect of 1-MCP have been widely studied.

Duan et al. (2004) observed that 1-MCP (1 μL/L) treatment inhibited pulp browning in "Zhonghuashuotao" peaches. Liguori et al. (2004) observed that higher concentration of 1-MCP can enhance the shelf life of stone fruit, including nectarines and melting flesh peaches. Treatment of 1-MCP at a concentration of 5 μL/L for 20 h time duration was optimum for inhibition of softening. 1-MCP treatment slowed the softening of fruit in a time-dependent manner and extended the time period for the fruits to become over-soft. Similarly, Liu et al. (2015) also observed that treatment of fruits of "Yuhualu" peach with 1-MCP (5 μL/L) for 24 h has delayed the respiration rate and ethylene production and enhanced the quality of fruits. Although each phenolic compounds has differential response to 1-MCP treatment, the treatment has retarded the onset of peak value for all the phenolic compounds over storage.

1-MCP as low as 0.1 μL/L can also be used to prevent mango fruit softening (Wang et al. 2006a, b). Treatment with 5.0 μL/L 1-MCP (6 h) can extend the shelf life of mangoes up to 12 days. The eating quality of these mangoes that are stored for 12 days was similar to the non-treated fruits. However, Sakhale et al. (2018) observed that 2000 ppb 1-MCP treatment (for 24 h) showed the best results in terms of various physical parameters. Vilas-Boas and Kader (2007) observed that 1-MCP treatment can reduce the ethylene production, but the respiration rate was not influenced in mangoes. Jiang and Joyce (2005) advocated that 1-MCP can be applied along with polyethylene bags to increase the shelf life of mango fruits under ambient temperature. However, Faasema et al. (2014) noted that the effectiveness of

1-MCP to delay the ripening of mangoes and enhance the shelf life can be affected by the storage temperature of treated fruits. Li et al. (2020) observed that 1-MCP have positive effect on quality-related parameters of mango fruit. The treatment also significantly induces the activity of various antioxidant enzymes and inhibited the activity of cell-wall-degrading enzymes like pectin esterase (PE), endopolygalacturonase (endo-PG), exopolygalacturonase (exo-PG), and endo-1,4-β-d-glucanase (EGase).

Owing to the rapid ripening rate and higher vulnerability to pathogens, peaches have a shorter postharvest shelf life. However, refrigeration or cold storage of peaches to increase the shelf life has another inherent problem, viz., vulnerability to chilling injury (Lurie and Crisosto 2005). In this regard, Jin et al. (2011) observed that 1-MCP (0.5 μL/L) treatment has dual effect in preventing chilling injury and maintaining quality of fruit in cold-stored "Baifeng" peaches. They also observed that 1-MCP treatment has resulted in the inhibition of the activity of polyphenol oxidase (PPO) and peroxidase (POD) and enhancement of antioxidant enzyme (SOD, CAT, and APX) activities. Lower respiration rate, ethylene production, ion leakage, and malondialdehyde (MDA) content were observed in the treated peach fruit as compared to the control. Recently, Liu et al. (2018) observed that 1-MCP (5 μL/L) application, along with intermittent warming (20 °C), has effectively extenuated the chilling injury and maintained superior fruit quality in peach cv. Yuhualu. This combined treatment has also maintained highest antioxidant activity and phenolic content. As evident, various concentrations of 1-MCP ranging between 0.5 and 5 μL/L were observed to enhance the shelf life and storability of peach fruits, but 5 μL/L was reported as the best concentration by many researchers.

Short shelf life of sweet cherry affects its marketability, consumer preference, and export. Being a non-climacteric stone fruit, the effect of 1-MCP (as it binds to the ethylene receptors in the plant cell) has been studied less as compared to the climacteric fruits. Some researchers claim 1-MCP to be relatively ineffective for extending cherry fruits' shelf life, while others reported contrasting results.

Gong et al. (2002) reported that 1-MCP treatment did not altered the fruit quality in terms of color and stem browning in sweet cherry (cv. "Bing" and "Rainier"). Exogenous application of ethylene enhanced stem browning irrespective of treatment with 1-MCP. The exogenously applied ethylene-induced changes in sweet cherry occurred via a process independent of ethylene receptors to which 1-MCP binds. Similarly, Mozetic et al. (2006) also observed that 1-MCP did not decrease the color change in sweet cherry cv. Lambert Compact. Among the various concentrations (0, 180, and 360 nL/L) used, the highest one resulted in significant reduction of cherry rot as compared to the untreated fruits. However, they had not noticed any significant change in anthocyanin and hydroxycinnamic acid profiles in fruits treated with 1-MCP.

In contrast to the above, Sharma et al. (2010) recommended that preharvest treatment of enhanced freshness formulation (EFF) (a hexanal formulation) in combination with postharvest treatment of hexanal and 1-MCP may increase the shelf life and quality of fruits of "Bing" sweet cherry. These treatments also enhanced the anthocyanin levels and activities of antioxidants like superoxide dismutase (SOD) and ascorbate peroxidase (APX). In another study, Yang et al. (2011) also supported the use of 1-MCP for improving the shelf life of sweet cherry. As compared to the control fruits, they observed a significant reduction in endogenous ethylene production, polyphenol oxidase activity, and malondialdehyde content in 1-MCP-treated fruits 60 days after cold storage. Among the three concentrations (1.0, 1.5, and 2.0 μL/L) of 1-MCP, 1.5 μL/L 1-MCP was noted as the best for maintaining fruit quality.

Fan et al. (2000) observed that treatment of apricot fruits with 1-MCP at a concentration of 1 μL/L (4 h at 20 °C) has reduced the respiration rate and ethylene production and exhibited less color change. The treated fruits maintained firmness and titratable acidity when stored at either of the temperature (0 or 20 °C). Dong et al. (2002) advocated that the application of 1-MCP at proper stage of maturity is important to increase the shelf life of apricot fruit. They noted that application of 1-MCP post-storage has improved the quality and delayed fruit ripening. Even at low concentration, 1-MCP reduced the flesh browning in apricot fruit. However, De Martino et al. (2006) have observed that the application of 1-MCP declined the production of ethylene and CO_2, irrespective of the time of application. It has also prevented the loss of tissue integrity of apricot fruits, which resulted in delayed ripening and maintained the fruit firmness.

Change in fruit color can also be controlled by the application of 1-MCP. Valero and Serrano (2010) observed a concentration-dependent effect of 1-MCP. Apricot fruits treated with 1-MCP at a concentration of 0.5 μL/L maintained a stable yellow color as compared to the concentration of 0.3 μL/L, after 21 days of cold storage. However, there was change in color from yellow to dark orange in the untreated fruits. Wu et al. (2015) noted that 1-MCP treatment maintained the membrane integrity of apricot fruits with alleviated lipid peroxidation and increased activity of antioxidants. They noted that chlorine dioxide can also be used to enhance the shelf life along with maintaining good quality of apricot fruits. Dong et al. (2002) advocated that 1-MCP can be used to increase both storage period and shelf life of "Royal Zee" plums. Both respiration rate and ethylene production of plum cv. "Royal Zee" were reduced by the 1-MCP treatment after both short- (10 days) and long-term (30 days) storage.

Valero et al. (2003) reported that the application of 1-MCP can enhance storability of plums up to 4 weeks of cold storage, while the control fruits were stored only up to 1 week. Pre-climacteric stage harvested plum fruits (cv. "President"), when treated with 1-MCP (0.3 and 0.5 mL/L), has resulted in ethylene inhibition, irrespective of the concentration. The treated plums exhibited delayed color change with a diminished ripening index. The application of 1-MCP also prevented the production of ethylene in both climacteric (cv. Santa Rosa) and suppressed climacteric type (cv. Golden Japan) (Martinez-Romero et al. 2003). However, the efficacy of 1-MCP was observed as concentration-dependent in Santa Rosa but concentration-independent in Golden Japan.

10.2.2 Methyl Jasmonate

Methyl jasmonate (MeJA), a volatile organic compound, is derived from jasmonic acid. It is used in plant defense system in response to various stresses and diverse developmental stages, including germination, root growth, flowering, fruit ripening, and senescence (Cheong and Do Choi 2003; Wasternack 2007; Ziosi et al. 2008). MeJA can also be applied at postharvest stage to reduce chilling injury and maintain the fruit quality. MeJA treatment at early and late fruit developmental stages can also downregulate the genes related to ethylene biosynthesis and fruit softening (Ruiz et al. 2009).

MeJA treatment can be used to lower the chilling injury and decay in mangoes. The reduction in the percent ion leakage of tissue was observed to have positive correlation with chilling tolerance. MeJA treatment also enhanced the total soluble solids (TSS) of mangoes (González-Aguilar et al. 2000). MeJA at 10^{-4} M concentration was observed as the most effective for the reduction of chilling injury in mango fruits stored at 5 °C and enhancing the shelf life. However, the concentration 10^{-5} M was effective in enhancing the coloration of mangoes (González-Aguilar et al. 2001). Apart from reduction of chilling injury, MeJA treatments also have significant effect on the reduction of malondialdehyde (MDA) content and electrolyte leakage (EL) of mango pulp and peel during cold storage (Junmatong et al. 2012). During cold storage, MeJA treatment can also lead to increase in the β-carotene content of mangoes (Boonyaritthongchai et al. 2017).

Among the temperate stone fruits, peach is highly vulnerable to chilling injury during cold storage. Meng et al. (2009) suggested that MeJA can be used to maintain the quality of peaches under low-temperature storage. Higher peroxidase activity coupled with lower phenolic compounds content has resulted in reduced chilling injury in low-temperature stored fruits. Based on the results, Jin et al. (2009) reported that a combine treatment of hot air and MJ vapor can be used to control the chilling injury and maintain the quality of peaches during low-temperature storage. They observed that, among the treatments, peach fruits treated with 1 μmol/L MJ vapor (38 °C for 12 h) with heat treatment at 38 C for 12 h followed by treatment with 1 μmol/L MJ vapor at 20 °C for 24 h had maintained the highest quality and exhibited the lowest chilling injury symptoms. Apart from higher antioxidant activities, they also observed a higher level of total phenolic and vitamin C contents. Recently, Yu et al. (2016) also reported similar findings in peach. They highlighted that hot air and MeJA treatments can effectively alleviate chilling injury of peach fruits during low-temperature storage. They noted the highest sucrose content and lowest chilling injury in the hot-air-treated fruits. Hot air, apart from enhancing the antioxidant enzyme activity, also increases the accumulation of heat shock proteins and maintains the membrane integrity, thus enhancing the shelf life and storability of peaches under the cold storages (Cao et al. 2010; Wang et al. 2014a, b). Zapata et al. (2014) observed that preharvest treatment of 0.5 mM MeJA at 63, 77, and 98 days after full blossom delayed the postharvest ripening process in plums. Ethylene production, respiration rate, and softening of fruits were also reduced significantly.

10.2.3 Salicylic Acid

Salicylic acid (SA), a natural colorless crystalline phenolic compound, functions as a plant hormone. Apart from playing important roles in various stages of plant growth and development, it also enhances the local and systemic resistance in fruits against various pathogens (Chan et al. 2007; Wang et al. 2006a, b). In addition to this, the application of SA also resulted in the inhibition of respiration rate and ethylene production and delayed senescence. Thus, it helps in increasing the fruit quality and shelf life (Chan et al. 2007; Tareen et al. 2012).

Junmatong et al. (2012) reported that SA (0.1 mM) reduced the chilling injury of mango during cold storage up to 35 days, while 1 mM extended the shelf life up to 42 days. SA treatment did not affect fruit firmness and skin color of the ripe mangoes.

Tareen et al. (2012) treated SA at various concentrations (0, 0.5, 1.0, 1.5, or 2.0 mmol/L) to study its efficacy on postharvest life of "Flordaking" peach. Treatments were done immediately after harvest. Although all concentrations of SA resulted in higher antioxidant enzyme activity of the fruits during 5 weeks of storage, 2.0 mmol/L concentration had the highest activity. It (2.0 mmol/L SA) also enhanced fruit firmness and weight but decreased the juice pH and maintained the fruit quality intact. However, Wang et al. (2006a, b) observed that SA at a concentration of 1 mM was significant to increase the firmness and decrease the chilling injury and thiobarbituric acid-reactive substance content of fruits as compared to untreated. Based on the findings, they concluded that SA can be used to alleviate the chilling injury symptoms of peaches during low-temperature storage through induction of higher antioxidant enzyme activity and heat shock proteins.

Like the peaches, SA can also be used to increase fruit quality and the storability of apricots. Wang et al. (2014a, b) studied the effect of SA (1.0 and 2.0 mM) treatment on the apricot (cv. Dahuang) fruit quality and antioxidant activity. The treated fruits exhibited higher PAL activity and hydrogen oxygen content. However, activity of ascorbate peroxidase and catalase was lowered, while superoxide dismutase and peroxidase showed the opposite trend in the SA-treated fruits. SA treatment also retarded the ripening process and quality loss as compared to the untreated fruits. Ezzat et al. (2017) concluded that salicylic acid and/or methyl jasmonate can be used to enhance the storability of apricot fruits. Combined treatment using fruit dipping method for 15 min into a solution of 0.2 mmol/L MeJA and 2 mmol/L SA for 15 min resulted in significant reduction in the fruit weight loss and fruit softening over the whole storage period.

10.2.4 Calcium Chloride

Calcium, a major element of fruit, not only protects the cell wall from various cell-wall-degrading enzymes but also leads to stabilization of cell wall (White and Broadley 2003). It also takes part in signal transduction at intracellular level in various physiological processes in the plant system. Apart from an essential element, calcium also plays a major role in keeping the postharvest quality of fruits and vegetables (Kirkby and Pilbeam 1984). It helps in maintaining membrane integrity and cell turgor and retarding membrane catabolism. Calcium application can be advocated at either pre- or postharvest stage to maintain the overall quality of fruit (Muzzaffar et al. 2018). Fruit ripening and senescence can also be reduced by the application of calcium (Ca^{2+}) (Ferguson 1984). Ca^{2+} can be used for maintaining the postharvest quality, extending both the shelf life and storability of various stone fruits (Lacey et al. 2009; Lester and Grusak 2001).

Singh et al. (1993) reported that preharvest sprays of calcium either in the form of calcium chloride or calcium nitrate can delay the ripening and had improved quality during storage as compared to the control "Dashehari" mangoes. However, 0.6% $CaCl_2$ was observed as the most effective treatment and can extend the storage life up to 10 days. $CaCl_2$ (3%) in combination with gum arabic (10%) have reduced oxidative damage, cell membrane changes, and decaying of mangoes and showed better fruit quality. The treatment also resulted in the inhibition of loss of ascorbic acid and phenolics (Khaliq et al. 2015, 2016). Preharvest application of $CaCl_2$ (0.5%) and $CaNO_3$ (1%) in Amrapali and Dashehari mangoes increased various fruit quality attributes and the shelf life for more than 7 days (Singh et al. 2017). Mohamed and El-khalek (2017) reported that preharvest treatment of mango cv. "Zebda" with calcium chloride (2% and 4%) has lowered the fruit deterioration and improved the fruit quality and storability.

Tzoutzoukou and Bouranis (1997) observed that the application of calcium at preharvest stage in the form of $CaCl_2$ increased Ca content of "Bebekou" apricot fruit up to 30–76%. Ethylene production was significantly lowered, and respiration rate was suppressed as compared to the control. Calcium-treated fruits were 70% firmer, and a positive correlation between the Ca content and fruit firmness was observed. However, Antunes et al. (2003) observed that the application of calcium chloride at postharvest stage can also improve the storage life in various cultivars of apricot.

Gupta et al. (2011) observed that "Earligrande" peach, when treated with 6% $CaCl_2$ solution (for 10 min) after the harvest at optimum stage, can be stored for 3 weeks under low temperature (0–2 °C, 85–90% RH) storage facility. The treatment also maintained fruit firmness and quality by reducing the spoilage of fruits and minimizing the physiological weight loss. However, a treatment consisting of combination of both $CaCl_2$ and salicylic acid can also be recommended for maintaining the fruit quality and reducing the chilling injury of honey peaches during 20 days of low-temperature storage, followed by 3 days of storage at ambient temperature (Gang et al. 2015). Although a single $CaCl_2$ application was effective for maintaining fruit quality, the combined treatment was observed as more effective in decreasing the chilling injury and polyphenol oxidase activity and lowering the respiration rate.

Hydro-cooling is an important operation to enhance the storability and shelf life of cherry fruits. The application of $CaCl_2$ (0.2–2.0%) by mixing it to the hydro-cooling water increased the cherry (cv. "Lapins" and "Sweetheart") fruit firmness and reduced pitting. It also increased the cherry fruit tissue Ca content which resulted in inhibited fruit senescence, enhanced total antioxidant capacity, and maintained quality. This application of $CaCl_2$ along with hydro-cooling water was also observed to enhance the shipping quality of cherries (Wang et al. 2014a, b).

Rahman et al. (2016) recommended that the application of 4% $CaCl_2$ solution can maintain the quality attributes of peach fruits for 30 days of storage. The harvested peaches cv. Texas A 69 were stored at low temperature (8–10 °C, 80–85% RH) after treatment with $CaCl_2$ solution (2% and 4%, 10 min). The treatment has resulted in significant reduction in weight loss (4.98%) and disease incidence (2.08%). However, there was increase in the fruit firmness (2.21 kg/cm^2). Based on the results, they observed that the fruit quality of peaches during storage is influenced by the application of $CaCl_2$ solution and storage duration.

10.2.5 Oxalic Acid

Oxalic acid (OA) is widely distributed in various plants and is normally accepted as a safe organic compound (Kim et al. 2008; Liang et al. 2009; Wang et al. 2009). Both the pre- and postharvest application of OA can be adopted to extend the shelf life and improve the quality of stone fruits at postharvest stages for commercial use (Razavi and Hajilou 2016).

OA (5 mM) treatment for 5 min can delay the mango (cv. Zill) fruit ripening and reduce fruit decay during storage. Physiologically, it resulted in reduced ethylene production, thus delaying the ripening of the fruits (Zheng et al. 2007). Razzaq et al. (2015) reported that oxalic acid treatment can reduce the softening of "Samar Bahisht Chaunsa" mango fruits during storage. The treatment of oxalic acid also resulted in higher antioxidative enzymes. Valero et al. (2011) reported that the treatment of sweet cherries with OA (1 mM) delayed the postharvest ripening process and maintained the fruit firmness and quality attributes for longer duration as compared to control. Martinez-Espla et al. (2014) observed that preharvest treatment of OA can also be explored for enhancing the postharvest quality of sweet cherry fruits.

Oxalic acid treatment can also effectively enhance the shelf life of plums (Wu et al. 2011). Plum (cv. "Damili") fruits, when treated with OA (5 mmol/L) for 3 min, resulted in lowered ethylene production. The softening of fruits was also delayed. The decline in the activity of polygalacturonase (PG) and pectin methylesterase (PME), i.e., delaying or slowing of pectin solubilization/degradation, can be associated with the inhibition of fruit softening (Wu et al. 2011).

Razavi and Hajilou (2016) recommended that preharvest treatments (15 days before harvest) with OA can be used to maintain fruit quality and pulp firmness following the postharvest storage of "Anjirymaleki" peach fruit. The treated fruits also maintained higher antioxidant (SOD, POD, and CAT) enzyme activity of the fruits. Internal browning during the storage can also be reduced by OA (5 mM solution for 10 min) treatment in peaches (Jin et al. 2014). The application of OA can be advocated to increase the chilling tolerance and decrease the chilling injury in peach fruits during the cold storage.

10.2.6 Melatonin

Melatonin, an indoleamine hormone, plays various physiological functions in plants (Gao et al. 2018; Farooq et al. 2018). Being ubiquitous in nature, melatonin has been identified in nearly all organs and tissues of plants (Jemima et al. 2011; Reiter et al. 2015). It is involved in various plant developmental processes, including growth, differentiation, ripening, and senescence, and performs protective functions against various abiotic stresses (Reiter et al. 2015; Tan 2015; Zhang et al. 2015). The application of melatonin to increase the postharvest shelf life and storability is well studied in stone fruits like peach.

Postharvest application of melatonin can delay the senescence and maintain the peach fruit quality. Peach fruits (cv. "Shahong" and "Qinmi") were treated with melatonin at a concentration of 0.1 mmol/L, followed by storage at ambient temperature for 7 days (Gao et al. 2016). The treated fruits of both cultivars exhibited a slowed senescence which was evident from the reduction in weight loss and respiration rate and lowered decay incidence. The treated fruits maintained higher firmness and TSS and increased activity of antioxidant enzymes (superoxide dismutase, peroxidase, catalase, and ascorbate peroxidase) as compared to the control. Recently, Gao et al. (2018) also observed that chilling injury symptoms in peach fruits (stored at 1 °C for 28 days) can be delayed by the treatment of 0.1 mM melatonin. A higher ratio of unsaturated to saturated fatty acids in peach fruit was maintained by the application of melatonin. Higher fruit firmness in the treated ones showed that melatonin can be used to alleviate the chilling injury of cold-stored peach fruits which shall be helpful in increasing the storage marketing window.

Melatonin can be used to enhance the postharvest shelf life of mango fruit also. Liu et al. (2020) reported that the treatment of melatonin 0.5 mM for 1 h can delay ripening, ethylene production, and softening of "Guifei" mango fruit. Abscisic acid accumulation was also suppressed in the treated fruit. However, Rastegar et al. (2020) reported that the enhanced shelf life of melatonin-treated mangoes during storage can be attributed to the higher activity of catalase and peroxidase enzymes.

10.2.7 Putrescine

Polyamines are a group of physiologically important compounds playing major role in fruit development and senescence (Valero et al. 1999). Polyamines, such as putrescine, act as antisenescent agents by antagonizing the effects of ethylene (Paksasorn et al. 1995). Available reports show that exogenous putrescine application has delayed the process of ripening and maintained the fruit firmness for longer duration in fruits such as apricots, lemons, apples, and pomegranates (Aly et al. 2019; Fawole et al. 2020; Valero et al. 1999).

Malik et al. (2002) observed that the treatment of putrescine at preharvest stage is more effective as compared to postharvest treatment of "Kensington Pride" mango fruits. They concluded that 1 mmol/L putrescine can be used to delay mango fruit ripening, while 2 mmol/L can enhance the shelf life and quality of "Kensington

Pride" mango fruit. Jawandha et al. (2012) reported that postharvest treatment of mangoes with putrescine at 2.0 mmol/L can be used to maintain good TSS and acid blend along with higher shelf life. Various research findings revealed that putrescine treatment can enhance antioxidant enzyme activities and maintain higher storability in mangoes (Razzaq et al. 2014; Wannabussapawich and Seraypheap 2018).

Zokaee Khosroshahi and Esna-Ashari (2007) observed that putrescine treatment resulted in the reduction of ethylene production and increase in flesh firmness of apricot fruit. They observed that the fruit weight loss negatively correlated with the concentration of putrescine. Martinez-Romero et al. (2002) observed that putrescine (1 mM) treatment can increase the firmness and reduce the bruising zones in already mechanically damaged fruits of "Mauricio" apricot. The reduction in color change, respiration rate, and ethylene production was also observed in the putrescine-applied fruits. Serrano et al. (2003) also recommended that postharvest treatment of 1 mM putrescine can increase the storability of plum fruits as it hindered ripening processes. The treatment of putrescine (1 mM) has a significant effect in increasing the shelf life and extending the storage duration. It also delayed the ripening and retained the fruit quality of various plum cultivars.

In "Angelino" plum, both pre- and postharvest putrescine applications have resulted in the lowering of production of ethylene during the storage as compared to control fruits. However, spraying of higher putrescine concentrations (2.0 and 1.0 mM) at the preharvest stage was more effective in hindering the production of ethylene as compared to the lower concentrations and applications at postharvest stage. The prestorage application decreased the fruit softening of plums during low-temperature storage through suppressing the biosynthesis of ethylene and reducing the activities of various fruit-softening enzymes including PE, EGase, *exo*-PG, and *endo*-PG (Khan et al. 2007). Later on, Khan et al. (2008) and Khan and Singh (2010) reported that, in plum, preharvest (1 week before the harvest) application of 2.0 mM putrescine was more beneficial in retarding ripening of fruit and can be used to increase the storability up to 6 weeks with maintaining quality of fruit. Recently, Abbasi et al. (2019) reported that spraying 1–3 mM putrescine during fruit growth can reduce chilling injury in peach fruit and retain fruit quality up to 6 weeks in cold storage.

10.2.8 Nitric Oxide

Nitric oxide, a bioactive molecule, plays major roles during signal transduction in various physiological processes such as development, stomatal regulation, and hormone signaling (Guo et al. 2003; Huang et al. 2003; Pagnussat et al. 2003; Perazzolli et al. 2006). It has been widely explored to enhance the shelf life of various climacteric and non-climacteric fruits including kiwi, avocado, and strawberry (Leshem and Pinchasov 2000; Shuhua et al. 2005; Wills et al. 2000; Ya'acov et al. 1998).

Zhu et al. (2006) observed a reduction in the production of ethylene and activity of 1-aminocyclopropane-1-carboxylic acid (ACC) oxidase in peach fruits, following the application of nitric oxide (5 and 10 μL/L). Flores et al. (2008) also observed that the treatment of peach (cv. "Rojo del Rito") fruits with nitric oxide at the concentration of 5 μL/L (4 h, at 20 °C) significantly reduced ethylene production and respiration rate. It was also observed that disintegration of the cell membranes was also lower in the treated plants. Nitric oxide can also be used to reduce pathogen infection. Peach fruits treated with nitric oxide have reduced incidence of brown rot disease. This can indirectly increase the shelf life of fruits (Gu et al. 2014).

Postharvest nitric oxide fumigation can be used to delay the softening and retard color development in mango fruits (Zaharah and Singh 2011b). Nitric oxide fumigation resulted in the inhibition of activities of ACS and ACO which resulted in the inhibition of ethylene biosynthesis in mangoes (Zaharah and Singh 2011a). Postharvest treatment of nitric oxide in the form of sodium nitroprusside (SNP) can improve the quality and increase the shelf life of mango fruits during storage through activation of ROS scavenging activities of antioxidant enzymes (Barman et al. 2014; Ren et al. 2017).

10.2.9 Hexanal

The action of phospholipase D initiates and plays a major role in the process of membrane degradation during ripening (Paliyath and Droillard 1992). Hexanal, a volatile C6 aldehyde, which occurs naturally, inhibits the phospholipase D action (El Kayal et al. 2017b). This property of hexanal can be exploited for maintaining the quality and increasing shelf life and storability of various fruits, vegetables, and flowers (Anusuya et al. 2016; Ashitha et al. 2019; El Kayal et al. 2017a; Paliyath and Subramanian 2008; Paliyath et al. 1999; Cheema et al. 2014).

Sharma et al. (2010) reported that preharvest spray (twice, at 7 and 15 days before harvest) of hexanal formulation (enhanced freshness formulation, EFF) in sweet cherry (cv. Bing) enhanced better coloration and firmness as compared to the untreated cherries. This quality enhancement was observed even after 30 days of harvest (stored at 4 °C). A higher chroma value was noticed which indicated enhanced redness in the treated fruits.

Enhanced freshness formulation (EFF) having hexanal (0.02% v/v) as the key ingredient can also be used to enhance the nectarine shelf life. Preharvest sprays (5 L per tree) of EFF (2%) at 15 and 10 days before harvest of "Fantasia" nectarines resulted in increased shelf life along with delaying the occurrence of chilling injury symptoms such as internal browning and mealiness/woolliness by 1 week (Kumar et al. 2018). Kaur et al. (2020) observed that 1600 μM EFF can reduce the respiration rate and pectin methyl esterase activity in mangoes. Both the lower and higher doses were not effective as the 1600 μM. Hexanal treatment resulted in the increase of firmness, TSS, and acidity and maintains the acceptable palatability of fruits up to 28 days of storage. Jincy et al. (2017) reported that, apart from increasing the antioxidant enzyme activity, hexanal treatment can also decrease membrane damage and ethylene evolution in mangoes.

10.3 Conclusion

Stone fruits have received attention and consumer demand across the globe. However, shorter shelf life creates a major hindrance in their prolong storage and marketability period. Pre- or postharvest application of chemicals like 1-MCP, salicylic acid, methyl jasmonate, oxalic acid, calcium chloride, putrescine, and melatonin has resulted in enhanced shelf life and increased storability of stone fruits. Chemical treatments also resulted in maintaining the fruit quality for a longer duration as compared to the non-treated fruits. However, it is well evident that there is a need for stage- and concentration-specific application of various chemicals to obtain optimum results in terms of extended shelf life and maximum storability.

References

Abbasi, N. A., Ali, I., Hafiz, I. A., Alenazi, M. M., & Shafiq, M. (2019). Effects of putrescine application on peach fruit during storage. *Sustainability, 11*(7), 2013.

Aly, M. A., Ezz, T. M., Abd El-Gawad, M. G., & Buazizah, M. H. S. (2019). Enhancement quality and storability of "Anna" apple fruits by some pre-harvest foliar applications. *Middle East Journal of Agriculture Research, 8*(1), 66–81.

Antunes, M. D. C., Correia, M. P., Miguel, M. G., Martins, M. A., & Neves, M. A. (2003). *The effect of calcium chloride postharvest application on fruit storage ability and quality of 'Beliana' and 'Lindo' apricot (Prunus armeniaca L.) cultivars*. In International conference on quality in chains. An integrated view on fruit and vegetable quality (pp. 721–726).

Anusuya, P., Nagaraj, R., Janavi, G. J., Subramanian, K. S., Paliyath, G., & Subramanian, J. (2016). Pre-harvest sprays of hexanal formulation for extending retention and shelf-life of mango (*Mangifera indica* L.) fruits. *Scientia Horticulturae, 211*, 231–240.

Ashitha, G. N., Preetha, P., & Varadharaju, N. (2019). Effect of postharvest hexanal treatment on shelf life of guava. *International Journal of Current Microbiology and Applied Science, 8*(7), 1057–1067.

Barkai-Golan, R., & Follett, P. A. (2017). Irradiation for quality improvement of individual fruits. In R. Barkai-Golan & P. A. Follett (Eds.), *Irradiation for quality improvement, microbial safety and phytosanitation of fresh produce* (pp. 55–107). San Diego, CA: Academic.

Barman, K., Asrey, R., Pal, R. K., Jha, S. K., & Bhatia, K. (2014). Post-harvest nitric oxide treatment reduces chilling injury and enhances the shelf-life of mango (*Mangifera indica* L.) fruit during low-temperature storage. *The Journal of Horticultural Science and Biotechnology, 89*(3), 253–260.

Benichou, M., Ayour, J., Sagar, M., Alahyane, A., Elateri, I., & Aitoubahou, A. (2018). Postharvest technologies for shelf life enhancement of temperate fruits. In S. A. Mir, M. A. Shah, & M. M. Mir (Eds.), *Postharvest biology and technology of temperate fruits* (pp. 77–100). Cham: Springer.

Blankenship, S. M., & Dole, J. M. (2003). 1-Methylcyclopropene: A review. *Postharvest Biology and Technology, 28*(1), 1–25.

Boonyaritthongchai, P., Chimvaree, C., Buanong, M., Uthairatanakij, A., & Jitareerat, P. (2017). Effect of methyl jasmonate on physical and chemical properties of mango fruit cv. Nam Dok Mai. *Horticulturae, 3*(1), 18.

Cao, S., Hu, Z., Zheng, Y., & Lu, B. (2010). Synergistic effect of heat treatment and salicylic acid on alleviating internal browning in cold-stored peach fruit. *Postharvest Biology and Technology, 58*(2), 93–97.

Chan, Z., Qin, G., Xu, X., Li, B., & Tian, S. (2007). Proteome approach to characterize proteins induced by antagonist yeast and salicylic acid in peach fruit. *Journal of Proteome Research, 6*(5), 1677–1688.

Cheema, A., Padmanabhan, P., Subramanian, J., Blom, T., & Paliyath, G. (2014). Improving quality of greenhouse tomato (*Solanum lycopersicum* L.) by pre-and postharvest applications of hexanal-containing formulations. *Postharvest Biology and Technology, 95*, 13–19.

Cheong, J. J., & Do Choi, Y. (2003). Methyl jasmonate as a vital substance in plants. *Trends in Genetics, 19*(7), 409–413.

De Martino, G., Vizovitis, K., Botondi, R., Bellincontro, A., & Mencarelli, F. (2006). 1-MCP controls ripening induced by impact injury on apricots by affecting SOD and POX activities. *Postharvest Biology and Technology, 39*(1), 38–47.

Dong, L., Lurie, S., & Zhou, H. W. (2002). Effect of 1-methylcyclopropene on ripening of 'Canino' apricots and 'Royal Zee' plums. *Postharvest Biology and Technology, 24*(2), 135–145.

Duan, Y. Q., Feng, S. Q., Zhao, Y. M., & Ma, Q. J. (2004). Effect of 1-methylcyclopropene (1-MCP) treatment on flesh cell ultrastructure of refrigerator-stored peaches. *Scientia Agricultura Sinica, 37*, 2039–2042.

Eckert, J. W., & Ogawa, J. M. (1988). The chemical control of postharvest diseases: Deciduous fruits, berries, vegetables and root/tuber crops. *Annual Review of Phytopathology, 26*(1), 433–469.

El Kayal, W., El-Sharkawy, I., Dowling, C., Paliyath, G., Sullivan, J. A., & Subramanian, J. (2017a). Effect of preharvest application of hexanal and growth regulators in enhancing shelf life and regulation of membrane-associated genes in strawberry. *Canadian Journal of Plant Science, 97*(6), 1109–1120.

El Kayal, W., Paliyath, G., Sullivan, J. A., & Subramanian, J. (2017b). Phospholipase D inhibition by hexanal is associated with calcium signal transduction events in raspberry. *Horticulture Research, 4*(1), 1–9.

Ezzat, A., Ammar, A., Szabo, Z., Nyeki, J., & Holb, I. J. (2017). Postharvest treatments with methyl jasmonate and salicylic acid for maintaining physico-chemical characteristics and sensory quality properties of apricot fruit during cold storage and shelf-life. *Polish Journal of Food and Nutrition Sciences, 67*(2), 159–166.

Faasema, J., Alakali, J. S., & Abu, J. O. (2014). Effects of storage temperature on 1-methylcyclopropene-treated mango (*Mangnifera indica*) fruit varieties. *Journal of Food Processing and Preservation, 38*(1), 289–295.

Fan, X., Argenta, L., & Mattheis, J. P. (2000). Inhibition of ethylene action by 1-methylcyclopropene prolongs storage life of apricots. *Postharvest Biology and Technology, 20*(2), 135–142.

Farooq, S., Mir, M. M., Ganai, S. A., Maqbool, T., Mir, S. A., & Shah, M. A. (2018). *Postharvest biology and technology of peach*. In Postharvest biology and technology of temperate fruits (pp. 169–199). Cham: Springer.

Fawole, O. A., Atukuri, J., Arendse, E., & Opara, U. O. (2020). Postharvest physiological responses of pomegranate fruit (cv. Wonderful) to exogenous putrescine treatment and effects on physico-chemical and phytochemical properties. *Food Science and Human Wellness*. https://doi.org/10.1016/j.fshw.2020.02.007.

Ferguson, I. B. (1984). Calcium in plant senescence and fruit ripening. *Plant, Cell and Environment, 7*(6), 477–489.

Flores, F. B., Sanchez-Bel, P., Valdenegro, M., Romojaro, F., Martinez-Madrid, M. C., & Egea, M. I. (2008). Effects of a pretreatment with nitric oxide on peach (*Prunus persica* L.) storage at room temperature. *European Food Research and Technology, 227*(6), 1599.

Gang, C., Li, J., Chen, Y., Wang, Y., Li, H., Pan, B., & Odeh, I. (2015). Synergistic effect of chemical treatments on storage quality and chilling injury of honey peaches. *Journal of Food Processing and Preservation, 39*(6), 1108–1117.

Gao, H., Lu, Z., Yang, Y., Wang, D., Yang, T., Cao, M., & Cao, W. (2018). Melatonin treatment reduces chilling injury in peach fruit through its regulation of membrane fatty acid contents and phenolic metabolism. *Food Chemistry, 245*, 659–666.

Gao, H., Zhang, Z. K., Chai, H. K., Cheng, N., Yang, Y., Wang, D. N., & Cao, W. (2016). Melatonin treatment delays postharvest senescence and regulates reactive oxygen species metabolism in peach fruit. *Postharvest Biology and Technology, 118*, 103–110.

Gong, Y., Fan, X., & Mattheis, J. P. (2002). Responses of 'Bing' and 'Rainier' sweet cherries to ethylene and 1-methylcyclopropene. *Journal of the American Society for Horticultural Science, 127*(5), 831–835.

González-Aguilar, G. A., Buta, J. G., & Wang, C. Y. (2001). Methyl jasmonate reduces chilling injury symptoms and enhances colour development of 'Kent' mangoes. *Journal of the Science of Food and Agriculture, 81*(13), 1244–1249.

González-Aguilar, G. A., Fortiz, J., Cruz, R., Baez, R., & Wang, C. Y. (2000). Methyl jasmonate reduces chilling injury and maintains postharvest quality of mango fruit. *Journal of Agricultural and Food Chemistry, 48*(2), 515–519.

Gu, R., Zhu, S., Zhou, J., Liu, N., & Shi, J. (2014). Inhibition on brown rot disease and induction of defence response in harvested peach fruit by nitric oxide solution. *European Journal of Plant Pathology, 139*(2), 369–378.

Guo, F. Q., Okamoto, M., & Crawford, N. M. (2003). Identification of a plant nitric oxide synthase gene involved in hormonal signaling. *Science, 302*(5642), 100–103.

Gupta, N., Jawandha, S. K., & Gill, P. S. (2011). Effect of calcium on cold storage and post-storage quality of peach. *Journal of Food Science and Technology, 48*(2), 225–229.

Huang, X., Bors, W., & Durner, J. (2003). *The role of nitric oxide in jasmonic acid signalling in Arabidopsis thaliana*. In Free radical research (Vol. 37, pp. 8–9). Oxon: Taylor and Francis.

Jawandha, S. K., Gill, M. S., Singh, N., Gill, P. P. S., & Singh, N. (2012). Effect of post-harvest treatments of putrescine on storage of mango cv. Langra. *African Journal of Agricultural Research, 7*(48), 6432–6436.

Jemima, J., Bhattacharjee, P., & Singhal, R. S. (2011). Melatonin-a review on the lesser known potential nutraceutical. *International Journal of Pharmaceutical Sciences and Research, 2*(8), 1975.

Jiang, Y., & Joyce, D. C. (2005). Effects of 1-methylcyclopropene alone and in combination with polyethylene bags on the postharvest life of mango fruit. *Annals of Applied Biology, 137*(3), 321–327.

Jin, P., Shang, H., Chen, J., Zhu, H., Zhao, Y., & Zheng, Y. (2011). Effect of 1-methylcyclopropene on chilling injury and quality of peach fruit during cold storage. *Journal of Food Science, 76*(8), S485–S491.

Jin, P., Zheng, Y., Tang, S., Rui, H., & Wang, C. Y. (2009). A combination of hot air and methyl jasmonate vapour treatment alleviates chilling injury of peach fruit. *Postharvest Biology and Technology, 52*(1), 24–29.

Jin, P., Zhu, H., Wang, L., Shan, T., & Zheng, Y. (2014). Oxalic acid alleviates chilling injury in peach fruit by regulating energy metabolism and fatty acid contents. *Food Chemistry, 161*, 87–93.

Jincy, M., Djanaguiraman, M., Jeyakumar, P., Subramanian, K. S., Jayasankar, S., & Paliyath, G. (2017). Inhibition of phospholipase D enzyme activity through hexanal leads to delayed mango (*Mangifera indica* L.) fruit ripening through changes in oxidants and antioxidant enzymes activity. *Scientia Horticulturae, 218*, 316–325.

Junmatong, C., Uthaibutra, J., Boonyakiat, D., Faiyue, B., & Saengnil, K. (2012). Reduction of chilling injury of 'Nam Dok Mai No. 4' mango fruit by treatments with salicylic acid and methyl jasmonate. *Journal of Agricultural Science, 4*(10), 126.

Kaur, K., Kaur, G., & Brar, J. S. (2020). Pre-harvest application of hexanal formulations for improving post-harvest life and quality of mango (*Mangifera indica* L.) cv. Dashehari. *Journal of Food Science and Technology*. https://doi.org/10.1007/s13197-020-04464-9.

Khaliq, G., Mohamed, M. T. M., Ali, A., Ding, P., & Ghazali, H. M. (2015). Effect of gum arabic coating combined with calcium chloride on physico-chemical and qualitative properties of mango (*Mangifera indica* L.) fruit during low temperature storage. *Scientia Horticulturae, 190*, 187–194.

Khaliq, G., Mohamed, M. T. M., Ghazali, H. M., Ding, P., & Ali, A. (2016). Influence of gum arabic coating enriched with calcium chloride on physiological, biochemical and quality responses of mango (*Mangifera indica* L.) fruit stored under low temperature stress. *Postharvest Biology and Technology, 111*, 362–369.

Khan, A. S., & Singh, Z. (2010). Pre-harvest application of putrescine influences Japanese plum fruit ripening and quality. *Food Science and Technology International, 16*(1), 53–64.

Khan, A. S., Singh, Z., & Abbasi, N. A. (2007). Pre-storage putrescine application suppresses ethylene biosynthesis and retards fruit softening during low temperature storage in 'Angelino' plum. *Postharvest Biology and Technology, 46*(1), 36–46.

Khan, A. S., Singh, Z., Abbasi, N. A., & Swinny, E. E. (2008). Pre-or post-harvest applications of putrescine and low temperature storage affect fruit ripening and quality of 'Angelino' plum. *Journal of the Science of Food and Agriculture, 88*(10), 1686–1695.

Kim, K. S., Min, J. Y., & Dickman, M. B. (2008). Oxalic acid is an elicitor of plant programmed cell death during *Sclerotinia sclerotiorum* disease development. *Molecular Plant-Microbe Interactions, 21*(5), 605–612.

Kirkby, E. A., & Pilbeam, D. J. (1984). Calcium as a plant nutrient. *Plant, Cell and Environment, 7*(6), 397–405.

Kumar, S. K., El Kayal, W., Sullivan, J. A., Paliyath, G., & Jayasankar, S. (2018). Pre-harvest application of hexanal formulation enhances shelf life and quality of 'Fantasia' nectarines by regulating membrane and cell wall catabolism-associated genes. *Scientia Horticulturae, 229*, 117–124.

Lacey, K., Hancock, N., & Ramsey, H. (2009). Measuring internal maturity of citrus. Farm note, Department of Agriculture and Food, Government of Western Australia, 354.

Leshem, Y. A. Y., & Pinchasov, Y. (2000). Non-invasive photoacoustic spectroscopic determination of relative endogenous nitric oxide and ethylene content stoichiometry during the ripening of strawberries *Fragaria anannasa* (Duch.) and avocados *Persea americana* (Mill.). *Journal of Experimental Botany, 51*(349), 1471–1473.

Lester, G. E., & Grusak, M. A. (2001). Postharvest application of chelated and non-chelated calcium dip treatments to commercially grown honey dew melons: Effects on peel attributes, tissue calcium concentration, quality, and consumer preference following storage. *HortTechnology, 11*(4), 561–566.

Li, L., Li, C., Sun, J., Sheng, J., Zhou, Z., Xin, M., & Li, J. (2020). The effects of 1-methylcyclopropene in the regulation of antioxidative system and softening of mango fruit during storage. *Journal of Food Quality, 2020*, 1–11.

Liang, Y., Strelkov, S. E., & Kav, N. N. (2009). Oxalic acid-mediated stress responses in *Brassica napus* L. *Proteomics, 9*(11), 3156–3173.

Liguori, G., Weksler, A., Zutahi, Y., Lurie, S., & Kosto, I. (2004). Effect of 1-methylcyclopropene on ripening of melting flesh peaches and nectarines. *Postharvest Biology and Technology, 31*(3), 263–268.

Liu, H., Cao, J., & Jiang, W. (2015). Changes in phenolics and antioxidant property of peach fruit during ripening and responses to 1-methylcyclopropene. *Postharvest Biology and Technology, 108*, 111–118.

Liu, S., Huang, H., Huber, D. J., Pan, Y., Shi, X., & Zhang, Z. (2020). Delay of ripening and softening in 'Guifei' mango fruit by postharvest application of melatonin. *Postharvest Biology and Technology, 163*, 111136.

Liu, H., Jiang, W., Cao, J., & Ma, L. (2018). A combination of 1-methylcyclopropene treatment and intermittent warming alleviates chilling injury and affects phenolics and antioxidant activity of peach fruit during storage. *Scientia Horticulturae, 229*, 175–181.

Lurie, S., & Crisosto, C. H. (2005). Chilling injury in peach and nectarine. *Postharvest Biology and Technology, 37*(3), 195–208.

Lurie, S., & Pesis, E. (1992). Effect of acetaldehyde and anaerobiosis as postharvest treatments on the quality of peaches and nectarines. *Postharvest Biology and Technology, 1*(4), 317–326.

Malakou, A., & Nanos, G. D. (2005). A combination of hot water treatment and modified atmosphere packaging maintains quality of advanced maturity 'Caldesi 2000' nectarines and 'Royal Glory' peaches. *Postharvest Biology and Technology, 38*(2), 106–114.

Malik, A. U., Singh, Z., & Dhaliwal, S. S. (2002, August). *Exogenous application of putrescine affects mango fruit quality and shelf life*. In XXVI international horticultural congress: Issues and advances in postharvest horticulture (Vol. 628, pp. 121–127).

Martinez-Espla, A., Zapata, P. J., Valero, D., Garcia-Viguera, C., Castillo, S., & Serrano, M. (2014). Preharvest application of oxalic acid increased fruit size, bioactive compounds, and antioxidant capacity in sweet cherry cultivars (*Prunus avium* L.). *Journal of Agricultural and Food Chemistry, 62*(15), 3432–3437.

Martinez-Romero, D., Dupille, E., Guillen, F., Valverde, J. M., Serrano, M., & Valero, D. (2003). 1-Methylcyclopropene increases storability and shelf life in climacteric and non-climacteric plums. *Journal of Agricultural and Food Chemistry, 51*(16), 4680–4686.

Martinez-Romero, D., Serrano, M., Carbonell, A., Burgos, L., Riquelme, F., & Valero, D. (2002). Effects of postharvest putrescine treatment on extending shelf life and reducing mechanical damage in apricot. *Journal of Food Science, 67*(5), 1706–1712.

Meng, X., Han, J., Wang, Q., & Tian, S. (2009). Changes in physiology and quality of peach fruits treated by methyl jasmonate under low temperature stress. *Food Chemistry, 114*(3), 1028–1035.

Mohamed, M. A. A., & El-khalek, A. (2017). Pre-harvest potassium silicate, chitosan and calcium chloride application improves Mango fruits (Zebda cv.) quality and storability. *Egyptian Journal of Horticulture, 44*(1), 17–32.

Mozetic, B., Simcic, M., & Trebse, P. (2006). Anthocyanins and hydroxycinnamic acids of Lambert Compact cherries (*Prunus avium* L.) after cold storage and 1-methylcyclopropene treatment. *Food Chemistry, 97*(2), 302–309.

Muzzaffar, S., Bhat, M. M., Wani, T. A., Wani, I. A., & Masoodi, F. A. (2018). *Postharvest biology and technology of apricot*. In Postharvest biology and technology of temperate fruits (pp. 201–222). Cham: Springer.

Pagnussat, G. C., Lanteri, M. L., & Lamattina, L. (2003). Nitric oxide and cyclic GMP are messengers in the indole acetic acid-induced adventitious rooting process. *Plant Physiology, 132*(3), 1241–1248.

Paksasorn, A., Hayasaka, T., Matsui, H., Ohara, H., & Hirata, N. (1995). Relationship of polyamine content to ACC content and ethylene evolution in Japanese apricot fruit. *Journal of the Japanese Society for Horticultural Science, 63*(4), 761–766.

Paliyath, G., & Droillard, M. J. (1992). The mechanisms of membrane deterioration and disassembly during senescence. *Plant Physiology and Biochemistry, 30*(6), 789–812.

Paliyath, G., Pinhero, R. G., Yada, R. Y., & Murr, D. P. (1999). Effect of processing conditions on phospholipase D activity of corn kernel subcellular fractions. *Journal of Agricultural and Food Chemistry, 47*(7), 2579–2588.

Paliyath, G., & Subramanian, J. (2008). *Phospholipase D inhibition technology for enhancing shelf life and quality*. In Postharvest biology and technology of fruits, vegetables, and flowers (pp. 240–245). Wiley Blackwell.

Perazzolli, M., Romero-Puertas, M. C., & Delledonne, M. (2006). Modulation of nitric oxide bioactivity by plant haemoglobins. *Journal of Experimental Botany, 57*(3), 479–488.

Rahman, M. U., Sajid, M., Rab, A., Ali, S., Shahid, M. O., Alam, A., & Ahmad, I. (2016). Impact of calcium chloride concentrations and storage duration on quality attributes of peach (*Prunus persica*). *Russian Agricultural Sciences, 42*(2), 130–136.

Rastegar, S., Khankahdani, H. H., & Rahimzadeh, M. (2020). Effects of melatonin treatment on the biochemical changes and antioxidant enzyme activity of mango fruit during storage. *Scientia Horticulturae, 259*, 108835.

Razavi, F., & Hajilou, J. (2016). Enhancement of postharvest nutritional quality and antioxidant capacity of peach fruits by preharvest oxalic acid treatment. *Scientia Horticulturae, 200*, 95–101.

Razzaq, K., Khan, A. S., Malik, A. U., Shahid, M., & Ullah, S. (2014). Role of putrescine in regulating fruit softening and antioxidative enzyme systems in 'Samar Bahisht Chaunsa' mango. *Postharvest Biology and Technology, 96*, 23–32.

Razzaq, K., Khan, A. S., Malik, A. U., Shahid, M., & Ullah, S. (2015). Effect of oxalic acid application on Samar Bahisht Chaunsa mango during ripening and postharvest. *LWT-Food Science and Technology, 63*(1), 152–160.

Reiter, R. J., Tan, D. X., Zhou, Z., Cruz, M. H. C., Fuentes-Broto, L., & Galano, A. (2015). Phytomelatonin: assisting plants to survive and thrive. *Molecules, 20*(4), 7396–7437.

Ren, Y., He, J., Liu, H., Liu, G., & Ren, X. (2017). Nitric oxide alleviates deterioration and preserves antioxidant properties in 'Tainong' mango fruit during ripening. *Horticulture, Environment, and Biotechnology, 58*(1), 27–37.

Ruiz, K. B., Bressanin, D., Ziosi, V., Costa, G., Torrigiani, P., Tadiello, A., & Bonghi, C. (2009). *Early jasmonate application interferes with peach fruit development and ripening as revealed by several differentially expressed seed and mesocarp genes*. In XI International symposium on plant bioregulators in fruit production (Vol. 884, pp. 101–106).

Sakhale, B. K., Gaikwad, S. S., & Chavan, R. F. (2018). Application of 1-methylcyclopropene on mango fruit (Cv. Kesar): potential for shelf life enhancement and retention of quality. *Journal of Food Science and Technology, 55*(2), 776–781.

Serrano, M., Martinez-Romero, D., Guillen, F., & Valero, D. (2003). Effects of exogenous putrescine on improving shelf life of four plum cultivars. *Postharvest Biology and Technology, 30*(3), 259–271.

Sharma, M., Jacob, J. K., Subramanian, J., & Paliyath, G. (2010). Hexanal and 1-MCP treatments for enhancing the shelf life and quality of sweet cherry (*Prunus avium* L.). *Scientia Horticulturae, 125*(3), 239–247.

Shuhua, Z., Jie, Z., & Huairui, S. (2005). Effects of nitric oxide (NO) on ripening and senescence of strawberry. *Scientia Agricultura Sinica, 38*, 1418–1424.

Singh, V., Pandey, G., Sarolia, D. K., Kaushik, R. A., & Gora, J. S. (2017). Influence of pre harvest application of calcium on shelf life and fruit quality of Mango (*Mangifera indica* L.) cultivars. *International Journal of Current Microbiology and Applied Sciences, 6*(4), 1366–1372.

Singh, B. P., Tandon, D. K., & Kalra, S. K. (1993). Changes in postharvest quality of mangoes affected by preharvest application of calcium salts. *Scientia Horticulturae, 54*(3), 211–219.

Sisler, E. C., Blankenship, S. M., & Guest, M. (1990). Competition of cyclooctenes and cyclooctadienes for ethylene binding and activity in plants. *Plant Growth Regulation, 9*(2), 157–164.

Sisler, E. C., & Serek, M. (1999). Compounds controlling the ethylene receptor. *Botanical Bulletin of Academia Sinica, 40*, 1–7.

Tan, D. X. (2015). Melatonin and plants. *Journal of Experimental Botany, 66*(3), 625–626.

Tareen, M. J., Abbasi, N. A., & Hafiz, I. A. (2012). Postharvest application of salicylic acid enhanced antioxidant enzyme activity and maintained quality of peach cv. 'Flordaking' fruit during storage. *Scientia Horticulturae, 142*, 221–228.

Thompson, A. K., & Bishop, D. (2016). Controlled atmosphere technology. Elsevier, reference module in food sciences. https://doi.org/10.1016/B978-0-08-100596-5.21136-0.

Tzoutzoukou, C. G., & Bouranis, D. L. (1997). Effect of preharvest application of calcium on the postharvest physiology of apricot fruit. *Journal of Plant Nutrition, 20*(2–3), 295–309.

Valero, D., Diaz-Mula, H. M., Zapata, P. J., Castillo, S., Guillen, F., Martinez-Romero, D., et al. (2011). Postharvest treatments with salicylic acid, acetylsalicylic acid or oxalic acid delayed ripening and enhanced bioactive compounds and antioxidant capacity in sweet cherry. *Journal of Agricultural and Food Chemistry, 59*(10), 5483–5489.

Valero, D., Martínez-Romero, D., Serrano, M., & Riquelme, F. (1999). *Polyamine roles on the post-harvest of fruits: A review*. In Recent research developments in agricultural and food chemistry (Vol. 3, Part I, pp. 39–55).

Valero, D., Martınez-Romero, D., Valverde, J. M., Guillen, F., & Serrano, M. (2003). Quality improvement and extension of shelf life by 1-methylcyclopropene in plum as affected by ripening stage at harvest. *Innovative Food Science and Emerging Technologies, 4*(3), 339–348.

Valero, D., & Serrano, M. (2010). *Postharvest biology and technology for preserving fruit quality*. Boca Raton: CRC Press.
Vilas-Boas, E. V. D. B., & Kader, A. A. (2007). Effect of 1-methylcyclopropene (1-MCP) on softening of fresh-cut kiwifruit, mango and persimmon slices. *Postharvest Biology and Technology, 43*(2), 238–244.
Wang, L., Chen, S., Kong, W., Li, S., & Archbold, D. D. (2006a). Salicylic acid pretreatment alleviates chilling injury and affects the antioxidant system and heat shock proteins of peaches during cold storage. *Postharvest Biology and Technology, 41*(3), 244–251.
Wang, B. G., Jiang, W. B., Liu, H. X., Lin, L., & Wang, J. H. (2006b). Enhancing the post-harvest qualities of mango fruit by vacuum infiltration treatment with 1-methylcyclopropene. *The Journal of Horticultural Science and Biotechnology, 81*(1), 163–167.
Wang, Q., Lai, T., Qin, G., & Tian, S. (2009). Response of jujube fruits to exogenous oxalic acid treatment based on proteomic analysis. *Plant and Cell Physiology, 50*(2), 230–242.
Wang, K., Shao, X., Gong, Y., Xu, F., & Wang, H. (2014a). Effects of postharvest hot air treatment on gene expression associated with ascorbic acid metabolism in peach fruit. *Plant Molecular Biology Reporter, 32*(4), 881–887.
Wang, Y., Xie, X., & Long, L. E. (2014b). The effect of postharvest calcium application in hydro-cooling water on tissue calcium content, biochemical changes, and quality attributes of sweet cherry fruit. *Food Chemistry, 160*, 22–30.
Wannabussapawich, B., & Seraypheap, K. (2018). Effects of putrescine treatment on the quality attributes and antioxidant activities of 'Nam Dok Mai No. 4' mango fruit during storage. *Scientia Horticulturae, 233*, 22–28.
Wasternack, C. (2007). Jasmonates: An update on biosynthesis, signal transduction and action in plant stress response, growth and development. *Annals of Botany, 100*(4), 681–697.
White, P. J., & Broadley, M. R. (2003). Calcium in plants. *Annals of Botany, 92*(4), 487–511.
Wills, R. B. H., Ku, V. V. V., & Leshem, Y. Y. (2000). Fumigation with nitric oxide to extend the postharvest life of strawberries. *Postharvest Biology and Technology, 18*(1), 75–79.
Wu, B., Guo, Q., Wang, G. X., Peng, X. Y., & Che, F. B. (2015). Effects of different postharvest treatments on the physiology and quality of 'Xiaobai' apricots at room temperature. *Journal of Food Science and Technology, 52*(4), 2247–2255.
Wu, F., Zhang, D., Zhang, H., Jiang, G., Su, X., Qu, H., & Duan, X. (2011). Physiological and biochemical response of harvested plum fruit to oxalic acid during ripening or shelf-life. *Food Research International, 44*(5), 1299–1305.
Ya'acov, Y. L., Wills, R. B., & Ku, V. V. V. (1998). Evidence for the function of the free radical gas-nitric oxide (NO•)- as an endogenous maturation and senescence regulating factor in higher plants. *Plant Physiology and Biochemistry, 36*(11), 825–833.
Yang, Q., Wang, L., Li, F., Ma, J., & Zhang, Z. (2011). Impact of 1-MCP on postharvest quality of sweet cherry during cold storage. *Frontiers of Agriculture in China, 5*(4), 631–636.
Yu, L., Liu, H., Shao, X., Yu, F., Wei, Y., Ni, Z., & Wang, H. (2016). Effects of hot air and methyl jasmonate treatment on the metabolism of soluble sugars in peach fruit during cold storage. *Postharvest Biology and Technology, 113*, 8–16.
Zaharah, S. S., & Singh, Z. (2011a). Mode of action of nitric oxide in inhibiting ethylene biosynthesis and fruit softening during ripening and cool storage of 'Kensington Pride' mango. *Postharvest Biology and Technology, 62*(3), 258–266.
Zaharah, S. S., & Singh, Z. (2011b). Postharvest nitric oxide fumigation alleviates chilling injury, delays fruit ripening and maintains quality in cold-stored 'Kensington Pride' mango. *Postharvest Biology and Technology, 60*(3), 202–210.
Zapata, P. J., Martinez-Espla, A., Guillen, F., Diaz-Mula, H. M., Martinez-Romero, D., Serrano, M., & Valero, D. (2014). Preharvest application of methyl jasmonate (MeJA) in two plum cultivars. 2. Improvement of fruit quality and antioxidant systems during postharvest storage. *Postharvest Biology and Technology, 98*, 115–122.
Zhang, N., Sun, Q., Zhang, H., Cao, Y., Weeda, S., Ren, S., & Guo, Y. D. (2015). Roles of melatonin in abiotic stress resistance in plants. *Journal of Experimental Botany, 66*(3), 647–656.

Zheng, X., Tian, S., Gidley, M. J., Yue, H., & Li, B. (2007). Effects of exogenous oxalic acid on ripening and decay incidence in mango fruit during storage at room temperature. *Postharvest Biology and Technology, 45*(2), 281–284.

Zhu, S., Liu, M., & Zhou, J. (2006). Inhibition by nitric oxide of ethylene biosynthesis and lipoxygenase activity in peach fruit during storage. *Postharvest Biology and Technology, 42*(1), 41–48.

Ziosi, V., Noferini, M., Fiori, G., Tadiello, A., Trainotti, L., Casadoro, G., & Costa, G. (2008). A new index based on vis spectroscopy to characterize the progression of ripening in peach fruit. *Postharvest Biology and Technology, 49*(3), 319–329.

Zokaee Khosroshahi, M. R., & Esna-Ashari, M. (2007). Post-harvest putrescine treatments extend the storage-life of apricot (*Prunus armeniaca* L.) 'Tokhm-sefid' fruit. *The Journal of Horticultural Science and Biotechnology, 82*(6), 986–990.

Packaging and Storage of Stone Fruits

11

K. Rama Krishna, J. Smruthi, and S. Manivannan

Abstract

Stone fruits respire at a higher rate and are soon perishable when kept at room temperature. Apart from the deterioration caused due to ripening activity, the stone fruits are also notified as delicate fruits due to their smooth peel which gets bruised immediately by impact or friction. The management of these two major factors can help in the regulation of physiological, biochemical and pathological activities of stone fruits. Packaging and storage are considered as major criteria to be taken into consideration in managing fruit deterioration. The recent scientific developments in the field of packaging and storage have brought new innovative process and recommendations which can enhance the storage/shelf life of stone fruits as well as maintain the quality for a longer period. The commodity packed gets the better acceptance by the consumer as it reduces the bruising caused at various stages of the supply chain and provides the produce at convenient sizes and packs required, whereas the storage ultimately keeps the fruits with maximum quality and availability for an extended period. The packaging and storage can be complementary to each other when the stakeholder at each stage of the supply chain knows what type of package is more suitable for a certain type of storage conditions and vice versa. In this chapter, we tried to pool up the scientific information and recommendations which can suit different stakeholders in using a varied type of packaging and storages for stone fruits. The packaging and storage of stone fruits such as mango, plum, peach, apricot, sweet

K. R. Krishna (✉) · S. Manivannan
Department of Horticulture, School of Life Sciences, Central University of Tamil Nadu, Thiruvarur, TN, India

J. Smruthi
Division of Food Science and Postharvest Technology, ICAR-Indian Agriculture Research Institute, New Delhi, India

M. M. Mir et al. (eds.), *Production Technology of Stone Fruits*,
https://doi.org/10.1007/978-981-15-8920-1_11

cherry, litchi, almond and dates, covering all the three major climate zones, are discussed in this chapter.

Keywords

Modified atmospheric packaging · Biodegradable packaging · Controlled atmospheric storage · Hypobaric storage · Export

11.1 Introduction

Stone fruits are valued for their nutrition as they are loaded with high antioxidants (Redondo et al. 2017). But like all other horticultural crops, stone fruits are highly perishable crops with a very short shelf life. The postharvest management of the stone fruits is of utmost importance as it is the only possible way to enhance its shelf life without compromising the quality parameters. Among different management aspects followed, the packaging and storage of produce become inevitable unit operations that directly affect the marketability of the stone fruits. A study conducted by the FAO (2014) revealed that the food loss in the postharvest chain of horticultural crops is estimated to be 45%. Adequate packaging and storage can help in reducing these losses. Yet, these losses remain a persistent challenge despite having solutions that are recognized globally (Ahmad and Siddiqui 2015). This may be due to the lack of awareness of the solutions and information to the stakeholders along the supply chain. In this chapter, we have tried to provide the information, suggestions and specifications laid done for packaging and storage of stone fruits (mango, plum, peach, apricot, sweet cherry, litchi, almond, dates) for both the national and international stakeholders.

11.1.1 Packaging of Stone Fruits

The packaging is considered as one of the most important aspects of the postharvest chain, i.e. from growers to consumers. The International Packaging Institute (IPI) defined packaging as the enclosure of items, packages or products in the bag, box, tray, bottle, can, cup, wrapped pouch or any other form of container to perform single or more functions as follows: containment, preservation, protection, utility, communication and performance (Siddiqui 2015). Packaging becomes a vital component for export as well as for local marketing of the stone fruits. It provides ease of handling and counted containers of uniform size. The standard-size packaging of the stone fruits reduces the need for repeated weighing, thereby faster handling and shipping/transporting of the containment to the destination. There are several types of packaging material commonly used for domestic marketing and export of the stone fruits (Table 11.1). These include the packages fabricated from paper and paper products (e.g. compressed cardboard, corrugated cardboard or fibreboard) (Marsh and Bugusu 2007; Raheem 2013), wood and wood products (Lee et al. 2008) and plastics (Marsh and Bugusu 2007). For retail and local marketing, apart from

Table 11.1 Packaging materials used by different stakeholders in India

Stakeholders	Type of packaging used	Capacity (kg)	Remarks
Producers	Bamboo basket/plastic crates/ bulk break	20, 25 and 30	Mostly bulk sales
	Gunny bags	5–50	
	Wooden boxes	5–15	
	Loose	–	
Preharvest contractors	Bulk/plastic crates/bamboo basket	20, 25 and 30	80–90% sales is in bulk. Remaining 10–20% is in packed form
Wholesalers	Bulk break/wooden boxes/ cardboard cartons	5, 10, 20	
Processors	Bulk break	–	–
Exporters	Cardboard carton boxes, thermo cool boxes, CFC boxes, foldable plastic boxes, etc.	3, 4.5, 5, 10, 20	As per specifications of exporter requirement
Retailers	Bamboo basket/bulk/push carts/cycles/kavadi	No specific size	Mostly loose or small plastic package of 1/2 or 1 kg
Cooperatives	Plastic crates	10, 20	–

above packaging material, suppliers and distributors are also using gunny bags made out of jute or plastic. With the advancement of technology in the field of packaging, innovations have evolved in the past two decades such as smart or intelligent packaging (Ghaani et al. 2016), biodegradable packaging (Ivonkovic et al. 2017) and modified atmospheric packaging (Ben-Yehoshua et al. 2005). Therefore, modern packaging not only contains, protects and gives value addition but also helps in taking decisions in choosing the correct commodity, assuring quality and are environmentally safe. The packaging material varies according to the crop, destination intended to be transported, availability of the packaging material and type of function it needs to perform. Generally, the stone fruits have soft tissue that gets bruised when the fruit experiences damage due to fall or shake or compact pressure while handling and transporting the produce. In this context, proper packaging material plays a significant role in reducing this damage or loss of produce due to impact damage. The use of plastic films as packing material should be chosen according to the permeability of the films to gases and humidity (Mangaraj et al. 2009; Castellanos and Herrera 2017) and the rate of respiration of the produce. The polymers used for packaging of most of the stone fruits in MAP conditions are polyethylene, polyvinyl chloride, polypropylene and polyethylene terephthalate (Mangaraj et al. 2009). Preferably, the thickness of the packaging film should be between 15 and 100 μm to be viable for mechanical and commercial reasons (Varoquaux et al. 2002).

11.1.2 Storage of Stone Fruits

The storage of fruit crops especially stone fruits is of great importance as there are several factors during the storage which directly influence the quality of fruits. The respiration rate and perishability of the fruit further get accelerated by the microenvironment of storage such as temperature, relative humidity and gas composition. Hence, the management of these vital components is important to enhance the shelf life and to maintain the better quality of stone fruits. The most important is the temperature of the produce and storage which determines the various physiological processes which in turn are responsible for hastening or delaying senescence in fruits. The harvested fruits are subjected to the pre-cooling, the major unit operation from which the respiratory heat is removed and to stabilize the metabolic activities in stone fruits. There are various methods through which the pre-cooling is administered in the fresh commodities (Elansari et al. 2019). The fruits reach their ultimate consumer after passing through various channels of operations after harvesting, transit and storage (wholesale and retail centres). During this chain of operations, fruits are exposed to various environmental conditions which could be either detrimental or beneficial.

The temperature at which fruits should be stored depends on their physiology. Stone fruits of temperate climate are usually stored at temperature just above the freezing point, and care should be taken to avoid chilling injury/mealiness-prone temperature as it's a major concern of quality deterioration, while the stone fruits of tropical climate are stored at temperatures above the critical chilling injury temperature (Patel et al. 2016). The postharvest handling is majorly temperature-dependent, as it determines the rate/speed of ripening, cellular breakdown and death (senescence). Apart from that, temperature hastens the development of various postharvest losses due to decay as high temperature and humidity are congenial for the growth of deteriorating microorganisms, especially fungi. The high temperature during the harvest and postharvest period results in a high physiological loss in weight (PLW) and shrinkage of fruits which drastically reduces the marketability of fruits (Gill et al. 2017). Similarly, the high PLW and respiration rate will eventually lead to senescence well in advance. The low temperature is also detrimental to fruit as it causes physiological maladies such as freezing and chilling injury (CI). The freezing injury occurs when fruits are stored below the freezing point (0 °C), whereas chilling injury can arise to a wide range of temperature above freezing point (Patel et al. 2016), and it deteriorates the quality of fruits and renders it unfit for consumption. For example, the temperature between 2.2 and 7.7 °C is proved to be the chilling temperature zone for peaches and nectarines, and it is always administered to store between 0 and 1 °C to avoid the mealiness in peaches (Crisosto et al. 1999). Globalization offered a wide variety or diversity of food to one's plate in any part of the world, and here lays the importance of temperature management of fruits so as to avail these temperate fruits in tropical and subtropical areas and vice versa to rejoice in its fresh form. Fruits should inevitably be pre-cooled well in advance before the storage to avoid the drastic reduction of marketability. The low temperature and high temperature are indefinitely detrimental to fruits, so it's important to store them under the optimum temperature which is again the commodity-specific (Gross et al. 2016). The physiology of fruit is complex, and it varies between different crops and even between the varieties of the same crop. The temperature-related disorders are reversible for a short point of time during storage, but after that, it became invariably irreversible. So it's very important to administer the management

Table 11.2 The effects of CO_2 and O_2 levels on crops while in storage

Modified CO_2 levels	Modified O_2 levels
Decreased discolouration levels	Reduced respiration rate
The decreased synthetic reaction in climacteric fruits	Reduced substrate oxidation
Inhibition of some enzymatic reactions	Delayed ripening of climacteric fruits
Decreased production of some organic volatiles	Prolonged storage life
Modified metabolism of some organic acids	Reduced rate of production of ethylene
Inhibition of the effects of ethylene	Reduced degradation rate of soluble pectin
Retarded fungal growth	Formation of undesirable odours and flavours
Production of off-flavours	Altered texture
Development of physiological disorders	Development of physiological disorders

activity on time. Intermittent warming of the stored commodity is an emerging management strategy; apart from that, temperature conditioning is also recommended (Ruoyi et al. 2005; Xi et al. 2012).

The modification of gaseous composition around the stone fruits is one of the most effective methods of storage to extend the shelf life. The controlled atmospheric (CA) storage is commercial and widely accepted worldwide. The gaseous components such as O_2, CO_2 and ethylene around the fruits are being controlled and monitored throughout the storage in controlled atmospheric storage (Ben-Yehoshua et al. 2005). The gas composition in CA storage varies according to the crop. It may also vary widely within the same species. The gas composition should be checked preciously throughout the storage. The undesirable gas composition may even have a deleterious effect on the crops (Table 11.2 (adopted from Thompson 1998)).

11.2 Packaging and Storage of Stone Fruits

11.2.1 Mango (*Mangifera indica*, Anacardiaceae)

11.2.1.1 General Packaging

An important operation before packaging of the harvested mangoes is de-sapping immediately, followed by washing, sorting and grading. For the local market and transportation of mangoes within the developing countries, the fruits are generally bulk packed maybe in multilayer, using wooden crates, thermoformed containers, gunny bags, bamboo baskets and plastic woven poly sacks of different sizes (Manalili 2011). For padding/cushioning between layers of mango fruits, dry grass, wheat straw or paddy, paper shredding or leaves of fruit tree are used. Sometimes, newspaper is used as a lining material in different packages. Plastic films as lining material are more advantageous than paper as they can hold more relative humidity in packaging, thereby less shrivelling of mangoes. But, these traditional methods do not integrate with the modern supply chain as the bulk packaging has various logistic issues (loading, unloading, pre-cooling, environmental and food safety) and do

Table 11.3 CFB specifications for mango export (capacity six no.)

Cultivar	External dimensions	Style of box	No. of plies	Type of flute	Grammage
Alphonso	225 × 170 × 100	RSC-0201,	3.0,	B (narrow)	250/150/150
Kesar	225 × 170 × 90	Telescopic-0300,	Lid-3, Tray-3	Lid-B	250/150/150
Dashehari	200 × 200 × 95	Telescopic-0306,	Lid-3, Tray-3	(narrow),	250/150/150
Banganpali	230 × 180 × 100	Telescopic-0312	Lid-3, Tray-3	Tray-B	250/150/150
Chausa	230 × 185 × 100			(narrow) Lid-B (narrow), Tray-B (narrow) Lid-B (narrow), Tray-B (narrow)	

Source: APEDA, Govt. of India. For other capacities such as 9, 12, 15 and 18 no., follow the link https://apeda.gov.in/apedawebsite/trade_promotion/Specifications-of-Packing.htm

Table 11.4 Wooden crate specifications for the domestic market (source: APEDA)

Type of packaging	Cultivar	Dimensions	Capacity (kg)
Wooden crates	Ratnagiri	45 × 30 × 30	16–18
	Malihabad	21.6 × 21.6 × 42	10–11

not fit into the modern cold chain system (Defraeye et al. 2015). Nowadays, the wooden crates are replaced with cardboard fibre boxes (CFB) which are more versatile in size, shape and functionality. Specification of the CFB for export and wooden crates for export markets is given in Tables 11.3 and 11.4.

11.2.1.2 Modified Atmospheric Packaging (MAP)

With the evolution of semipermeable films, MAP is considered a viable alternative for controlled atmospheric storage. The polymeric films when wrapped around the fruits create a modified atmosphere in a passive way, where the concentration of CO_2 increases and O_2 decreases around the fruit (Singh et al. 2013). The 'Baneshan' mangoes were pre-packed in a high-molecular-high-density polyethylene (HM-HDPE) film that has shown reduced respiration, weight loss and ripening (Narayana 1989). The effect of MAP is more advantageous when it is combined with other treatments such as pre-fungicide applications prior to storage. The 'Nam Dok Mai' mango cultivar was treated with hot benomyl (1000 ppm) solution at 55 °C for 5 min and seal packed in the polyvinyl chloride (PVC) film (0.01 mm) and stored at 13 °C. It was able to prolong the storage life of mangoes for 4 weeks without any adverse effect on the quality of the stone fruit (Sosnrivichai et al. 1992). The polymeric films having micro-perforation showed longer shelf life by regulating the gas composition exchange, especially O_2. Polyolefin (SM60M) film (contains micro-perforations of eight holes of 0.4 mm diameter/in^2) was used as packaging material for mango (cv. 'Tommy Atkins' and 'Keitt') and stored at 14 °C. The fruits had a storage life of 2–3 weeks and additional shelf life of 1 week at 17 °C in polyolefin film wrapped onto polystyrene trays (Rodov et al. 1994). The mangoes (cv. 'Alphonso') harvested at mature hard green condition, packed in micro-perforated D-955 film, can be stored for 1 month at 8 °C without any chilling injury (CI) symptoms even after 1 week at ambient condition (Rao et al. 2003). The mangoes (cv. 'Banganapalli' and 'Alphonso') could be stored for 5 weeks when the fruits were individually shrink-wrapped and stored at 8 °C. The fruits had no CI symptoms and ripened normally with good quality even after 1 week at ambient conditions (Rao and Shivashankara 2004). Xtend®, a micro-perforated polyethylene film, was used as a carton liner that leads to the storage of 'Keitt' and 'Tommy Atkins' mangoes for 3 weeks at 12 °C (Pesis et al. 2000). Due to the demand for biodegradable and eco-friendly packaging, the plastic films used as MAP can be replaced with biodegradable films. For example, the 'Alphonso' mangoes when stored in chitosan-covered wax-lined cartons, at ambient conditions (27 °C, 65% RH), were successfully stored for 20 days without any off-flavour and microbial infection (Srinivasa et al. 2002).

11.2.1.3 Storage

Evaporative Cool Storage

One of the low-cost storage structures which can be adapted on-farm for short-term storage is evaporative cool storage. The mango cv. 'Baneshan' and 'Amrapali' shelf life was extended over 1 week when the fruits were stored using evaporative cool chamber (Narayana 1989). The zero energy cool chamber (ZECC) developed as an on-farm evaporative cooling structure by Roy and Pal (1989) was able to prolong the shelf life of green mango for 3–4 days and ripe mango for 9 days when compared to ambient condition. Tefera et al. (2007) had reported that the evaporative cooling storage (14.33–19.26 °C and 70.15–82.4% RH) was able to extend the storage life of mangoes from 3 days to 28 days in comparison to the fruits stored at ambient conditions.

Low-Temperature Storage/Cold Storage at Different Levels of Handling of Fruits

The general recommended storage temperature of mango is 12–13 °C (54–56 °F) at 85–95% relative humidity (Mitra and Baldwin 1997). Depending on the stage of maturity at harvest, cultivar/variety and season of the harvest of mangoes, the storage temperature can vary between 7 °C and 13 °C. The sensitivity of mango cultivar to a temperature lower than 10 °C depends upon the stage of maturity of the fruit, the cultivar and the extent of the period the fruits are stored (Narayana et al. 2012). The green mango fruits can be stored between 10 °C and 15 °C, while ripe mango fruits can tolerate below 10 °C also (Nunes et al. 2007). The temperature range of 10–12 °C is maintained in the temporary holding storage rooms before loading onto truck trailers or marine containers. When it comes to different cooling methods used to control the storage temperature, forced-air cooling is recommended over room cooling method, as forced-air cooling rapidly cools the surface of fruit reducing extreme water loss (reduces the water vapour gradient across the fruit skin), thus slowing down the water movement out of the fruit due to transpiration (Brecht et al. 2010). Care should be taken to maintain the relative humidity around 85–95% if the fruits are stored for a longer period. It is also advised to pre-cool the reefer containers, i.e. cooler than 12 °C so that the fluctuations of temperature are minimum after the transfer of mangoes to the reefer containers. The dock staging area should be sufficiently refrigerated to maintain a temperature of 12–15 °C. The temperature of retail stores for sales of mango should not be around 10 °C. Perhaps, the majority of mango cultivars are prone to chilling injury if they are stored for a long period at a temperature below 10 °C (Nunes et al. 2007). Therefore, at retail conditions where mangoes are intended for longer display, they should always be held at an ambient room temperature of 21–22 °C in the store. Other general recommendations for low-temperature storage are as follows:

- At 12 °C and maintaining 85% RH, the mango fruits can be stored for 2–3 weeks (Mercantilia 1989).
- At 13.3 °C and 85–90% RH, the mangoes can be stored for 14–25 days (SeaLand 1991) (Table 11.5).

Table 11.5 Cultivar-wise recommended storage temperatures of mango

Mango cultivar	Recommended temperature	Relative humidity	Storage period	References
Other countries				
Alphonso	13 °C	90%	2–3 weeks	Snowdon (1990)
Caraboa	7.2–10 °C	85–90%	17–24 days	Pantastico (1975)
Ceylon	10 °C	–	3 weeks	Thompson (1971)
Haden	12–14 °C	90%	2 weeks	Snowdon (1990)
Irwin	10 °C	85–90%	3 weeks	Lutz and Hardenburg (1968)
Julie	11–12 °C	90%	2 weeks	Snowdon (1990)
Keitt	12–14 °C	90%	2 weeks	
Pico	7.2–10 °C	85–90%	17 days	Pantastico (1975)
Zill	10 °C	90%	3 weeks	
India				
Alphonso	7–9 °C	90%	7 weeks	Snowdon (1990)
Bangalore	5.5–7 °C	90%	7 weeks	
Chausa	8 °C	85–90%	3–4 weeks	NHB[a]
Dashehari	7.2–7.9 °C	85–90%	35 days	Mann and Singh (1976)
Langra	14 °C	85–90%	3–4 weeks	NHB[a]
Mallika	12 °C	85–90%	3–4 weeks	
Neelum	5.5–9 °C	90%	5–6 weeks	Snowdon (1990)
Raspuri	5.5–9 °C	90%	5–6 weeks	
Safeda	5.5–7 °C	90%	7 weeks	

[a]National Horticulture Board (NHB): http://nhb.gov.in/report_files/mango/mango.htm

Controlled Atmospheric Storage (CA)

General recommended CA storage temperatures in mango are as follows:

- 12 °C with 5% CO_2 and 5% O_2 for 20 days (Hatton and Reeder 1966)
- 10–15 °C with 5% CO_2 and 5% O_2 (Kader 1986)
- 10–15 °C with 5–10% CO_2 and 3–5% O_2 (Kader 1989)
- 13 °C with 5% CO_2 and 5% O_2 (SeaLand 1991)
- 10 °C with 90% RH in 10% CO_2 + 5% O_2 (Lawton 1996)
- 10–14 °C and 85–90% RH with 5–10% CO_2 and 2–5% O_2 for 1–4 weeks (Burden 1997)
- 10 °C with 10% CO_2, 3% O_2 and 87% N_2 for 2 weeks (Kim et al. 2007)

However, Bender et al. (2000) recommended CA storage conditions for 'Tommy Atkins', 'Keitt', 'Haden' and 'Kent' as follows:

- Mature green: 12 °C with 25% CO_2 and 3–4% O_2 for 3 weeks.
- Tree ripe: 8 °C with 25% CO_2 and 3–4% O_2 (or) 5 °C with 10% CO_2 and 5% O_2 (Table 11.6).

Table 11.6 Controlled atmospheric storage of some mango cultivars

Cultivar	Temperature	CO_2	O_2	Storage period	References
Alphonso	13 °C	5%	5%	4 weeks	Rao and Rao (2008)
	8 °C	5%	5%	45 days	Niranjana et al. (2009)
	(when treated with hot water treatment of 55 °C for 5 min)				
Amelie	11 °C	5%	5%	4 weeks	Medlicott and Jeger (1987)
	10–12 °C	5%	5%	28 days	Kane and Marcellin (1979)
Banganapalli	13 °C	3%	5%	5 weeks	Rao and Rao (2008)
Caraboa	8 °C	10%	6%	4 weeks	Bleinroth et al. (1977)
Carlotta	10–11 °C	5%	2%	6 weeks	Medlicott and Jeger (1987)
Chok Anan	15 °C	0%	5%	28 days	Shukor et al. (2000)
Delta R2E2	13 °C	6%	3%	38 days	Lalel et al. (2005)
Haden	10–11 °C	5%	2%	6 weeks	Medlicott and Jeger (1987)
Irwin (mature)	8–12 °C	5%	5%	28 days	Maekawa (1990)
Irwin (tree ripe)	5–15 °C	5–10%	5%	30 days	Nakamura et al. (2004)
Kent	12 °C	10%	5%	29 days	Lizana and Ochagavia (1997)
Kensington Pride	13 °C	5%	3%	26 days	Johnson et al. (1990)
	13 °C	4–6%	2–4%	30–35 days	Lalel et al. (2004)
Rad	13 °C	4%	6%	25 days	Noomhorm and Tiasuwan (1995)
Tommy Atkins	12 °C	5%	5%	31 days	Lizana and Ochagavia (1997), Bender et al. (2000)

Hypobaric or Low-Pressure Storage
In hypobaric storage, the pressure inside the storage is reduced to remove excess oxygen, thereby creating CA storage. This method not only removes oxygen but also diffuses out the ethylene from the tissue of fruits, thus prolonging the storage life of fruits (Narayana et al. 2012). When hypobaric storage was used for mango (cv. 'Haden'), the ripening of the fruits was four times slower at 150 mmHg than in air (Burg 1975). These fruits ripened normally with good acceptability as well as less decay. Similarly, 'Irwin', 'Tommy Atkins' and 'Kent' were stored in low-temperature storage maintaining 76–152 mmHg at 13 °C and 98–100% RH for 3 weeks. The fruits took a longer time to complete the full ripening, had higher acceptability after removal from storage and were kept at ambient conditions (Spalding and Reeder 1977). Ilangantileke and Salocke (1989) reported that the mango's (cv. 'O Krong') fruit life when pre-cooled at 15 °C, waxed and maintained at 60–160 mmHg at 13 °C was extended for 4 weeks in storage with normal ripening and without any adverse effect on quality. Hypobaric storage has proven to lengthen the shelf life of mango cultivars, reduce CI symptoms, control fruit fly and reduce development of anthracnose and stem rot (Apelbaum et al. 1977; Spalding and Reeder 1977; Jackman et al. 1988; Davenport et al. 2006). But the use of hypobaric storage is limited to high-value crops, as the construction of hypobaric storage is very expensive due to the low-pressure requirement (Wills et al. 2007) (Table 11.7).

11.2.2 Plum and Peach (*Prunus domestica* and *Prunus persica*, Rosaceae)

Plums and peaches are climacteric and highly perishable. They produce a quite high amount of ethylene (Kader 2002) and respire moderately (Rizvi and Perdue 1981). Further, the harvested fruits show high loss of nutrients without proper treatments, packing or storage.

11.2.2.1 General Packaging

Packaging reduces wastages by protecting the fruits from surface dirt, moisture loss and shrinking, undesirable changes in the physical appearance, mechanical damages deteriorating physiological conditions and entomological and pathological spoilage. There are several types of packaging employed in plums and peaches. Films, bags, crates and boxes made up of polymers are the most commonly used packing materials. Apart from that, there are different types of packaging employed

Table 11.7 Hypobaric storage of some mango cultivars

Cultivars	Pressure (mmHg)	Temperature (°C)	Storage period	References
Pairi, Haden and Maya	100 and 75	13	25 and 35 days, respectively	Apelbaum et al. (1977)
Rad	100	15	45 days	Chen (1987)
Keitt	20	13.3	45 days	Burg (2004)

in these two fruit crops like coloured wraps, modified atmospheric packing (MAP), antimicrobial packaging (AMP), active packaging (AP) and biodegradable packaging (BP). These packages are mainly to prolong the shelf life of the product, and it is reliant on the barrier properties of packaging material to gases like CO_2, O_2 and ethylene.

11.2.2.2 Modified Atmospheric Packaging (MAP)

The recommended gas mixtures for peach under MAP are 1–2% O_2, 3–5% CO_2 and 93–96% N_2 (Scetar et al. 2010). The shelf life of late-maturing peach ('Jesca Clone') was studied using cinnamon essential oil incorporated into plastic packaging (PET trays). The active label package was able to reduce the decay of fruits to 13% vs. 86% compared to non-active packaging. The fruits recorded low weight loss (3.4%) and high firmness (15.9 N) at the end of 12 days of storage at ambient condition (Prado et al. 2011). Mir et al. (2018) suggested ethylene absorber with four perforations gave the best result in increasing shelf life when peach fruits of the cultivar 'Shan-I-Punjab', harvested at colour break stage and packed in thermocol trays wrapped with LDPE film. Peach (cv. 'Shan-I-Punjab') was packed in paper mould trays and wrapped with cryovac heat shrinkable RD-106. The fruits were stored for 9 days with good quality at supermarket (18–20 °C, 90–95 RH) condition (Mahajan et al. 2015).

A recent trend on the packaging is aimed at biodegradable packaging materials as it is environmentally safe. Accordingly, Seglina et al. (2013) used polylactic acid (PAL) boxes, polypropylene and cardboard boxes to extend the shelf life of diploid plum cultivar 'Kometa'. The plum fruits can be stored in good quality for 8 days at 4 °C by using PLA boxes. The quality of fruits was maintained due to selective gas and moisture barrier properties of polylactic material. The packaging also ensured minimal weight loss and promoted normal plum ripening with good sensory attributes.

Coloured wraps like transparent, yellow and white polyethylene films were tried (Khan et al. 2013) in plums to extend the shelf life. The fruits wrapped with transparent film showed reduced weight loss and decay index, and had higher TSS and acidity, whereas maximum ascorbic acid was retained in the fruits wrapped with yellow films. The fruits wrapped with a transparent film had higher overall sensory acceptability (Table 11.8).

11.2.2.3 Storage

Cold Storage

The fruits of plum and peach are climacteric, respire and ripen at a rapid rate and deteriorate within 2–3 days at room temperature. Low-temperature storage becomes imperative for extended storage life and quality of the fruit. The storage temperature of 0–1 °C (above freezing point) with 90–95% RH is recommended for extended storage life of plum and peach (Wang et al. 2016; Lurie and Crisosto 2005). At this temperature, the fruits can be stored for 2–3 weeks with good quality. Both the fruits are prone to chilling injury (flesh translucency, internal browning, mealiness, pulp

Table 11.8 Recommended modified atmospheric packaging (MAP) of different varieties of plum and peach

Cultivar	Package material	Thickness of the film	Storage temperature and RH	Storage duration	References
Plum					
Angeleno	PET trays wrapped with polyethylene film to form pallet bag unit	100 μm	1 °C, 90–95% RH	60 days	Peano et al. (2017)
Ariddo di Core	Coextrusion of PET, EVOH and PE (Corapack, Italy)	90 and 65 μm	1 °C, 90–95% RH	21 days	Giuggioli et al. (2016)
Black Amber	LDPE (LifeSpan™)	30 μm	0–1 °C, 90–92% RH	56 days	Singh and Singh (2013), Singh et al. (2013)
Black Beauty	PE	30 μm	0 °C, 90–92% RH	50 + 7 days	Erkan and ESKİ (2012)
Peach					
Douradao	LDPE bags	50–60 μm	1 °C, 90–95% RH	28 days	Santana et al. (2011)
Royal Glory	Hot water treated (46 °C) and packed in PE bags (LDPE+HDPE+LDPE)	35 μm	1 °C, 90–95% RH	1–2 weeks	Malakou and Nanos (2005)
Shan-I-Punjab	Shrinkable RD-106	15 μm	28–30 °C, 60–65% RH	4 days	Mahajan et al. (2015)

bleeding, flavour loss and failure to ripen) when stored at 2.2–7.6 °C (Singh 2010; Crisosto et al. 1999). Therefore, the temperatures just above freezing point are best for storage of plum and peach.

Controlled Atmospheric (CA) Storage

CA storage (raised CO_2 and lowered O_2) supported by low temperature and high relative humidity has reported extending the fruit quality of plum and peach for a longer time. However, the gas concentration, storage temperature, cultivar type, maturity, pretreatments and duration of storage influence the final quality of the fruit (Saltveit 2003). Steffens et al. (2014) recommended CA condition for storing 'Laetitia' plums is 2 kPa O_2 + 2 kPa CO_2, since it allows for a slower apparent ripening of the fruit (lower evolution of the skin red colour) and low intensity of internal breakdown, whereas higher CO_2 concentration (above 5%) may lead to deleterious effects on the fruit. The plum (cv. 'Jubileum', 'Victoria' and 'Opal') fruits were CA stored at 2 °C with 5, 15 and 25% CO_2 for 3 weeks. All the CO_2 concentrations lead to the development of bicarbonate-like taste that resulted due to accumulation of ethanol (Wang and Vestrheim 2003).

Peaches (cv. 'Eldorado') grown in conventional and organic systems were subjected to CA storage at 1.5 °C and 90% relative humidity (RH) with 2% O_2 and 15% CO_2. It presented higher uniformity and intense characteristic taste up to 30 days to the conventional system and up to 15 days to the organic system (Cantillano et al. 2010). Peaches (cv. 'Flavorcrest' and 'Red Tape') were stored up to 45 days with good quality when CA storage condition of 5% CO_2 and 2% O_2 at 0–0.5 °C with 90–95% RH was opted (Akbudak and Eris 2003). The sliced peaches (cv. 'Fay Elberta') were preserved with good quality up to 7 days at 5 °C with CA gas composition of 2% O_2 and 12% CO_2 (Wright and Kader 1997). The peach (cv. 'Rich Lady') fruits had most acceptable sensory score when stored at 2 °C with CA condition of 3 kPa O_2 and 10 kPa CO_2 for 15 days and then kept at 20 °C for 1 day (Ortiz et al. 2008).

11.2.3 Apricots (*Prunus armeniaca*, Rosaceae)

11.2.3.1 General Packaging

Apricots are marketed immediately after harvested, and their availability for longer time and distance depends on proper packaging and storage. The packaging of apricots is either carried out at filed or in specific postharvest handling centres. Prior to packaging, the postharvest operations such as cleaning, sorting, grading, waxing and chemical treatments become necessary to fetch a better price. Apricot fruits are harvested/picked into plastic baskets or picking bags. The fruits harvested are generally packed in CFB and reusable containers made of wood, bamboo or plastic. The graded apricot fruits are packed in 5 kg wooden boxes for transport and marketing (Sharma and Krishna 2017). The variation in size of apricot fruits in a package should not be more than 5% of the total count and may differ no more than 6 mm when measured from the widest portion of the fruit cross-section (Muzzaffar et al. 2018). The polymers such as polystyrene trays and flow pack made of polyvinyl chloride wrapped in transparent plastic films with or without holes can form an attractive packaging for apricots. Generally, the apricots are single- and two-layer tray packed or volume filled in about 10 kg net (Carlos and Kader 1999).

11.2.3.2 Modified Atmospheric Packaging (MAP)

MAP combined with low-temperature storage can maintain the quality of apricot fruits for a longer time. In MAP, the atmospheric gases surrounding the fruit are maintained by either the passive or active way. The apricots (cv. Canino) can be stored for 35 days at 0 °C when fruits inside MA packaging were able to produce passively 13–15 kPa CO_2 and 10 kPa O_2. The fruits did not show any decay or browning after removal from storage for 4 days at 20.8 °C (Kosto et al. 2002). Ayhan et al. (2010) found that the apricot cv. 'Kabaasi' stored at 4 °C, packed in either biaxially oriented polypropylene film (BOPP) or cast polypropylene film (CPP), could be successfully stored for 28 days. Similar storage life was also obtained when the 'Kabaasi' cultivar of apricot was coated with 5% NatureSeal® and actively packed in biaxially oriented polypropylene film (5% O_2 and 10% CO_2)

and was able to be stored up to 28 days at 4 °C (Muftuoğlu et al. 2012). The apricot (cv. Aprikoz) was packed in a plastic box covered with stretch film (12 and 16 μm) and stored at 0 °C with 90–95% RH that was able to extend the storage life up to 35 days than control (Koyuncu et al. 2010). The apricot fruits (cv. Tom Cot®) were stored at 1 °C with 90–95% RH using different packaging material. A Multilayer film (obtained from coextrusion of PET, ethylene vinyl alcohol (EVOH) and polyethylene (PE)) of thickness 90 μm and 65 μm and biodegradable film (Mater-Bi, Novamont, Italy) of 25 μm thick were able to prolong the storage life of apricot fruits up to 21 days (Peano et al. 2014). MAP in combination with pretreatments and cold temperature forms an excellent synergy to extend the storage life of apricot fruits. The apricot (cv. Shahroudi) pretreated with salicylic acid (0.1 mM), packed in polyethylene trays wrapped with cellophane film (20 μm), were able to have an extend storage life up to 19 days at 0.5 °C (80% RH) with good sensory quality (Moradinezhad and Jahani 2016). The dried apricots are pretreated with sulphur-dioxide before storage to retain the quality for a longer period. These can be avoided without compromising on quality by using MAP. The sun-dried apricots (cv. 'Hacihaliloglu') without sulphur-dioxide treatment can be stored at 5 °C for 30 weeks when the packaging was done using polyethylene trays laminated with polyvinyl chloride infused with a gas composition of 10% CO_2, 2% O_2 and 1% N (Elmaci et al. 2008).

11.2.3.3 Storage

Ambient Storage

The apricot's perishability in the air at near-optimum temperature and RH was found to be very high with less than 2 weeks of potential storage life (Kader 1993). The shelf life of apricot at ambient conditions is not more than 3–5 days and at cold storage was 2–4 weeks depending on variety (Ishaq et al. 2009; Ghasemnezhad and Shiri 2010). At 20 °C and 60% RH, the shelf life may be only for 1–2 days (Mercantilia 1989).

Cold Storage

- −0.5 to 0 °C and 85–95% RH: the storage life was for 1–2 weeks (Hardenburg et al. 1986).
- −1 to 0 °C and 90–95% RH: the storage life was for 1–4 weeks (Snowdon 1990).
- −0.5 to 0 °C and 90–95% RH: the storage life was for 3–4 weeks (Carlos and Kader 1999).
- −0.5 to 0 °C and 90–95% RH: the storage life was for 7–21 days (Camelo 2004).
- 'Aprikoz' variety: at 0 °C and 90% RH, the storage life was for 14 days (Koyuncu et al. 2010).
- 'Kabaasi' variety: at 4 °C, the storage life was for 7 days (Muftuoğlu et al. 2012).
- −1 to 4 °C: the storage life was for 2 weeks (Wills and Golding 2016).

Controlled Atmospheric (CA) Storage

The apricot fruits when stored in an altered gas environment (CA) for a certain period and removed to ambient temperature were edible for longer periods (El-Ramady et al. 2015). Generally, an altered atmosphere with 2–3% CO_2 and 2–3% O_2 is considered optimum for apricot storage (Muzzaffar et al. 2018). According to Kader et al. (1989), the minimum levels of CO_2 and O_2 concentration tolerated by apricots are 2% each. Deleterious effects such as flesh browning (Hardenburg et al. 1986) and loss of flavour or development of odd flavour (Carlos and Kader 1999; Folchi et al. 1995) were seen when the apricot fruits are stored in CA condition with higher CO_2 (˃5%) and lower O_2 (<1%). For short-term storage, late-harvested apricot fruits were benefited from low O_2 (1%) at 15 °C than early harvested fruits (Botondi et al. 2000). The CA requirement may vary between varieties. The apricots (cv. 'Aprikoz') were exposed to both MA and CA condition. It was noticed that in CA condition, the fruits had better external appearance and taste than the fruits in MA condition (Koyuncu et al. 2010). It was recommended that the pre-storage treatment of apricots with 20% CO_2 for only 2 days may reduce decay incidence during storage and/or transportation in CA or air (Carlos and Kader 1999) (Table 11.9).

Hypobaric Storage

When subjected to low-pressure storage of 102 mmHg, the storage life of apricot fruits was able to extend from 53 days to 90 days in cold storage. Hypobaric condition not only retarded the colour development while in storage but also retained quality of the fruits such as carotenoids, sugars and acid levels as that of cold storage (Haard and Salunkhe 1975). When apricot (cv. 'Xinjiang Kuqa') fruits are stored in decompressed storage of 50 kPa at 0 °C, there was inhibition in the increase of polygalacturonase (PG), pectinmethylesterase (PE) and cellulose activity of the stored fruits; thereby, fruits maintained higher firmness and extended storage life (Wang et al. 2012).

11.2.4 Sweet Cherry (*Prunus avium*, Rosaceae)

11.2.4.1 General Packaging

Cherries are very sensitive to bruising, and extra care should be taken for proper handling while packaging, storing and transporting. Generally, sweet cherries are packed in meshes and transparent plastic bags or boxes and clamshells with proper ventilation. The unit size may vary from country to country, for example, 5.3–9 kg

Table 11.9 Recommended CA storage conditions for apricot

Temperature	CO_2	O_2	Storage period	References
(cv. Royal) 0 °C	2.5–3%	2–3%	–	Hardenburg et al. (1986)
−0.5 °C	1.5%	1.5%	–	Truter et al. (1994)
0–5 °C	2–3%	2–3%	–	Bishop (1996)
−0.5 °C	1.5%	1.5%	2 weeks	Vander Merwe (1996)

in the USA and 2–5 kg in Italy or 500 g trays to 250–500 g plastic bags. Before storage and shipping, the cartons are labelled, palletized and shrink-wrapped.

11.2.4.2 Modified Atmospheric Packaging (MAP)

MAP was proven to be an effective way of maintaining the quality of sweet cherries during storage and marketing. Both active and passive methods of MAP are extensively used to extended the storage life and maintain the quality of cherries. An equilibrium-modified atmosphere is attained in the package by selective permeability of package films to gases and respiration of the fruits (Zoffoli and Rodriguez 2009). Micro-perforated films performed better than the traditional perforated containers in maintaining the quality of the sweet cherries (Chiabrando et al. 2019). Even biodegradable trays and films were found to be suitable for extended storage of sweet cherries (Giacalone and Chiabrando 2013). Antimicrobial films were developed by using two layers of LDPE, where sodium metabisulphite ($Na_2S_2O_5$) and calcium sulphite ($CaSO_3$) were sandwiched between the two layers of LDPE. This antimicrobial film released SO_2, thereby inhibiting the fungal pathogens and increasing the storage life up to 30 days at 0 °C (Yu et al. 2012).

The quality parameters such as colour and firmness of the fruit, greenness of the stem, sugars and acidity of the cherries were well maintained when MA packaged and stored at low temperature (Kahlke et al. 2009; Khorshidi et al. 2011). The colour of the cherries was well maintained due to the increase in anthocyanin content during storage period (Conte et al. 2009). Relative humidity (90–95%) plays a key factor for maintaining the greenness of the stem in sweet cherries (Alonso and Alique 2006). In case when wet cherries are freshly MA packed which may lead to accumulation of free water and condensation in a package which might risk fruit splitting or decay, it is recommended to insert a dry paper liner into the MAP to control in-package RH (Wills and Golding 2016). The change of gas composition inside the MAP with higher CO_2, lower O_2 and low temperature might have reduced the rate of respiration leading to the lowered metabolic activity of the fruit, thereby maintaining the quality of the sweet cherries. But, there was off-flavour development due to the production of ethanol and acetaldehyde accumulation when the gas composition was abnormally altered, i.e. low O_2 concentration of $\leq 1.5\%$ and high concentration of CO_2 of $\geq 12\%$ (Goliáš et al. 2007). The use of gas composition (CO_2 and O_2) in the MAP varies from variety to variety (Table 11.10).

11.2.4.3 Storage

Refrigerated Storage

The recommended refrigerated/cold storage of sweet cherries is −1 to 0.5 °C (Camelo 2004) where the cherries can be stored for 2–4 weeks under 95% RH (Gross et al. 2016). Generally, sweet cherries don't have extended shelf life and are prone to colour loss, firmness loss, browning of the stem, fruit pitting, bruise and decay at ambient temperature. At retail conditions, the sweet cherries are recommended to be stored at 4 °C (Alique et al. 2003).

Table 11.10 Recommended modified atmospheric packaging (MAP) of different varieties of sweet cherry

Variety	Storage temp.	Packaging material	Thickness of the film/model	Gas composition		Storage period	References
				CO_2	O_2		
Ferrovia	6 °C	Life + PET containers sealed with perforated Topaz 421 film	21 μm	8 kPa	14 kPa	15 days	Chiabrando et al. (2019)
	0 °C	Oriented polypropylene bags	20 μm	4%	10%	28 days	Conte et al. (2009)
	0.5 °C	Plastic bags	–	10%	10%	30 days	Massignan et al. (2006)
0900 Ziraat	0 °C	Parka (1% biofilm) sealed on Xtend® trays	815-CH 97/a	–	–	21 days	Aglar et al. (2017)
	0 °C	Xtend bags	CH-49	16.3%	3.6%	21 days	Küçükbasmac et al. (2008)
Skeena	0 °C	Breatheway®	363–106-A	8–10%	6.5–7.5%	6 weeks	Wang et al. (2015)
Black Star	4 °C	Polylactic acid (PLA) tray thermos sealed with biodegradable film	Film: Mater-Bi 174	31.5%	0.2%	35 days	Giacalone and Chiabrando (2013)
Napoleon	0 °C	PP trays sealed with BOPP film	21 μm	5%	5%	28 days	Esturk et al. (2012)
Sweet Heart	0–1.5 °C	FF 604 (Chile), Sunvac (Australia), View Fresh (USA)	–	6–9 kPa	10–13 kPa	42 days	Zoffoli and Rodriguez (2009)
Hedelfingen	3 °C	LDPE (LifeSpan)	L-204	7–8%	4–5%	4 weeks	Padilla-Zakour et al. (2004)
	2 °C	Polyolefin	PD-900	–	–	6 weeks	Skog et al. (2002)
Navalinda	4 °C	Polypropylene punnets sealed with MAP 3.0 micro-perforated film	35 μm	3–4%	–	8 days	Alique et al. (2003)
Merton Bigarreau	0 °C	Polystyrene trays wrapped with PP film	PP90	–	–	8 weeks	Celikel et al. (2003)
Lapins	3 °C	LDPE (LifeSpan)	L-204	7–8%	4–5%	4 weeks	Padilla-Zakour et al. (2004)
	0 °C	Breatheway® (USA)	363-106-A	8–10%	6.5–7.5%	6 weeks	Wang et al. (2015)

Controlled Atmospheric (CA) Storage

The major cause of quality loss in sweet cherries is due to rotting when stored at 0 °C or on subsequent shelf life at 10 °C or 20 °C. To increase the storage life and maintain the quality of sweet cherries, several researchers have tried different combination of CO_2 and O_2 concentration using CA storage. The minimum levels of CO_2 and O_2 tolerated by sweet cherries are 15% and 2%, respectively (Kader 1989) (Table 11.11).

The effect of CA storage with high CO_2 and low O_2 concentration has proven to be beneficial for control of decay in sweet cherries. It was recommended that O_2 levels can be further reduced below 10% and/or by increasing CO_2 above 15%, thereby reducing decay along with reduced pitting, toughening and maintained titratable acidity in 'Van' and 'Sam' variety of sweet cherry (Wang and Vestrheim 2002). Under CA condition, there was inhibition of skin darkening (Luchsinger et al. 2005), reduced decay/microbial spoilage (Serradilla et al. 2012), reduced stem browning (Gasser et al. 2004), maintained vitamin C, inhibition of the enzyme activity such as peroxidase and polyphenol oxidase, reduced malondialdehyde (Tian et al. 2004) and maintained anthocyanin content (Remon et al. 2003). High CO_2 (>15%) and low O_2 (<1%) are not recommended as sweet cherry fruits suffered from off-flavour development and loss of quality (Gross et al. 2016; Goliáš et al. 2007). But, cold storage conditions with high CO_2 (20%) were recommended for short-term storage of sweet cherry without loss of fruit quality (Dziedzic and Błaszczyk 2019).

Hypo and Hyperbaric Storage

The sweet cherry cv. 'Bing' was able to be stored for 93 days under hypobaric/reduced atmosphere of 102 mmHg at 0 °C. The hypobaric condition was able to delay the chlorophyll and starch breakdown as well as the formation of carotenoids. But, the hypobaric condition was able to maintain greenness of the stem/pedicel of the fruit only up to 60 days of storage (Salunkhe and Wu 1973). Borecka (1986) reported that the sweet cherry cv. 'Emperor Francis' when stored at 2 °C with the hypobaric condition of 5 kPa was tasted best and stored well for 30 days with good quality. The sweet cherries were exposed to the short hypobaric condition of 0.50 atm for 4 h that was able to effectively control grey mould and brown rot (Romanazzi et al. 2001). Similar results were obtained when sweet cherries are treated with 1% chitosan and 0.50 atm and stored for 14 days at 0 °C (Romanazzi et al. 2003). This combination was able to control the decay caused due to *Alternaria*, *Rhizopus* and green rot.

Table 11.11 CA storage recommendation for some varieties of sweet cherry

Variety	Storage temp.	Gas composition		Storage period	References
		CO_2	O_2		
0900 Ziraat	0 °C	25%	5%	60 days	Akbudak et al. (2008)
Bing	1 °C	10%	2%	21 days	Luchsinger et al. (2005)
Lapis	1 °C	10%	5%	40 days	Tian et al. (2004)
Sweet Heart	1 °C	2%	5%	6 weeks	Remon et al. (2003)

The hyperbaric condition was evaluated on sweet cherries by Romanazzi et al. (2008). The sweet cherry cv. 'Ferrovia' was pretreated for 24 h under hyper pressure of 1140 (approx. 1.5 atm) and later stored at 0 °C for 14 days followed by 7 days at 20 °C. The short hyperbaric treatment was able to reduce the incidence of grey mould, blue rot and brown rot effectively when compared to the fruits stored at 1 atm for 4 h.

11.2.5 Litchi (*Litchi chinensis*, Sapindaceae)

11.2.5.1 General Packaging

Being a non-climacteric fruit, the harvest of litchi commences once it's fully mature and ripe, which in turn reduces the marketability period of this juicy aril. Utmost care should be taken while harvesting and after harvesting, as fruits start browning once it is harvested or separated from the mother plant. The browning in litchi is mainly attributed to the activities of polyphenol oxidase and anthocyanin degradation by desiccation. Litchi is a subtropical fruit whose harvest coincides with summer that aggravates the short shelf life of this miracle fruit (Purbey et al. 2019). So to ensure the fruits reach its ultimate consumer in its best-intended quality, proper storage and packaging are inevitable. It is advocated that harvesting of the fruits early in the morning and carrying it to packhouse within 2 h can assure the best quality. As a first step towards the postharvest operation, fruits are pre-cooled to 10 °C to remove the excess field and respiratory heat from the produce.

The litchi fruits are packed for various utilities intended to be sold in the local market and long-distance and overseas market. So the selection of material for packaging also differs based on utility. For the local or domestic markets nearby, litchi fruits are packed in either bamboo baskets or wooden crates lined with leaves or papers. To maintain the freshness of produce, immense care should be given while packing for the distant and overseas market; a good CFB box for distant transit should be light in weight, shallow and rigid enough to protect the fruits from any kind of pressure and should have good ventilation. The wholesale package box size varies between 2–2.5 kg and 5 kg. The litchis of uniform size and of the same variety and grade should be packed. The packaging should be done in such a way that the visible part in the package must represent the entire contents of the packed fruits. Any unpleasant and foreign matters with odours are not desirable. The consumer packages are normally packed in market-preferred consumer punnets of variable sizes (Purbey et al. 2019). Recently, Israel's agriculture department has developed a unique packaging method called Extend Cast™ for retail packaging which promises to keep fruit fresh with attractive red colour and obviates the use of sulphur dioxide (Table 11.12).

11.2.5.2 Modified Atmospheric Packaging (MAP)

The litchi fruits retained good quality up to 12 days when packed using MAP with a storage temperature of 8 ± 2 °C with 85% RH (Gosh et al. 1998). When compared between polyethylene (PE) and polyvinyl chloride (PVC) film for MA packaging of

Table 11.12 Litchi packaging dimension for CFB (Purbey et al. 2019)

Capacity	Dimension (cm)	Number of ply
Consumer pack (2 kg)	26 × 18 × 13	3
Consumer pack (4 kg)	40 × 18 × 20	3
Cold chain transport package	40 × 30 × 15	3

litchi fruits, the PVC films have given best results as it effectively reduced the weight loss without condensation of water droplets inside the package. PVC film package also reduced the pericarp browning when stored at 5 °C with an extended storage life up to 36 days (Fontes et al. 1999). The litchi fruits are stored better when polyethylene film (150 gauge) package was incorporated with sulphur dioxide pads. The fruits showed no discolouration and were stored up to 2 weeks at 2 °C (Ragnoi 1989). Under retail marketing conditions, the litchi fruits harvested in the early morning (6 AM), kept in perforated polymer bags and packed in cardboard fibre boxes had maintained the better quality with reduced browning and had maximum anthocyanin content. The fruits were marketable up to 5 days at 32 °C with 49% RH (Purbey et al. 2018). Pre-cooling in combination with MA packaging of litchi fruits had reduced pericarp browning and extended storage life. It is recommended to hydrocool (pre-cooling) the destalked litchi fruits at 0–1 °C for 12–15 min and pack it in LDPE plastic bags (30 μm) followed by cold storage at 5 °C (Munbodh 2015). This can extend the storage life of litchi cv. 'Taiso' up to 3 weeks. The minimally processed litchi cv. 'Mauritius' packed inside non-perforated clamshell trays performed better without any decay incidence and were stored up to 9 days with good quality (Mphahlele et al. 2020).

11.2.5.3 Storage

Refrigerated Storage

As the litchi fruits can't be kept for a longer time under ambient condition especially in summers of tropical and subtropical condition, proper storage at optimum temperature is inevitable. Fruits can be stored for about 30 days if stored under 2–4 °C with 85–90% RH. For short-term storage of less than 2 weeks, a temperature of 7 °C is satisfactory with 90–95% RH (Molla et al. 2017).

Controlled Atmospheric (CA) Storage

The CO_2 and O_2 threshold limits which may cause injury to the litchi fruits are >15% and <1%, respectively (Gross et al. 2016). CA storage containing 3–5% O_2 and 5% CO_2 at 5–7 °C is recommended (Kader 1998). Off-flavour development was noticed when the CO_2 concentration is between 10 and 15%. Extended storage life of 22 days was seen when litchi cv. 'Mauritius' was stored under the CA condition of 5% CO_2 and 3% O_2 (Vilasachandran et al. 1997). Relative humidity of 90% and CA storage with the gas composition of 3–5% CO_2 and 3–5% O_2 at 1 °C were able to lower the losses of total anthocyanin content and reduce the activity of polyphenol oxidase, thereby maintaining quality and controlled pericarp browning of the fruits (Jiang and Fu 1999). Mahajan and Goswami (2004) recommended the CA storage with 3.5% CO_2 and 3.5% O_2 at 2 °C with 92–95% RH for litchi cv. 'Bombay' up to 56 days with good quality. The negative changes in litchi cv. 'Gola', i.e. weight loss, pericarp browning, membrane leakage and malondialdehyde content, were well controlled when the fruits are stored under CA condition of 1% O_2 and 5% CO_2 at 5 °C with a storage life of 35 days (Ali et al. 2016).

11.2.6 Almond (*Prunus amygdalus*, Rosaceae)

11.2.6.1 General Packaging

Almonds should be harvested once the fruits start to dehisce (splitting of the hull) so as to avoid fungal and pest infestation especially navel orange worm. Mechanical harvesting is followed in nuts, and once it's dried in the orchard (ranges from few days to 2 weeks under the sun), almonds must be picked to ensure the best-quality nuts without any concealed damage. The concealed damage is detrimental as it causes discolouration of nuts and, in severe case, it imparts unpalatable flavour. So it is important not to expose nuts to high moisture and temperature. In order to avoid the aflatoxin contamination by *Aspergillus flavus*, care should be taken during and after harvest, and a thorough examination of nuts is recommended during the sorting operation.

The packaging of almond requires a careful and thorough understanding of crop characteristics like high protein and fat content. For the overseas market, in-shell almonds are packed in sacks, and the naturally shelled almonds are packed in fibre bulk bins or cartons, depending on the product (in-shell or shelled) and volume. The cut almonds are highly prone to damage by oxygen, moisture and light, so extreme care should be taken by packaging in cartons with a plastic liner in order to prevent the moisture entry/migration (Ziaolhagh 2013). Roasted almonds must be protected from oxygen by either vacuum packaging or packing with inert gases like nitrogen. Vacuum-packed bag foil is typically used for roasted almonds processed in developed countries and shipped in 25 pound (11.3 kg) cartons.

11.2.6.2 Modified Atmospheric Packaging (MAP)

To ensure quality maintenance of peeled almonds over a longer time, i.e. up to 546 days, it is recommended to use metallized film (PET + PEBD + AI + Surlyn) package infused with nitrogen (150 mmHg) and refrigerated at 4 °C (Senesi et al. 1991). The multilaminate film made of polyamide (PA) + polyethylene (PE) was able to store almond kernels up to 20 months under vacuum at ambient temperature (Ziaolhagh 2013). Fresh green almonds (kernel fully developed) cv. 'Mazetta' are stored up to 120 days when they were MA packed in polyethylene film bags (ZOEPAC) and stored at 2 °C (Ben Abda et al. 2018).

11.2.6.3 Storage

Low-Temperature Storage

After processing and drying, nuts are stored in bins at a temperature of 10 °C and 65–70% RH. The moisture plays an important role in deciding the shelf life of nuts; almond contains about 10–15% moisture on a fresh weight basis; and for improvement in the stability and good shelf life, it should be reduced to 5–8% (Wu et al. 2019). For long-term storage, cool temperature of less than 0 °C is optimal, but a higher temperature may not stimulate insect activity and may work well to control moisture migration (and also minimize lipid oxidation). In general, almonds are shelf-stable and can be stored up to 2 years if the environmental conditions are conducive. But in the case of tropical and subtropical conditions, where the environmental conditions are abusive, the shelf life is limited. At 0 °C, the in-shell almonds can be stored up to 20 months, while at 10 °C, they can be stored up to 16 months and only for 8 months at 20 °C (Gross et al. 2016). Low-temperature storage is better for shelled almonds and can retain quality up to 6 months. Care should be taken not to mix shelled almonds with other commodities in storage as they can readily absorb odours due to their high lipid content.

Controlled Atmospheric (CA) Storage

Almonds can tolerate more than 12% of the CO_2 without affecting its quality (Gross et al. 2016). In-shell almonds were found to be more stable than the kernels alone. The fresh almond flavour in the kernels can be maintained by elevated CO_2 and low O_2 atmosphere. CA atmosphere of 1% O_2 and 9–9.5% CO_2 was able to control the insect infestation along with maintaining the flavour of the kernels (Guadagni et al. 1978). The O_2 levels as low as 0.5% were able to remove the stability difference between in-shell nuts and shelled kernels (Kader 1996).

11.2.7 Dates (*Phoenix dactylifera*, Arecaceae)

11.2.7.1 General Packaging

The harvesting of dates depends on its stage like khalal, rutab and tamar and market demand or consumer preference. The khalal stage is considered as fresh fruit stage, and it's the first to harvest among the three. The varieties with low tannin are preferred at this stage as high tannin and varieties which are astringent are not fit for consumption. The second stage, rutab, is considered as the favourite and most delicate stage of date. The high-perishability nature of this stage makes it difficult to handle and transport despite high market demand. Owing to the high-perishability nature of the dates (rutab), utmost care should be taken while packing. The tamar stage is considered as non-perishable because of its low moisture content, and it is ideally marketed as dried dates in the market (Siddiq and Greiby 2014). Most of the date varieties are harvested at tamar stage. The packhouse operations such as fumigation, washing, storage, refrigeration, hydration and dehydration are carried out to ensure the availability of best-quality fruits to the ultimate consumers. The material changes accordingly with retail and bulk packing.

For the bulk transport, the fruits are packed in 5 kg or 15 lbs. corrugated fibreboard boxes, and these fruits are either marketed in open markets or for customized purchase in case of bulk packing. Dry dates are packed in large reinforced cartons especially for export purpose. The fruit may be handed over to be repacked in the countries where it is to be marketed and where retail packing will be carried out according to the customer requirements. For the package of rutab-stage dates, use of plastic liner in the box helps in minimizing the water loss. The consumer package or retail package has got great importance in recent years as they have to be adopted and styled according to consumer demand. These consumer packages may include transparent film bags or trays overwrapped with films of various sizes and shapes. The package with resealing property is of great demand in market, and dates are also packed in transparent plastic tubs. The information for the client is usually on the lid. Round fibreboard cans having metallic top and bottom (500–1000 g), rigid transparent plastic containers (200–300 g) and plastic bags (50–60 g) usually made of PET polyethylene are common packaging material for dates (Gross et al. 2016).

11.2.7.2 Modified Atmospheric Packaging (MAP)

'Barhee' dates harvested at khalal stage were packed in polyethylene bags with active modified atmosphere of 5% CO_2 and stored at 4 °C. The dates showed acceptable results in quality and extended the storage life up to 28 days, whereas CO_2 concentration of 30% had deleterious quality changes on dates harvested at khalal stage (Mortazavi et al. 2010). Date cv. 'Khalas' was harvested at khalal stage, packed in non-perforated PE sealed onto plastic tray with active MA and stored at 0 °C for 27 days. The MA packaging with CO_2 concentration of 10–20% and O_2 concentration of 17–20.8% was able to reduce the ripening (dark red colour), retain firmness, maintain yellow colour and have higher sensory score (Al-Eid et al. 2012). At 20 °C, MAP was more effective than the vacuum packaging. The date cv. 'Deglet Noor' harvested at tamar stage was packed in BPP film bags (30 μm) with active CO_2 gas concentration of 60 kPa and stored at 20 °C up to 30 days without any adverse changes in physical and chemical attributes (Jemni et al. 2016). The active MAP condition of 5% O_2, 20% CO_2 and 75% N_2 was able to reduce the weight loss (0.45%) of date cv. 'Barhi' when packed in three-layered LDPE (80 μm) and stored at 5 °C. But, when date fruits were passively MA packed, they had higher firmness, colour and sensory attributes than the active MA packaging. In passive MAP, the fruit weight loss was 3.2–3.4% and stored up to 30 days (Al-Kaabi 2020).

11.2.7.3 Storage

Low-Temperature Storage

Khalal and rutab-stage dates should be stored at 0 °C at 85–95% RH, whereas tamar-stage dates are stored at 0 °C with only 65–75% RH (Aleid and Kader 2013). At 0 °C, dates can be retained in storage without any quality losses up to 1 year. With the increase in temperature and moisture, the physiological and pathological deterioration also increases in dates. But, dates are least affected by chilling and freezing injury owing to its high sugar content. The longer shelf life can be achieved by storing dates at low temperature. Dates cv. 'Deglet Noor' (soft and firm) can be kept for more than 1 year at −17.5 °C while withstanding for 1 month at 27 °C, 8 months at 5 °C and 3 months at 15 °C (Rygg et al. 1953), whereas partially dried and dry dates can be stored up to 1 year at 0 °C and 20 °C, respectively, without significant quality losses at 70–75% RH (Gross et al. 2016).

Controlled Atmospheric (CA) Storage

A plastic chamber holding controlled atmosphere of 60–80% CO_2 and 4–7% O_2 at ambient temperature had an equal effect in controlling pest (*nitidulid* beetles) and maintained the quality of dates as that stored at −18 °C up to 4.5 months (Navarro et al. 2001). Dates cv. 'Barhi' was CA stored at 0 °C with 20% CO_2. The dates were well stored up to 26 weeks with maintained fruit colour, firmness and tannins. There was also reduced degradation of caffeoylshikimic acid (CSA, a major phenolic compound) (Al-Redhaiman 2005).

11.3 Conclusion

Packaging and storage play a key role in enhancing the storage life and quality of stone fruits. Due to the fast perishability of the fruits, packaging and storage become a vital component in the supply chain for longer availability and distribution of the fruits. The advancement in science and technology has put forth various improved packaging and storage technologies for different stakeholders across the supply chain. Here, we have tried to pool up the scientific information and recommendations on packaging and storage of stone fruits spread across the world. The chapter gives insight into the successful utilization of the information for extending the quality and availability of stone fruits using packaging and storage. The best suitable package and storage recommendations for different stone fruits and their varieties/cultivars discussed go a long way for planning further scientific studies, market strategies, reducing losses and wastage in varied segments of the supply chain.

References

Aglar, E., Ozturk, B., Guler, S. K., Karakaya, O., Uzun, S., & Saracoglu, O. (2017). Effect of modified atmosphere packaging and 'Parka' treatments on fruit quality characteristics of sweet cherry fruits (*Prunus avium* L.'0900 Ziraat') during cold storage and shelf life. *Scientia Horticulturae, 222*, 162–168.

Ahmad, M. S., & Siddiqui, M. W. (2015). *Postharvest quality assurance of fruits*. Cham: Springer.

Akbudak, B., & Eris, A. (2003). Effect of some postharvest treatments and controlled atmosphere (ca) storage on basic quality criteria of peaches and nectarines. *Acta Horticulturae, 599*, 297–303.

Akbudak, B., Tezcan, H., & Eris, A. (2008). Determination of controlled atmosphere storage conditions for '0900 Ziraat' sweet cherry fruit. *Acta Horticulturae, 795*, 855–859.

Aleid, S. M., & Kader, A. A. (2013). *Dates: Postharvest science, processing technology and health benefits*. Hoboken: John Wiley & Sons.

Al-Eid, S. M., Barber, A. R., Rettke, M., Leo, A., Alsenaien, W. A., & Sallam, A. A. (2012). Utilisation of modified atmosphere packaging to extend the shelf life of Khalas fresh dates. *International Journal of Food Science & Technology, 47*(7), 1518–1525.

Ali, S., Khan, A. S., Malik, A. U., & Shahid, M. (2016). Effect of controlled atmosphere storage on pericarp browning, bioactive compounds and antioxidant enzymes of litchi fruits. *Food Chemistry, 206*, 18–29.

Alique, R., Martínez, M. A., & Alonso, J. (2003). Influence of the modified atmosphere packaging on shelf life and quality of Navalinda sweet cherry. *European Food Research and Technology, 217*(5), 416–420.

Al-Kaabi, H. J. (2020). Effect of modified atmosphere packaging on quality of Barhi dates at Khalal stage. *International Journal of Food Studies, 9*, 125–134.

Alonso, J., & Alique, R. (2006). Sweet cherries. In Y. H. Hui, J. Barta, M. P. Cano, T. W. Gusek, J. S. Sidhu, & N. K. Sinha (Eds.), *Handbook of fruits and fruit processing* (pp. 359–367). Oxford: Blackwell Publishing.

Al-Redhaiman, K. N. (2005). Chemical changes during storage of 'Barhi' dates under controlled atmosphere conditions. *HortScience, 40*(5), 1413–1415.

Apelbaum, A., Zauberman, G., & Fuchs, Y. (1977). Sub atmospheric pressure storage of mango fruits. *Scientia Horticulturae, 7*(2), 153–160.

Ayhan, Z., Eştürk, O., & Müftüoğlu, F. (2010). Effects of coating, modified atmosphere (MA) and plastic film on the physical and sensory properties of apricot. *Acta Horticulturae, 876*, 143–150.

Ben Abda, J., Ben Amara, M., Bahri, N., & Abdelli, S. (2018). Effect of storage temperature and modified atmosphere packaging on quality of fresh green 'Mazetta' almonds. *Acta Horticulturae, 1194*, 921–926.

Bender, R. J., Brecht, J. K., Sargent, S. A., & Huber, D. J. (2000). Mango tolerance to reduced oxygen levels in controlled atmosphere storage. *Journal of the American Society for Horticultural Science, 125*, 707–713.

Ben-Yehoshua, S., Beaudry, R. M., Fishman, S., Jayanty, S., & Mir, N. (2005). Modified atmosphere packaging and controlled atmosphere storage. In S. Ben-Yehoshua (Ed.), *Environmentally friendly technologies for agricultural produce quality*. Boca Raton, FL: Taylor and Francis.

Bishop, D. (1996). Controlled atmosphere storage. In C. J. V. Dellino (Ed.), *Cold and chilled storage technology*. London: Blackie.

Bleinroth, E. W., Garcia, J. L. M., & Yokomizo, Y. (1977). Conseracao de quatro variedades de manga pelo frio e em atmosfera controlada. *Coletanea de Instituto de Tecnologia de Alimentos, 8*, 217–243.

Borecka, H. (1986). The quality of sweet cherries stored under hypobaric conditions. *Acta Agrobotanica, 39*(2), 195–200.

Botondi, R., Crisa, A., Massantini, R., & Mencarelli, F. (2000). Effects of low oxygen short-term exposure at 15 C on postharvest physiology and quality of apricots harvested at two ripening stages. *The Journal of Horticultural Science and Biotechnology, 75*(2), 202–208.

Brecht, J. K., Sargent, S. A., Kader, A. A., Mitcham, E. J., Maul, F., Brecht, P. E., & Menocal, O. (2010). *Mango postharvest best management practices manual* (p. 78). Gainesville: University of Florida, Horticultural Sciences Department.

Burden, J. N. (1997). Postharvest handling for export. In S. Mitra (Ed.), *Postharvest physiology and storage of tropical and subtropical fruits* (pp. 1–20). Oxford: CAB International.

Burg, S. P. (1975). Hypobaric storage and transportation of fresh fruits and vegetables. In N. F. Haard & D. K. Salunke (Eds.), *Postharvest biology and handling of fruits and vegetables* (pp. 172–188). Westport, CT: AVI.

Burg, S. P. (2004). *Postharvest physiology and hypobaric storage of fresh produce*. Wallingford: CAB International.

Camelo, A. F. L. (2004). *Manual for the preparation and sale of fruits and vegetables: From field to market* (Vol. 151). FAO: Rome.

Cantillano, F. F., Treptow, R. D. O., & Schünemann, A. P. (2010). Controlled atmosphere storage of peaches cultivar 'Eldorado' grown in conventional and organic system. *Acta Horticulturae, 872*, 365–370.

Carlos, H. C., & Kader, A. A. (1999). *Apricots postharvest quality maintenance guidelines*. Davis: Department of Pomology, University of California, Davis.

Castellanos, D. A., & Herrera, A. O. (2017). Modified atmosphere packaging: Design and optimization strategies for fresh produce. *Postharvest Handling, 85*.

Celikel, F. G., Özelkök, S., & Burak, M. (2003). A study on modified atmosphere storage of sweet cherry. *Acta Horticulturae, 628*, 431–438.

Chen, R. C. (1987). *Effect of precooling and waxing treatment on mango under hypobaric storage*. Unpublished master's thesis, AIT, Bangkok, Thailand.

Chiabrando, V., Garavaglia, L., & Giacalone, G. (2019). The postharvest quality of fresh sweet cherries and strawberries with an active packaging system. *Food, 8*(8), 335.

Conte, A., Scrocco, C., Lecce, L., Mastromatteo, M., & Del Nobile, M. A. (2009). Ready-to-eat sweet cherries: Study on different packaging systems. *Innovative Food Science and Emerging Technologies, 10*(4), 564–571.

Crisosto, C. H., Mitchell, F. G., & Ju, Z. (1999). Susceptibility to chilling injury of peach, nectarine, and plum cultivars grown in California. *HortScience, 34*, 1116–1118.

Davenport, T. L., Burg, S. P., & White, T. L. (2006). Optimal low-pressure conditions for long-term storage of fresh commodities kill Caribbean fruit fly eggs and larvae. *HortTechnology, 16*, 98–104.

Defraeye, T., Cronjé, P., Berry, T., Opara, U. L., East, A., Hertog, M., & Nicolai, B. (2015). Towards integrated performance evaluation of future packaging for fresh produce in the cold chain. *Trends in Food Science & Technology, 44*(2), 201–225.

Dziedzic, E., & Błaszczyk, J. (2019). Evaluation of sweet cherry fruit quality after short-term storage in relation to the rootstock. *Horticulture, Environment, and Biotechnology, 60*(6), 925–934.

Elansari, A. M., Fenton, D. L., & Callahan, C. W. (2019). *Precooling*. In Postharvest technology of perishable horticultural commodities (pp. 161–207). Cambridge: Woodhead Publishing.

Elmaci, Y., Altug, T., & Pazir, F. (2008). Quality changes in unsulfured sun dried apricots during storage. *International Journal of Food Properties, 11*(1), 146–157.

El-Ramady, H. R., Domokos-Szabolcsy, É., Abdalla, N. A., Taha, H. S., & Fári, M. (2015). *Postharvest management of fruits and vegetables storage*. In Sustainable agriculture reviews (pp. 65–152). Cham: Springer.

Erkan, M., & ESKİ, H. (2012). Combined treatment of modified atmosphere packaging and 1-methylcyclopropene improves postharvest quality of Japanese plums. *Turkish Journal of Agriculture and Forestry, 36*(5), 563–575.

Esturk, O., Ayhan, Z., & Ustunel, M. A. (2012). Modified atmosphere packaging of "Napoleon" cherry: Effect of packaging material and storage time on physical, chemical, and sensory quality. *Food and Bioprocess Technology, 5*(4), 1295–1304.

FAO. (2014). *Safe food: Global initiative on food loss and waste reduction*. Italy: Rome. www.fao.org/safe-food.

Folchi, A., Pratella, G. C., Tian, S. P., & Bertolini, P. (1995). Effect of low O_2 stress in apricot at different temperatures. *Italian Journal of Food Science, 7*, 245–253.

Fontes, V. L., De Moura, M. A., Vieira, G., & Finger, F. L. (1999). Influence of plastic films and temperature on postharvest pericarp browning in litchi (*Litchi chinensis*). *Revista Brasileira de Armazenamento, 241*, 56–59.

Gasser, F., Mattle, S., & Hohn, E. (2004). Cherry: Storage trials 2003. *Obst und Weinbau, 140*, 6–10.

Ghaani, M., Cozzolino, C. A., Castelli, G., & Farris, S. (2016). An overview of the intelligent packaging technologies in the food sector. *Trends in Food Science & Technology, 51*, 1–11.

Ghasemnezhad, M., & Shiri, M. A. (2010). Effect of chitosan coatings on some quality indices of apricot (*Prunus armeniaca* L.) during cold storage. *Caspian Journal of Environmental Sciences, 8*(1), 25–33.

Giacalone, G., & Chiabrando, V. (2013). Modified atmosphere packaging of sweet cherries with biodegradable films. *International Food Research Journal, 20*(3), 1263–1268.

Gill, P. P. S., Jawandha, S. K., Kaur, N., & Singh, N. (2017). Physico-chemical changes during progressive ripening of mango (*Mangifera indica* L.) cv. Dashehari under different temperature regimes. *Journal of Food Science and Technology, 54*(7), 1964–1970.

Giuggioli, N. R., Sottile, F., & Peano, C. (2016). Quality indicators for modified atmosphere packaging (MAP) storage of high-quality European plum (*Prunus domestica* L.) cultivars. *Italian Journal of Food Science, 28*(3), 376–390.

Goliáš, J., Němcová, A., Čaněk, A., & Kolenčíková, D. (2007). Storage of sweet cherries in low oxygen and high carbon dioxide atmospheres. *Horticultural Science, 34*, 26–34.

Gosh, U., Bhatacharjee, A., Bose, P. K., Choudhuri, D. R., & Gangopadhyah, H. (1998). Physicochemical changes of litchi in modified atmospheric storage. *Indian Journal of Chemical Technology, 5*, 393–396.

Gross, K. C., Wang, C. Y., & Saltveit, M. E. (Eds.). (2016). *The commercial storage of fruits, vegetables, and florist and nursery stocks*. Washington, DC: United States Department of Agriculture, Agricultural Research Service.

Guadagni, D. G., Soderstrom, E. L., & Storey, L. (1978). Effect of controlled atmosphere on flavor stability of almonds. *Journal of Food Science, 43*, 1077–1080.

Haard, N. F., & Salunkhe, D. K. (1975). *Symposium: Postharvest biology and handling of fruits and vegetables (pp. 193)*. Westport, CT: AVI Publishing.

Hardenburg, R. E., Watada, A. E., & Wang, C. Y. (1986). The commercial storage of fruits, vegetables and florist and nursery stocks. *United States Department of Agriculture, Agricultural Research Service, Agriculture Handbook, 66*, 130.

Hatton, T. T., & Reeder, W. F. (1966). Controlled atmosphere storage of 'Keitt' mangoes. *Proceedings of the Caribbean Region of the American Society for Horticultural Science, 10*, 114–119.

Ilangantileke, S., & Salocke, V. (1989). *Low pressure storage of Thai mango*. Paper presented in 5th CA research conference, Wenatchee, WA (pp. 103–117).

Ishaq, S., Rathore, H. A., Masud, T., & Ali, S. (2009). Influence of postharvest calcium chloride application, ethylene absorbent and modified atmosphere on quality characteristics and shelf life of apricot (*Prunus armeniaca* L.) fruit during storage. *Pakistan Journal of Nutrition, 8*(6), 861–865.

Ivonkovic, A., Zeljko, K., Talic, S., & Lasic, M. (2017). Biodegradable packaging in the food industry. *Journal of Food Safety and Food Quality, 68*, 26–38.

Jackman, R. L., Yada, R. Y., Marangoni, A., Parkin, K. L., & Stanley, D. W. (1988). Chilling injury. A review of quality aspects. *Journal of Food Quality, 11*(4), 253–278.

Jemni, M., Chniti, F., Harbaoui, K., Ferchichi, A., & Artés, F. (2016). Partial vacuum and active modified atmosphere packaging for keeping overall quality of dates. *Journal of New Sciences Agriculture & Biotechnology, 29*(1), 1656–1665.

Jiang, Y. M., & Fu, J. R. (1999). Postharvest browning of litchi fruit by water loss and its prevention by controlled atmosphere storage at high relative humidity. *LWT-Food Science and Technology, 32*(5), 278–283.

Johnson, G. I., Sangchote, S., & Cooke, A. W. (1990). Control of stem end rot *Dothiorella dominicana* and other postharvest diseases of mangoes cultivar Kensington Pride during short and long term storage. *Tropical Agriculture, 67*, 183–187.

Kader, A. A. (2002). *Postharvest technology of horticultural crops* (3rd ed.). Oakland, California, Publication no. 3311.

Kader, A. A. (1986). Biochemical and physiological basis for effects on controlled and modified atmospheres on fruits and vegetables. *Food Technology, 40*, 99–104.

Kader, A.A. (1989). *A summary of CA requirements and recommendations for fruit other than pome fruits*. In International controlled atmosphere conference fifth, proceedings, Wenatchee, Washington. Volume 2. Other commodities and storage recommendations (pp. 303–328).

Kader, A. A. (1993). Postharvest handling. In J. E. Preece & P. E. Read (Eds.), *The biology of horticulture—An introductory textbook* (pp. 353–377). New York: Wiley.

Kader, A. A. (1996). In-plant storage. In W. C. Micke (Ed.), *Almond production manual* (pp. 274–277). Davis, CA: Division of Agriculture and Natural Resources, University of California. Pub. no. 3364.

Kader, A. A. (1998). *Produce facts-lychee. Perishables handling quarterly, no. 94*. Davis, CA: University of California.

Kader, A. A., Zagory, D., Kerbel, E. L., & Wang, C. Y. (1989). Modified atmosphere packaging of fruits and vegetables. *Critical Reviews in Food Science & Nutrition, 28*(1), 1–30.

Kahlke, C. J., Padilla-Zakour, O. I., Cooley, H. J., & Robinson, R. L. (2009). Shelf-life and marketing window extension in sweet cherries by the use of modified atmosphere packaging. *New York Fruit Quarterly, 17*, 21–24.

Kane, O., & Marcellin, P. (1979). Effects of controlled atmosphere on the storage of mangoes (varieties 'Amelie' and 'Julie'). *Fruits, 34*, 123–129.

Khan, M. S., Zeb, A., Rahatullah, K., Ihsanullah, I., Ahmed, N., & Ahmed, S. (2013). Storage life extension of plum fruit with different colored packaging and storage temperatures. *Journal of Environmental Science, Toxicology and Food Technology, 7*(3), 86–93.

Khorshidi, S., Davarynejad, G., Tehranifar, A., & Fallahi, E. (2011). Effect of modified atmosphere packaging on chemical composition, antioxidant activity, anthocyanin, and total phenolic content of cherry fruits. *Horticulture, Environment, and Biotechnology, 52*, 471–481.

Kim, Y., Brecht, J. K., & Talcott, S. T. (2007). Antioxidant phytochemical and fruit quality changes in mango (*Mangifera indica* L.) following hot water immersion and controlled atmosphere storage. *Food Chemistry, 105*(4), 1327–1334.

Kosto, I., Weksler, A., & Lurie, S. (2002). Modified atmosphere storage of apricots. *Alon Hanotea, 56*, 173–175.

Koyuncu, M. A., Dilmaçünal, T., & Özdemir, O. (2010). Modified and controlled atmosphere storage of apricots. *Acta Horticulturae, 876*, 55–66.

Küçükbasmac, F., Özkaya, O., Ağar, T., Saks, Y., & Tefen, S. L. (2008). Effect of retail-size modified atmosphere packaging bags on postharvest storage and shelf-life quality of '0900 Ziraat' sweet cherry. *Acta Horticulturae, 795*, 775–780.

Lalel, H. J. D., Singh, Z., & Tan, S. C. (2004). Biosynthesis of aroma volatile compounds and fatty acids in 'Kensington Pride' mangoes after storage in a controlled atmosphere storage at different oxygen and carbon dioxide concentrations. *Journal of Horticultural Science and Biotechnology, 79*, 343–353.

Lalel, H. J. D., Singh, Z., & Tan, S. C. (2005). Controlled atmosphere storage affects fruit ripening and quality of 'Delta R2E2' mango. *Journal of Horticultural Science and Biotechnology, 80*, 551–556.

Lawton, A. R. (1996). *Cargo care*. Cambridge: Cambridge Refrigeration Technology. (these recommendations were given as general guidelines and CRT accept no responsibility for their use).

Lee, D. S., Yam, K. L., & Piergiovanni, L. (2008). *Food packaging science and technology*. Boca Raton: CRC Press.

Lizana, L. A., & Ochagavia, A. (1997). Controlled atmosphere storage of mango fruits (*Mangifera indica* L.) cvs Tommy Atkins and Kent. *Acta Horticulturae, 455*, 732–737.

Luchsinger, L., Mardones, C., & Leshuk, J. (2005). Controlled atmosphere storage of 'Bing' sweet cherries. *Acta Horticulturae, 667*, 535–537.

Lurie, S., & Crisosto, C. H. (2005). Chilling injury in peach and nectarine. *Postharvest Biology and Technology, 37*(3), 195–208.

Lutz, J. M., & Hardenburg, R. E. (1968). *The commercial storage of fruits, vegetables, and florist and nursery stocks*. USDA Agric. Handbook No. 66. 94 pp.

Maekawa, T. (1990). On mango CA storage and transportation from subtropical to temperate regions in Japan. *Acta Horticulturae, 455*, 732–737.

Mahajan, B. V. C., Dhillon, W. S., Kumar, M., & Singh, B. (2015). Effect of different packaging films on shelf life and quality of peach under super and ordinary market conditions. *Journal of Food Science and Technology, 52*(6), 3756–3762.

Mahajan, P. V., & Goswami, T. K. (2004). Extended storage life of litchi fruit using controlled atmosphere and low temperature. *Journal of Food Processing and Preservation, 28*(5), 388–403.

Malakou, A., & Nanos, G. D. (2005). A combination of hot water treatment and modified atmosphere packaging maintains quality of advanced maturity 'Caldesi 2000' nectarines and 'Royal Glory' peaches. *Postharvest Biology and Technology, 38*(2), 106–114.

Manalili, N. M. (2011). *Appropriate food packaging solutions for developing countries*. Rome.

Mangaraj, S., Goswami, T. K., & Mahajan, P. V. (2009). Applications of plastic films for modified atmosphere packaging of fruits and vegetables: A review. *Food Engineering Reviews, 1*(2), 133.

Mann, S. S., & Singh, R. N. (1976). The cold storage life of 'Dashehari' mangoes. *Scientia Horticulturae, 5*(3), 249–253.

Marsh, K., & Bugusu, B. (2007). Food packaging—Roles, materials, and environmental issues. *Journal of Food Science, 72*, 39–55.

Massignan, L., Lovino, R., de Cillis, E. M., & Santomassi, F. (2006). Cold storage of cherries with a combination of modified atmosphere and controlled atmosphere. *Rivista di Frulticolture edi Ortofloricoltura, 68*, 63–66.

Medlicott, A. P., & Jeger, M. J. (1987). *The development and application of postharvest treatments to manipulate ripening of mangoes*. In R. A. Prinsley & G. Tucker (Eds.), Mangoes. A review (pp. 56–77). Common Wealth Science Council.

Mercantilia. (1989). *Guide to food transport fruit and vegetables (pp 247)*. Copenhagen: Mercantilia Publishers.

Mir, A. A., Bandral, J. D., & Sood, M. (2018). Effect of active packaging on quality and shelf life of peach fruits. *The Pharma Innovation Journal, 7*(4), 1076–1082.

Mitra, S. K., & Baldwin, E. A. (1997). Mango. In *Postharvest physiology and storage of tropical and subtropical fruits* (pp. 85–122). New York: CAB International.

Molla, M. M., Rahman, E., Khatun, A., Islam, M. F., Uddin, M. Z., Ullah, M. A., & Miaruddin, M. (2017). Color retention and extension of shelf life of litchi fruit in response to storage and packaging technique. *Technology, 12*(5), 322–331.

Moradinezhad, F., & Jahani, M. (2016). Quality improvement and shelf life extension of fresh apricot fruit (*Prunus armeniaca* cv. Shahroudi) using postharvest chemical treatments and packaging during cold storage. *International Journal of Horticultural Science and Technology, 3*(1), 9–18.

Mortazavi, S. M. H., Arzani, K., & Arujalian, A. A. (2010). Modified atmosphere packaging of date fruit (*Phoenix dactylifera* L.) Cultivar 'Barhee' in khalal stage. *Acta Horticulturae, 882*, 1063–1069.

Mphahlele, R. R., Caleb, O. J., & Ngcobo, M. E. (2020). Effects of packaging and duration on quality of minimally processed and unpitted litchi cv.'Mauritius' under low storage temperature. *Heliyon, 6*(1), e03229.

Muftuoğlu, F., Ayhan, Z., & Esturk, O. (2012). Modified atmosphere packaging of Kabaaşı apricot (*Prunus armeniaca* L. 'Kabaaşı'): Effect of atmosphere, packaging material type and coating on the physicochemical properties and sensory quality. *Food and Bioprocess Technology, 5*(5), 1601–1611.

Munbodh, R. S. (2015). Effect of hydro cooling and packaging on the shelf life of cold stored destalked litchis cultivar Taiso. *University of Mauritius Research Journal, 21*, 420–437.

Muzzaffar, S., Bhat, M. M., Wani, T. A., Wani, I. A., & Masoodi, F. A. (2018). Postharvest biology and technology of apricot. In *Postharvest biology and technology of temperate fruits* (pp. 201–222). Cham: Springer.

Nakamura, N., Rao, D. V. S., Shilina, T., & Nawa, Y. (2004). Respiration properties of tree-ripe mango under CA condition-review. *Japan Agricultural Research Quarterly, 38*, 221–226.

Narayana, C. K., Rao, D. S., & Roy, S. K. (2012). Mango production, postharvest physiology and storage. In M. Siddiq (Ed.), *Tropical and subtropical fruits: Postharvest physiology, processing and packaging* (pp. 259–276). Ames, IA: John Wiley & Sons.

Narayana, C. K. (1989). *Studies on postharvest behaviour of mango (Mangifera indica L.) cv. 'Baneshan'*. PhD thesis, Indian Agricultural Research Institute, New Delhi.

Navarro, S., Donahaye, J. E., Rindner, M., & Azrieli, A. (2001). *Storage of dates under carbon dioxide atmosphere for quality preservation*. In Proc. Int. Conf. Controlled Atmosphere and Fumigation in Stored Products, Exclusive Printing Services, Clovis, CA, USA (pp. 231–239).

Niranjana, P., Gopalakrishna, R. K. P., Sudhakar, R. D. V., & Madhusudhan, B. (2009). Effect of controlled atmosphere storage (CAS) on antioxidant enzymes and DPPH-Radical scavenging activity of mango (*Mangifera indica* L.) CV. Alphonso. *African Journal of Food, Agriculture, Nutrition and Development, 9*(2), 779–792.

Noomhorm, A., & Tiasuwan, N. (1995). Controlled atmosphere storage of mango fruit, *Mangifera indica* L. cv. 'Rad.'. *Journal of Food Processing & Preservation, 19*, 271–281.

Nunes, M. C. N., Emond, J. P., Brecht, J. K., Dea, S., & Proulx, E. (2007). Quality curves for mango fruit (cv. Tommy Atkins and Palmer) stored at chilling and non-chilling temperatures. *Journal of Food Quality, 30*(1), 104–120.

Ortiz, A., Lara, I., Graell, J., López, M. L., & Echeverría, G. (2008). Sensory acceptance of ca-stored peach fruit. Relationship to instrumental quality parameters. *Acta Horticulturae, 796*, 225–230.

Padilla-Zakour, O. I., Tandon, K. S., & Wargo, J. M. (2004). Quality of modified atmosphere packaged 'Hedelfingen' and 'Lapins' sweet cherries. *HortTechnology, 14*(3), 331–337.

Pantastico, E. B. (Ed.). (1975). *Postharvest physiology, handling and utilisation of tropical and sub tropical fruits and vegetables*. Westport, CT: AVI Publishing.

Patel, B., Tandel, Y. N., Patel, A. H., & Patel, B. L. (2016). Chilling injury in tropical and subtropical fruits: A cold storage problem and its remedies: A review. *International Journal of Science, Environment and Technology, 5*(2), 1882–1887.

Peano, C., Giuggioli, N. R., & Girgenti, V. (2014). Effects of innovative packaging materials on apricot fruits (cv Tom Cot®). *Fruits, 69*(3), 247–258.

Peano, C., Giuggioli, N. R., Girgenti, V., Palma, A., D'Aquino, S., & Sottile, F. (2017). Effect of palletized MAP storage on the quality and nutritional compounds of the Japanese Plum cv. Angeleno (*Prunus salicina* Lindl.). *Journal of Food Processing and Preservation, 41*(2), e12786.

Pesis, E., Aharoni, D., Aharon, Z., Ben-Arie, R., Aharoni, N., & Fuchs, Y. (2000). Modified atmosphere and modified humidity packaging alleviated chilling injury in mango fruit. *Postharvest Biology & Technology, 19*, 93–101.

Prado, M. P., Rodriguez-Lafuente, A., & Nerin, C. (2011). Active label-based packaging to extend the shelf-life of "Calanda" peach fruit: Changes in fruit quality and enzymatic activity. *Postharvest Biology and Technology, 60*(3), 211–219.

Purbey, S., Pongener, A., Marboh, E., & Lal, N. (2019). Advances in packaging of litchi fruit to maintain the quality. *Current Journal of Applied Science and Technology, 38*(1), 1–11.

Purbey, S. K., Pongener, A., Kumar, V., & Nath, V. (2018). Effect of time of harvest and packaging on quality and shelf life of litchi fruit. *Acta Horticulturae, 1211*, 65–70.

Raheem, D. (2013). Application of plastics and paper as food packaging materials—An overview. *Emirates Journal of Food and Agriculture, 25*, 177–188.

Rao, D. V. S., & Shivashankara, K. S. (2004). *Individual shrink wrapping of mango fruits for alleviation of chilling injury and extension of storage life*. Paper presented at 1st Indian Horticultural Congress, Pusa Campus, New Delhi.

Rao, D. V. S., Gopalakrishna, R. K. P., & Niranajana, P. (2003). *Alleviation of chilling injury of mango by CA storage and MA packing*. Paper presented at 5th international food convention, Mysore. 189 p.

Redondo, D., Arias, E., Oria, R., & Venturini, M. E. (2017). Thinned stone fruits are a source of polyphenols and antioxidant compounds. *Journal of the Science of Food and Agriculture, 97*(3), 902–910.

Remon, S., Marquina, P., Peiro, J. M., & Oria, R. (2003). Storage potential of sweetheart cherry in controlled atmospheres. *Acta Horticulturae, 600*, 763–769.

Rizvi, S. S., & Perdue, R. R. (1981). Requirements for foods packaged in polymeric films. *Critical Reviews in Food Science & Nutrition, 14*(2), 111–134.

Rodov, V., Ben-Yehoshua, S., Fishman, S., Gotlieb, S., Fierman, T., & Fang, D. Q. (1994). Reducing decay and extending shelf life of bell-peppers and mangoes by modified atmospheres packaging. *Australian Centre for International Agricultural Research Proceedings, 50*, 416–418.

Romanazzi, G., Nigro, F., & Ippolito, A. (2003). Short hypobaric treatments potentiate the effect of chitosan in reducing storage decay of sweet cherries. *Postharvest Biology and Technology, 29*(1), 73–80.

Romanazzi, G., Nigro, F., & Ippolito, A. (2008). Effectiveness of a short hyperbaric treatment to control postharvest decay of sweet cherries and table grapes. *Postharvest Biology and Technology, 49*(3), 440–442.

Romanazzi, G., Nigro, F., Ippolito, A., & Salerno, M. (2001). Effect of short hypobaric treatments on postharvest rots of sweet cherries, strawberries and table grapes. *Postharvest Biology and Technology, 22*(1), 1–6.

Roy, S. K., & Pal, R. K. (1989). *A low cost zero energy cool chamber for short term storage of mango*. In III international mango symposium 291 (pp. 519–524).

Ruoyi, K., Zhifang, Y., & Zhaoxin, L. (2005). Effect of coating and intermittent warming on enzymes, soluble pectin substances and ascorbic acid of *Prunus persica* (cv. Zhonghuashoutao) during refrigerated storage. *Food Research International, 38*(3), 331–336.

Rygg, G. L., Furr, J. R., Nixon, R. W., & Armstrong, W. W. (1953). Factors affecting the spoilage of dates at room temperature. *Date Growers' Institute Report, 30*, 10–14.

Saltveit, M. E. (2003). Is it possible to find an optimal controlled atmosphere? *Postharvest Biology and Technology, 27*, 3–13.

Salunkhe, D. K., & Wu, M. T. (1973). Effects of sub-atmospheric pressure storage on ripening and associated chemical changes of certain deciduous fruits. *Journal of the American Society for Horticultural Science, 98*, 113–116.

Santana, L. R. R. D., Benedetti, B. C., Sigrist, J. M. M., & Sato, H. H. (2011). Effects of modified atmosphere packaging on ripening of'Douradão'peach related to pectolytic enzymes activities and chilling injury symptoms. *Revista Brasileira de Fruticultura, 33*(4), 1084–1094.

Scetar, M., Kurek, M., & Galic, K. (2010). Trends in fruit and vegetable packaging—A review. *Croatian Journal of Food Technology, Biotechnology and Nutrition, 5*(3–4), 69–86.

SeaLand. (1991). *Shipping guide to perishables*. Iselim, NJ: SeaLand Services.

Segliņa, D., Olšteine, A., Krasnova, I., & Juhņeviča, K. (2013, August). *Use of packaging materials for extending the shelf life of diploid plum variety 'Kometa'*. In Proceedings of the Latvian Academy of Sciences. Section B. Natural, Exact, and Applied Sciences (Vol. 67, No. 2, pp. 174–178). Sciendo.

Senesi, E., Rizzolo, A., & Sarlo, S. (1991). Effect of different packaging conditions on peeled almond stability. *Italian Journal of Food Science, 3*, 209–218.

Serradilla, M. J., Martín, A., Ruiz-Moyano, S., Hernández, A., López-Corrales, M., & De Guía Córdoba, M. (2012). Physicochemical and sensorial characterisation of four sweet cherry cultivars grown in Jerte Valley (Spain). *Food Chemistry, 133*(4), 1551–1559.

Sharma, R. R., & Krishna, H. (2017). *Apricot*. In Textbook of temperate fruits (pp. 88). CBS Publishers & Distributors Pvt Ltd.

Shukor, A. R. A., Razali, M., & Omar, D. (2000). Post storage respiratory suppression and changes in chemical compositions of mango fruit after storage in low-oxygen atmosphere. *Acta Horticulturae, 509*, 467–470.

Siddiq, M., & Greiby, I. (2014). Overview of date fruit production, postharvest handling, processing, and nutrition. In M. Siddiq, S. M. Aleid, & A. A. Kader (Eds.), *Dates: Postharvest science, processing technology and health benefits* (1st ed., pp. 1–28). Chichester: John Wiley and Sons.

Siddiqui, M. W. (Ed.). (2015). *Postharvest biology and technology of horticultural crops: Principles and practices for quality maintenance*. Boca Raton: CRC Press.

Singh, S. P., & Singh, Z. (2013). Controlled and modified atmospheres influence chilling injury, fruit quality and antioxidative system of Japanese plums (Prunus salicina Lindell). *International Journal of Food Science and Technology, 48*(2), 363–374.

Singh, S. P. (2010). *Postharvest oxidative stress in plums: Mechanism and implications for storage and fruit quality*. PhD thesis, Curtin University, WA, Australia.

Singh, Z., Singh, R. K., Sane, V. A., & Nath, P. (2013). Mango-postharvest biology and biotechnology. *Critical Reviews in Plant Sciences, 32*(4), 217–236.

Skog, L. J., Schaefer, B. H., & Smith, P. G. (2002). *On-farm modified atmosphere packaging of sweet cherries*. In XXVI international horticultural congress: Issues and advances in postharvest horticulture (Vol. 628, pp. 415–422).

Snowdon, A. L. (1990). *A colour atlas of postharvest diseases and disorders of fruits and vegetables. Volume 1. General introduction and fruits* (p. 302). London: Wolfe Scientific.

Sosnrivichai, S., Gomolmanee, S., Boonyakiat, D., Uthaibutra, J., Boon-Long, P., & Gemma, H. (1992). Seal packaging by plastic film as a technique for limiting fungal decay of mangoes. *Acta Horticulturae, 296*, 23–32.

Spalding, D. H., & Reeder, W. F. (1977). Low pressure (hypobaric) storage of mangos. *Journal American Society for Horticultural Science, 102*, 367–369.

Srinivasa, P. C., Baskaran, R., Ramesh, M. N., Prashanth, K. V. H., & Tharanathan, R. N. (2002). Storage studies of mango packed using biodegradable chitosan film. *European Food Research and Technology, 215*, 504–508.

Steffens, C. A., do Amarante, C. V. T., de Oliveira Alves, E., Brackmann, A., Corrêa, T. R., & Espindola, B. P. (2014). Storage of 'Laetitia' plums (Prunus salicina) under controlled atmosphere conditions. *African Journal of Biotechnology, 13*(32), 3239–3243.

Ragnoi, S. (1989). Development of the market for Thai lychee in selected European countries (Doctoral dissertation, MSc thesis, Silsoe College, Cranfield Institute of Technology).

Rao, S. D. V., & Rao, G. K. P. (2008). Controlled atmosphere storage of mango cultivars 'Alphonso' and 'Banganapalli' to extend storage-life and maintain quality. *The Journal of Horticultural Science and Biotechnology, 83*(3), 351–359.

Tefera, A., Seyoum, T., & Woldetsadik, K. (2007). Effect of disinfection, packaging, and storage environment on the shelf life of mango. *Biosystems Engineering, 96*, 201–212.

Thompson, A. K. (1971). *The storage of mango fruit. Tropical agriculture trinidad* (pp. 63–70). Hoboken: John Wiley & Sons.

Thompson, A. K. (1998). *Controlled atmosphere storage of fruits and vegetables*. Oxon: CAB International.

Tian, S. P., Jiang, A. L., Xu, Y., & Wang, Y. S. (2004). Responses of physiology and quality of sweet cherry fruit to different atmospheres in storage. *Food Chemistry, 87*(1), 43–49.

Truter, A. B., Combrink, J. C., & Von Mollendorff, L. J. (1994). Controlled-atmosphere storage of apricots and nectarines. *Deciduous Fruit Grower, 44*(11), 421–427.

Van der Merwe, J. A. (1996). Controlled and modified atmosphere storage. In J. C. Combrink (Ed.), *Integrated management of post-harvest quality* (pp. 104–112). Paarl: Paarl Printers.

Varoquaux, P., Gouble, B., Ducamp, M. N., & Self, G. (2002). Procedure to optimize modified atmosphere packaging for fruit. *Fruits, 57*(5–6), 313–322.

Vilasachandran, T., Sargent, S. A., & Maul, F. (1997). Controlled atmosphere storage shows potential for maintaining postharvest quality of fresh litchi. In A. A. Kader (Ed.), *7th International controlled atmosphere research conference, University of California, Davis, July 13–18, 1997, Abstract 54*. Davis, CA: University of California.

Wang, J., Pan, H., Wang, R., Hong, K., & Cao, J. (2016). Patterns of flesh reddening, translucency, ethylene production and storability of 'Friar' plum fruit harvested at three maturity stages as affected by the storage temperature. *Postharvest Biology and Technology, 121*, 9–18.

Wang, L., & Vestrheim, S. (2002). Controlled atmosphere storage of sweet cherries (*Prunus avium* L.). *Acta Agriculturae Scandinavica, 52*(4), 136–142.

Wang, L., & Vestrheim, S. (2003). Controlled atmosphere storage of Norwegian grown plums (Prunus domestica L.). *Acta Agriculturae Scandinavica, B, 53*(1), 33–37.

Wang, P., Gao, H., Zhou, Y., Fang, X., & Mao, J. (2012). Effects of hypobaric storage on softening and cell wall metabolism of Xinjiang kuqa apricot fruits. *Transactions of the Chinese Society of Agricultural Engineering, 28*(16), 254–258.

Wang, Y., Bai, J., & Long, L. E. (2015). Quality and physiological responses of two late-season sweet cherry cultivars 'Lapins' and 'Skeena' to modified atmosphere packaging (MAP) during simulated long distance ocean shipping. *Postharvest Biology and Technology, 110*, 1–8.

Wills, R. B. H., & Golding, J. B. (2016). *Postharvest. An introduction to the physiology and handling of fruit and vegetables* (p. 72). Wallingford: CAB International.

Wills, R. B. H., McGlasson, W. B., Graham, D., & Joyce, D. C. (2007). *Postharvest; an introduction to the physiology and handling of fruit, vegetables and ornamentals* (pp. 83–99). Wallingford: CABI Publishers.

Wright, K. P., & Kader, A. A. (1997). Effect of controlled-atmosphere storage on the quality and carotenoid content of sliced persimmons and peaches. *Postharvest Biology and Technology, 10*(1), 89–97.

Wu, J., Lin, X., Lin, S., Chen, P., Huang, G., Peng, P., et al. (2019). California almond shelf life: Changes in moisture content and textural quality during storage. *Transactions of the ASABE, 62*(3), 661–671.

Xi, W. P., Zhang, B., Shen, J. Y., Xu, C. J., & Chen, K. S. (2012). Intermittent warming alleviated the loss of peach fruit aroma-related esters by regulation of AAT during cold storage. *Postharvest Biology and Technology, 74*, 42–48.

Yu, X. M., Xu, W. C., & Li, D. L. (2012). Effect of an antimicrobial packaging on postharvest quality of sweet cherry. *Applied Mechanics and Materials, 200*, 249–253.

Ziaolhagh, H. (2013). Effect of packaging on shelf life of almond kernels. *International Journal of Agriculture and Crop Sciences, 5*(1), 15.

Zoffoli, J. P., & Rodriguez, J. (2009). Effect of active and passive modified atmosphere packaging of sweet cherry. *Acta Horticulturae, 1020*, 115–119.

Hi-Tech Stone Fruit Industry, Issues, and Approaches

12

Mohammad Maqbool Mir, T. Angmo, Umar Iqbal, Gh. Hassan Rather, M. U. Rehman, Rifat Bhat, Amit Kumar, Nowsheen Nazir, Ashaq H. Pandit, and M. Amin Mir

Abstract

Stone fruits are an important temperate fruit crops and play a significant role in the food and nutritional security. Due to the increase in temperature, stone fruit cultivation is decreasing, because most of the cultivated stone fruit crop appeared to be unable to receive enough chilling in most winters. The increase in the temperature acts as a treat to its profitable cultivation. So, it became an imperative to go for hi-tech cultivation in stone fruit cultivation. Hi-tech cultivation is a precise production technique for the efficient use of inputs at the appropriate time and quantity for maximization of yield and quality of different fruit crops. The production potential of stone fruit can be enhanced through hi-tech approaches which include high-density planting (HDP), micro-irrigation, fertigation, integrated nutrient management (INM), and integrated pest management (IPM). But there are many limitations in promoting hi-tech cultivation in stone fruit industry. These issues can be overcome by using recent technologies integrated with indigenous knowledge and will have a far-reaching consequence to see that hi-tech horticulture reaches a new high.

Keywords

Hi-tech · Stone fruits · Issues · Approaches

M. M. Mir (✉) · T. Angmo · U. Iqbal · G. H. Rather · M. U. Rehman · R. Bhat · A. Kumar · N. Nazir · A. H. Pandit · M. A. Mir
Division of Fruit Science, Sher-e-Kashmir University of Agricultural Sciences & Technology of Kashmir, Shalimar, Srinagar, Jammu and Kashmir, India

M. M. Mir et al. (eds.), *Production Technology of Stone Fruits*,
https://doi.org/10.1007/978-981-15-8920-1_12

12.1 Introduction

Stone fruits are member of the Rosaceae, subfamily Prunoideae. The term stone fruits is used to denote fruits of the *Prunus* species, viz., peach, nectarine, plum, cherry, and apricot, which contain hard stony seed in the flesh. The fruit that is one-seeded is classified as drupe. The growth of fruit is described as double sigmoidal with rapid growth during cell division, followed by pit hardening (period of slow growth), and then a period of rapid cell enlargement (Flore 1994). Stone fruits are fleshy fruits containing a hard pit that protects the embryo, which is characteristic of all species of the genus *Prunus*. In most species, the juicy mesocarp is the edible part of the fruit (Infante 2012). The area under stone fruits is rapidly increasing in temperate parts of the world due to its rich nutritional and economic value. The interest in growing these crops under subtropical regions during the last three decade has also increased due to the availability of high-quality and low-chilling cultivars (Sharpe et al. 1990). Nutritionally, stone fruits are good sources of vitamins and minerals (Table 12.1), and there is an increasing interest in their potential value due to the presence of phenolic compounds with antioxidant properties (Nirgude et al. 2018).

But stone fruit cultivation is decreasing because most cultivated stone fruit crops appeared to be unable to receive enough chilling in most winters due to the increase in temperature. The Indian Meteorological Department monitors the climatic variability and has divulged that temperatures are increasing in Kashmir and some part of Jammu areas where temperate fruits are mainly grown, with substantial increase in the maximum temperature of 0.05 °C/year. Due to the increase in temperature and decrease in rainfall, apricot and cherries are fast disappearing from some areas of Kashmir Valley (Lal et al. 2018). The yield and quality of stone fruits are fast deteriorating in the valley due to the general rise in temperature and less availability of water. From the last few years, there has been distinguishable slow growth in production and productivity in rain-fed Kashmir's Karewas areas (Lal et al. 2018). Fruits like cherry, apple, plum, peach, and apricot are damaged heavily due to unusual hailstorms and windstorms in the summer (Lal et al. 2018). There has been marked change in the pattern of snowfall in Kashmir in recent years, which is affecting all the stone and pome fruits. It has been noticed that snowfall and flowering are coinciding in some years, leading to great loss in quantity and quality (Datta 2013).

Due to these issues of changes in the climate, it's become an imperative to go for hi-tech cultivation in stone fruit crops other than the conventional cultivation.

The term hi-tech cultivation refers to the precise production techniques for the efficient use of inputs at the appropriate time and quantity for maximization of yield and quality of different fruit crops. It is an adoption of any technology which is modern, less environment-dependent, and capital intensive and has the capacity to improve productivity and quality of horticultural crops (Chadha 2001). Hi-tech cultivation is an environment-friendly, intensive technique to capacitate farmers for obtaining high productivity and quality products to fetch more money. Hi-tech fruit crop cultivation is a chain system of culturing fruits, with a proper linkage from selection of seed/variety to the finished product through the process of modern

Table 12.1 Vitamin and mineral content of stone fruit (mg 100 g^{-1} edible portion)

	Vitamins				Minerals					
Crop	Vit. C	Thiamin	Riboflavin	Niacin	K	Na	Ca	Mg	Fe	Zn
Apricot										
Bavinity	7	0.02	0.03	1.4	350	1	16	9	0.3	0.1
Trevatt	16	0.03	0.04	1.1	320	3	15	9	0.3	0.2
Cherry										
Eagle	21	0.04	0.02	0.4	200	0	14	7	0.2	0.1
Lambert	16	0.02	0.03	0.6	240	0	28	7	0.3	0.1
Ron's	18	0.03	0.03	0 0.6	280	1	30	12	0.4	0.1
Nectarine										
Goldmine	13	0.02	0.04	1.1	240	1	11	6	0.2	0.1
Nectared	11	0.02	0.03	1.4	230	0	8	8	0.1	0.1
Redgold	–	0.02	0.04	–	240	0	7	7	0.1	0.1
P6	–	0.02	0.03	1.1	220	1	6	6	0.1	0.1
Peach										
Blackburn	14	0.01	0.05	0.9	190	1	9	6	0.4	0.1
Cardinal	11	0.02	0.03	1.0	170	2	4	6	0.1	0.1
Elberta	7	0.01	0.03	1.0	190	2	6	6	0.1	0.1
Fragar	10	0.02	0.03	1.0	170	2	5	6	0 0.1	0.1
Halehaven	16	0.01	0.03	1.1	180	1	5	7	0.1	0. 1
J. H. Hale	8	0.01	0.03	1.1	220	2	9	7	0 0.1	0.1
Maygold	8	0.02	0.03	–	–	–	–	–	–	–
Redhaven	6	0.01	0.05	1.1	180	2	7	7	0.2	0.1
Plum										
Blood	6	0.03	0.04	0.8	170	2	8	6	0.2	0.1
Greengage	5	0.05	0.04	0.6	140	1	6	6	0.2	0.1
Mariposa	5	0.04	0.04	0.7	190	2	7	7	0.4	0.1
Narrabeen	4	0.02	0.05	0.3	120	1	6	4	0 0.1	0.1
Santa Rosa	4	0.03	0.05	0.2	140	3	6	6	0.2	0.1
Wilson	6	0.04	0.05	0.9	160	0	7	6	0.2	0.1

Wills et al. (1983)

technique of crop production through postharvest management techniques (Bhattacharyya et al. 2017).

The production potential of stone fruit can be enhanced through hi-tech approaches. These methods involve high-density planting (HDP), improved varieties and rootstock, training, and pruning, soil and leaf analysis-based integrated nutrient management, integrated pest management (IPM), protected cultivation, fertigation, micro-irrigation, disease and weed management, appropriate mechanization and essential innovations such as preharvest bagging, and the use of growth regulators. The detailed descriptions of each adopted technology are given below:

12.2 High-Density Planting

HDP is defined as planting at a density in excess of that planted under normal spacing. In other words, it is the planting of more plants than the optimum through the manipulation of tree size (Singh et al. 2017). High-density planting accommodates a maximum number of plants per unit area and produces a maximum yield of quality fruits. It was originated in Europe at the end of 1960 for temperate fruits (Singh et al. 2017; Usha et al. 2015). Majority of temperate fruit in Australia, America, and New Zealand are grown under this system. In HDP, attempt is made to have less structural branches and more fruiting branches. The branches on the plant should be so arranged that they should be having a minimum shade on other branches (Singh et al. 2017). HDP technologies have also been developed in stone fruits, and it produces a maximum number of fruit per unit area and produces good-quality fruit as compared to the traditional planting system. The main objective of HDP is to make use of horizontal and vertical space per unit time and to harness maximum possible returns per unit of input of resources. High-density planting is differentiated into four categories:

Low-density planting (LDP), 100 plants/ha—There is no use of dwarfing rootstocks and required minimum training and pruning, and trees come into commercial production potential after 10–15 years.

Medium-density planting (MDP), 250–500 plants/ha—It requires proper pruning and gives high-yielding and high-quality fruit.

High-density planting (HDP), >1000 plants/ha—HDP requires rigorous training and pruning, dwarfing rootstocks, and chemicals to maintain optimum growth; both yield and expenses are higher; and the establishment and maintenance of HDPs require technical backup.

Ultrahigh-density planting (UHDP), >2000–5000 trees/ha—This requires severe pruning and training, proper canopy management, chemical assistance, and nutrient management, which are essential; this also requires technical backup.

Superhigh-density planting (SHDP) (meadow), >10,000–40,000 or more plants/ha—Severe top pruning is practiced similar to mowing of grass land; tree yields after 1 or 2 years after planting; this requires heavy uses of growth regulators as well as judicial canopy managements (Usha et al. 2015).

12.3 Advantages of High-Density Planting

High density orcharding also facilitates more efficient use of inputs, easy and early harvest, distance between plants, apart from higher yield and net economic returns per unit area in the initial years. Advantages also include reduced labor costs, increase fruit quality, and early cropping.

12.4 Component of High-Density Planting

- Use of dwarfing rootstock and interstock.
  Dwarfing rootstock and interstock had a significant effect on vigor and structural development of fruit trees.
- Use of dwarf scion.
- Efficient training and pruning.

Available scientific evidence supports the fact that training of young orchard trees and pruning in the bearing trees increase fruit yield and quality and prolong the economic life of trees. In fact, training and pruning have become a prerequisite for high-density and mechanized orcharding. Pruning restricts the vegetative growth, thus keeping the trees in manageable size and form. It allows good airflow and adequate light penetration inside the tree canopy. It also minimizes disease risk and assists in good fruit coloration. Like all other pome fruits, stone fruit trees should also be pruned to develop a strong, well-balanced framework of scaffold branches. Unwanted branches should be cut back early to avoid the necessity of large cuts in later years. The open center or modified leader system is a pruning system which is best suited to stone fruit trees.

- Use of Growth Retardant.

Growth regulators such as ethephon, daminozide, ancymidol, B-9 (phosphon D), cycocel, chlormequat, and paclobutrazol are extensively used to reduce shoot growth by 30% to 0% and increased return bloom in the succeeding years. Tying down the branches to allow them to grow to 45° angle from the main stem is the standard practice to control tree size (Usha et al. 2015).

12.4.1 Micro-irrigation and Fertigation

Micro-irrigation has emerged as an appropriate technology to meet the increasing demand for irrigation, especially for the horticultural crops, as this method has proved to have an efficiency of 95%. It ensures higher quality of crops, increases yield, consumes less energy and water, uses less chemical and fertilizer, reduces leaching and run off, produces less weeds and less soil compaction (Saxena and Rao 2019). Increased yield, reduced harvesting time, and economy are the main reasons for the adoption of this system (micro-irrigation), mainly in high value horticultural crops. In addition to yield increase, drip irrigation also improves the quality of produce and reduces crop susceptibility to disease. Drip irrigation is also used in glass houses, protected greenhouses, or for the off-season crops (Dashora 2017). Fertigation is one of the recent techniques of applying fertilizers through drip irrigation systems (Bussi et al. 1991; Srinivas 2004) which permits the application of various fertilizer formulations directly at the site of roots and thus improves fertilizer use efficiency (Singh et al. 2005). Fertigation is a method of fertilizer

Table 12.2 Fertilizer use efficiency in fertigation (%)

Nutrient	Fertilizer use efficiency (%) Soil application	Drip +Fertigation
Nitrogen	30–50	95
Phosphorous	20	45
Potassium	60	80

Biswas (2010)

application in which fertilizer is incorporated within the irrigation water by the drip system. In this system, the fertilizer solution is evenly distributed in the irrigation. The efficiency is more due to the high availability of nutrients. In this method, water-soluble fertilizers as well as liquid fertilizer are used. The Table 12.2 shows the fertilizer use efficiency being increased from 80% to 90% by using this method (Biswas 2010).

12.5 Advantages of Fertigation

- Through fertigation, nutrients and water are supplied near the active root zone which results in greater absorption by the crops. Nutrient and water losses due to volatization and evaporation are very minimum or completely absent because the nutrient solution is supplied through network of pipe lines.
- There is a possibility of getting 25–50% higher yield as water and fertilizers are supplied evenly to the crops without wastage.
- Water use efficiency is found to be increased up to 90% (Manohar et al. 2001), and fertilizer use efficiency is also higher which helps to save nutrients up to 80% (Sandal and Kapoor 2015; Srivastava and Malhotra 2017).
- Combining water and liquid fertilizers with insecticides and herbicides (Kumar 1999) saves labor, machinery, and time of application separately.
- Cultivation of stone fruit trees under light soil condition always poses problem; hence, fertigation offers growing of these crops under such soil condition and also minimizing soil compaction by avoiding the involvement of heavy traffic of equipment as in the conventional method of fertilizer application, thereby maintaining and improving the physical, biological, and chemical nature of soil (Haynes and Swift 1987).
- In conventional approaches, overfertilization and over-irrigation result in high intensity of weeds and pathogens, whereas fertigation facilitates reduced weed population and contact time of pathogen with the tree (Battilani 2008).

Advances in micro-irrigation techniques have emerged as greater application of fertigation in stone fruits. Stone fruit crops play an important role in the food and nutritional security. These fruits are mostly perennial in nature and require proper irrigation and fertilization for optimum growth, yield, and quality. In recent past, many researchers, after using this technique in different stone fruits, have pointed out the advantages of fertigation than other means of conventional fertilization. In peach, fertigation significantly improved the tree growth, yield, and physical and chemical properties of fruits (Bryla et al. 2003; Banyal et al. 2014). The valuable effects of fertigation on the growth of tree, yield attributes, quality characters, leaf and fruit nutrient content, nutrient use efficiency, and chemical traits were observed in nectarines (Krige and Stassen 2008; Singh et al. 2015). Fertigation treatments have been found to be significantly superior in terms of vegetative and reproductive growth, yield, nutrient, and other quality characteristics in cherry (Neilsen et al. 2010). In apricot, it improves growth, quality attributes, and leaf nutrients of the fruit. It also increases water and fertilizer use efficiency in apricot (Bybordi 2013). Under fertigation, water and fertilizers are supplied at the right time and required levels, thus over-feeding is totally avoided; fertigation also helps in meeting the physiological needs of the trees at different stages of growth (Shirgure 2013).

12.5.1 Integrated Nutrient Management

Integrated nutrient management (INM) refers to the maintenance of soil fertility and of plant nutrient supply at an optimum level for sustaining the desired productivity through optimization of benefits from all possible sources of organic, inorganic, and biological components in an integrated manner. In simple words, integrated nutrient management (INM) is the use of organic resources (manures, crop residues, etc.) together with inorganic fertilizers to optimize crop nutrition.

The basic concept of integrated nutrient management system (INMS) is to improve soil productivity through a balanced use of fertilizers combined with biological sources of plant nutrients and to reduce inorganic (fertilizer) input cost also. Integrated nutrient supply/management (INS) aims at adjustment and maintenance of soil fertility and of plant nutrient supply to an optimum level for sustaining the desired crop productivity through optimization of benefit from all possible sources of plant nutrients in an integrated manner (Roy and Ange 1991).

12.6 Components of INM

Soil sources: Mobilizing unavailable nutrients and using appropriate crop varieties and cultural practices.

Mineral fertilizer: Coated urea, super granules, and direct use of locally available rock phosphate in acid soils, muriate of potash (MOP), single super phosphate (SSP), and micronutrient fertilizers.

Organic sources: Organic manures like farm yard manure (FYM), human excreta, crop waste, residues, sewage sludge, rural compost, press mud, and industrial by-products. Other potential nutrients of organic sources like nonedible oil cakes and wastes from food processing industries are also included.

Biological sources: Biofertilizers are the carrier-based preparation designed to improve the soil fertility and help plant growth by their increased number and biological activity in the rhizosphere. Microorganisms that are involved in

Table 12.3 Types of microorganisms involved in biological nitrogen and phosphate fixation and solubilization of other nutrients

S. No.	Groups	Examples
N_2-fixing biofertilizers		
1.	Free-living	*Azotobacter*, *Beijerinckia*, *Clostridium*, *Klebsiella*, *Anabaena*, *Nostoc*
2.	Symbiotic	*Rhizobium*, *Frankia*, *Anabaena azollae*
3.	Associative symbiotic	*Azospirillum*
P-solubilizing biofertilizers		
1.	Bacteria	*Bacillus megaterium* var. *phosphaticum*, *Bacillus subtilis*, *Bacillus circulans*, *Pseudomonas striata*
2.	Fungi	*Penicillium* sp., *Aspergillus awamori*
P-mobilizing biofertilizers		
1.	Arbuscular mycorrhiza	*Glomus* sp., *Gigaspora* sp., *Acaulospora* sp., *Scutellospora* sp., *Sclerocystis* sp.
2.	Ectomycorrhiza	*Laccaria* sp., *Pisolithus* sp., *Boletus* sp., *Amanita* sp.
3.	Ericoid mycorrhizae	*Pezizella ericae*
4.	Orchid mycorrhiza	*Rhizoctonia solani*
Biofertilizers for micronutrients		
1.	Silicate and zinc solubilizers	*Bacillus* sp.
Plant growth-promoting rhizobacteria		
1.	Pseudomonas	*Pseudomonas fluorescens*

Singh et al. (2017)

biological nitrogen and phosphate fixation and solubilization of other nutrients are given in Table 12.3.

Biofertilizers can fix N in the range of 20–200 kg/ha/year, solubilize P in the range of 30–50 kg/ha/year, and mobilize P, Fe, Zn, and Mo to a varying extent. Biofertilizers are associated in the production of plant growth-promoting (root-colonizing) bacteria, including the phosphorus-solubilizing *Pseudomonas* spp. and nitrogen-fixing *Azospirillum*, and are known to produce growth hormones which often lead to increased shoot and root growth. They suppress the growth of pathogenic microorganisms by productions of antibiotics and bacteriocins (Singh et al. 2017).

12.6.1 Integrated Pest Management

Integrated pest management (IPM) is an effective and environmentally sensitive approach to pest management that depends on a combination of techniques such as biological control, modification of cultural practices, habitat manipulation, and use of resistant varieties. IPM programs use current, comprehensive information on the life cycles of pests and their interaction with the environment. This information, in combination with available pest control methods, is used to manage pest damage by the most economical means and with the least possible hazard to people, property, and the environment.

For stone fruit, developing an IPM program involves more than putting together a pest spray program. It involves a proactive approach to growing, beginning with site and cultivar selection, and an understanding of cultural practices that will help to delay, reduce, or eliminate potential problems. It also involves an understanding of the life cycle of the pest (whether it may be insect or mite), knowing insect vectors, being able to identify beneficial, understanding their life cycles, understanding how the environment impacts potential pest problems and plant health, understanding how different pesticides work on the ground and the proper timing, understanding and utilizing alternative management methods, and understanding economic and injury thresholds (IPM guideline, Los and Concklin 2013).

12.6.2 Protected Cultivation

Protected cultivation is defined as the cropping technique wherein the microclimate that is surrounding the plant body is controlled partially or fully as per the requirement of the plant species which is grown during their period of growth. Protected cultivation is actually an invention which achieves higher nutrient and water use efficiencies, increases photosynthetic efficiency, and reduces respiratory loss, which are the advantages of protected cultivation.

Activities like greenhouse construction, shade net, and mulching is worth adopting. Greenhouses are used mainly for growing of fruit nurseries, hardening tissue culture plants, and growing high-value crops. In the greenhouse, temperature and humidity are kept under control by using micro-irrigation and fogging system as per the need of the crops which are to be grown in the greenhouse (Singh et al. 2017). Greenhouse reflects back 43% of the net solar radiation incident and allows the transmittance of the "photosynthetically active solar radiation" in the range of 400–700 Nm wavelengths which increases the photosynthetic efficiency of crops grown inside the green house (Roy et al. 2018). Moderately short to moderately long wavelength (15–390 Nm) ultraviolet rays damaging the crops are absorbed by the glass or polyethylene sheet used as cladding material for greenhouse/poly house. For this reason, excellent crop growth and high yield are generally realized in crops grown under such protected structure (Roy et al. 2018).

12.7 Advantages of Protected Cultivation

- Under unfavorable agroclimatic conditions, crops can be grown.
- In comparison to those under open field conditions, crop yield can be several times higher in protected cultivation.
- Produce quality of is superior.
- Higher input use efficiencies are achieved.

- Management of insect pest, diseases, weed, etc. is easier under protected cultivation.
- Income per unit area increases.
- It puts up an excellent opportunity to produce fruits for export.

The types protected structures and design depend mainly on the availability of material, purpose of protected cultivation, climate of region, and available market for quality produce. For protected cultivation, climate-controlled greenhouse, semi-climate-controlled greenhouse, naturally ventilated or low-cost greenhouse, walk-in tunnels, insect-proof net house, and plastic low tunnels can be used (Dashora 2017).

12.7.1 Issues Under Hi-Tech Promotion

Some of the important issues in the promotion of high-value crops are as follows:

1. Unsuccessful efforts in bringing more and more small holders under high-value system of production.
2. Transfer of technology has been a serious issue in fruit crops as compared with other horticultural crops, and lowland production technology is highly transferable from one major region to another. This transfer of technology is not easy in fruit crops because of the substantial climate difference; it is not easily understandable why it is so difficult to transfer technology from one zone to another.
3. Lack of access to institutional credit: informal sources have continued to play a major role in the credit market due to lack of access to adequate formal credit.
4. Lack of institutional development for input support, training, extension, and capacity building.
5. Issue in marketing linkages for ensuring remunerative price of produce on sustainable basis.
6. Issue in enabling policy interventions:
   (a) Lack of encouragement in aggregation of farmers into farmer groups such as Farmer Interest Groups (FIGs)/Farmer Producer Organizations (FPOs) and Farmer Producer Companies (FPCs) to bring economy of scale and scope.
   (b) Lack of support in skill development and creation of employment generation opportunities for youth of rural areas in horticulture and postharvest management, especially in the cold chain sector.
7. High-value fruit crops are perishable in nature, and measures toward diversification call for simultaneously addressing critical infrastructure needs like quick transport facility and cold storages.
8. Absence of specific risk mitigation measures to off-set high production and risks related to market.
9. In fertigation of stone fruits, there are several issues such as the following:
   (a) Huge initial investment: This is the major factor restraining the growth of the market. The initial investment cost for installing fertigation system for

water-soluble fertilizer application is high and is hence less suitable for low-value crops considering that the returns on investment may be low.
   - (b) Critical nutrient application doses: Nutrient application is one of the most factors in addition to fertilizer dose among different horticultural practices, which influence the growth and yields of crops. Nutrient application in fertigation is critical, as overdosage may cause yield losses.
   - (c) Nonavailability of water-soluble fertilizers or liquid fertilizers at reasonable rates has been hindrances in large-scale adoption of this technology. Thus, fertigation has been largely limited to greenhouse-grown crops or high-value crops like grapes (Dashora 2017).
   - (d) Nonavailability of liquid fertilizers at reasonable price in remote or hilly areas (Nirgude et al. 2018).
10. Constraints in IPM were as follows:
   - (a) Institutional constraints:
   - Integrated pest management requires an interdisciplinary, multifunctional approach to solving problems of pest. Lack of coordination between different institutes and implementation agencies leads to difficulty in institutional integration.
   - (b) Informational constraints:
   - Farmers lacked the skills necessary to practiced integrated pest management. To instill confidence in the farmers in many cases, the field-level extension workers are not trained sufficiently in IPM (lack of trained IPM consultants).
   - Lack of information on pesticide resistance and on how to use biologically based methods, e.g., natural enemies (Brunner 1994).
   - (c) Sociological constraints:
   - Many farmers considered IPM practices to be risky as compared to the use of chemical pesticide because many farmers expressed their lack of faith in IPM.
   - To develop and adopt IPM methods, there is a lack of incentives.
   - (d) Political constraints:
   - The continuance subsidy on pesticide by the government for political reason and its tie up with the government provided credit for crop production; this acts as a leading constraint to farmers' acceptance of IPM.
   - Vested interest associated with pesticide trade.
   - (e) Economic constraints:
   - There is a lack of resources which directed toward a biologically based IPM.
   - Market demand for blemish-free produce.
   - Lack of funds to give training to farmers and extension workers on the use of IPM.
11. Constraints in HDP are as follows:
   - (a) Poor or inadequate availability of quality planting material, varieties, and dwarfing rootstocks.
   - (b) In HDP, there is a lack of standardization of production technology in various fruit crops, particularly stone fruits.

(c) As compared to the traditional system of planting, high-density planting needs higher initial establishment cost which common orchardist cannot afford.
(d) HDP needs for a more professional and scientific competency for management compared to the conventional planting at wider spacing.
(e) Tree/plant canopy management right from establishment stage.
(f) For HDP, nonavailability of complete package and use of mechanization (Mishra and Goswami 2016).

12.8 Hi-Tech Fruit Cultivation: Way Forward

Hi-tech fruit crop cultivation is emerging as a dynamic tool for increasing productivity of crops, particularly stone fruit cultivation, which in turn can be used as an initiative in the most challenging task of doubling the farmers' income (DFI). The key elements to carry forward the goal of hi-tech fruit crop cultivation may include the following:

1. Improved soil and water management practices through hi-tech interventions.
2. Development of high-yielding varieties/hybrids, conservation of existing germplasm, and exploitation of underutilized plant and land use with development of new varieties.
3. Application of biotechnology and nanotechnology.
4. Precision farming oriented to targeted yield, crop- and region-specific nutrient management, and irrigation resource conservation.
5. Incorporating postharvest technology will not only reduce postharvest losses but will also impart value addition.
6. Capacity-building through training and practical demonstration of improved technologies.

To keep pace with the modern information-based decisions, it is necessary that real-time data is recorded from the crops and communicated to the decision makers immediately as a profitable hi-tech horticulture entrepreneurship. Considerable advances have been made in the developed nations with respect to new types of sensor technologies, automated irrigation control technologies, and decision-support tools. These have significantly improved the efficiency and reliability of the technologies and also decreased purchase and operating costs (Lea-Cox 2012). The use of wireless sensor-based irrigation network (WSIN) has potential benefits in terms of reduced use of water and decreased CO_2 emissions. But, more importantly, it will help in considerably reducing the application rate of nutrients (up to 40%) and also the runoff rates of applied nutrients (up to 40%) (Majsztrik et al. 2013). The cutting-edge technologies that can be harnessed for hi-tech horticulture include sensor networks, cloud computing, augmented reality, unmanned air vehicles, and control area network (Ahrary and Ludena 2015). Therefore, the integrated use of such technologies together with indigenous knowledge will have a far-reaching consequence to see that hi-tech horticulture reaches a new high (Bhattacharyya et al. 2017).

12.9 Conclusion

Stone fruits are an important temperate fruit crops. The production potential of stone fruits can be enhanced through hi-tech approaches which include high-density planting, fertigation, promotion of mechanization, integrated nutrient management, canopy management, and integrated pest management. There are many limitations in promoting hi-tech cultivation in stone fruit industry. These issues can be overcome by the use of recent technologies based on scientific lines integrated with indigenous knowledge and will have a far-reaching consequence to see hi-tech horticulture reaches a new way. Furthermore, there is a bad need of value addition of stone fruit crops due to their perishable nature which include establishment of CA stores, scientific packaging, and transit of produce safely to the consumer's yard.

References

Ahrary, A., & Ludena, D. A. R. (2015). *A cloud-based vegetable production and distribution system*. In R. Neves-Silva et al. (Eds.), Intelligent Decision Technologies, Smart Innovation, Systems and Technologies—Proceedings of the 7th KES International Conference (pp. 11–20).

Banyal, S. K., Sharma, D., & Jarial, K. (2014). Effect of nitrogen fertigation on yield and fruit quality of peach (*Prunus persica* L.) under low hill conditions of Himachal Pradesh. *Current Advances in Agricultural Sciences, 6*(2), 158–160.

Battilani, A. (2008). Manipulating quality of horticultural crops with fertigation. *Acta Horticulturae, 792*, 47–60.

Bhattacharyya, T., Haldankar, P. M., Patil, V. K., Salvi, B. R., Haldavanekar, P. C., Pujari, K. H., & Dosani, A. A. (2017). Hi-tech horticulture: Pros and cons. *Indian Journal of Fertilisers, 13*(12), 46–58.

Biswas, B. C. (2010). Fertigation in Hi tech agriculture. *Fertilizer Marketing News, 41*(10), 4–8.

Brunner, J. F. (1994). Integrated pest management in tree fruit crops. *Food Reviews International, 10*(2), 135–157.

Bryla, D. R., Trout, T. J., & Ayars, J. E. (2003). Growth and production of young peach trees irrigated by furrow, micro jet, surface drip or subsurface drip systems. *Hort Science, 38*(6), 1112–1116.

Bussi, C., Huguet, J. G., & Defrance, H. (1991). Fertilization scheduling in peach orchard under trickle irrigation. *Journal of Hort Science, 66*(4), 487–493.

Bybordi, A. (2013). Quantitative and qualitative effects of nutrient applications and irrigation methods on apricot. *Middle-East Journal of Scientific Research, 14*(3), 423–431.

Chadha, K. L. (2001). *Hi-tech horticulture in India. Policy paper 13*. New Delhi: National Academy of Agricultural Sciences.

Dashora, L. K. (2017). Hi tech horticulture practices. Winter school compendium on Hi-tech intervention in fruit production towards hastening productivity, nutritional quality and value addition at Agriculture University Kota, College of Horticulture & Forestry, Jhalawar, Rajasthan, India.

Datta, S. (2013). Impact of climate change in Indian Horticulture—A review. *International Journal of Science, Environment and Technology, 2*(4), 661–671.

Flore, J. A. (1994). Stone fruit. In B. Schaffer & P. C. Anderson (Eds.), *Handbook of environmental physiology of fruit crops* (Vol. 1, pp. 233–270). Boca Raton: CRC Press.

Haynes, R. J., & Swift, R. S. (1987). Effect of trickle fertigation with three forms of nitrogen on soil pH, levels of extractable nutrients below the emitter and plant growth. *Plant and Soil, 102*, 211–221.

Infante, R. (2012). Harvest maturity indicators in the stone fruit industry. *Stewart Postharvest Review, 1*, 3.

Krige, G. T., & Stassen, P. J. C. (2008). Mineral nutrient distribution and requirement of pulse drip fertigated 'Donnarine' nectarine trees. *Acta Horticulturae*, (772), 355–360.

Kumar, A. (1999). *Pressurized irrigation system towards enhanced water use efficiency*. In Souvenir XXXIVth annual convention. Indian Society of Agricultural Engineers (pp. 26–44).

Lal, S., Singh, D. B., Sharma, O. C., Mir, J. I., Sharma, A., Raja, W. H., Kumawat, K. L., & Rather, S. A. (2018). Impact of climate change on productivity and quality of temperate fruits and its management strategies. *International Journal of Advance Research in Science and Engineering, 7*.

Lea-Cox, J. D. (2012). Using wireless sensor networks for precision irrigation scheduling. In M. Kumar (Ed.), *Problems, perspectives and challenges of agricultural water management* (pp. 233–258). Rijeka: InTech Press.

Los, L., & Concklin, M. (2013). *IPM guidelines for insects and diseases of stone fruits. University of Connecticut IPM program*. Storrs, CT: College of Agriculture and Natural Resources, UConn.

Majsztrik, J. C., Price, E. W., & King, D. M. (2013). Environmental benefits of wireless sensor-based irrigation networks: Case-study projections and potential adoption rates. *Hort Technology, 23*, 783–793.

Manohar, K., Khan, R., Kariyanna, M. M., & Sreerama, R. (2001). *An overview of status, potential and research accomplishment of drip irrigation in Karnataka*. In: Proceedings of International Conference on Micro and Sprinkler Irrigation System (pp. 69–79).

Mishra, D. S., & Goswami, A. K. (2016). High density planting in fruit crops. *Hort Flora Research Spectrum, 5*, 3.

Neilsen, G. H., Neilsen, D., Kappel, F., Toivonen, P., & Herbert, L. (2010). Factors affecting establishment of sweet cherry on Gisela 6 rootstock. *Hort Science, 45*(6), 939–945.

Nirgude, V., Misra, K. K., Singh, P. N., Singh, A. K., & Singh, N. (2018). NPK fertigation of stone fruit crops: A review. *International Journal of Chemical Studies, 6*(2), 3134–3142.

Roy, R. N., & Ange, A. L. (1991). In Integrated plant nutrient systems (IPNS) and sustainable agriculture. *Proc. FAI Annual Seminar*, FAI, New Delhi, pp SV/1–1/1–12.

Roy, S. S., Sharma, S. K., Ansari, M. A., Prakash, N., & Ngachan, S. V. (2018). *Hi-tech protected nursery for production of quality planting material of fruit and vegetable crops* (Vol. 1, pp. 169–202). Conservation Agriculture for Advancing Food Security in Changing Climate.

Sandal, S. K., & Kapoor, R. (2015). Fertigation technology for enhancing nutrient use and crop productivity: An overview. *Himachal Journal of Agricultural Research, 41*(2), 114–121.

Saxena, C. K., & Rao, K. V. R. (2019). Microirrigation for high water productivity in horticultural crops. *Phytochemistry of Fruits and Vegetables*, 179–199.

Singh, D., Sharma, S. D., & Kumar, P. (2015). Nitrogen fertigation for nectarines (*Prunus persica* var. *nucipersica*): Lateral and vertical nutrient acquisition and cropping behaviour in rainfed agro-ecosystem. *Indian Journal of Agricultural Sciences, 85*(11), 1440–1447.

Singh, J., Bhatnagar, P., Meena, C. B., & Mishra, A. (2017). Hi -Tech interventation in fruit production for hastening productivity, quality and value addition. Winter school compendium on Hi-tech intervention in fruit production towards hastening productivity, nutritional quality and value addition at Agriculture University Kota, College of Horticulture & Forestry, Jhalawar, Rajasthan, India.

Singh, Y., Singh, C. S., Singh, A., Singh, A. K., & Singh, K. (2005). Fertigation: A key for hi-tech. agriculture. *Agriculture Water Management, 52*, 128–149.

Shirgure, P. S. (2013). Citrus fertigation: A technology of water and fertilizers saving. *Scientific Journal of Crop Science, 2*(5), 56–66.

Sharpe, R. H., Sherman, W. B., & Martsolf, J. D. (1990). Peach cultivars in Florida and their chilling requirements. *Acta Horticulturae, 279*, 191–197.

Srinivas, K. (2004). Fertigation in horticultural crops. In K. L. Chadha, B. S. Ahloowalia, K. V. Prasad, & S. K. Singh (Eds.), *Crop improvement and production technology of horticultural crops* (Vol. 1, pp. 506–525). New Delhi: Proceedings of the First Indian Horticulture Congress.

Srivastava, A. K., & Malhotra, S. K. (2017). Nutrient use efficiency in perennial fruit crops: A review. *Journal of Plant Nutrition*.

Usha, K., Thakre, M., Goswami, A. K., & Deepak, N. G. (2015). *Fundamental of fruit production*. New Delhi: Division of Fruits and Horticultural Technology, Indian Agricultural Research Institute.

Wills, R. B., Scriven, F. M., & Greenfield, H. (1983). Nutrient composition of stone fruit (*Prunus* spp.) cultivars: Apricot, cherry, nectarine, peach and plum. *Journal of the Science of Food and Agriculture, 34*, 1383–1389.

Growth and Supply Chain of Stone Fruits in the World: An Indian Outlook

13

S. H. Baba

Abstract

This study envisages findings of an investigation of status and supply chain of important stone fruits in the world with special reference to India. This chapter clearly brought out that stone fruits have an important role in human food and can be consumed in many forms. The USA, Turkey, China, Spain and Iran are the few countries which have acquired better ranks in terms of maximum area under stone fruits or their production. India is relatively better placed with respect to the area under plums and peaches; however, the yield level of important stone fruits in the country is less compared to major producers. Trends in the area, production and yields of important stone fruits in the world vis-à-vis India have been discussed in this section. This chapter provides an insight into the economic feasibility in production and supply chain of selected stone fruits. It further discussed the various problems perceived by growers in cultivation of these fruits and on the basis of findings; this section concludes with a few broader policy suggestions for sustaining stone fruit economy in their niche areas.

Keywords

Stone fruits · Growth · Value chain · India · World

S. H. Baba (✉)
Division of Social Sciences, FoFy, Sher-e-Kashmir University of Agricultural Sciences and Technology of Kashmir, Shalimar, Srinagar, Jammu and Kashmir, India
e-mail: drshbaba@skuastkashmir.ac.in

M. M. Mir et al. (eds.), *Production Technology of Stone Fruits*,
https://doi.org/10.1007/978-981-15-8920-1_13

13.1 Introduction

India has made a rapid progress in the production of horticultural crops (NHB 2005) and has emerged as the second largest fruit producer globally. India is producing 89,018 MT of fruits from an area of about 6188 hectares (Anonymous 2016a). Apple, citrus, banana, grapes, mango and stone fruits including apricot, almond, peach, plum, apricot, nectarine and cherry are important fruits growing in the country. Cultivation of fruit crops is economically more viable compared to cereals, and per rupee return to investment was reported to be significantly higher across the world. The cultivation of fruits contributed significantly in accelerating the growth of agricultural sector and has significant impact on the country's economy. The comparative advantage at natural niches has favoured fruit cultivation across the country, and farmers in the past have been laying new orchards wherever natural conditions, institutional role and socioeconomic conditions were favourable. In consideration of viability of the fruits, the Government of India has prioritized this sector and has constituted National Horticulture Mission to double the fruit and vegetable production.

The idiom 'stone fruit' was used to indicate fruits belonging to *Prunus* species, and peaches, nectarines, plums, cherries and apricots are few of them. The genus *Prunus* and family Rosaceae, is considered to be the source of apricot, almond, peach, nectarine, cherry and plum. *Prunus* originated in Central Asia with secondary origins in Europe, North America and Eastern Asia (Watkins 1995). The introduction of cherry, apricot and plum in the Mediterranean basin was linked with the invasion of Alexander the Great (356–323 BCE), who dominated Persia and then continued east through Afghanistan, Turkistan, Pakistan and north-western India up to the Indus River. Almonds (*Prunus dulcis*) grow wild throughout Central Asia and Southwest Asia from Turkey and Syria to the Caucasus, the Hindu Kush mountains and deserts of Tian Shan (Zohary and Roy 1975). The history of cultivation of apricot (*Prunus armeniaca*) goes back to 5000 years in China with the first ascription to Emperor Yu (2205–2198 BCE; 658 BCE), and superior orchards were described in 406–250 BCE (Faust et al. 1998). Cherry consists of three main categories, viz. the sweet cherry, the tart cherry and the ground cherry. The origins of these species overlap and include Central Europe and areas around the Black Sea with the sweet cherry as far east as central Russia. Among *Prunus* species, plums have wide diversity and have been cultivated on three continents (Okie and Weinberger 1996; Faust and Suranyi 1999), and there are links between the major subgenera (Watkins 1995). Archaeological finds of plums in Europe date to Neolithic times (Faust and Suranyi 1999). The culture of peach dates to 2000 BCE, and there are now thousands of cultivars in China (Wang 1985).

Presently, these crops are commercially grown in the temperate zone where strong light, clear sky, long seasons and warm temperature prevail mainly in low and mid hills with altitudinal range of 1000–2000 m above mean sea level (Ghosh 2001). Stone fruits have an important role in human food and can be consumed as fresh, dried or processed products. These fruits are economically and nutritionally essential, and they constitute the most popular fruit consumed worldwide (Xiaoyong

Zhao et al. 2015; Poonam et al. 2011; Faust and Timon 1995). There has been a steady increase in the demand for stone fruits and their processed products owing to the increasing health and nutritional awareness of consumers. The phytochemicals in stone fruits include anthocyanins, carotenoids and phenolic acids (Tomas-Barberan et al. 2001; Weinert et al. 1990; Senter and Callahan 1991; Tourjee et al. 1998; Gil et al. 2002; Cevallos-Casals et al. 2006; Vicente et al. 2009; Scordino et al. 2012). Some stone fruits have high antioxidant activity (Wang et al. 1996; Prior et al. 1998; Cevallos-Casals et al. 2006). Further, few fruits (like plum) have high catechin content that suppresses cancer cells and other ailments (Chung et al. 2001; Kampa et al. 2000; Damianaki et al. 2000; Sartippour et al. 2001). Besides nutritional values, stone fruits are capital- and labour-intensive and help in generation of employment and additional income through its strong backward and forward linkages. Most of the stone fruits are commonly consumed as processed products including jam, ready-to-serve (RTS) drink or dried products and are regularly consumed year round; that has raised its demand among all the consumer groups. In the recent few years, stone fruit cultivation, plum, peach and apricot in particular, has become common feature in the temperate and subtropical setting of Northern Indian. Stone fruits are cultivated in nearly 0.31 million ha area with annual production of 2.52 million tonnes (Gangwar and Singh 1998) though Himalayan states have also shown an increase in both area and production of stone fruits (Gangwar et al. 2005; Baba et al. 2012; Baba 2018).

13.2 Status of Stone Fruits

There are a number of fruits which are under stone fruit category; however, we have considered almond, peach, plum, cherries and apricot in this study. The time-series data, with effect from 1965 to 2018, pertaining to area and production of these stone fruits were obtained from the official website of the Food and Agriculture Organization (FOA) (http://www.fao.org). The figures around the world imply the aggregate of all the producing countries of the world, and the yield of fruits around the world indicates the average yield of fruits produced in different countries. Stone fruits form important crops of cropping pattern across the world. Currently, important stone fruits occupy an area of over 7000 hectares throughout the world of which major proportion of area is under plum (2649 hectares) followed by almond (http://www.fao.org). An attempt has been made to see the distribution of stone fruits across major producers in the world, and this is presented in this section.

13.2.1 Countries with Maximum Area Under Stone Fruits

If we talk about the area under important stone fruits, Spain has allotted an area of over 6.6 lakh hectares under almond followed by the USA and Turkey. The total area under India is estimated at 0.07 lakh hectare under almond and appears at the

18th rank after Pakistan. The area under almond falls exclusively under a temperate territory, Jammu and Kashmir. We have seen a decline of area under this fruit owing to climate change; the farmers fail to irrigate their almond crop even during crucial day owing to the less availability of irrigation water at highlands which form traditional habitat for almond in this region. As far as apricot is concerned, Turkey tops the list of apricot-growing countries in terms of area under it, allocating an area of about 1.26 lakh hectares. Iran and Uzbekistan come next in the list, while the area under this fruit in India is only 0.06 lakh hectares. Within India, apricot is localized in cold-arid region of union territory of Ladakh. In this region, this crop is challenged by the infestation of codling moth for which the fresh export of this fruit is completely prohibited. Turkey has allotted 0.84 lakh hectares of area under cherries and appears at top in terms of area under this fruit followed by the USA, Chile and Syria. India has only very small area under cherries and appears at the last among 25 top countries of the world in terms of area. In terms of the area under peach including nectarine and plum, China appears at the top of the list, while India has better position in terms of area allocation under these stone fruits. India appears at eighth and seventh place with respect to area under peach and plum (http://www.fao.org). There is a need to demand assessment of stone fruits in the trading partners of India, and accordingly, productive capacities need to be allotted for the production of these crops in India (Fig. 13.1).

13.2.2 Countries with Maximum Production of Stone Fruits

As far as the production of important stone fruits is concerned, the USA appeared as the top producer of almond in the world which may be due to much higher yield level of this fruit compared to Spain. Spain is consistent with its maximum area allocation and appears next in the list of major almond producers in the world followed by Iran and Morocco. The total area under India is estimated at 0.07 lakh tonnes under almond and appears at the 23rd rank in the world. Owing to the decline in yield level and area under this fruit, India has slipped down in the list of production of almond crop. As far as apricot is concerned, Turkey tops the list in apricot production, producing about 7.5 lakh tonnes. Iran and Uzbekistan come next in the list, while India appears nowhere in the list of 25 major apricot-producing countries owing to much less productivity of this fruit. Turkey is producing 6.4 lakh tonnes of cherries and appears at the top in terms of production of this fruit followed by the USA, Uzbekistan and Chile. The appearance of Uzbekistan at the third place among top cherry-producing countries is because of maximum yield of this fruit in this country. Although India has only very small area under cherries and appears at the last among top countries of the world, India appears nowhere in the list, and this scenario is due to much less productivity of this fruit in India which emphasizes upon dissemination of production technologies in its niche areas. In terms of the production of peach including nectarine and plum, China appears at the top of the list, while India has better position in terms of production of these stone fruits. India

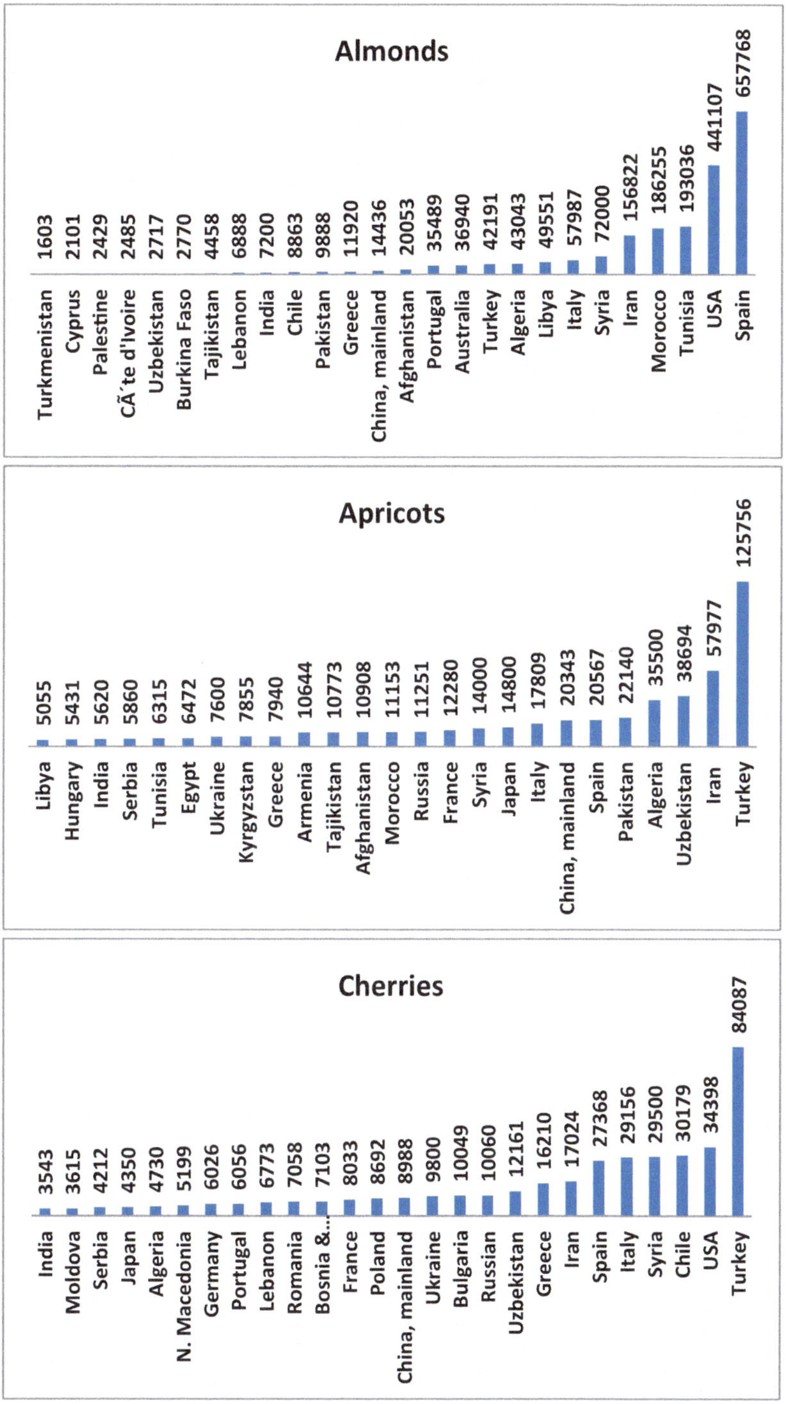

Fig. 13.1 Area under stone fruits in major producing countries of the world (ha). Source: Authors' calculation on FAO data (http://www.fao.org)

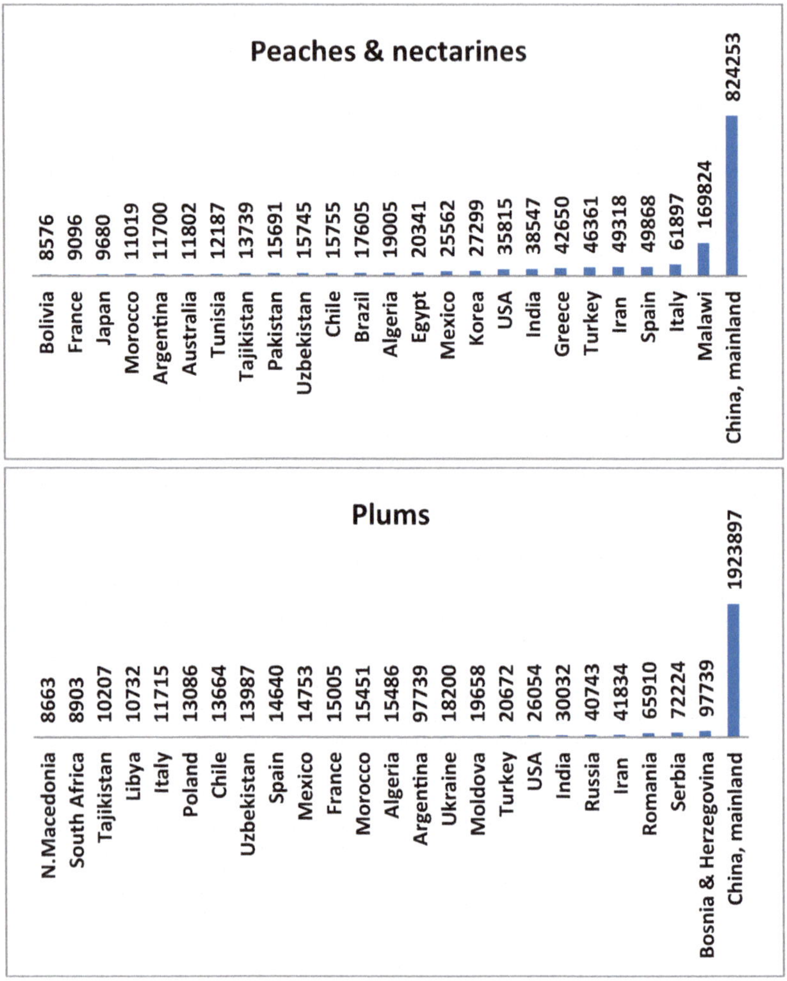

Fig. 13.1 (continued)

appears at ninth and seventh place with respect to production of peach and plum (Fig. 13.2).

13.2.3 Trends in the Area, Production and Yield of Stone Fruit in the World

Although India is not the maximum producer of important stone fruits in the world, this country has a potential of expanding more area under these fruits, and production could be increased manifolds if research and development efforts are focused to improve yield levels of these crops. The available production technologies have

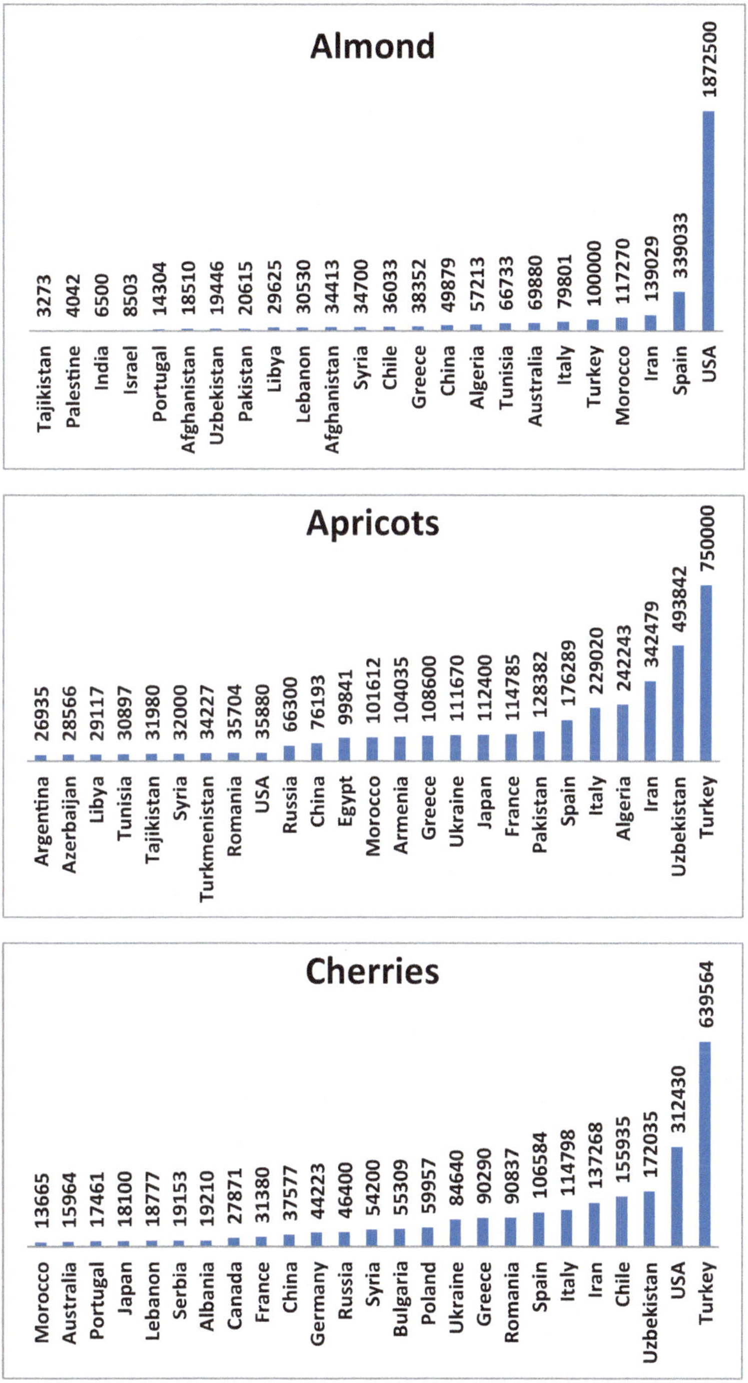

Fig. 13.2 Production of stone fruits in major producing countries of the world (tonnes). Source: Authors' calculation on FAO data (http://www.fao.org)

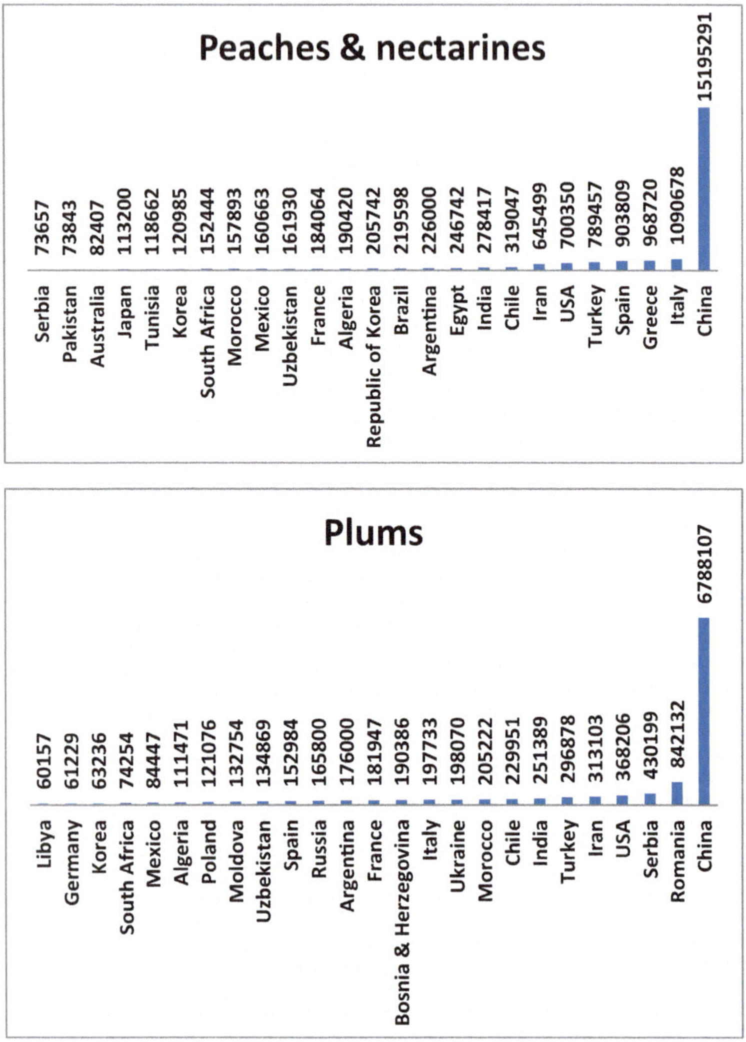

Fig. 13.2 (continued)

to be disseminated with less lag for which extension network has to be strengthened and streamlined. In this backdrop, an attempt has been made in this section to examine the growth of stone fruits in the world as aggregate, and in India, the trend for area and production of stone fruits has been examined for India to have an idea about average scenario of production and yield in the world. Moreover, lesson from this analysis would help to develop a road map for the encouragement of stone fruit economy in the developing nations and in India in particular.

Since the early 1960s, the area under important stone fruits has shown a significant increase with few fluctuations in some years. The expansion of area under stone fruits was more pronounced for plum. The area under peach including nectarine has also increased though it is still around 2 percent of world area. Other stone fruits in India constitute only 1 percent of world area under those crops. Peach and cherries have exhibited a favourable growth across the world and in India (Table 13.1) though there has been 3.50- and 5.85-fold increase in peach and plum, respectively, in India which is appreciable.

The world has harvested 24,453 tonnes of peach in 2018 which is relatively higher among stone fruits produced in the world. Plum comes next in terms of production around the world (Table 13.2). Apricot production is less than one-third of plum production around the world, while cherries and almond comes after this stone fruit. The production scenario of stone fruits in India seems to be in tandem with world production of these fruits. India produces 1.14 and 1.99 percent of peach and plum, respectively, while the production of other stone fruits in India is less than one percent. Plum production in India has exhibited a drastic increase, and the increase in peach and cherries in India is relatively better than the world's increase in production. The expansion of area under plum and peach and the increase in production of these crops reflect the steady increase of global demand for these fruits.

The yield levels of different stone fruits have exhibited a distinct pattern over the years throughout the world vis-à-vis India though it is based significantly upon bulkiness of the produce. While the yield of almond has shown a good progress across the world, it has been gloomy trend in the case of plum and cherries. On the other hand, the yield level of plum in India is double the global yield; however, the yield of other stone fruits in the country has been significantly low. Considering the constraints in bringing more area under the cultivation of these fruits, the emphasis would really be on the enhancement of yield levels, and the demand for production technologies would apparently be warranted. Studies have exhibited huge technological gaps of important fruit crops including stone fruits in the niche area in India (Baba et al. 2012, 2014). Extension network has to be streamlined to disseminate available technologies to harness the gain from technology adoption (Table 13.3).

In order to observe the pattern of growth, the entire reference period of the data was divided into two sub-periods, viz. period I (1965–1991) and period II (1992–2018), simply for clarity of the growth (Table 13.4). The estimates of compound growth rates exhibited a diverse scenario, i.e. in some cases, area is expanding at faster pace towards later period; in some cases, production is assuming more growth rate; and in few other cases, the growth seems more in period II which is the more favourable scenario and is desired. In the world average figures, we can see that while the pace of growth of area under almond has decreased over the years, the production and yield levels have increased at faster pace implying that the production of this crops is increasing owing more to yield gains though the area expansion has also contributed to this increase. The increasing rate of growth in period II of apricot seems to have been the most desired scenario across the world. The declining area and production of apricot in India during period I has been reversed, and despite deceleration of yield, production has appreciably increased over the years. The scenario of cherries seems to be discouraging where the area has been decreasing at increasing rate; the negative growth of yield has restricted the growth of

Table 13.1 Area under some important stone fruits in the world and India (000 ha)

Year	Almonds			Apricots			Peaches and nectarines			Plums			Cherries		
	World	India	India % of world	World	India	India % of world	World	India	India % of world	World	India	India % of world	World	India	India % of world
1965	789.8	–	–	230.0	3.0	1.3	592.6	11.0	1.9	448.4	3.5	0.8	102.4	1.35	1.32
1970	918.0	–	–	277.6	3.6	1.3	690.0	11.0	1.6	605.3	4.7	0.8	154.3	1.40	0.91
1975	1093.1	9.67	0.88	270.5	3.0	1.1	745.0	11.0	1.5	580.5	5.5	0.9	171.3	1.50	0.88
1980	1222.0	16.33	1.34	294.1	2.0	0.7	754.4	14.7	1.9	631.1	7.0	1.1	167.7	1.20	0.72
1985	1349.1	17.40	1.29	314.0	1.6	0.5	996.3	15.5	1.6	1261.4	10.0	0.8	290.5	1.30	0.45
1990	1446.7	19.20	1.33	316.5	2.0	0.6	1331.0	15.5	1.2	1517.6	12.5	0.8	292.5	1.30	0.44
1995	1486.9	19.32	1.30	397.6	2.5	0.6	1381.2	18.3	1.3	1679.6	14.1	0.8	314.0	1.63	0.52
2000	1649.7	18.06	1.09	435.5	3.5	0.8	1274.5	21.5	1.7	2184.5	14.2	0.7	335.7	2.81	0.84
2005	1698.5	15.55	0.92	480.6	4.0	0.8	1506.8	33.5	2.2	2254.8	20.0	0.9	355.6	3.06	0.86
2010	1717.3	17.65	1.03	564.4	4.8	0.9	1537.0	37.0	2.4	2411.2	25.0	1.0	396.6	3.23	0.81
2015	1747.5	7.13	0.41	559.5	5.0	0.9	1626.7	37.9	2.3	2500.6	27.7	1.1	414.0	3.64	0.88
2018	2041.3	7.10	0.35	548.8	5.6	1.0	1712.4	38.5	2.3	2649.0	30.0	1.1	442.0	3.54	0.80
Fold Δ	2.58	0.73	0.39	2.39	1.87	0.79	2.89	3.50	1.21	5.91	8.58	1.45	4.32	2.62	0.61

Source: Authors' calculation on FAO data (http://www.fao.org)

Table 13.2 Production of some important stone fruits in the world and India (000 tonnes)

Year	Almonds, with shell			Apricots			Peaches and nectarines			Plums			Cherries		
	World	India	India % of world	World	India	India % of world	World	India	India % of world	World	India	India % of world	World	India	India % of world
1965	605.1	–	–	1362.0	9.2	0.68	5816.1	48.0	0.83	4815.9	14.0	0.29	1112.7	3.40	0.31
1970	676.7	–	–	1630.3	10.8	0.66	6371.9	48.5	0.76	6144.3	19.0	0.31	1460.0	3.90	0.27
1975	752.4	1.81	0.24	1546.6	9.5	0.61	6542.2	48.0	0.73	4979.5	22.0	0.44	1359.8	4.00	0.29
1980	919.7	1.86	0.20	1734.6	6.5	0.37	7535.1	78.0	1.04	5992.3	28.0	0.47	1279.1	2.00	0.16
1985	1145.6	2.59	0.23	2028.9	4.8	0.24	7752.5	65.0	0.84	6558.4	40.0	0.61	1529.9	3.00	0.20
1990	1315.4	2.21	0.17	2189.1	6.0	0.27	9397.4	70.0	0.74	6110.9	50.0	0.82	1397.3	3.20	0.23
1995	1047.6	6.57	0.63	2088.7	7.3	0.35	10924.8	89.0	0.81	6191.7	58.0	0.94	1627.0	4.40	0.27
2000	1452.2	10.90	0.75	2863.0	9.9	0.35	13281.5	120.0	0.90	8404.9	78.0	0.93	1898.8	7.80	0.41
2005	1824.6	14.33	0.79	3626.0	11.3	0.31	18041.5	203.3	1.13	9830.4	146.2	1.49	1843.6	8.77	0.48
2010	2569.8	12.51	0.49	3303.5	13.6	0.41	20540.5	244.0	1.19	10699.9	200.0	1.87	1998.3	9.55	0.48
2015	2480.6	7.06	0.28	3960.1	14.2	0.36	23870.9	264.8	1.11	11546.4	228.9	1.98	2236.9	11.09	0.50
2018	3209.9	6.36	0.20	3871.1	16.0	0.41	24453.4	278.4	1.14	12608.7	251.4	1.99	2563.1	10.95	0.43
Fold Δ	5.30	3.52	0.82	2.84	1.73	0.61	4.20	5.80	1.38	2.62	17.96	6.86	2.30	3.22	1.40

Source: Authors' calculation on FAO data (http://www.fao.org)

Table 13.3 Yield of some important stone fruits in the world and India (q/ha)

Year	Almonds		Apricots		Peaches and nectarines		Plums		Cherries	
	World	India	World	India	World	India	World	India	World	India
1965	7.7	–	59.2	30.7	98.1	43.6	107.4	40.0	108.7	25.2
1970	7.4	–	58.7	30.0	92.4	44.1	101.5	40.4	94.6	27.9
1975	6.9	1.87	57.2	31.7	87.8	43.6	85.8	40.0	79.4	26.7
1980	7.5	1.14	59.0	32.5	99.9	53.1	95.0	40.0	76.3	16.7
1985	8.5	1.49	64.6	30.0	77.8	41.9	52.0	40.0	52.7	23.1
1990	9.1	1.15	69.2	30.0	70.6	45.2	40.3	40.0	47.8	24.6
1995	7.0	3.40	52.5	29.0	79.1	48.6	36.9	41.1	51.8	27.1
2000	8.8	6.04	65.7	28.6	104.2	55.9	38.5	54.9	56.6	27.8
2005	10.7	9.22	75.4	28.2	119.7	60.6	43.6	73.2	51.8	28.7
2010	15.0	7.08	58.5	28.3	133.6	65.9	44.4	80.0	50.4	29.6
2015	14.2	9.90	70.8	28.3	146.7	69.9	46.2	82.7	54.0	30.5
2018	15.7	8.96	70.5	28.4	142.8	72.2	47.6	83.7	58.0	30.9
Fold Δ	2.05	4.79	1.19	0.93	1.46	1.66	0.44	2.09	0.53	1.23

Source: Authors' calculation on FAO data (http://www.fao.org)

production on ground of area expansion. In India, cherries have exhibited a favourable growth in area, production and yield levels. The growth of yield and production of plum and peach at yet faster pace towards period II throughout the world as well as in India together is desired though the role of technologies could not be ruled out for the sustenance of economy of stone fruits in view of the fact that the major increase in production was due to area expansion and the gain on account of yield increase would help to achieve scaling up the production manifolds.

13.2.4 Forecasted Value of Area, Production and Productivity of Stone Fruits

The estimated linear equation for area/production/yield for important stone fruits considered in this along with the coefficient of determination has been presented in Table 13.5. The positive regression coefficient of all equation except production of peach and plum indicated that cultivation of stone fruits could be taken up on yet more commercial scale and could be considered for upliftment of farming community. The area, production and productivity of stone fruits were predicted for the year 2030 by maintaining the current trends and employing linear function. The forecasted area, production and productivity of fruits in the world as a whole and in India revealed that there may be a significant increase in area and production of these fruits if the current trends will be maintained. The forecast indicated that there could be an expansion of area under almond by over 7 percent; however, due to the increase in area under non-bearing orchards, there will be an increase of only 2.46 percent in yield level and 1.39 percent decline in production of the entire world. While the estimated value for apricot in the world has revealed an increase of area

Table 13.4 Estimates of compound growth rates (%)

		World			India		
Fruit	Item	Period I	Period II	All	Period I	Period II	All
Almond	Area	2.60*	0.98*	1.55*	3.78*	−2.73*	0.08
		0.12	***0.07***	***0.07***	***0.48***	***0.79***	***0.34***
	Yield	0.56*	2.80*	1.53*	9.76*	0.39	6.11*
		0.28	***0.29***	***0.13***	***2.42***	***1.91***	***0.84***
	Production	3.17*	3.79*	3.08*	5.97*	3.12	6.02*
		0.28	***0.28***	***0.10***	***2.60***	***1.94***	***0.81***
Apricot	Area	1.02*	1.73*	1.89*	−3.52*	3.58*	1.25*
		0.20	***0.10***	***0.07***	***0.42***	***0.15***	***0.29***
	Yield	0.75*	0.96*	0.31*	−0.15*	−0.14*	−0.19*
		0.24	***0.21***	***0.09***	***0.06***	***0.02***	***0.02***
	Production	1.77*	2.69*	2.20*	−3.67*	3.45*	1.06*
		0.19	***0.23***	***0.08***	***0.43***	***0.13***	***0.29***
Cherries	Area	3.88*	1.41*	2.67*	−0.49	3.49*	2.31*
		0.36	***0.05***	***0.12***	***0.25***	***0.34***	***0.18***
	Yield	−3.27*	0.09	−1.44*	−1.96*	0.58*	0.51*
		0.42	***0.14***	***0.16***	***0.70***	***0.03***	***0.21***
	Production	0.61	1.50*	1.24*	−2.45*	4.07*	2.82*
		0.23	***0.15***	***0.08***	***0.82***	***0.34***	***0.35***
Peach	Area	2.89*	0.91*	2.06*	1.73*	3.71*	2.91*
		0.32	***0.13***	***0.11***	***0.25***	***0.30***	***0.12***
	Yield	−1.21*	2.73*	0.90*	0.49	1.68*	0.94*
		0.29	***0.12***	***0.16***	***0.30***	***0.05***	***0.09***
	Production	1.68*	3.64*	2.95*	2.22*	5.39*	3.85*
		0.12	***0.14***	***0.08***	***0.32***	***0.32***	***0.16***
Plum	Area	4.43*	2.02*	3.89*	5.31*	3.72*	4.09*
		0.56	***0.15***	***0.17***	***0.30***	***0.23***	***0.11***
	Yield	−3.61*	0.65*	−2.16*	−0.98*	3.31*	1.74*
		0.56	***0.13***	***0.21***	***0.38***	***0.24***	***0.19***
	Production	0.82*	2.67*	1.72*	4.33*	7.03*	5.83*
		0.18	0.16	0.09	0.20	0.31	0.13

Figures in italic and bold are standard errors of growth rates; * Denotes significance at 0.05 or better probability level
Source: Authors' analysis on FAO data (http://www.fao.org)

and production by over 19 percent, its yield may increase by 3.47 percent only. On the other hand, there would be a decline of 13.42 percent in apricot production in India, and this decline was due to the contraction of area under this fruit. If the current trend prevails, there would be an increase of about 20 percent in the area under cherries; however, the forecasted values imply about 56.12 percent decline in yield of this fruit. The Indian trends indicated increase of 12.3 percent in area and 9.33 percent increase in production of cherries. The estimated future values for peach in the world and India indicate that although on current trend the area under this fruit is expected to increase, the expected increase in production of this fruit is offset to

Table 13.5 Estimates of linear regression and forecasted value for 2030

Fruit	Items	World					India				
					Forecasting value for 2030					Forecasting value for 2030	
		b_0	b_1	R^2	Value	% change	b_0	b_1	R^2	Value	% change
Almond	Area (ha)	20,253	849,387	0.96	2,186,086	7.09	15,910	16.72	0.0034	16,847	137.41
	Yield (hg/ha)	161	5504	0.71	16,111	2.46	−129	199	0.6487	10,997	22.67
	Production (tonnes)	44,064	257,042	0.88	3,165,268	−1.39	605	279	0.5284	16,218	154.95
Apricot	Area (ha)	7120	183,735	0.92	653,659	19.10	45	2043	0.36	5010	−10.85
	Yield (hg/ha)	208	59,249	0.19	72,981	3.47	−55	30,824	0.71	27,222	−4.12
	Production (tonnes)	55,453	969,237	0.88	4,629,164	19.58	111	6512	0.29	13,816	−13.42
Cherries	Area (ha)	6546	96,526	0.95	528,574	19.60	50	683	0.77	3982	12.39
	Yield (hg/ha)	−1086	97,106	0.58	25,449	−56.12	125	22,642	0.17	30,877	−0.08
	Production (tonnes)	20,999	1,089,507	0.83	2,475,426	−3.42	165	1085	0.72	11,972	9.33
Peach	Area (ha)	21,801	553,307	0.89	1,992,145	16.33	623	4310	0.87	45,437	17.87
	Yield (hg/ha)	1027	75,189	0.46	142,955	0.11	514	39,445	0.67	73,400	1.62
	Production (tonnes)	367,270	2,221,056	0.89	26,460,850	8.21	4717	−6272	0.84	305,035	9.56
Plum	Area (ha)	49,637	132,711	0.95	3,408,734	28.68	504	1	0.93	33,289	10.85
	Yield (hg/ha)	−1474	104,283	0.68	7030	−85.23	980	25,149	0.69	89,819	7.30
	Production (tonnes)	135,440	3,884,722	0.86	12,823,736	1.71	4494	−38,538	0.82	258,091	2.67

Source: Authors' analysis on FAO data (http://www.fao.org)

a significant extent by the decline of its yield levels. If the current trend for area of plum prevails, then its area may expand by 11 percent in India, and there will be an increase in forecasted yield, but the total production is estimated to increase by only 3 percent. Contrary to this, the aggregate yield of plum throughout the world is estimated to decline by about 85 percent which is really a cause of concern for the sustenance of this fruit crop. Although the forecasted values are only indicative of the evolving scenario, it emphasized upon R&D efforts to innovate for the enhancement of yield level of stone fruits. Encouragement of adoption of production technologies and improvement of resource use efficiency would have long-term positive influence on the upscaling of production of stone fruits.

13.3 Economic Perspective: Micro-evidences

In view of the considerable demand of stone fruits across the globe or among a particular group of consumers, it became imperative to investigate into their economic feasibility in their niche areas. For this important piece of investigation, this study chose a temperate and a cold-arid region which form niche for the growth of these crops. Ladakh region for cold-arid agroclimatic setting and Kashmir region for temperate setting were purposively selected to have an economic outlook of stone fruits in India.

13.3.1 Scenario in the Niche Area (Jammu and Kashmir)

It could be seen from Table 13.6 that the area under important stone fruits in the erstwhile Jammu and Kashmir state constituted about 7 percent total states' fruit area though this proportion has decline by 2.5 percent since 2007. Similarly, the production of stone fruit as percentage of total fruit production in the state has decline from 2.9 percent in 2007 to just 1.9 percent in 2016. The decline was more pronounced in the yield levels of these fruits. While almond and apricot occupy major part of the area under stone fruits, apricot, plum and peach constitute major proportion of stone fruit production.

An attempt was made to quantify determinants of change in area under stone fruits employing Nerlovian model. Regression estimates revealed that lagged prices, lagged area under these fruits and production were significant determinants of area expansion under important stone fruits in the study area. The estimates of adjusted R^2 indicated that the variables specified in the models gave best fit to the estimates (Table 13.7). The estimates highlighted that production improvement would encourage farmers to allocate more area under fruits. However, it again emphasizes upon productivity improvement through adoption of input/production technologies. The price incentives would also improve returns to farmers, which should kept pace with the increase in input prices.

Table 13.6 Area, production and yield of stone fruits in the study area (Jammu and Kashmir, India)

	2007					2016					Difference		
	Area		Production		Yield	Area		Production		Yield	Area	Production	Yield
Fruit	000 ha	% of total	000 tonnes	% of total	(q/ha)	000 ha	% of total	000 tonnes	% of total	(q/ha)	(000 ha)	(000 tonnes)	(q/ha)
Almond	16.4	54.0	11.3	24.1	6.9	7.1	31.2	6.4	14.9	9.0	−9.3	−4.9	2.10
Apricot	4.8	15.7	12.8	27.3	26.7	6.2	27.1	12.8	30.0	20.8	1.4	0.1	−5.88
Peach	2.2	7.3	4.3	9.1	19.1	2.6	11.5	5.9	13.8	22.6	0.4	1.7	3.59
Plum	3.8	12.6	7.8	16.7	20.3	4.0	17.7	9.3	21.9	23.2	0.2	1.5	2.85
Cherries	3.1	10.3	10.6	22.7	33.9	2.8	12.4	8.3	19.4	29.3	−0.3	−2.3	−4.59
Total	30.4	100.0	46.7	100.0	15.4	22.7	100.0	42.7	100.0	18.8	−7.7	−4.0	3.43

Source: Anonymous (2007, 2016b)

Table 13.7 Estimates of regression model

Variables	b_0	$Area_{-1}$	Production	Price	$Price_{-1}$	Adj. R^2	Fcal
Coefficient	−2386	0.57[a]	0.03[a]	9.93	3.06[a]	0.6461	133.0[a]
		(0.06)	(0.00)	(14.7)	(0.84)		

Figures within parentheses indicate standard error
[a]Denotes significance at 0.05 or better probability level

Table 13.8 Cost of cultivation of different-age apricot orchards (%)

Age of orchard	Labour	Manuring/fertilization	Others	Total variable cost (Rs/kanal)
1st	58.8	12.3	28.9	2341
2nd	86.0	10.5	3.5	554
3rd	84.7	11.8	3.5	657
4th	85.0	12.1	2.9	815
5th to 10th	82.9	14.3	2.9	1001
10th to 15th	84.7	13.1	2.1	1440
1fifth to 25th	85.6	12.5	1.9	1812
> 25th	83.4	14.6	2.0	2385

Source: Authors' calculation data collected in a field survey, 2018

13.3.2 Economic Feasibility of Stone Fruit Cultivation

13.3.2.1 Apricot

Cost of Apricot Cultivation

Within apricot-dominated area, this stone fruit alone occupied about 54 percent of gross of cropped area due to natural niche that favours its successful cultivation under cold-arid agroclimatic setting of Ladakh. The cropping intensity in the study area was only 100.34 percent owing to the fact that this region received harsh climatic conditions for the most part of the year that prevents multiple cropping of major crops growing in this region. Apricot is cultivated more or less organic in cold-arid region of Ladakh, India. All the farmers were found applying farmyard manure (FYM) available at their farms and only small quantity of urea on their orchard (Table 13.8). About Rs 2341 are required for establishment of new one kanal (1 kanal = 1/20 hectare) orchard of apricot, but later, this cost declines by about one-fourth in the second year and gradually increased with the age of orchard. In the old age, orchards about Rs 2400 per kanal are invested in variable inputs. Across different age groups, the share of labour cost accounted for 59–86 percent of total cost on variable items. It could be seen from the table that the labour cost alone accounted for major share of total variable costs. The favourable return to farm fixed resources (RFFR) ratio is also indicative of better economic feasibility of apricot cultivation compared to other crops (Table 13.9). Accordingly, it is better to shift more area towards apricot if other supplementary factors required for its better performance are made available around apricot-growing belt of this region.

Table 13.9 Economic feasibility of apricot

Crop	Total variable costs (Rs/ha)	Gross return (Rs/ha)	RFFR
Apricot	27,510	47,690	1.73

Source: Authors' calculation data collected in a field survey, 2018

Marketing System of Apricot

It was seen that the overall marketed surplus was more than 60 percent of the total apricot production per farm. Apricot produce is harvested either with fastening green nets underneath foliage and fruit part of the tree to collect the delicate fruits that may drop during harvesting or without nets that reduced fruit loss up to the extent of 30 percent. However, only 17.5 percent of farmers in the study area were found making use of these nets. The proportion of marketed surplus was little higher in Halman variety (a sweet variety) followed by Khantay (a bitter wild variety to be used for oil extraction).

Major proportion of apricot surplus was observed to be sold after drying (88.5%) it in sun, and the rest (11.5%) goes into the market as fresh. The marketing of apricot in Ladakh is not well organized. The commonly encountered channels identified in the marketing of apricot and its products in the region have been presented in Table 13.10. A majority of the farmers sold their surplus of dried apricot to contractors (village traders). As high as 16 and 12 percent of farmers sold surplus of dry fruits directly to wholesalers at Jammu and Srinagar markets, respectively.

There is no channel of fresh apricot marketing with respect to markets outside Ladakh owing to restrictions on shifting apricot as fresh to prevent transmission of codling moth infestation in other regions. Out of total average marketable surplus available per farm, 21 percent was sold as fresh. Out of total fresh sales, over 71 percent of surplus was sold through retailers as reported by 66.5 percent of farmers. There are two modernized supply channels (III and IV) of apricot marketing in which fresh apricot sold by producers to processing units was value added and sold as apricot nectar (or ready-to-serve (RTS) drink) as in Channel III or apricot jam as in Channel IV. Though modernized channels prevent losses to a better extent and ensure higher absolute returns to producers, fresh apricots which were sold through these channels constituted just10 percent of total surplus.

Price Spread in Marketing of Fresh Apricot Through Traditional/ Modernized Channels

In Channel I, producer comes to daily local mandis, made stall of their produce and sold the fresh apricot directly to the consumers. Producers incurred all the expenses to take their marketable surpluses to the consumers. In this channel, producers realized only 58.58 percent of consumer's price as net return; the lower return in channel was due to the inclusion of postharvest losses (Table 13.11). In Channel II, retailer exists as one of the intermediaries between producer and consumer. Here, the producers sold their produce to the retailers in the daily local mandi. At the

Table 13.10 Channels commonly perused by farmers in apricot marketing (Percent)

Channels	Functionaries									Farmer	Surplus
Sale as dry											
I	Farmer	→	Wholesaler (S)	→	Retailer (S)	→	Consumer			11.9	13.5
II	Farmer	→	Wholesaler (J)	→	Retailer (J)	→	Consumer			16.3	16.4
III	Farmer	→	Retailer (L)	→	Consumer					9.4	7.3
IV	Farmer	→	Contractor	→	Wholesaler (S)	→	Retailer (S)	→	Consumer	27.5	26.3
V	Farmer	→	Contractor	→	Wholesaler (J)	→	Retailer (J)	→	Consumer	31.5	30.8
Others										3.5	5.7
Sale as fresh											
I	Farmer	→	Consumer							21.3	10.0
II	Farmer	→	Retailer	→	Consumer					66.5	71.3
III	Farmer	→	Processing unit	→	Wholesaler	→	Retailer	→	Consumer	5.5	8.3
IV	Farmer	→	Processing unit	→	Wholesaler	→	Retailer	→	Consumer	4.1	7.3
Others										2.6	3.3

Note: *S* = Srinagar (wholesale market at 400 km distance), *J* = Jammu (distance at 450 km) and *L* local
Channels III and IV in fresh apricot marketing are w.r.t. ready-to-serve (RTS) drink and jam making
Source: Authors' calculation data collected in a field survey, 2018

Table 13.11 Price spread in marketing of fresh apricot (% of consumers' price)

Particulars	Channels			
			Modernized channel	
	I	II	III	IV
Net price received by farmer	58.58 (6.74)	36.77 (4.78)	2.00 (8.66)	5.01 (8.66)
Consumers' price	100.0 (11.5)	100.0 (13.0)	100.0 (432.0)	100.0 (172.8)

Note: Figures within parentheses indicate absolute amount in rupees
Source: Authors' calculation data collected in a field survey, 2018

Table 13.12 Price spread in marketing of dried apricot (% of consumers' price)

Particulars	Channels				
	I	II	III	IV	V
Net price received by farmer	62.36 (123.5)	64.31 (130.6)	60.96 (94.5)	40.13 (79.5)	39.14 (79.5)
Consumers' price	100.0 (198)	100.0 (203)	100.0 (155)	100.0 (198)	100.0 (203)

Note: Figures within parentheses indicate absolute amount in rupees
Source: Authors' calculation data collected in a field survey, 2018

retailer's level, spoilage of produce forms another cost component (Table 13.12). The producers received about 37 percent of consumer's price; however, it was observed that the returns to farmer and margins of retailer were badly affected when spoilage costs were considered. Channels III and IV are modernized chains of apricot supply wherein processors invested huge amount on value addition of fresh produce purchased from producers. Processors value-added fresh apricots in the form of either apricot ready-to-serve (RTS) drink or apricot jam and sold it off to consumers through wholesalers and retailers (Table 13.11). Although the net price received by the producers in this channel constituted only meagre proportion of consumers' price, this price was relatively more in absolute terms compared to other channels. To sum up, producer received higher returns (in absolute terms) in the channels through which they pushed their produce to processing units, yet despite this fact, only 10 percent of fresh produce was sold through Channels III and IV.

Price Spread in Marketing of Dried Apricot

The price spread in various marketing channels of dried apricot has been presented in Table 13.12. In all the five channels of dried apricot marketing, producers dried the apricot in their own traditional way and spent on labour and ingredients. It could be concluded that producers received higher proportion of consumers' price as net return in channels with lower number of intermediaries. Producers were found to receive higher returns (both in absolute and in proportionate terms) in first two channels because they sold off their produce directly to the wholesalers.

Table 13.13 Cost of cultivation of different-age almond orchards (%)

Age (years)	Labour	Manuring/ fertilization	Pesticides	Others	Total variable costs (Rs/kanal)
1st	50.2	10.2	0.0	39.7	1950
2nd	73.5	26.3	0.0	0.2	293
3rd	80.5	19.3	0.0	0.3	550
4th	81.4	18.4	0.0	0.2	869
5th to 10th	59.2	19.1	21.6	0.1	1437
10th to 15th	56.3	25.5	17.6	0.6	1843
1fifth to 25th	58.7	25.6	15.5	0.2	2582
> 25th	62.5	25.4	11.7	0.4	3499

Source: Authors' calculation data collected in a field survey, 2018

Table 13.14 Economic feasibility of almond and other competing crops

Crop	Total variable costs (Rs/ha)	Gross return (Rs/ha)	RFFR
Almond	46,557	122,049	2.62

Source: Authors' calculation data collected in a field survey, 2018

13.3.2.2 Almond

It was observed that about 58 percent of the total cropped area was allocated to almond fruit. The paddy and vegetables were mostly grown under irrigated conditions and occupied about 25.79 and 1.13 percent of cropped area, respectively. The cropping intensity in this region was just 112.6 percent indicating thereby that farmers were generally growing two crops in a year on a small portion of cropped area. Out of the total area of 14.85 kanals under almond in the study area, 13.14 kanals were alone under Kagzi variety. Wont, a wild variety of almond, was also visible in parts of the study area, though their proportion in total area was very meagre.

Cost of Almond Cultivation

The structure of variable costs involved in the cultivation of different-age almond on one kanal of land has been estimated and presented in Table 13.13. The cost on variable inputs keeps on increasing with the age of orchard except in the first year of orchard establishment. Since the new orchard is to be laid on new soil, it requires a handsome amount to be spend on land improvement, seedlings, digging/filling of pits and other items that resulted in higher cost on variables in this period compared to other years. In all the age groups, the cost on labour constituted higher proportion of total variable costs followed by manuring and fertilization. The cost on pesticides was almost absent up to fourth year of almond orchard, but later, it also constituted a good proportion of total variable cost though in >25-year orchards it was about 12 percent.

Almond was more profitable in the study area as indicated by RFFT ratio which indicated that one rupee investment on variable cost of almond cultivation would generate 2.62 rupees as gross returns (Table 13.14). This ratio seems more favourable for this crop compared to other competing resources in the region.

Marketing Arrangements

On an average, each farm has 692 quintal of almond produce of which the overall marketed surplus constitutes 81 percent. Loss at field stage accounted for the major proportion of total losses, and as high as 114 q/farm goes as waste at farm owing to scarcity of skilled labour at harvesting and traditional methods of harvesting/dehulling. As high as 54 percent of total surplus of almond is marketed as nut, and the rest is value added and sold as kernels. In papery variety (thin shelled), over 73 percent of surplus was sold after converting them into kernels due to the fact that it is easy to extract kernels out of delicate shells of this variety. Contrary to this, only 13 percent of wont (wild) variety surplus could be value added and sold as kernel. In view of increasing demand of kernels and import of this product from other countries of the world, there is need to replace wont variety with papery variety for achieving higher possible surplus of highly accepted.

Although almond is sold either as kernel or as nuts, there are no different channels for their marketing. The marketing of almonds is not well organized; the commonly encountered channels identified in its marketing have been presented in Table 13.15. A majority of the farmers sold their surplus to preharvest contractors. Out of total average marketable surplus available per farm, 80.5 percent was sold out through these contractors. As high as 12 and 11 percent of farmers sold surplus directly to wholesalers at Jammu and Srinagar markets, respectively.

Price Spread

The price spread in various marketing channels of almond as nut as well as kernel has been presented in Table 13.16. In all the five channels of almond marketing, producers received higher proportion of consumer's price in the channel where he sold his produce directly to retailer. Presence of preharvest contractors in marketing channel of almond was found to deprive farmers from real benefits, and they received relatively lower price for their produce when they sold standing fruit to these intermediaries. It could be concluded that producers received higher proportion of consumers' price as net return in channels with lower number of intermediaries.

13.3.2.3 Cherry

Cropping pattern in the study farming systems revealed dominance of cherry in terms of area allocation. In addition to cherry, common crops grown under this were paddy, mustard, fodder, etc. The paddy is mostly grown under irrigated conditions and occupied about 9.3 percent of cropped area. Vegetables were cultivated on 7.98 percent of cropped area, and the cropping intensity in this region was over 112 percent.

Cost of Cherry Cultivation

The cost of cultivation of cherry has been worked out and presented in Table 13.18. Cherry fruit is propagated mainly through suckers, which are cut from main mother plant and maintained at new farms. The nursery of cherry involves good amount for

Table 13.15 Marketing channels and net price received by farmers (%)

Channel	Functionaries									Produce	Farmers
I	Farmers	→	Wholeseller (S)	→	Retailer (S)	→			Consumers	11	6
II	Farmers	→	Wholeseller (J)	→	Retailer (J)	→			Consumers	12	8
III	Farmers	→	Retailers	→	Consumers					7	4
IV	Farmers	→	Contractor	→	Wholeseller (S)	→	Retailer (S)	→	Consumers	40	43
V	Farmers	→	Contractors	→	Wholeseller (J)	→	Retailer (J)	→	Consumers	29	37.5
VI	Others									1	1.5
										(100.0)	(100.0)

Note: S = Srinagar (wholesale market at 55 km distance) and J = Jammu (distance at 315 km)
Source: Authors' calculation data collected in a field survey, 2018

Table 13.16 Price spread in marketing of almond

Channel	Nut			Kernel		
	Net price received by farmers		Price paid by consumer (Rs/kg)	Net price received by farmers (Rs/kg)		Price paid by consumer (Rs/kg)
	Amount (Rs/kg)	%				
I	149	67.8	205	195	60.9	320
II	151	67.1	210	205	61.2	335
III	161	73.7	205	215	67.2	320
IV	73	35.6	205	132	41.3	320
V	73	34.8	210	132	39.4	335

Source: Authors' calculation data collected in a field survey, 2018

Table 13.17 Cost of cultivation of different-age cherry orchards (%)

Age group (year)	Labour	Manures/ fertilizers	Pesticides	Others	Total variable costs (Rs/ kanal)
1st	40.8	52.9	0.0	6.2	2266
2nd	41.6	31.3	0.0	27.1	295
3rd	84.7	15.3	0.0	0.0	733
4th	66.5	30.4	3.1	0.0	518
5th to 10th	65.9	26.6	7.6	0.0	977
10th to 15th	56.6	31.9	11.5	0.0	1469
1fifth to 25th	58.6	31.5	9.9	0.0	1935
> 25th	59.0	32.8	8.2	0.0	2425

Source: Authors' calculation data collected in a field survey, 2018

Table 13.18 Economic feasibility of cherry (Rs/ha)

Crop	Total variable costs	Gross return	RFFR
Cherry	52,546	171,877	3.27

Source: Authors' calculation data collected in a field survey, 2018

its establishment at new orchards. Expenses were increased in improvement of land for orchard, digging of pits, fencing and planting, seedlings, etc. In the first year of its establishment, like other fruits, as high as Rs 2266 per kanal is invested out of which major proportion was spent on labour and fertilization (Table 13.17). However, at this stage, no pesticides were sprayed on crops. In 2nd-, 3rd- and fourth-year orchards, the costs fluctuated but are very less compared to year of establishment from fifth year onwards; the total cost increases with age of orchards, and the cost on labour in each age category increases in proportion with total cost. The cost on pesticides also adds to the total costs and accounted for some percentage of total variable costs, though its share declined in later years.

Economic feasibility analysis of cherry fruit and other competing crops revealed that it is relatively more remunerative (RFFR = 3.27). The cost structure of cherry would increase by folds if it is estimated for only bearing-age orchards (Table 13.18).

Marketing Arrangements

The overall marketed surplus constitutes about 69 percent of its total production per farm. Loss at field in the form of fruit drops accounted for the major proportion of total losses. It was observed that the losses at field were in the form of rudimentary fruits and drop bruise and due to infestation of pests. Losses also accrue to the fruit at picking stage due to harvesting injuries, owing to its perishable nature and physical injuries during packing. The proportion of marketable surplus was relatively higher in the case of makhmali variety followed by mishri and gool variety. Measures to prevent these losses need to be taken to prevent various field losses.

The marketing of cherry is not well organized; the commonly encountered channels identified in its marketing have been presented in Table 13.19. A majority of the farmers sold their surplus to contractors. Out of total average marketable surplus available per farm, 71 percent was sold out through these contractors. As high as 17 and 9 percent of farmers sold surplus directly through retailers and wholesalers, respectively.

Price Spread

The price spread in various marketing channels of cherry has been presented in Table 13.20. In all the four channels of cherry marketing, producers received higher proportion of consumer's price in the channel where he sold his produce directly to retailer. Presence of contractors in marketing channel of cherry was found to deprive farmers from real benefits, and they received relatively lower price for their produce when they sold standing fruit to these intermediaries. Poor capital availability of farmers made them to dispose off their produce through contractors. Encouragement of cooperative coupled with institutional loans would definitely improve their returns by way of direct rules. It could be concluded that producers received higher proportion of consumers' price as net return in channels with lower number of intermediaries.

13.3.2.4 Plum

The common crops grown under this plum-dominated region were plum, paddy, mustard, fodder, vegetable, etc. It was observed that about 48.6 percent of the total cropped area was allocated to plum fruit. The cropping intensity in this region was just 113.41 percent indicating thereby that farmers were generally growing one crop in a year and they could multiply their system on a small portion of cropped area.

Cost of Plum Cultivation

Like other fruits, plum production is also a cost- and labour-intensive activity. It was observed that only 1.5 percent of farmers have their own nurseries of plum and majority of them purchase seedlings from private nurseries or from block agricultural offices. The establishment of new orchard of plum requires investment in land improvement and its levelling and fencing. Moreover, pits are dig out to plant new seedlings; therefore, in the first year, a higher proportion of total cost was spent on labour and planting material. In the second year, only Rs 280.00 per kanal was spent

Table 13.19 Marketing channels and net price received by farmers (percent)

Channel	Functionaries								Farmer	Produce
I	Farmers	→	Retailer (Srg)	→	Consumer				17	17
II	Farmers	→	W.s (Srg)	→	Ret (Srg)	→	Consumer		9	17
III	Farmers	→	Contractor	→	W.s (Srg)	→	W.s (Srg)	→	Consumer	32	29
IV	Farmers	→	Contractor	→	W.s (Delhi)	→	Ret (Delhi)	→	Consumer	39	33
	Others									3	4

Note: *W.s* wholesaler
Source: Authors' calculation data collected in a field survey, 2018

Table 13.20 Price spread in marketing of cherry

Channels	Price received by farmer		Price paid by consumer (Rs/kg)	
	Amount (Rs/kg)	%		
I	69	62.7	110	
II	67	60.9	110	
III	45	40.9	110	
IV	45	21.4	210	

Source: Authors' calculation data collected in a field survey, 2018

Table 13.21 Cost of cultivation of different-age plum orchard (%)

Age of orchard	Labour	Fertilizer	Pesticides	Others	Total variable costs (Rs/kanal)
1st	45.1	9.8	0.0	45.1	2993
2nd	70.3	29.7	0.0	0.0	280
3rd	87.7	12.3	0.0	0.0	881
4th	82.6	15.7	1.7	0.0	999
5th to 10th	79.0	16.2	4.8	0.0	1361
10th to 15th	71.5	21.6	6.8	0.0	1902
1fifth to 20th	74.7	18.5	6.8	0.0	2160
> 20th	77.5	17.4	5.2	0.0	3062

Source: Authors' calculation data collected in a field survey, 2018

Table 13.22 Economic feasibility of plum

Crop	Total variable costs (Rs/ha)	Gross return (Rs/ha)	RFFR
Plum	42,625	92,071	2.16

Source: Authors' calculation data collected in a field survey, 2018

on variable costs; this cost however increased gradually with age of orchard (Table 13.21). In all bearing- and non-bearing-age orchards, cost on labour constituted major proportion of total cost on variable inputs, followed by fertilizers and pesticides. It was observed that the share of expenditure on fertilizers and pesticides declined in more than 20-year orchards. Farmers of this region applied more urea than recommended, while other fertilizers or manures used much below scientific recommendations. The technological gap with respect to muriate of potash (MOP) was relatively more. This scenario emphasized upon strengthening of extension to educate farmers in this regard.

The cost and return structure of average-age orchard of plum crop revealed that one rupee expenditure on variable inputs would yield 2.16 as gross revenue which is almost double than paddy and mustard. In proportion of total variable costs, about 70 percent are alone spent on labour in plum which is higher compared to other crops (Table 13.22).

Marketing Arrangements

Since no part of surplus plum can be hoarded, then overall marketed surplus was about 76 percent of its total production. It was observed that the losses in the form

Table 13.23 Marketing channels and net price received by farmers (%)

Channel	Functionaries	Produce	Farmer
I	Farmer-Retailer-Consumer	4	7
II	Farmer-Wholeseller(oss)/CA-Retailer-Consumer	8	11
III	Farmer-Preharvest contractor-Wholeseller (L)-Retailer-Consumer	52	49
IV	Others	5	3

Note: *oss* = outside the state, *CA* = commission agent, *L* = local
Source: Authors' calculation data collected in a field survey, 2018

of bruise damages and due to infestation of pests account for a good proportion of total spoilages. Loss also accrues to the fruit at picking stage due to harvesting injuries owing to its high perishability, loss of fruits due to slopes, fissures in fruits and physical injuries during packing. These fruits were neither marketed nor consumed in any form.

The marketing of plum is not well organized; the commonly encountered channels identified in its marketing have been presented in Table 13.23. A majority of the farmers sold their surplus to preharvest contractors. Out of total average marketable surplus available per farm, 52 percent was sold out through these contractors, while as high as 8% of farmers sold surplus directly to wholesalers.

Price Spread

The price spread in various marketing channels of plum has been presented in Table 13.24. In all the six channels of plum marketing, producers received higher proportion of consumer's price in the channel where he sold his produce directly to retailer. Presence of preharvest contractors in marketing channel of pear was found to deprive farmers from real benefits, and they received relatively lower price for their produce when they sold standing fruit to these intermediaries. It could be concluded that producers received higher proportion of consumers' price as net return in channels with lower number of intermediaries.

13.3.2.5 Peach

Cropping pattern under peach-based farming systems indicated that the common crops grown under this region were peach, paddy, fodder, vegetable, etc. About 49.5 percent of the total cropped area was allocated to peach. The cropping intensity in this region was 153.5 percent indicating thereby that farmers were generally growing one crops in a year.

Cost of Peach Cultivation

Peach cultivation is a cost- and labour-intensive activity. It was observed that majority of farmers purchase seedlings from block agricultural office. The establishment of new orchard of peach involves Rs 3573/kanal investment in land improvement, digging of pits and planting of seedlings. In the second year, only Rs 330.00 per kanal was spent on variable costs; this cost however increased gradually with age of orchard (Table 13.25).

Table 13.24 Price spread in marketing of plum

Channel	Net price received		Consumers price (Rs/box)
	Amount (Rs/box)	% of consumer's rupee	
C1	115	54.2	212.0
C2	143	67.4	212.3
C3	112	34.5	324.6
C4	122	30.0	406.4

Source: Authors' calculation data collected in a field survey, 2018

Table 13.25 Cost of cultivation of different-age peach orchard (%)

Age of orchard	Labour	Fertilizer	Pesticides	Others	Total variable costs (Rs/kanal)
1st	43.0	8.8	0.0	48.2	3573
2nd	68.3	30.7	0.0	1.0	330
3rd	84.0	14.3	0.0	1.7	899
4th	80.5	17.8	1.3	0.4	1111
5th to 10th	78.0	16.2	5.3	0.5	1561
10th to 20th	70.5	21.6	7.8	0.0	2102
> 20th	75.5	18.4	6.2	0.0	3472

Source: Authors' calculation data collected in a field survey, 2018

Table 13.26 Economic feasibility of peach and other competing crops

Crop	Total variable costs (Rs/ha)	Gross return (Rs/ha)	RFFR
Peach	36,095	88,071	2.44

Source: Authors' calculation data collected in a field survey, 2018

The per rupee return over variable cost from peach crop revealed that one rupee expenditure on variable inputs would yield 2.44 (Table 13.26).

Marketing Arrangements

The utilization pattern of peach reported by the sample farmers indicated that the overall marketed surplus was about 72 percent of its total production, and the rest goes as field loss, family consumption or gifts. Losses at field were in the form of bruise damages and due to infestation of pests. Loss also accrues to the fruit at picking stage due to harvesting injuries owing to its high perishability, loss of fruits due to slopes, fissures in fruits and physical injuries during packing. The marketing of peach is unregulated and is predominantly in the hands of market functionaries. The commonly encountered channels identified in its marketing have been presented in Table 13.27. A majority of the farmers sold their surplus to preharvest contractors. Out of total average marketable surplus available per farm, 22 percent surplus was sold out through wholesalers.

Price Spread

The price spread in various marketing channels of peach, presented in Table 13.28, revealed that producers receive good proportion of consumer's price in the channel

Table 13.27 Marketing channels and net price received by farmers (%)

Channel	Functionaries	Produce	Farmer
I	Farmer-retailer-consumer	7	12
II	Farmer-Wholerseller/CA-Retailer-Consumer	28	22
III	Farmer-Preharvest contractor-Wholeseller-Retailer-Consumer	59	59
IV	Others	6	7

Note: *CA* = commission agent
Source: Authors' calculation data collected in a field survey, 2018

Table 13.28 Price spread in marketing of peach

Channel	Net price received		Consumers price (Rs/box)
	Amount (Rs/box)	% of consumer's rupee	
I	160	67.3	240
II	168	61.2	237
III	142	33.8	355
IV	122	30.8	362

Source: Authors' calculation data collected in a field survey, 2018

where he sold his produce directly to retailer. Presence of preharvest contractors in marketing channel of peach makes farmers loss a major share to them, and they received relatively lower price for their produce when they sold standing fruit to these intermediaries.

13.4 Problems and Policy Suggestions

The responses of farmers regarding different problems in stone fruit-growing regions were obtained and the results summarized in Fig. 13.3. Farmers were asked about the problems related to production, and majority of the farmers expressed less favourable weather conditions as major problem hindering improvement of fruit production. Small/fragmented/sloppy holdings were other critical problems faced by the farmers in the study area. These problems hinder performance of stone fruits as it affects capital investment and infrastructural development. As high as 72.5 percent of farmers highlighted the incidence of insect/pests a major problem in production of stone fruits. Poor irrigation facilities in this region have been concern of over 68.3 percent of farmers in this region. Besides, there were a number of other problems reported by farmers that impart disinterest in them in cultivation of stone fruits.

There were a number of problems as perceived by farmers relating to marketing of stone fruit that discourage them to manage more area under this fruit crop. Among different marketing-related problems, perishability was considered to be a major market-based problem. Non-availability of markets and more distance to markets

Fig. 13.3 Problems perceived by farmers in stone fruit cultivation (%). Source: Authors' calculation data collected in a field survey, 2018

were reported by about 88 percent of farmers. Among other problems, higher transportation charges and uneven market prices were highlighted by over 61 percent of farmers. Besides, there are a number of other problems related to marketing that constrained expansion of area/production of stone fruits.

With regard to institutional problems, most of the problems were related with credit. Stone fruit growers reported no specific insurance scheme as safety net. This problem was highlighted by 81.3 percent of farmers. Majority of farmers expressed the cumbersome loaning procedure as the main institutional problem. Farmers in the study area perceived this problem more intensely. Inadequate extension staff was yet another problem reported by a good proportion of farmers.

13.5 Conclusion and Policy Suggestions

13.5.1 Bridging Technological Gaps

Technological gaps in different agroclimatic regimes need to be bridged wherein seed replacement rates to the desired level are essential. The coordination in dissemination of technologies of different institutions and farming community involved is imperative.

13.5.2 Innovations

Innovations in the form of location and farm-size-specific machines and farming tools are required as substitute to a good proportion of agricultural labourers.

Further, farm equipment should be female-labour-friendly in view of increasing role of females in agriculture in this region.

13.5.3 Access to Information and Extension and Capacity Development

Given the scenario that only 5 percent of farmers have access to extension agencies, there is a need to streamline the extension system of the agricultural system. Efforts are to be made to take technologies and interventions to the field without any time lag. There is a need to bring reforms in the approaches linking farmers to the scientific community by way of strengthening market-led extension rather than production based. Application of ICT and extension network with mass media support would have an immense utility. It is imperative for capacity development among human capital associated with agricultural sector. Knowledge management coupled with capacity building forms an important domain area for the development of agricultural sector of the state.

13.5.4 Linking Production to Markets Through Firm Value Chain

There is a need of integrating production with marketing through value addition and other postharvest management practices to secure livelihood of agricultural labour by engaging them for most part of the year. Encouragement of micro-agricultural enterprises like mushroom cultivation, home-based processing units and apiculture at farm level around each production centre could also provide them employment in farm-based off-farm jobs.

13.5.5 Orchard Management

Emphasis is on the orchard management on scientific lines, capacity development in plant health management, canopy management, orcharding system, etc.; moreover, the pollination management has also an important role to play.

13.5.6 Provision of Logistics

Establishment of focal points with provisions of all possible infrastructures/facilities for soil/pesticide testing, value addition, scientific packaging, quality control, etc. around the production centres is inevitable for agricultural growth. In addition, institutions have a role to play in the form of database creation and provision of important logistics within market yard and in the transit to ensure safe delivery of farm produce.

13.5.7 Expansion of Storage Capacities

The available cold store facilities for the storage of perishable surpluses of fresh fruits in the state are meagre. The institutional role in the encouragement of establishment of CA stores/warehousing with provision of various services like hypothecation and insurance to provide risk cover to the farmers becomes prominent.

13.5.8 Emphasis on Marketing Aspect (National Agricultural Markets (NAM), Spot Markets, etc.)

The dominance of small/marginal farmers and existence of chain of functionaries in marketing of horticultural crops often put farmers in distress. eNAM would increase their access to markets through warehouse-based sales and thus obviate the need to transport his produce to the mandi. Bulk buyers, processors, exporters, etc. benefit from being able to participate directly in trading at the local mandi/market level through the NAM platform, thereby reducing their intermediation costs. The gradual integration of all the major mandis in the states into eNAM will ensure common procedures for issue of licences, levy of fee and movement of produce. The NAM will also facilitate the emergence of value chains in major agricultural commodities across the country and help to promote scientific storage and movement of agri goods.

13.5.9 Market Intelligence to Benefit the Poor

Strengthening of market intelligence for ensuring delivery of agricultural produce at the place of its demand with effective integration of domestic markets and diffusion of price signals is urgently required. There is a need to advance loan to small/marginal farmers for marketing their produce, and group lending for these activities should have a provision. The fixation of minimum support price (MSP) for all the crops covering cost of production and ensuring minimum profit of 50 percent of this cost are essential. Further establishment of 'village procurement centres' and organization of 'small farmers' fair' would have crucial role in the development of resource-poor farmers.

13.5.10 Contract Farming and FPOs

In view of regressive fragmentation of holding and emergence of marginal/small farmers, the emphasis is on the collectivization of farming and marketing activities. The famers are to be encouraged to act as a business unit to empower them as important market actors. Strengthening of cooperatives and creation of FPOs and registration and linkages with financial institutions including NABARD are required to give commercial orientation to the unorganized farming community. The cooperative societies should be revitalized to increase the bargaining strength of growers. There is a need to inculcate more professionalism among different stakeholders and skilled management in cooperative organizations.

13.5.11 Emphasis on Agripreneurship

A shift from agriculture to agribusiness is an essential pathway to revitalize Indian agriculture and to make more attractive and profitable venture. Applying the thought and practice of entrepreneurship in the field of horticulture generates wide range of economic benefits. The managerial, technical and innovative skills of entrepreneurship applied in the field of agriculture may yield positive results, and a well-trained agripreneur may become a role model to all such disheartened farmers. Entrepreneurship in horticulture would help to (a) reduce the burden of agriculture, (b) generate employment opportunities for rural youth, (c) control migration from rural to urban areas, (d) increase national income, (e) support industrial development in rural areas (f) and reduce the pressure on urban cities.

13.5.12 Input Supply Support

There is a need for an effective regulation of pesticide/fertilizer trade in view of the availability of spurious/substandard inputs in the market. Establishment of input checkposts at each production centre equipped with chemical testing facilities is needed, and it should be mandatory that each imported container of pesticides should undergo registration at this checkpost with sample-based testing. Cost on pesticides being a major share of total cost of apple cultivation, effective measures like dissemination of integrated pest management (IPM) modules should be adopted to prevent the disease and insect/pest incidence in apple crop.

13.5.13 Development Subsidies

Development subsidies rather than input subsidies are advocated for the creation of long-term asset having long productive life. These subsidies are expected to create stock of capital in public-private domain productive services which may flow for productivity gains of agricultural capacities.

13.5.14 Regional Enterprise Planning and Place for Cash Crops

Regional specialization of production centres on the basis of comparative advantage will enhance technical and allocative efficiency in production processes. There is a need to exploit niches in agroecosystems for regional specialization of production centres in commodities based upon comparative advantage.

Besides, there is a need to establish special agricultural zones based on climate/physiographic factors and niches, production and supply of quality seed/planting material/improved breeds of animals and poultry and development of seed banks/stores. The possibility of participatory breeding has to be taken vigorously.

References

Anonymous. (2016a). National Horticulture Board, New Delhi.

Anonymous. (2016b). *Digest of statistics*. Directorate of Statistics & Economics, Government of Jammu & Kashmir, India.

Anonymous. (2007). *Digest of statistics*. Directorate of Statistics & Economics, Government of Jammu & Kashmir, India.

Baba, S. H. (2018). *S&T intervention in agricultural and allied sectors for strengthening livelihood security in Kashmir Division*. Final technical report of DST sponsored research project, Sher-e-Kashmir University of Agricultural Sciences & Technology of Kashmir, Shalimar Campus, Srinagar, J&K-India.

Baba, S. H., Wani, M. H., & Malik, H. A. (2012). *Fruit economy linkages and role in rural upliftment and employment generation in J&K. Final Technical Report of ICAR-HTM sponsored research project*. Division of Agricultural Economics & Marketing, Sher-e-Kashmir University of Agricultural Sciences & Technology of Kashmir, Shalimar Campus, Srinagar, J&K-India.

Baba, S. H., Wani, M. H., Zargar, B. A., & Malik, H. A. (2014). Imperatives for sustenance of agricultural economy in the mountains: A prototype from Jammu & Kashmir. *Agricultural Economics Research Review, 27*(2), 243–257.

Cevallos-Casals, B., Byrne, D. H., Okie, W. R., & Cisneros-Zevallos, L. (2006). Selecting new peach and plum genotypes rich in phenolic compounds and enhanced functional properties. *Food Chemistry, 96*, 273–280.

Chung, L. Y., Cheung, T. C., Kong, S. K., Fung, K. P., Choy, Y. M., Chan, Z. Y., & Kwork, T. T. (2001). Induction of apoptosis by green tea catechins in human prostate cancer DU 145 cells. *Life Sciences, 68*, 1207–1214.

Damianaki, A., Bagogeorgou, E., Kampa, M., Notas, G., Hatzoglou, A., Panagiotou, S., Gemetzi, C., Kouroumalis, E., Martin, P. M., & Castanas, E. (2000). Potent inhibitory action of red wine polyphenols on human breast cancer cells. *Journal of Cellular Biochemistry, 78*, 42941.

Faust, M., & Suranyi, D. (1999). Origin and dissemination of plum. *Horticultural Review, 23*, 179–231.

Faust, M., Suranyi, D., & Nyujto, F. (1998). Origin and dissemination of apricot. *Horticultural Review, 22*, 225–266.

Faust, M., & Timon, B. (1995). Origin and dissemination of peach. *Horticultural Reviews, 17*, 331–379.

Gangwar, L. S., & Singh, S. (1998). Economic evaluation of Nagpur mandarin cultivation in Vidarbha region of Maharashtra. *Indian Journal of Agricultural Economics, 53*(4), 648–653.

Gangwar, L. S., Ilyas, S. M., Singh, D., & Kumar, S. (2005). An economic evaluation of kinnow mandarin cultivation in Punjab. *Agricultural Economics Research Review, 18*(1), 71–80.

Ghosh, S. P. (2001). Temperate fruit production in India. *Acta Horticulture*, 565–132.

Gil, M., Tomas-Barberan, F., Hess-Pierce, B., & Kader, A. (2002). Antioxidant capacities, phenolic compounds, carotenoids, and vitamin A contents of nectarine, peach, and plum cultivars from California. *Journal of Agricultural and Food Chemistry, 50*, 4976–4982.

Kampa, M., Hatzoglou, A., Notas, G., Damianaky, A., Bakogeorgou, E., Gemetzi, C., Kouroumalis, E., Martin, P. M., & Castanas, E. (2000). Wine antioxidant polyphenols inhibit the proliferation of human prostate cancer cell lines. *Nutrition and Cancer, 37*, 223–233.

NHB (Indian Horticulture Database). (2005). National Horticulture Board, Gurgaon, Haryana.

Okie, W. R., & Weinberger, J. H. (1996). Plums. In J. Janick & J. N. Moore (Eds.), *Fruit breeding* (Vol. 1, pp. 559–607). New York: Wiley.

Poonam, V., Raunak, G., Kumar, C. S., Reddy, L., Jain, R., Sharma, S. K., Prasad, A. K., & Parmar, V. S. (2011). Chemical constituents of the Genus Prunus and their medicinal properties. *Current Medicinal Chemistry, 18*, 3758–3824.

Prior, R. L., Cao, G., Martin, A., Sofic, E., Mcewen, J., Obrien, C., Lischner, N., Ehlenfeldt, M., Kalt, W., Krewer, G., & Mainland, C. (1998). Antioxidant capacity as influenced by total phenolic and anthocyanin content, maturity, and variety of Vaccinium species. *Journal of Agricultural and Food Chemistry, 46*(7), 2686–2693.

Sartippour, M. R., Heber, D., Ma, J., Lu, Q., Go, V. L., & Nguyen, M. (2001). Green tea and its catechins inhibit breast cancer xenografts. *Nutrition and Cancer, 40*(2), 149–156.

Scordino, M., Sabatino, L., Muratore, A., Belligno, A., & Gagliano, G. (2012). Phenolic characterization of Sicilian yellow flesh peach (Prunus persica L.) cultivars at different ripening stages. *Journal of Food Quality, 35*, 255–262.

Senter, S. D., & Callahan, A. (1991). Variability in the quantities of condensed tannins and other major phenols in peach fruit during maturation. *Journal of Food Science, 56*, 1585–1587.

Tomas-Barberan, F. A., Gil, M. I., Cremin, P., Waterhouse, A. L., Hess-Pierce, B., & Kader, A. A. (2001). HPLC-DAD-ESIMS analysis of phenolic compounds in nectarines, peaches, and plums. *Journal of Agricultural and Food Chemistry, 49*, 4748–4760.

Tourjee, K. R., Barrett, D. M., Romero, M. V., & Gradziel, T. M. (1998). Measuring flesh color variability among processing clingstone peach genotypes differing in carotenoid composition. *Journal of the American Society for Horticultural Science, 123*, 433–437.

Vicente, A. R., Manganaris, G. A., Sozzi, G. O., & Crisosto, C. H. (2009). Nutritional quality of fruits and vegetables. In W. J. Florkowski, R. L. Shewfelt, B. Brueckner, & S. E. Prussia (Eds.), *Postharvest handling: A systems approach* (2nd ed., pp. 58–106). New York: Academic. ISBN: 978-0-12-374112-7.

Wang, H., Cao, G., & Prior, R. L. (1996). Total antioxidant capacity of fruits. *Journal of Agricultural and Food Chemistry, 44*(3), 701–705.

Wang, Y. (1985). Peach growing and germplasm in China. *Acta Horticulture, 175*, 3–55.

Watkins, R. (1995). Cherry, plum, peach, apricot and almond: Prunus spp. (Rosaceae). In J. Smartt & N. W. Simmonds (Eds.), *Evolution of crop plants* (2nd ed., pp. 423–429). Essex: Longman Scientific & Technical.

Weinert, I., Solms, J., & Escher, F. (1990). Diffusion of anthocyanins during processing and storage of canned plums. *Lebensmittel Wissenschaft Technologie, 23*, 396–399.

Zhao, X., Zhang, W., Yin, X., Su, M., Sun, C., Li, X., & Chen, K. (2015). Phenolic composition and antioxidant properties of different peach [Prunus persica (L.) Batsch] cultivars in China. *International Journal of Molecular Sciences, 16*, 5762–5778.

Zohary, D., & Roy, P. S. (1975). Beginning of fruit growing in the Old World. *Science, 187*, 319–327.

Diseases of Stone Fruit Crops

14

N. A. Khan, Z. A. Bhat, and M. A. Bhat

Abstract

Stone fruits (peach, plum, cherry, apricot and nectarines) are widely cultivated throughout the temperate regions of the world. There are various production impediments in stone fruit cultivation of which diseases are considered as the major limiting factors which affect the yield and quality of the fruits, thus making their cultivation less remunerative. Leaf curl, shot hole, brown rot, powdery mildew, scab, rust, cankers, root and collar rot, bacterial spot, crown gall and plum pox are important diseases of these fruit crops which can be managed through an integrated approach involving use of cultural, chemical and biological measures, besides using resistant cultivars. Economic importance, symptoms, causal organisms, disease development and management of major diseases of stone fruits are described.

Keywords

Causal organism · Diseases · Disease development · Stone fruits · Symptoms · Management

14.1 Introduction

Stone fruit crops such as peach, plum, cherry, apricot and nectarines are widely distributed throughout the temperate regions of the world. In India, they are mainly cultivated in the northern states, comprising Jammu and Kashmir (J&K), Himachal Pradesh (HP), Uttarakhand and some parts of Punjab. Stone fruit crops, in general,

N. A. Khan · Z. A. Bhat (✉) · M. A. Bhat
Division of Plant Pathology, Sher-e-Kashmir University of Agricultural Sciences and Technology of Kashmir, Shalimar, Srinagar, Jammu and Kashmir, India

M. M. Mir et al. (eds.), *Production Technology of Stone Fruits*,
https://doi.org/10.1007/978-981-15-8920-1_14

and cherry and peaches, in particular, are prone to a number of diseases caused by fungi, bacteria and viruses, affecting leaves, fruits, twigs, blossoms and roots, which inflict huge economic losses to fruit growers and also increase the cost of management. The success of any plant disease control programme is primarily governed by a crop production system which fulfils the essential goals of pest management. Economic losses due to disease vary with climatic conditions, cultivar susceptibility and orchard or nursery management practices. The monoculturing of susceptible cultivars especially under temperate environmental conditions promotes severe disease development. Under such situations, any lapses or interruptions in the disease management process, sometimes, facilitate epiphytotics. Thus, adoption of appropriate and effective protection strategies to restrict such losses becomes unavoidable. Therefore, cultivation of stone fruit crops for optimal yield and quality demands improved technical support for management of various diseases besides making availability of quality planting material with short gestation period and low to medium initial investments.

14.2 Leaf Curl

The disease has worldwide occurrence particularly in Africa, Australia, China, Europe, Japan, North and South America, New Zealand and Portugal, attacking peach, cherry, plum, apricot and nectarine. Although on peaches the disease was reported earlier in 1821 from England, Berkeley in 1860 for the first time gave a detailed description of the organism causing the disease. In India, Sydow and Butler (1916) and Ferraris (1928) reported the occurrence of the disease from Himachal Pradesh and Kashmir, respectively. Padwick (1945) recorded the disease in almond nurseries from Kashmir. The disease chiefly infects leaves and young shoots, causing severe leaf defoliation, while fruit infection is less frequent. Different species of *Taphrina* have been reported to invade different stone fruit crops. *Taphrina communis* and *T. pruni* cause plum pocket on plum, while *T. cerasi* causes leaf curl and witches' broom on cherries. In India, the disease is more severe on peaches and nectarines reducing fruit set, fruit quality and fruit yield and weakens the tree vigour through premature defoliation.

14.2.1 Symptoms

Disease symptoms appear on young leaves as yellow to reddish areas in early spring. Progression of the diseased areas causes thickening and puckering of leaf blades along the midrib (Plate 14.1). Puckered areas develop red or purple colour, become brittle and occasionally may develop white coating of spores. Deformed leaves curl, abscise and defoliate prematurely. After defoliation, dormant buds become active, and new flesh of leaves emerges which also succumbs to the disease. In current years, twigs can also be infected which become thickened and may exude a gum-like substance. Severely infected twigs die. Fruit infection is rare which is characterized by irregular, raised, wrinkled, reddish brown lesions (Koul 1967).

Plate 14.1 Red-coloured puckered leaf blade along the midrib

14.2.2 Causal Organism

Taphrina deformans (Burk.) Tulasne

The fungus produces naked asci on the upper leaf surface which break through the cuticle of distorted leaf and appear as powdery grey felt-like area on the malformed leaf. Asci are unitunicate, cylindrical to clavate, round or truncate at the apex with a stalk cell and measure 17–56 × 7–15 μm in size. The ascospores bud frequently within the ascus to produce blastospores or conidia which are round, oval or elliptical in shape and measure 3–7 μm in diameter (Pscheidt 1995).

14.2.3 Disease Development

The disease is more destructive in areas with prolonged periods of cool and moist weather. The fungus overwinters as ascospores and conidia mostly on the bud scales (Plate 14.2).

Ascospores produced on upper surface of hypertrophied leaves and other tree parts are discharged forcibly onto young tissues in early spring and germinate to bud conidia. Bud conidia reproduce on twigs and shoot tips which on germination penetrate the developing leaves and other organs directly through the cuticle or

Plate 14.2 Nectarine plant severely infected with leaf curl disease

through stomata and invade the tissues by producing intercellular mycelium. Budding of conidia occurs at relative high humidity of 95%. Periods of cool, humid weather during early bud development favour disease development. The optimum temperature for fungal growth is 20 °C, and the maximum is between 26 °C and 30 °C. Ascospores and bud conidia can survive for several months under hot, dry conditions (Anderson 1956) (Plate 14.3).

14.2.4 Management

The disease can be effectively managed by integrating both cultural and chemical practices. Collection and destruction of overwintered diseased leaves can give an effective disease control, besides spraying the trees at pre-bud burst, fruitlet and 15–20 days after fruitlet stage with different fungicides. Copper oxychloride 50 WP at 0.3% and carbendazim 50 WP at 0.05% have been found highly effective against the disease (Sharma et al. 1987, 1988; Bhardwaj and Ved 1995). Other fungicides reported effective against the disease include captan 50 WP at 0.3%, dodine 65 WP at 0.06% and chlorothalonil 70 WP at 0.3% (Ponti et al. 1993).

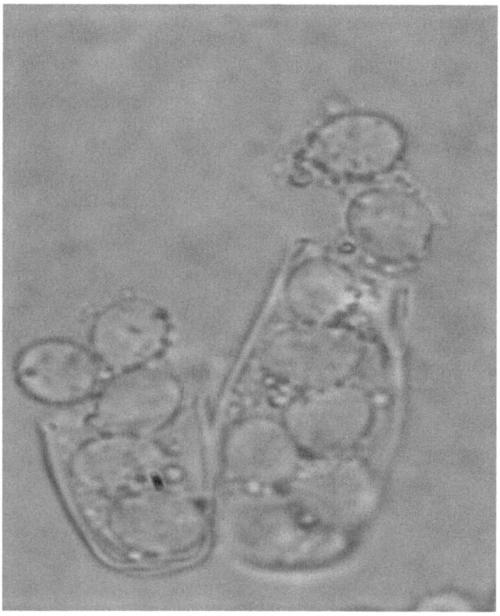

Plate 14.3 Ascospores of the fungus releasing from the ascus

14.3 Shot Hole

The disease also known as *Coryneum* blight, California peach blight and pustular spot has worldwide occurrence, affecting all the cultivated *Prunus* species. Leveille (1843) from France for the first time reported its occurrence on peach, while Berkeley (1864) recorded the disease from England. In India, Koul (1967) and Munjal and Kulshrestha (1968) for the first time observed the disease on almond from Kashmir. Earlier, Singh (1943) reported its occurrence from Kumaon hills *on Prunus persica, P. avium* and *P. armeniaca* under the name *Clasterosporium* sp. The disease is more severe on apricot and cherry, besides almond, peach and plum.

14.3.1 Symptoms

Disease symptoms are frequently observed on leaves, twigs and fruits but rarely on blossom. On leaves, the disease is initiated as small dark brown lesion with reddish margins. In warm, dry climate, the lesions enlarge rapidly, dry out and abscise resulting in shot holes, giving the leaf a ragged appearance. Such early dehiscence largely eliminates the formation of sporodochia on the lesions. On fruit, deep purple raised lesions develop mostly on the upper side, which become rough and corky on coalescence and may secrete gummy exudates. Severe fruit infection causes fruit cracking at infection sites. On twigs, small, raised purplish spots develop which expand to necrotic cankers oozing gum droplets. The leaves, twigs, and unopened buds get blighted. The diseased buds become darker in colour and are occasionally covered with gummy exudates (Putoo and Razdan 1991) (Plate 14.4).

Plate 14.4 Ragged appearance of peach leaf due to abscission of disease spots

14.3.2 Causal Organism

Wilsonomyces carpophilus (Lev.) Adaskaveg, Ogawa and Butler
The earlier generic names of the fungus include *Clasterosporium*, *Coryneum* and *Stigmina*.

The fungus produces olivaceous brown to black dot-like sporodochial conidiomata, on necrotic twig cankers and occasionally on fruit. Conidiophores are subhyaline to light brown in colour, simple to irregularly branch proliferating sympodially and bearing a solitary conidium and measure 17–45 × 5–11 μm in size. Conidia are sub-hyaline to golden brown in colour, thick walled, ellipsoidal or fusoid with apical cell ovate and basal cell truncate usually with three to five transverse double-walled septa, slightly constricted at each septum and measure 20–90 × 7–16 μm in size (Ellis 1959; Koul 1967).

14.3.3 Disease Development

The shot-hole fungus overwinters as conidia within infected buds and on twig lesions. In early spring, conidial germination proceeds disease development. Conidia are not easily detached from the conidiophores by moving air, but are readily removed by water. The incubation period ranges from 2 to 3 days at 20–28 °C

Plate 14.5 Raised purple-red disease lesions on apricot fruit

with 90–100% relative humidity (Gupta et al. 1972) (Plate 14.5). Periodic showers, high relative humidity (70–80%) and optimum temperature (19–22 °C) favour disease development (Kosogrova 1976).

14.3.4 Management

The disease can be effectively managed by spraying the trees at leaf-fall in autumn and before bud burst and fruitlet in spring. Further, spray can be conducted 15–20 days after fruitlet depending upon disease severity. Fungicides like captan 50 WP at 0.3%, copper oxychloride 50 WP at 0.3%, mancozeb 75 WP at 0.3%, carbendazim 50 WP at 0.05% and thiophanate methyl 70 WP at 0.05% have been found effective against the disease. Copper oxychloride should be sprayed immediately after leaf-fall to reduce the disease carry-over, but its application on apricot should be avoided beyond pink bud stage (Anderson 1956; Gupta et al. 1972; Angelov 1980; Ogawa et al. 1995a, b).

14.4 Rust

The disease is worldwide in occurrence particularly in countries like Australia, Brazil, China, France, Japan, New Zealand and the United States, affecting all the cultivated *Prunus* species. The disease was first reported from Australia in 1890 (Dunegan 1938). In India, Koul (1967) for the first time reported its occurrence on almond from Kashmir, while Waraich and Khatri (1977) reported its occurrence on plum from Himachal Pradesh. In Kashmir, besides almond, the disease has also been observed on peach, plum and apricot (Mushtaq et al. 2014). Early season disease appearance results in premature leaf defoliation with reduced yields by causing abscission of immature prune fruit.

14.4.1 Symptoms

Disease symptoms are mostly observed on leaves and twigs but rarely on fruit. On leaves, the disease is initiated as angular, pale yellowish green spots on both the surfaces, which later change into bright yellow islands on the upper surface. Those on the lower surface develop typical orange-brown rust pustules bearing urediniospores. Black to dark brown teliospores develop in late season amongst the orange-brown urediniospores. Fruit infection usually occurs when the conditions are favourable for the fungus to cause infection at a relatively late stages of fruit maturity. Water-soaked green spots appear on the fruit surface which become sunken as the fruit growth continues. Later on, the spots become deeper green in colour and develop yellow boarders (Plate 14.6).

Urediniospores may or may not be present in fruit lesions. Twig infection characterized by water-soaked appearance and swelling is usually inconspicuous on all host species but is more frequently found on peach. Infected twigs rupture lengthwise to expose urediniospores. Although twig infection does not cause much damage, it can serve as important source of primary inoculum (Dunegan 1938; Koul 1967).

Plate 14.6 Almond leaf showing rust symptoms

14.4.2 Causal Organism

Stone fruit rust is caused by two species of fungi, viz., *Tranzschelia discolor* (Fuckel) Tranzschel and Litv. and *Tranzschelia pruni-spinosae* (Pers. ex. Pers) Dietel.

Tranzschelia discolor associated with cultivated stone fruit has worldwide occurrence, while *T. pruni-spinosae* is found on wild *Prunus* species. The two species of fungi are differentiated from each other on the basis of their teliospore morphology. Teliospores of *T. pruni-spinosae* are uniformly coarsely verrucose over both cells and are similar in shape, size and colour (Plate 14.7).

The teliospores of *T. discolor* differ in shape, size, colour and verrucose markings. They are pulverulent forming black clusters, bi-celled and chestnut brown in colour. The apical cell is coarsely verrucose and globoid, while basal cell is oblong or ovate-oblong, generally tapered towards the base and lighter in colour than the apical cell. The teliospore size of *T.* discolor ranges from 15–20 × 30–39 μm in size, while *T. pruni-spinosae* ranges from 18–27 × 30–39 μm. The urediniospores identical in both the species are borne on short stalks, intermixed with hyaline capitate paraphysis, ovate-clavate in shape and measure 20–48 × 12–16 μm in size. *Tranzschelia discolor* exists in a series of definable formae speciales. The names were *T. discolor* f.sp. *dulcis* for strains attacking almond, *T. discolor* f.sp. *persicae* for strains attacking peach and *T. discolor* f.sp. *domesticae* for strains attacking prunes and other hosts (Hawksworth et al. 1983; Bertrand 1995) (Plates 14.8 and 14.9).

Plate 14.7 Rust sori on undersurface of plum leaves containing urediniospore and teliospores

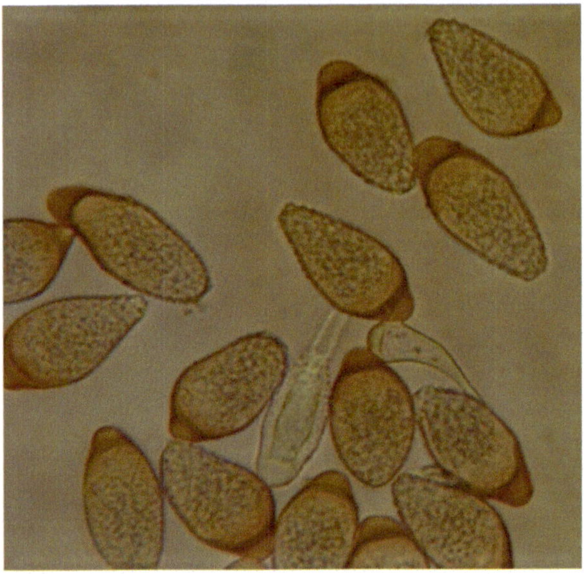

Plate 14.8 Uredinospores of the fungus

14.4.3 Disease Development

The fungus overwinters as mycelia in twigs or as urediniospores on twigs or on non-abscised leaves. However, *Anemone coronaria* has been reported to be the alternate host. The viability of urediniospores on twig surface is greatly reduced during the winter months. Urediniospores germinate over wide temperature range of 8–38 °C, the optimum being 13–26 °C. Thus, the availability of viable spores and moisture, rather than temperature, is considered the major limiting factor for infection. The duration of wet periods required for infection at various temperatures is unknown, although an 18 h wet period at 20 °C is adequate for heavy infection. The incubation period for leaf infection is 7–10 days (Bertrand 1995).

14.4.4 Management

The disease can be effectively managed with preventive fungicidal sprays. Trees sprayed with zineb 80 WP five times during the growing season retained the normal foliage with 35% increase in fruit yield (Decker and Buchanan 1975). Foliar application of propiconazole 12 EC and myclobutanil 10 WP at 25–50 ppm showed strong curative effect against *T. discolor* infections in plum leaves. The disease incidence was low in trees treated with captan 50 WP at 0.3% or with mancozeb 75 WP at 0.3% or with dithianon 70 WP at 0.3% and bitertanol 25 WP at 0.05% (Shabi et al. 1990). No attempts have been made to control this disease with biological

Plate 14.9 Teliospores of the fungus

methods; however, Koul (1967) showed that in the Kashmir Valley, *Darluca filum* parasitizes the rust fungus.

14.5 *Cercospora* Leaf Spot

Cercospora leaf spot has been reported from many cherry-growing areas of the world including France, Germany, Israel, Italy and the United States affecting both sweet and sour cherries. The pathogen is also reported worldwide on other *Prunus* species including almond, peach and plum. In India, Koul (1962) for the first time recorded its occurrence on sweet cherry from Kashmir. The disease is more severe on cherry than other stone fruits leading to yellowing and premature defoliation.

14.5.1 Symptoms

Disease is initiated as small circular light brown spots with dark red margins, which later develop into reddish brown lesions. The lesions are observed on both the leaf surfaces which are scattered and rarely coalesce. They may be few or sometimes so numerous as to cover the major portion of the leaf. The affected leaves develop pronounced chlorosis of the leaf tissue, as a result of which they appear either mottled or completely yellow. Sometimes, the necrotic lesions drop out, leaving a shot-hole symptom. Severely infected trees defoliate early in midsummer which stimulates new growth (leaves and blooms) in autumn, which causes further debilitation of tree vigour (Sztejnberg 1986).

14.5.2 Causal Organism

The disease is caused by the fungus *Cercospora circumscissa* Sacc. The teleomorph is reported as *Mycosphaerella cerasella* Aderhold. However, in India, *Cercospora rubrotincta* Ell. and Ev. has been reported on almond and sweet cherry. The lesions on leaves are covered with dark brown stromata from which conidiophores and conidia arise. Conidiophores (20–27 × 3–5 μm) are pale brown to brown, sparingly septate and geniculate. Conidia (28–118 × 2.7–5.0 μm) are initially hyaline but with maturity become olivaceous in colour. They are obclavate and straight to slightly curved in shape and are one to seven septate (Little 1987).

14.5.3 Disease Development

The fungus overwinters as substomatal stroma on diseased leaf debris. In spring, conidiophores and conidia of the pathogen develop from substomatal stroma of the fungus. Conidia serve as primary source of inoculum. Optimum temperature for disease development ranges from 20 to 25 °C. Conidia from current-season lesions are disseminated through wind and water splashes to initiate secondary cycles. High humidity, rain and dew favour disease development (Koul 1962; Sztejnberg 1995) (Plate 14.10).

14.5.4 Management

Spraying the trees three to four times at interval of 15–20 days starting from leaf bud burst stage with either mancozeb 75 WP at 0.3% or captan 50WP at 0.3%, or copper oxychloride 50WP at 0.3% or carbendazim 50WP at 0.05% or bitertanol 25 WP at 0.05% is effective in checking the disease. Application of captan before leaf-fall was found most effective. Collection and destruction of diseased leaf litter from orchard floor significantly reduce the primary inoculum in the spring (Sztejnberg 1978, 1986; Verma and Gupta 1979).

14.6 Cherry Leaf Spot

The disease has worldwide occurrence and is a serious disease of both sweet and sour cherries, particularly in areas with humid climate, like Australia, Canada, France, Italy, Germany, Japan and the United States. Karsten (1885) for the first time documented the occurrence of the disease from Finland on *Prunus padus* and named the causal fungus as *Cylindrosporium padi*, although Peck (1878) described

Plate 14.10 Cercospora spots on cherry leaves

it under the name *Septoria cerasina* from New York. In India, Nisar et al. (2014) for the first time reported its occurrence from Kashmir. The disease causes severe midsummer defoliation which reduces tree vigour leading to production of fewer blossoms and reduced fruit set in the successive seasons. The trees may die from winter injury, if severe cold follows midsummer defoliation.

14.6.1 Symptoms

The disease is initiated as small, reddish to purple red spots on upper leaf surface and occasionally on lower surface which on enlargement and coalescence result in the formation of small, brown necrotic patches. Initially, the outline of the spot is not well defined, but in later stages, the spots become somewhat circular in shape. Occasionally, the area around the spots remains green which gives the leaf a mottled appearance. Under humid weather conditions, white to light pink mildew-like growth appears on the corresponding side of the spots containing the asexual reproductive structures of the fungus. Affected leaves turn yellow, abscise and defoliate prematurely. Only a few spots per leaf can result in chlorosis. Upward leaf curling along the margins may also occur as a result of severe infection. Fruit infection is rare, but fruit pedicles may be infected under server epidemics. Premature defoliation results in reduced tree vigour and produces fewer blossoms in successive season (Jones 1995; Ellis 2008; Nisar et al. 2014) (Plate 14.11).

Plate 14.11 Severely infected chlorotic cherry leaf

14.6.2 Causal Organism

*Telo. Blumeriella jaapii (*Rehm) Arx
*Ana. Cylindrosporium padi (*Lib.) P. Karst. ex. Sacc

The fungus produces acervuli on the lower leaf surface containing conidia which are released as whitish to pinkish sticky mass. They are hyaline, bi-celled, elongated, curved or flexuous, with a tapering apex and measuring 55.5–72.7 μm in size. Microconidia produced only on the host in acervular stomata during August–September are hyaline and straight to allantoid and measures 4.5–6 × 1.5–2.0 μm (Nisar et al. 2014). Apothecia immersed in host tissue are reportedly flat to shallow, convex, globose to obpyriform and dark brown in colour. Asci are unitunicate, clavate with a long stout pedicellate base and measure 50–60 × 12–14 μm in size. Ascospores are usually unicellular or have a median septum, hyaline, ellipsoid to elongate-fusiform and often slightly curved and measure 30–50 × 3.5–4.5 μm in size (Jones 1995).

14.6.3 Disease Development

The fungus survives through the winter in infected fallen leaves. In spring, apothecia are produced in previous season's stroma. Apothecia and ascospores are formed at an optimum temperature of 17 °C and 13 °C, respectively. Ascospore discharge is highest at 16 °C to 30 °C which may occur immediately after rain fall from late bloom to about 6 weeks after petal fall. The fungus infects the leaves through stomata. Optimum conditions for lesion development are temperature of 15–20 °C coupled with rain fall and high humidity. Conidia are carried by rain splash and air currents. Secondary infection by conidia continues until leaf-fall in autumn (Garcia and Jones 1993; Ogawa et al. 1995a, b) (Plate 14.12).

14.6.4 Management

Since the fungus overwinters on fallen leaves serving as primary source of inoculum, then orchard sanitation involving collection and destruction of fallen leaves helps in reducing the disease inoculum. Various fungicides like chlorothalonil 70 WP at 0.3%, dodine 65 WP at 0.06%, copper oxychloride 50 WP at 0.3%, bitertanol 25 WP at 0.05%, difenoconazole 25 EC at 0.03%, penconazole 10 EC at 0.04% and hexaconazole 5 EC at 0.05% (Jones et al. 1993) have been found effective in checking the disease. The fungicides should be applied after petal fall and at 10-day intervals till harvest with a postharvest application after 2–3 weeks.

Plate 14.12 Whitish felt-like patches on undersurface of cherry leaf

14.7 Scab

The disease also known as freckles or black spot is an important disease of peaches. Besides peaches, the disease also affects almond, apricot and nectarine but rarely plum and cherry. Peach scab fungus was first time described in 1877 after collecting the fruit specimen from Australia and was named as *Cladosporium carpophilum* (Anderson 1956). In India, Baruah and Bora (1980), Khosla et al. (2009) and Nasreen et al. (2017) reported its occurrence on peach and almond from Mysore, Himachal Pradesh and Kashmir, respectively. The disease is worldwide in occurrence and is more severe in areas with warm humid climate.

14.7.1 Symptoms

The disease occurs on leaves, twigs and fruits. Small, greenish to olivaceous circular spot appears on the fruit surfaces, more frequently near the stem end. The spots enlarge and become dark olivaceous to black in colour. In severe infection, individual spots merge together and form a uniform velvety blotch. The fruit cracks and becomes abnormal in shape. Twig infection occurs on current season's growth where raised, circular to oval, greenish lesions of size 3–6 mm develop, which later turn brownish with raised purple to dark brown boarder. On leaves, angular to circular pale green areas develop on the lower leaf surface, which on sporulation turn olive green. In severe disease infection, lesions coalesce to cause leaf chlorosis resulting in premature defoliation. Long and narrow lesions are mostly formed along the leaf midrib (Nasreen et al. 2017) (Plate 14.13).

14.7.2 Causal Organism

Telo. Venturia carpophila E.E. Fisher
Ana. Cladosporium carpophilum (Thuemen)

The fungus initially produces hyaline, septate and branched hyphae which later thicken, become olivaceous in colour and form pseudoparenchymatous layer from which conidiophores arise. The conidiophores are short and erect, rarely branched, one to several septate and elongated at the base. Conidia are borne at the apex of conidiophores either singly or in chains. They are usually unicellular, occasionally bi-celled with a slight constriction at the septum, hyaline to slight olive green and fusoid to ovate in shape and measure 15–17 × 4–5 μm in size (Hendrix 1995; Khosla et al. 2009). The perfect stage of the fungus has not been reported till now, but since the imperfect stage on the twigs produces abundant conidia for initial infection, the perfect stage would be of little importance in the life cycle of the pathogen.

Plate 14.13 Olive green to black scab lesions along the midrib on lower leaf surface

14.7.3 Disease Development

The fungus overwinters as mycelia and chlamydospores in lesions and on twigs. Primary infection occurs from the conidia produced in the spring, 2–6 weeks after the shuck-split stage of development. Under favourable weather conditions, infection begins to occur at about shuck fall. Conidial germination occurs at a temperature range of 15–30 °C, the optimum temperature being 25–30 °C at relative humidity of 94–100%. Conidial liberation from the lesions occurs relatively at high relative humidity of 100%, which decreases with the decrease in humidity (Plate 14.14).

The pathogen has remarkably long incubation periods of 25–45 days and 40–70 days for disease appearance on leaves, twigs and fruit (Lawrence and Zehr 1982).

14.7.4 Management

The disease is mainly controlled through the use of fungicidal sprays, however, pruning of the trees helps to allow good penetration of sunlight and also increases air circulation which helps in disease control. Low-lying areas should not be selected as planting sites. Timely application of fungicides starts from calyx split and every 2 weeks thereafter for a total of four to five sprays. Various fungicides like chlorothalonil 70 WP at 0.3%, wettable sulphur 80 WP at 0.2%, carbendazim 50 WP at

Plate 14.14 Dark olivaceous velvety blotches over the peach fruit surface causing fruit cracking

0.05%, bitertanol 25 WP at 0.05% and captan 50 WP at 0.3% have been found effective in checking the disease (Hendrix 1995).

14.8 Brown Rot

Brown rot also known as grey mould or ripe rot is a major disease of all commercially grown *Prunus* species and is worldwide in distribution. Besides fruit rot under field and postharvest conditions, blightening of affected blossoms, spurs, twigs and small branches is also observed. The fruit rot phase of the disease was first reported by Schweinitz in 1822 on peach from the United States, while the blossom and leaf blight phase was reported in 1886 on cherry (Anderson 1956). In India, Sharma and Koul (1988) reported its fruit rot phase, while Gupta and Byrde (1988) recorded its blossom and twig blight phase from Himachal Pradesh.

14.8.1 Symptoms

The fungus initially invades the blossom which turns grey to dark brown, resulting in blossom blight. The fungus spreads through the peduncle and reaches to branches resulting in twig blight, followed by formation of elliptical-fusoid canker which usually develops from blighted twigs or fruit spurs with massive gum formation. The canker slowly girdles the twig and blights the portion distal to the cankered area. Leaves on such shoots turn tan to brown and remain attached instead of abscising. Usually, cankers remain restricted to twigs and do not expand. Fruit rot, the most destructive phase of the disease, is common on mature fruit than on immature fruit, where small, circular light brown spots develop on fruit surface, which expand rapidly and, under favourable conditions, engulf the entire fruit. Under wet and humid conditions, ashy-grey tufts of fungus develop over the lesion surface. Usually, both mature and immature fruits with brown rot tend to remain attached to the tree because the toxin released from the decaying tissue causes shoot dieback which prevents the formation of abscission layer between the fruit and peduncle (Sharma and Koul 1988; Gupta and Byrde 1988; Ogawa et al. 1995a, b) (Plate 14.15).

14.8.2 Causal Organism

Brown rot disease is caused by closely related more than one species of fungi belonging to genus *Monilinia.* The species include the following:

Monilinia fructicola (Wint.) Honey
Monilinia fructigena (Aderh. and Ruhl.) Honey
Monilinia laxa (Aderh. and Ruhl.) Honey

Species identification based on conidial morphology is somewhat difficult, as all the three species produce hyaline, lemon-shaped conidia in monilioid chains with meagre variation in their conidial size. However, *M. fructigena* produces largest conidia of the three species averaging 13 × 22 μm in size. The apothecia of *M. fructigena* and *M. fructicola* are similar, light brown in colour, 3 mm in size at maturity, cupulate and stipitate, while the apothecia of *M. laxa* and *M. fructigena* are rare. Asci of *M. fructigena* are 112–180 × 9–12 μm in size, while those of *M. laxa* and

Plate 14.15 Ashy-grey tufts of the fungus on mature rotten peach

M. fructicola are 121–188 × 7.5–11.8 and 102–215 × 3–12 µm, respectively. Ascospores are hyaline and ellipsoidal in *M. fructigena* and ellipsoidal to oval in *M. fructicola.* The ascospore size ranges from 6–15 × 4–8 µm for *M. fructicola* and 7–19 × 4.5–8.5 µm for *M. laxa* (Batra 1979; Howitt and Leach 1939). Macroconidia are produced in abundance by *M. fructicola,* less frequently by *M. fructigena* and rarely by *M. laxa* (Harada 1975; Byrde and Willetts 1977). Brown rot caused by *M. fructigena* reported from other countries of the world has not been reported from India.

14.8.3 Disease Development

The fungus overwinters on mummified fruits (either on ground or still on tree) and on twig and branch cankers as mycelium and conidia. Conidia are disseminated by wind and rain which germinate rapidly under favourable conditions. Mummies fallen on the ground and continuously exposed to moisture produce apothecia, while mummies hanging on tree fail to produce apothecia. The apothecia discharge ascospores during the bloom period but do not contribute to fruit infection. Later, the conidia produced on blighted blossom provide a source of infection for ripening fruit. Prolonged wet weather during blossom results in extensive blossom infection. Maximum blossom infection occurs at a temperature of 16 °C with four hours of wetness (Biggs and Northover 1988).

14.8.4 Management

Brown rot can be effectively managed by controlling the blossom blight phase of the disease. Removal and destruction of the infected plant parts including mummies and blighted and cankered twigs reduce the inoculum available for blossom and fruit infection. Pruning practices that promote good air circulation and sunlight penetration into the tree canopy help to prevent the conditions favouring infection and disease development. Good nutrition and proper soil moisture reduce the tree stress and thus increase tree tolerance to brown rot infection. Spraying the trees three to four times starting from bud swell stage at interval of 10–15 days with fungicides like carbendazim 50 WP at 0.05% or thiophanate methyl 70 WP at 0.05% or tebuconazole 29.5 EC at 0.05% or myclobutanil 10 WP at 0.03% or captan 50 WP at 0.3% or chlorothalonil 70 WP at 0.3% can give an effective control of blossom blight and fruit rot. To reduce postharvest decay, rapid cooling (cold storage) of the fruit is practised to delay brown rot development (Ogawa et al. 1995a, b).

Plate 14.16 Powdery mildew disease on apricot leaves

14.9 Powdery Mildew

The disease occurs on all the stone fruit crops but is more serious in nursery plantations where seedlings are infected in early growing season which remain stunted. The disease is worldwide in occurrence particularly in European countries. In India, Butler and Bisby (1931) reported the occurrence of the disease on almond from Kashmir caused by *Sphaerotheca*, while Koul (1967), Khan et al. (1975), Pandotra et al. (1968) and Sharma (1985) reported that the disease on almond and plum is caused by *Phyllactinia* and *Uncinula*, respectively (Plate 14.16).

14.9.1 Symptoms

The powdery mildew fungi infect leaves, young shoots and fruits. On young leaves, the disease is initiated as fine, netlike growth. As the infection advances, the leaf becomes covered with whitish patches of fungal growth. Severely infected leaves become chlorotic, puckered, roll upward, abscise and defoliate prematurely. White patches of the fungus also infect current season's young green shoots which may become curved at the tips and remain stunted. Buds may also be invaded by the fungus which fails to open or open improperly. White circular spots may also occur on fruit which may spread over fruit surface. The infected fruit turns dark brown with rusty appearance and becomes deformed (Koul 1967; Sharma 1985).

14.9.2 Causal Organism

The identification of powdery mildew on stone fruits is difficult because more than one species of powdery mildew fungi were found to occur on various stone fruits. Identification of these species is simplified if perfect state (ascocarp) is present. Species currently recognized on stone fruits include the following:

Sphaerotheca pannosa (Wallr. ex. Fr.) Lev
Podosphaera tridactyla (Wallr.) deBary
*Podosphaera clandestina (*Wallr. ex. Fr.) Lev
Podosphaera leucotricha (Ell. and Ev.) Salmon
Podosphaera clandestina

The conidia are produced in chains on short erect conidiophores which are elliptical to cylindrical, contain fibrosin bodies and measure 20–33 × 9.8–21.3 μm in size. Cleistothecia are globose to subglobose and black or brown in colour with five to twenty fasciculate appendages and measure 70–110 μm in dia. Asci are ovate and measure 61–123 × 36–79 μm in size. Ascospores are subreniform to elliptical, unicellular and hyaline and measure 20–35.7 × 12–18.7 μm in size.

14.9.2.1 *Podosphaera tridactyla*

The conidia are ellipsoidal to barrel shaped, borne on erect conidiophores and measure 20–32 × 13–18 μm in size. Cleistothecia are scattered to gregarious, brown to black in colour with one to eight fasciculate to sub-fasciculate appendages and measure 70–105 μm in dia. Asci are subglobose to ellipsoidal-ovoid and measure 50–90 × 50–80 μm in size. Ascospores are six to eight per ascus, hyaline to cream coloured and ellipsoidal-ovoid to subglobose and measure 16–30 × 9–20 μm in size.

14.9.2.2 *Sphaerotheca pannosa*

The conidia are produced in chains on short erect conidiophores and measure 20–33 × 12–19 μm in size. The cleistothecia are gregarious and brown to brown-black in colour with few septate appendages and measure 70–115 μm in dia. Asci are ellipsoidal-ovoid containing four to eight ascospores and measure 70–100 × 50–80 μm in size. Ascospores are hyaline, unicellular and cylindrical and measure 16–28 × 9–18 μm in size (Grove 1995).

14.9.3 Disease Development

The fungus overwinters as mycelia in the buds. On peach and apricot cleistothecia are rarely observed but are common on roses. Primary infections are, therefore, mainly through conidia produced from overwintering mycelia and occasionally by ascospores. Conidia and ascospores are carried by wind to green tissues. Under favourable temperature and moisture conditions, the spores germinate and infect the tissues through fine hyphae which absorb the nutrients through specialized globose

Plate 14.17 Powdery mildew disease on peach

structure haustoria. Optimum temperature for conidial germination lies between 21 and 27 °C (Weinhold 1961) (Plate 14.17).

14.9.4 Management

Clipping and proper destruction of diseased shoots and other diseased materials help in minimizing the disease inoculum. Three to four sprays of fungicides like wettable sulphur 80 WP at 0.2% or dinocap 48 EC at 0.1% or triadimefon 25 WP at 0.05% or flusilazole 40 EC at 0.02% or tebuconazole 250 EW at 0.05% or pyraclostrobin + boscalid 38 WG at 0.03% or hexaconazole 5 EC at 0.05% before bloom, at petal fall and 2 weeks later are effective in controlling the disease (De La Torre Almaraz and Ceballos Silva 1990; Dong et al. 1991; Huang et al. 1995).

14.10 Frosty Mildew

Frosty mildew on peach and apricot is found where these fruits are grown but is common in neglected orchards. The disease was first reported from the United States. In India, the disease was recorded from Himachal Pradesh (Sohi et al. 1964; Sharma and Paul 1986) (Plate 14.18).

Plate 14.18 Frosty mildew diseases on apricot

14.10.1 Symptoms

The disease is characterized by the appearance of white powdery patches on lower leaf surface, similar in appearance to powdery mildew. White patches may be discrete or cover the entire leaf surface. Corresponding to white powdery patches on the lower surface, pale yellow or pale green areas develop on the upper leaf surfaces, which latter become reddish in autumn (Higgins and Wolf 1937) (Plate 14.19).

14.10.2 Causal Organism

Ana. Cercosporella persica Sacc
Telo. Mycosphaerella pruni-persicae Deighton

Conidia borne singly on conidiophores accumulate in masses which are hyaline, vermicular to clavate and multiseptate and measure 17–86 × 2.5–7 μm. Perithecia produced during spring are black, erumpt and globose and measure 75–110 × 60–106 μm in diameter releasing hyaline, slightly curved, bi-celled ascospores which measure 12–20 × 2.5–3.5 m in size (Higgins and Wolf 1937).

Plate 14.19 Frosty mildew diseases on apricot

14.11 *Cryptosporiopsis* Blight

The disease is worldwide in occurrence particularly in countries experiencing warm and moist climate. The disease was first reported on peach from England in 1859 and later from the United States, Israel, Japan and India on different stone fruit crops. In India, Lee et al. (1981) reported its occurrence on almond from Kashmir, causing leaf spot and twig canker. The disease occurs in various phases, viz. leaf spot, dieback, twig canker and fruit spot phase. Of these, twig blight involving current season's growth has assumed serious proportions. Putoo (1980) reported its epidemic phase from various parts of Kashmir.

14.11.1 Symptoms

On current season's twigs, the disease is initiated as small, irregular, water-soaked chlorotic lesion around the base of leaf petiole. The chlorotic lesion elongates along the twig axis and forms sunken streaks which are initially purplish in colour, later becoming reddish pink with purplish margins and off-white centre. Twig girdling occurs at the point of infection which results in drooping, yellowing and drying of leaves above the girdled region. Complete girdling results in drying of affected twigs. On leaves, the disease is characterized by the appearance of small, circular chocolate brown spots which later turn creamy white and are surrounded by chocolate brown hollow. Small pinhead-like structures (acervuli) containing the conidia of the fungus breaking through the epidermis develop in the centre of the spots. Coalescence of these spots results in leaf blightening, and the blighted leaves remain clinged to the twigs throughout the season unless disturbed mechanically or through wind (Beig et al. 2006).

14.11.2 Causal Organism

Ana. Cryptosporiopsis corticola
Telo. Pezicula corticola (Jorgenson) Nannf

The fungus produces black-coloured acervuli either solitary or botryose which measures 591–765 × 212–267 μm in size. Conidia borne acrogenously on philaids of conidiogenous cells are unicellular, hyaline to slight yellowish in colour and oval to ellipsoidal in shape, with one end rounded and other slightly tapered with a truncate base, and measure 5.9–10.0 × 2.5 μm in size (Lee et al. 1981).

14.11.3 Disease Development

The pathogen overwinters as asexual fruiting bodies (acervuli) on twigs, pruned snag, leaf petioles, buds, etc. The spores ooze out in the month of May and are disseminated by wind and rain. The optimum temperature for pathogenic growth is 20–25 °C.

14.11.4 Management

Cryptosporiopsis blight can be managed through an integrated crop management strategy. Prune the diseased twigs, and dress wounds and all pruning cuts with a wound dresser. Spraying the trees twice in the month of May and June with either penconazole 10 EC at 0.05% or copper oxychloride 50 WP at 0.25% has been found effective in managing the leaf blight phase of the disease (Beig et al. 2006).

14.12 *Cytospora* Canker

Cytospora canker also known as perennial canker, *Valsa* canker or *Leucostoma* canker is an important canker disease of sweet cherries and other stone fruits. The disease is worldwide in occurrence especially in temperate zones of the world. *Cytospora* cankers on peach was first reported from New York in 1900 and later on cherry from the United States (Biggs 1995). The pathogen has wide host range; besides stone fruit crops, it also invades apple, pear, olive, willow and other flowering plants, reducing their productivity and also shortening the tree's longevity.

14.12.1 Symptoms

The canker symptoms on stone fruit crops depend on the part of the tree that is infected. Infection of small twigs appears as sunken, discoloured areas, often with alternating zonation lines, around the winter-killed buds or leaf scars. The infected tissues darken with age, and an amber-coloured gum oozes from them until the twig is killed. Twig or branch infection also produces leaf symptoms during the growing season. Leaves on infected branch often turn yellow, droop and may wilt and die. Dead twigs and branches are usually covered with pinhead-sized black structures serupting through dead bark which are the reproductive structures of the pathogen (Biggs 1989).

Cankers that form on the main trunk, branch crotches, scaffold limbs and older branches are the most conspicuous expression of infection. Cankers on these tree parts are initiated as small necrotic spot which enlarges lengthwise more, leading to formation of oval to elliptical brownish and sunken cankered area, often surrounded by raised callus tissue. Diseased bark turns dark in colour. The bark finally shrivels, becomes spongy and sloughs off, exposing blackened dead wood beneath. Numerous small pimples like pycnidia appear on the dead bark (Tekauz and Patrick 1974).

14.12.2 Causal Organism

Leucostoma cincta (Fr. ex. Fr.) Hohn
Syn. *Valsa cincta* (Fr. ex. Fr.)
Ana. *Leucocytospora cincta* (Sacc.)
Leucostoma persoonii Hohn.
Syn. *Valsa leucostoma* (Pers. ex. Fr.)
Ana. *Leucocytospora leucostoma* (Pers.) Hohn

Pycnidial stromata are formed on cankers and on killed twigs and branches which are black on the surface and grey to greyish brown internally. The pycnidial discs are white in case of *L. persoonii* and grey to brownish grey in *L. cincta*. Mature pycnidia under moist conditions extrude orange- to amber-coloured tendril or cirrus, containing the conidia of the fungus. The conidia of both the species are allantoid and hyaline and measure 5–10 × 1–2 μm in size. Perithecia are formed much later, often 2–3 years after pycnidial stromata appear. They may be formed within or underneath the pycnidial stromata. The perithecial stromata of *L. cincta* are round (1.6–2.8 mm in dia.), prominent and delimited by a black, circumscribing conceptacle (30–80 μm thick). Black ostioles of individual perithecia (10–30 per perithecial stromata) are arranged circinately around a central pycnidium. Individual perithecia are 200–350 μm in dia. Asci are clavate and sessile to subsessile and measure 45–80 × 7–12 μm in size. Ascospores are hyaline and unicellular allantoid and measure 15–30 × 4–8 μm in size. The perithecial stromata of *L. persoonii* are round (2–3 mm in dia.) and whitish possessing a circumscribing conceptacle similar to that of *L. cincta*. Asci are fusoid to clavate and sessile to subsessile and measure

35–45 × 7–8 μm in size. Ascospores are hyaline, unicellular and allantoid and measure 10–18 × 2–5 8 μm in size. Asci and ascospores of *L. cincta* are generally larger than those of *L. persoonii*. The teleomorph stage of *L. persoonii* is fairly common in nature on *Prunus* spp., whereas *L. cincta* is relatively rare on *Prunus* spp. and more common on *Malus* spp. (Sakuma et al. 1980: Tamura and Saito 1982: Proffer and Jones 1989).

14.12.3 Disease Development

The pathogen invades the host tissue through wounds or dead tissues. Mechanical and winter injuries and other stresses such as sunburnt tissues provide sites for infection (Kable et al. 1967). The pathogen can enter the node through leaf scars or dead buds during the dormant season where the disease may progress to older limbs and scaffold branches. Fungus overwinters through conidia produced in multichambered pycnidia on the cankered twigs and branches. The potential role of ascospores is uncertain. Conidia are most abundant under cool, moist conditions of late fall and early spring but are present throughout the year if the rainfall is sufficient. During wet weather, conidia oozes out in the form of long coiled threads which are splashed by rain or may be carried by insects. At temperatures 14–20 °C, pathogen can initiate more rapid canker expression (Willison 1936; Bertrand and English 1976).

14.12.4 Management

Leucostoma canker can be managed through an integrated crop management strategy. Management of cankers is based on preventive measures that minimizes winter injury, sunburn and insect damage; promotes optimum plant health; and facilitates rapid wound healing. Application of balanced doses of fertilizers promotes good tree vigour, Pruning of cankered branches and twigs and removing of cankers from tree trunks and large limbs during dry weather. Treating the wounds and all pruning cuts with a wound dresser. Spraying trees with carbendazim 50 WP or with thiophanate-methyl 70 WP at 0.05% immediately after pruning and before rains reduces the disease (Moller and Carter 1970; Biggs 1989; Jones and Aldwinkle 1990). Spraying the trees with methyl benzimidazole fungicides, viz. benomyl or Topsin M, during growing season and after the pruning or removal of cankers along with dressing the wound with paints containing Topsin M, polyoxin D or guzatine also gave good control of *Valsa* canker (Tamura 1984).

14.13 *Phytophthora* Root and Crown Rot

Root and crown rot are the most important soil-borne diseases causing considerable losses to stone fruit crops. These diseases are more disastrous than other aerial diseases, because they often result in the death of the tree. The diseases are worldwide in occurrence, and losses have been reported throughout the world affecting all ages of trees.

14.13.1 Symptoms

Above-ground symptoms of root and crown rot-affected trees exhibit insufficient extension of terminal growth and sparse, small and chlorotic foliage. Fruits remain highly coloured, undersized and sunburned. Shoots and scaffold branches show die-back symptoms. As the disease progresses, trees collapse and die suddenly in late spring or summer following years with excessively wet weather. Trees affected with root rot have only few feeder roots, and those remaining are often decayed with dark brown to black discoloration in the cortex and stele. The affected trees exhibit decay, which often extends to the crown and up to the graft union but seldom extends much above the graft union.

14.13.2 Causal Organism

Many *Phytophthora* spp. have been implicated as root and crown rot pathogens of stone fruit trees. *Phytophthora cactorum* (Lebert and Cohn), *P. syringae* (Kleb.) Kleb., *P. megaspermaa* Drechsler, *P. citricola* Sawada, *P. cryptogea* Pethyb and Laff and *P. cinnamomi* Rands have been usually found associated with the root and crown rot diseases of stone fruits in the temperate regions of the world (Taylor and Washington 1984; Wilcox and Mircetich 1985; Bielenin and Jones 1988).

14.13.3 Disease Development

Phytophthora is primarily a soil-borne pathogen that thrives in the soil. Roots are predisposed to infection when the soil is saturated or nearly so. The fungus produces zoospores within sporangia when soil moisture levels are high. Zoospores swim when free or standing water is present. On close contact to tree roots or crown, these spores lose flagella, encyst, germinate and initiate infection. Hyaline and aseptate mycelium of the fungus colonizes the roots. Sporangiophore-bearing sporangia are produced. Continuous crop of zoospores is produced for secondary infection. Perennation of the pathogen occurs through the production of oospores which are sexual spores. The pathogen survives in the soil for several years. Pathogen population is directly influenced by soil moisture and temperature. Soil temperature of 12–20 °C with pH 5 to 6 is best suited for the survival of the fungal propagules (Rana and Gupta 1984).

14.13.4 Management

Soil water management is the most important aspect for avoiding *Phytophthora* root and crown rot. Sites with good drainage system reduce the rate of disease development. Tree basin of the infected trees should be drenched with systemic fungicides like Ridomil or Fosetyl-Al or during rainy seasons to avoid disease spread (Taylor and Washington 1984). Dead bark near the crown or collor region should be scarified during the dormant season and applied with wound dressing paint or paste like Chaubattia or Bordeaux. Besides chemical management and other strategies, *Phytophthora* root and crown rots include careful soil water management and genetically resistant root stock. It is advisable to keep the graft union at least 30 cm above the soil line.

14.14 Bacterial Spot

The disease also known as bacterial short hole, black spot and bacterial leaf spot was initially described by Erwin F. Smith in 1903 on Japanese plum. The disease is worldwide in occurrence particularly in Australia, Canada, Japan, Italy, New Zealand, South Africa, the United States, Brazil, etc. The disease affects peach, nectarine, plum, apricot and prunes and is responsible for their severe yield losses by reducing the fruit quality and leaf vigour through leaf spots, defoliation and twig infection. It is more severe in areas with warm and humid environmental conditions during the cropping season.

14.14.1 Symptoms

Disease symptoms appear on leaves, twigs and fruits. On leaves, small, water-soaked, angular, grey lesions appear on the undersides of leaves, often located along the midrib, leaf tip or margin. As the lesions enlarge, they turn brown or black with purplish and necrotic centre, and the centres of spots can fall out, giving the leaf a tattered appearance. Heavily infected leaves turn yellow and lead to premature leaf-fall. Infected fruit is stained with brown to black pits and cracks which are mistaken for insect damage. Under humid conditions, gummy exudates may be seen on the fruit lesions. Elliptical cankers develop on twigs of current or one-year-old growth. Dieback symptom as a result of twig canker is more severe on plum and apricot than on peach.

14.14.2 Causal Organism

Xanthomonas campestris pv. *pruni* (Smith) Dye

The bacterium is gram-negative, rod shaped and motile and measures 0.2–0.4 × 0.8–1.0 μm in size.

14.14.3 Disease Development

The bacterium overwinters on cankered area on twigs and in buds. In spring, the bacterial population increases, spreads by rain splashes and insects and causes infection to young leaves, fruits and twigs under moist conditions through natural openings, leaf scars and wounds. Warm, rainy weather throughout the season is conducive to secondary infections.

14.14.4 Management

Chemical control of the disease is not feasible, so planting of highly susceptible cultivars should be avoided as the disease is more severe on some cultivars. Fertilization should be adequate to maintain good tree health. Dormant applications of fixed copper (copper oxychloride) may reduce bacterial populations. The antibiotic oxytetracycline at 500–700 ppm and fungicides like dodine and ziram have also been used with varying degrees of success.

14.15 Crown Gall

Crown gall is a widespread disease which affects large number of woody and herbaceous plants; however, it is more common on pome and stone fruits. The disease affects *Prunus* spp. both under nursery conditions and mature plants in the orchard.

14.15.1 Symptoms

The disease is characterized by the gall formation on root and stem near the soil line of affected plants, which vary in size from microscopic to more than 10 cm in diameter. These galls are initially white, round, fleshy swellings which latter turn tan to brownish during the dormant season.

14.15.2 Causal Organism

Agrobacterium tumefaciens (Smith and Townsend) Conn

Crown gall disease is caused by the biovar 2 of *A. tumefaciens*. The bacterium is gram-negative, aerobic, motile (by means of peritrichous flagella) and rod shaped and measures 0.6–1.0 × 1.5–3.0 μm in size.

14.15.3 Disease Development

The bacterium overwinters in soil and can survive in soil as saprophyte for several years. As soon as they come in contact with the susceptible plants, they enter the roots or stem near the soil line through wound. Infection of plants by *Agrobacterium* tumorigenic strains consists of the transfer of a fragment of bacterial Ti plasmid (T-DNA) into the plant cell and its incorporation into the plant genome. This transfer is controlled by virulence (vir) genes which are located on the Ti plasmid. Expression of T-DNA genes, which code for auxin and cytokinin synthesis, triggers uncontrolled cell division and growth resulting in tumour formation (Zhu et al. 2000). Tumour formation inhibits transport of water and nutrients which may partly girdle the bigger roots or crown and reduce plant growth.

14.15.4 Management

Chemical control of the disease is impractical especially in commercial orchards owing to its soil-borne nature. Sanitation and cultural practices which include using of certified disease-free planting material, avoiding injury to crowns and roots of the plant as the bacterium enters through wounds and avoiding use of infected bud/graft wood for propagation purpose have been effective in reducing the disease. Biological control of the disease has been achieved by the use of *Agrobacterium radiobacter* strain 84 that produces antibiotic agrocin 84. An improved strain K84 derived from strain 84 has provided more effective control of the disease.

14.16 Plum Pox

Plum pox also known as 'Sharka' disease was first detected in Eastern Europe in 1918 (Atanassov 1932), and since then, the disease spread to various other parts of the world. The disease affects and causes serious losses in plum, peach, nectarine and apricot by reducing the quality and quantity of the fruit. The disease can lead to 83–100% yield losses in highly susceptible cultivars (Kegler and Hartmann 1998; Waterworth and Hadidi 1998).

14.16.1 Symptoms

Symptoms of the disease are observed on leaves and fruits. Plum leaves show vein yellowing and light green to yellow ring formation which is restricted to few leaves per shoot. On unripe fruit, similar type of rings develops which later disappear as the fruit attains colour and maturity. Sunken spots are sometimes also observed on fruit. Red rings and spots can also occur on the stones. The affected fruits are low in sugar, tasteless and drop prematurely. In peach, symptoms on leaves consist of chlorotic vein clearing with twisted and distorted lamina. On peach fruits, similar symptoms as observed in plum, however, affected fruits are sometimes distorted.

In apricot, leaf symptoms are similar to plum but are less conspicuous. Affected fruits may be misshapen and flavourless with dry flesh around the stone, turn brown or become necrotic and may have rings on the surface of the seed.

14.16.2 Causal Organism

Plum pox virus belongs to genus *Potyvirus* group of viruses. The virus is characterized by its flexuous, filamentous particles of approx. 750 nm in length. The virus contains a single-stranded RNA with a molecular weight of 3.5×10^6. The virus is transmitted by aphids in non-persistent manner.

14.16.3 Disease Development

Infected plants are the major source of inoculums. The virus is transmitted by grafting and non-persistently by aphid vectors.

14.16.4 Management

Use of healthy and/or resistant/tolerant planting material is the only effective method to control the disease. Roughing out the infected plants as soon as the symptoms are observed and management of aphids which act as vectors of virus can help in preventing the spread of the disease.

14.17 Conclusion

Stone fruits such as peach, plum, cherry, apricot and nectarines are invaded by various fungal and bacterial pathogens. Amongst fungal diseases, the most important ones are peach leaf curl, *Stigmina* blight, powdery mildew, brown rot, peach scab, rust, leaf spots, canker and root diseases. Amongst the bacterial diseases, crown gall and bacterial blight are the most important ones. Stone fruit crop diseases are effectively managed by integrating various cultural, chemical and biological measures. Fungal fruit and foliar diseases are successfully controlled by timely spraying of fungicides. Canker and dieback diseases are managed by proper pruning and applying fungicidal paints or pasts. Diseases caused by fastidious bacteria are managed by applying antibiotics and removing infected plants.

References

Anderson, H. W. (1956). *Diseases of fruit crops* (501 pp). New York: McGrew–Hill Book Company.

Angelov, A. (1980). The effectiveness of breeding in fruit growing in Bulgaria. *Ovoshchavstvo, 59*, 12–16.

Atanassov, D. (1932). Plum pox. A new virus disease. *Annals of the University of Sofía, Faculty of Agriculture and Silviculture, 11*, 49–69.

Baruah, S. N., & Bora, K. N. (1980). A fungal disease of peach plant collected from Gauhati University Campus, Assam (*Venturia carpophila*). *Science and Culture, 46*, 264.

Batra, L. R. (1979). First authenticated North American record of *Monilinia fructigena* with notes on related species. *Mycotoxon, 8*, 476–484.

Beig, M. A., Nisar, A. K., & Dar, G. H. (2006). Disease status of almond blight in Kashmir and morpho-physiological studies on its causal pathogen *Cryptosporiopsis corticola. Applied Biological Research, 9*, 22–28.

Berkeley, M. J. (1864). Outlines of British Fungology. *Gardens Chronicle, 24*, 938.

Bertrand, P. F. (1995). Rust. In M. O. Joseph, I. Z. Eldon, W. B. Georg, F. R. David, U. Kiyoto, & K. U. Jerry (Eds.), *Compendium of stone fruit diseases* (pp. 23–24). St. Paul, MN: APS. Press.

Bertrand, P. F., & English, H. (1976). Release and dispersal of conidia and ascospores of *Valsa Leucostoma. Phytopathology, 66*, 987–991.

Bhardwaj, S. S., & Ved, R. (1995). Incidence of apricot leaf curl and its management. In V. K. Gupta & R. S. Sharma (Eds.), *Integrated disease management and plant health* (pp. 227–229). Jodhpur: Scientific Publishers.

Bielenin, A., & Jones, A. L. (1988). Prevalence and pathogenicity of *Phytophthora* spp. From sour cherry trees in Michigan. *Plant Disease, 72*, 473–476.

Biggs, A. R. (1989). Integrated approach to control Leucostoma canker of peach in ontario. *Plant Disease, 73*, 869–874.

Biggs, A. R. (1995). Leucostoma canker. In M. O. Joseph, I. Z. Eldon, W. B. Georg, F. R. David, U. Kiyoto, & K. U. Jerry (Eds.), *Compendium of stone fruit diseases* (pp. 28–30). St. Paul, MN: APS Press.

Biggs, A. R., & Northover, J. (1988). Influence of temperature and wetness duration on infection of peach and sweet cherry fruits by *Monilinia fructicola. Phytopathology, 78*, 1352–1356.

Butler, E. J., & Bisby, G. R. (1931). *The fungi of India*. ICAR Sci memo. no. 1.

Byrde, R. J. W., & Willetts, H. J. (1977). *The brown rot fungi of fruit: Their biology and control* (171pp). New York: Pergaman Press.

De La Torre Almaraz, R., & Ceballos Silva, A. P. (1990). Chemical control of peach disease at Acajete, Puebla. *Revista mexicana de fitopatología, 8*, 181–190.

Decker, R., & Buchanan, D. W. (1975). Effect of rust (*Tranzschelia discolor*) on the tree performance and fruit yield of 'Early Amber' peach and Sun gold nectarine. *Proceedings of the Florida State Horticultural Society, 86*, 333–335.

Del Schweinitz, L. D. (1822). Synopsis fungorum carolinae superioris secundum observationes. *Soc. Nat. Cur. Lips, 1*, 20–131.

Dong, C. O., Kim, S. B., Jang, H. I., Cho, M. D., & Lee, E. K. (1991). Studies on the ecological characteristics and control methods of peach powdery mildew. *Journal of Korean Horticultural Science, 32*, 191–198.

Dunegan, J. C. (1938). The rust of stone fruits. *Phytopathology, 28*, 417–426.

Ellis, M. B. (1959). *Clasterosporium and some allied Dematiaceae Phragmosporae. Mycological paper (Vol. 72)*. Surrey: CMI, Kew.

Ellis, M. A. (2008). *Cherry leaf spot. Fact sheet agricultural and natural resources*. Ohio State University (pp. 3). http://ohioline.osu.edu.

Ferraris, T. (1928). Agricultura ef fitopathologianel Kashmir. *Curiamo de Plante, 6*, 61–68.

Garcia, S. M., & Jones, A. L. (1993). Influence of temperature on apothecial development and ascospore discharge by *Blumeriella jaapii. Plant Disease, 77*, 776–779.

Grove, G. G. (1995). Powdery mildew. In M. O. Joseph, I. Z. Eldon, W. B. Georg, F. R. David, U. Kiyoto, & K. U. Jerry (Eds.), *Compendium of stone fruit diseases* (pp. 12–14). St. Paul, MN: APS Press.

Gupta, G. K., & Byrde, R. J. W. (1988). *Monilinia laxa* associated with blossom wilt of apricot and almond in Himachal Pradesh, India. *Plant Pathology, 37*, 591–593.

Gupta, G. K., Agarwal, R. K., & Dutt, K. (1972). Stigmina blight of peach and its control in Himachal Pradesh. *Advances in Horticultural Science, 9*, 49–51.

Harada, Y. (1975). Cultural studies on the *Monilinia* disease fungus of apples. V1. The effect of light on conidial germination, mycelial growth and sporulation. *Bulletin of the Faculty of Agriculture, Hirosaki University*, 25pp.

Hawksworth, D. L., Sutton, B. C., & Ainsworth, G. C. (1983). *Ainsworth and Bisby's dictionary of fungi* (7th ed.). Kew Surrey: Commonwealth Mycological Institute.

Hendrix, F. F. (1995). Scab. In M. O. Joseph, I. Z. Eldon, W. B. Georg, F. R. David, U. Kiyoto, & K. U. Jerry (Eds.), *Compendium of stone fruit diseases* (pp. 11–12). St. Paul, MN: APS Press.

Higgins, B. B., & Wolf, F. A. (1937). Frosty mildew of peach. *Phytopathology, 27*(6), 690–696.

Howitt, W. B., & Leach, L. D. (1939). Brown rot sclerotinia occurring in California and their distribution on stone fruits. *Phytopathology, 29*, 337–351.

Huang, J. W., Chen, J. H., Chung, W. C., & Yaung, S. H. (Eds.). (1995). Chemical control of powdery mildew on Japanese apricot. *Journal of Agriculture and Forestry*, 44: 13–18.

Jones, A. L. (1995). Cherry leaf spot. In M. O. Joseph, I. Z. Eldon, W. B. Georg, F. R. David, U. Kiyoto, & K. U. Jerry (Eds.), *Compendium of stone fruit diseases* (pp. 21–22). St. Paul, MN: APS Press.

Jones, A. L., & Aldwinkle, H. S. (Eds.). (1990). *Compendium of Apple and Pear diseases* (pp. 28–30). St. Paul, MN: APS Press.

Jones, A. L., Ehret, G. R., Garcia, S. M., Kesner, C. D., & Klein, W. M. (1993). Control of cherry leaf spot and powdery mildew on sour cherry with alternate-side application of Fenarimol, myclobutanil and tebuconazole. *Plant Disease, 77*, 703–706.

Kable, P. F., Fliegel, P., & Parker, K. G. (1967). Cytospora canker on sweet cherry in New York State association with winter injury and pathogenicity to other species. *Plant Disease Reporter, 51*, 155–157.

Karsten, P. A. (1885). Symbolae and mycoligiam Fennicam, XVI. *Meddelanden af Societas pro Fauna et Flora Fennica, 11*, 148–161.

Kegler, H., & Hartmann, W. (1998). Present status of controlling conventional strains of plum pox virus. In A. Hadidi, R. K. Khetarpal, & H. Koganezawa (Eds.), *Plant virus disease control* (pp. 616–628). St Paul, MN: APS Press.

Khan, M. W., Malik, K. A., & Khan, M. A. (1975). Perithecial stages of certain powdery mildews including some new records-111. *Indian Phytopathology, 28*, 199–201.

Khosla, K., Bhardwaj, S. S., & Gupta, A. K. (2009). Scab disease of peach and nectarine. *Journal of Mycology and Plant Pathology., 39*, 342.

Kosogrova, E. A. (1976). Dynamics of Clasterosporiosis of plum in northern Kazakhstan. *Zashchita Rastenii, 6*, 1344.

Koul, T. N. (1962). Occurrence of *Cercospora rubrotincta* on cherry in India. *Current Science, 31*, 349.

Koul, T. N. (1967). Diseases of stone-fruits in Kashmir. *Horticulturist, 2*, 52–58.

Lawrence, E. G., & Zehr, E. L. (1982). Environmental effects on the development and dissemination of *Cladosporium carpophilum* on peach. *Phytopathology, 72*, 773–776.

Lee, V. C., Lal, S. P., & Gupta, G. K. (1981). A new blight disease of almond in Kashmir valley and its control. *Pesticides, 15*, 22–25.

Leveille, J. H. (1843). Observations sur quelques champignons de la flore des environs de Paris. *Ann. Sci. Bot., 19*, 215.

Little, S. (1987). *Cercospora circumscissa descriptions of pathogenic fungi and bacteria, no. 911*. Kew, Surrey: Commonwealth Mycological Institute.

Moller, W. J., & Carter, M. V. (1970). Field evaluation of benomyl for control of limb dieback in apricots. *Australian Journal of Experimental Agriculture and Animal Husbandry, 10*, 488–489.

Munjal, R. L., & Kulshrestha, D. D. (1968). Some Dematiaceous Stigmina species. Hyphomycetes from India. V. Stigmina species. *Indian Phytopathology, 21*, 309–314.

Mushtaq, A. B., Nisar, A. K., & Nisar, A. Q. (2014). First record of rust caused by Tranzschelia discolor (Fuckel) Tranz. & Litv. on stone fruits in Jammu & Kashmir (India). *Applied Biological Research, 16*(2), 264–265.

Nasreen, F. K., Khursheed, A., Mohammad, H., Saba, B., Mushtaq, A. B., & Nisar, A. Q. (2017). First record of scab disease of almond by *Cladosporium carpophilum* in India. *Indian Phytopathology, 70*(3), 403–404.

Nisar, A. K., Mushtaq, A. B., Khurshaid, A., & Nisar, A. Q. (2014). First record of cherry leaf spot caused by *Cylindrosporium padi* (P.Karst.) from India. *Applied Biological Research, 16*(1), 122–123.

Ogawa, J. M., Zehr, E. I., & Biggs, A. R. (1995a). Brown rot. In M. O. Joseph, I. Z. Eldon, W. B. Georg, F. R. David, U. Kiyoto, & K. U. Jerry (Eds.), *Compendium of stone fruit diseases* (pp. 7–10). St. Paul, MN: APS Press.

Ogawa, J. M., Zehr, E. I., & Biggs, A. R. (1995b). Cherry leaf spot. In M. O. Joseph, I. Z. Eldon, W. B. Georg, F. R. David, U. Kiyoto, & K. U. Jerry (Eds.), *Compendium of stone fruit diseases* (pp. 21–22). St. Paul, MN: APS Press.

Padwick, G. W. (1945). *Notes on Indian fungi 111. Mycological paper no.12*. Surry: CMI, Kew.

Pandotra, V. N., Kachroo, J. L., & Sastry, K. S. M. (1968). Six powdery mildew from Jammu and Kashmir State. *Proceedings of the Indian Academy of Sciences, 67B*, 119–124.

Peck, C. H. (1878). Report of the botanist. *New York State Museum of Natural History, 29*, 29–82.

Ponti, I., Brunel, A., Spada, G., Tosi, C., Gafaffoni, M., Emiliani, G., & Cont, C. (1993). Activity of dodine against peach blister. *Inf. Agra., 49*, 75–79.

Proffer, T. J., & Jones, A. L. (1989). A new canker disease of apple caused by Leucostoma cincta and other fungi associated with cankers on apple in Michigan. *Plant Disease, 73*, 508–514.

Pscheidt, J. W. (1995). Peach leaf curl. In M. O. Joseph, I. Z. Eldon, W. B. Georg, F. R. David, U. Kiyoto, & K. U. Jerry (Eds.), *Compendium of stone fruit diseases* (p. 22). St. Paul, MN: APS Press.

Putoo, B. L. (1980). Almond blight. A new fungal disease in Kashmir. *Indian Journal of Mycology and Plant Pathology, 10*, LXV.

Putoo, B. L., & Razdan, V. K. (1991). Stigmina blight of cherry. *Plant Disease Research, 6*, 60.

Rana, K. S., & Gupta, G. K. (1984). Effect of fungicides on the viability of *Phytophthora cactorum* propagules in the soil. *Phytopathology, 110*, 245–250.

Sakuma, T., Mizuno, N., Koganezawa, H., & Miyagawa, H. (1980). Studies on apple Valsa canker. Seasonal differences in lesion appearance and pycnidial formation. *Bulletin of the Fruit Trees Research Station, Ser 7*, 101–108. (Morioka, Japan).

Shabi, E., Elisha, S., Birger, R., & Singer, G. (1990). Almond diseases and their control. *Aon Hanotea, 42*, 445–450.

Sharma, A. K. (1985). A new species of Phyllactinia from India. *Transactions of the British Mycological Society, 85*, 756–759.

Sharma, R. C., & Paul, Y. S. (1986). New disease of peach (*Prunus persica* L) from India. *Indian Journal of Plant Pathology, 4*, 182.

Sharma, R. L., & Koul, J. L. (1988). Occurrence of brown rot (*Monilinia* spp.) on stone fruits in Himachal Pradesh. *Plant Disease Research, 3*, 46–47.

Sharma, R. C., Jindal, K. K., & Gupta, V. K. (1987). *Managing fungal peach leaf curl through chemical and host resistance*. In Proceedings of 11th International Congress on Plant Protection held in Manila in October 1987 (pp. 51).

Sharma, R. C., Gupta, V. K., & Garg, R. C. (1988). Dormant application of fungicides for the management of peach leaf curl. *Indian Phytopathology, 41*, 259–260.

Singh, U. B. (1943). *Some diseases of fruits and fruit trees in Kumaon hills*. ICAR Misc. Bull. no. 51.

Smith, E. F. (1903). Observations on hitherto unreported bacterial diseases, the cause of which enters the plant through ordinary stomata. *Science, 17*, 456–457.

Sohi, H. S., Jain, S. S., Sharma, S. L., & Verma, B. R. (1964). New records of plant diseases from Himachal Pradesh. *Indian Phytopathology, 17*, 42–45.

Sydow, H., & Butler, E. J. (1916). Fungi Indiae Orientalis part IV. *Annals of Mycology, 14*, 177–220.
Sztejnberg, A. (1978). Fungicides controlling cherry leaf spot disease caused by the fungus *Cercospora circumscissa in* Israel. *Alon Hanotea, 32*, 341–344.
Sztejnberg, A. (1986). Etiology and control of cherry leaf disease in Israel caused by fungus *Cercospora circumscissa. Plant Disease, 70*, 349–351.
Sztejnberg, A. (1995). Cercospora leaf spot. In M. O. Joseph, I. Z. Eldon, W. B. Georg, F. R. David, U. Kiyoto, & K. U. Jerry (Eds.), *Compendium of stone fruit diseases* (pp. 23–27). St. Paul, MN: APS Press.
Tamura, O. (1984). Studies on the epidemiology and control of Japanese apple canker caused by *Valsa cerotosperma* (Tobe.ex.Fr.)Maire (=*V.mali*) Miyabe et Yamada. *Report of Hokkaido Prefectural Agricultural Experiment Station, 31*, 34–42.
Tamura, O., & Saito, I. (1982). Histopathological changes of apple bark infected by *Valsa cerato-sperma* during dormant and growing period. *Annals of the Phytopathological Society of Japan, 48*, 490–498.
Taylor, P. A., & Washington, W. S. (1984). Curative treatments for *Phytophthora cactorum* in peach trees using metalaxyl and fosetyl Al. *Australian Journal of Plant Pathology, 13*, 31–34.
Tekauz, A., & Patrick, Z. A. (1974). The role of twig infection on the incidence of perennial canker of peach. *Phytopathology, 64*, 683–688.
Verma, K. D., & Gupta, G. K. (1979). Studies on leaf spot of cherry (*Prunus avium*) caused by *Cercospora circumscissa* and its control. *Progressive Horticulture, 10*, 57–62.
Waraich, K. S., & Khatri, H. L. (1977). Occurrence of pink disease and rust on plum in India. *Indian Journal of Mycology and Plant Pathology, 7*, 202–206.
Waterworth, H. E., & Hadidi, A. (1998). Economic losses due to plant viruses. In A. Hadidi, R. K. Khetarpal, & H. Koganezawa (Eds.), *Plant virus disease control* (pp. 1–13). St Paul, MN: APS Press.
Weinhold, A. R. (1961). Temperature and moisture requirements for germination of conidia of *Sphaerotheca pannosa. Phytopathology, 51*, 699–703.
Wilcox, W. F., & Mircetich, S. M. (1985). Pathogenicity and relative virulence of seven *Phytophthora* spp.on Mahaleb and Mazzard cherry. *Pathology, 75*, 1451–1455.
Willison, R. S. (1936). Peach canker investigation–II. Infection studies. *Canadian Journal of Research, 14*, 27–44.
Zhu, J., Oger, P. M., Schrammeijer, B., Hooykaas, P. J. J., Farrand, S. K., & Winans, S. C. (2000). The bases of crown gall tumorigenesis. *Journal of Bacteriology, 182*, 3885–3895.

Integrated Pest Management of Stone Fruits

15

Bashir Ahmad Rather, M. Maqbool Mir, Umar Iqbal, and Shabir Ahmad Mir

Abstract

Stone fruits (apricot, cherry, peach, nectarine and plum) are subjected to many biotic stresses of which insect pests, viz. aphids, scales, mealy bugs, mites, borers, lepidopteran caterpillars and fruit flies, are responsible for many problems related to different plant parts and cause a considerable damage to these fruit crops. Close monitoring of insect pests for incidence and potential damage is considered one of the key steps for effective pest management in stone fruits. It facilitates information about the current pest status and crop condition and is also helpful in selecting the best possible combination of the pest management strategies. The execution of cultural and habitat management practices also plays a significant role in minimizing the pest stress on stone fruits, as they make the crop environment less favourable for the pests. Exploitation of breeding for pest-resistant varieties is a continuous process and is the most economic and environmentally sound method for insect pest management. Emphasizing biocontrol with the help of conservation and augmentation of natural enemies of pests such as predators, parasitoids, entomopathogenic nematodes, fungi and bacteria in IPM programmes will further help in reducing the pest population build-up and outbreak of secondary pests. The dormant or delayed dormant spray of horticultural mineral oils and need-based use of insecticides and acaricides in stone fruits can help to manage the pest populations below economically damaging levels

B. A. Rather (✉)
Entomology, High Mountain Arid Agriculture Research Institute (HMAARI), SKUAST-K, Leh, Ladakh, India

M. M. Mir · U. Iqbal
Division of Fruit Science, Sher-e-Kashmir University of Agricultural Sciences & Technology of Kashmir, Srinagar, Jammu and Kashmir, India

S. A. Mir
Department of Food Science and Technology, Government College For Women, Srinagar, Jammu and Kashmir, India

M. M. Mir et al. (eds.), *Production Technology of Stone Fruits*,
https://doi.org/10.1007/978-981-15-8920-1_15

when the pests cannot be controlled by other means. They are considered an important component of integrated pest management programmes.

Keywords

Stone fruits · Integrated pest management · Monitoring · Damage · Insect pests

15.1 Introduction

The word stone fruit refers to any of hundreds of fruit species belonging to genus *Prunus*. These include apricots, cherries, peaches, nectarines, plums, almonds and some interesting new interspecific hybrids such as the plumcot. They occupy a distinct place amongst the cultivated fruits because of their excellent dessert quality and widespread use in canning and processing industry. These are grown in wide range of climatic conditions from subtropical to temperate regions, but the fruits grown in the hills are superior to those produced in the plains. Drupe is the botanical term used to describe these fruits. A few of these crops are represented by several species such as sweet cherries (*Prunus avium* L.) and sour or tart cherries (*Prunus cerasus* L.) in case of cherries and the European plums (*Prunus domestica*) and the Asian or Japanese plums (*Prunus salicina* L.) in case of plums. The edible portion of stone fruits consists of fleshy exocarp and mesocarp tissues overlying a stony endocarp (the stone or pit) except for almonds and a few apricot species where the seed is consumed. A single seed is contained within this pit and fruit development seldom continues if this seed aborts (Looney and Jackson 1999).

Stone fruits are attacked by a number of root-, foliage-, stem- and fruit-feeding insects from the very beginning of their growth to fruit maturity. Most of these pests are common to peach, plum and apricot and occur in almost all the regions. Butani (1979) listed 80 species of insect pests of peach, 60 of plum and 30 of apricot. The different parts of these fruit crops like stem, leaves, flowers and fruits attract different categories of insect and non-insect pests and cause significant damage. They present a diverse challenge to integrated pest management (IPM), because of their growth habit being perennial and complexity in structural framework. This chapter will focus on the economically important insect pest of stone fruits. The pests are categorized into the following groups or categories (Table 15.1).

15.2 Description of Major Insect Pests of Stone Fruits

Peach, plum and apricot occupy a prominent place amongst the cultivated stone fruits because of their excellent dessert quality and widespread use in canning and processing industry throughout the world. However, peach occupies the major position amongst the stone fruit grown commercially in temperate to subtropical conditions. In general, peach attracts more pests in comparison to plum, cherry and

Table 15.1 Different categories of insect pests on stone fruits (Thakur and Gupta 2004)

Pest category	Hosts	Nature of damage	Name of pests
Foliage feeders	Peach, plum, apricot and cherry	Mesophyll stylet feeders and bulk foliage feeders	Different mite species: • European red mite, *Panonychus ulmi* Koch • Two spotted spider mite, *Tetranychus urticae* Koch Different aphid species: • Peach leaf curl aphid, *Brachycaudus helichrysi* Kalt • Black cherry aphid, *Myzus cerasi* (Fabr.) • Mealy plum aphid, *Hyalopterous pruni* Koch • Green peach aphid, *Myzus persicae* Sulz • Almond mealy bug, *Drosicha dalbergiae* Stebbing Bulk foliage feeders: • **Hairy caterpillar**, *Lymantria obfuscata* Wlk. • **June green beetles/chaffer beetles** (*Adoretus simplex* and *Holotrichia* spp.), green June beetles (*Cotinis nitida* L.), Japanese beetle (*Popillia japonica* Newman) • **Flea beetle**, *Altica himensis* Shukla • **Brown-tail moth**, *Euproctis* spp.
Structural feeders	Peach, plum, apricot and cherry	Superficial woody tissue and shoot feeders, wood-boring insect pests and root-system-boring insects	Different borer species: • Peach tree borer, *Synanthedon exitosa* (say) • Lesser peach tree borer, *Synanthedon pictipes* (Grote and Robinson) • Peach twig borer, *Anarsia lineatella* Zeller • Flat-headed borer, *Chrysobothris* spp. • Flat-headed cherry stem borer, *Sphenoptera lafertei* Thomson • Cherry stem borer, *Aeolesthes holosericea* fab. • Shot-hole/pinhole borer, *Scolytus nitidus* and *S. rugulosus* Muller Lecanium scales: • Plum scale or brown apricot scale, *Parthenolecanium corni* (bouche) • Peach scale, *Parthenolecanium persicae* • White peach scale, *Pseudaulacaspis pentagona* • San Jose scale, *Quadraspidiotus perniciosus* (Comst.)

(continued)

Table 15.1 (continued)

Pest category	Hosts	Nature of damage	Name of pests
Fruit and seed feeders	Cherry, apricot, plum and peach	Direct pests of buds, fruitlets and mature fruits	Different fruit fly species: • Cherry fruit fly, *Rhagoletis cingulata* (Loew) • Mediterranean fruit fly, *Ceratitis capitata* Wiedemann • Peach fruit fly, *Daucus zonatus* Saund. • Oriental fruit fly, *Daucus dorsalis* Hend. • **Oriental fruit moth**, *Grapholita molesta* (Busck) • **Apricot chalcid**, *Eurytoma samsonovi* Vassiliev • **Codling moth**, *Cydia pomonella* L. • **Plum curculio**, *Conotrachelus nenuphar* Herbst

apricot. Almost 80 insect pest species have been recorded infesting peach throughout the world. But the most common insect pests which hamper the quality of produce and yield include different species of aphids like peach leaf curl aphid, peach mealy aphid, peach green aphid, San Jose scale, white peach scale, fruit moth, peach fruit flies, flat-headed stem borer, bark-eating caterpillar, peach twig borer, grey weevils and cockchafer beetles (Horton et al. 2008). Cherry has its origin in a region between Caspian and Black seas, but presently, it is grown in all the temperate countries of the world. There are over 60 species of pests that inflict direct or indirect damage to cherry in different parts of the world. About 40 species have been reported as pests of cherry from India which causes injury to cherry. However, only a few species are serious pests of cherry, while most other species are polyphagous (Butani 1979). The economically important insect pests of stone fruits are discussed below.

15.2.1 Aphids

Aphids are small sap-feeding insects that can multiply rapidly but are considered to be occasional pests of stone fruits. Aphids feed on plants by injecting their needle-like mouthparts into the tissue and sucking out the plant juice. Nymphs and adults suck the sap from leaves, petioles, blossoms and fruits causing leaf curl and distortion and weakening of trees. The symptoms of aphids are seen in leaves, fruits and shoots. Leaves are curled downwards and sticky with honeydew secreted by the aphids. Honeydew may also drip onto the fruit causing russet spots and black sooty mould. Blossoms wither, and fruits do not develop or drop if formed. Sometimes, sooty fungus grows on honeydew exuded by aphids, thereby reducing the market value of fruits (Wilson 2014a). Heavy infestations can reduce fruit quantity and quality on mature trees, limit fruit set the following year and may kill young trees. Several species are particularly virulent with regard to apricots, such as hop aphids, mealy plum aphids and thistle aphids. Aphids can be green, black, pink or of mixed colour (Stoven and Bush 2020). They are active from February to October. During winter, they are found in the egg stage at the base of buds. With the flow of cell sap during the spring, the eggs hatch and the nymphs move onto the primordial leaves where they start sucking the sap. In about 4 weeks, the nymphs are fully fed and change into non-winged adults. There are no males. The females produce eggs without fertilization. These hatch inside the body of mothers. Thus, the females give birth to young ones instead of laying eggs. Each viviparous female produces about 50 young ones in her lifetime of 13 days and completes about three to four generations on the fruit plants (Atwal and Dhaliwal 2015). Winged males and females are produced with the increase in temperature during the season. They migrate to other alternative host plants and again start reproducing asexually and complete four to five generations on these hosts from June to October. The winged females are produced again in early November, and they migrate back to peach, plum and other fruit trees. The old foliage has been shed by that time; therefore, the females lay eggs at the base of buds. Egg laying is completed by the mid of December when the females die (Thakur and Gupta 2004). There are three important species of aphids reported on stone fruits, viz. peach leaf curl aphid or plum aphid, *Brachycaudus helichrysi* Kalt.; mealy plum aphid, *Hyalopterous pruni* Koch.; and green peach aphid, *Myzus persicae* Sulz. These three species of aphids are widely distributed and destructive pests of peach, plum, apricot and almond. They are prevalent both in the plains and in mountainous areas throughout the world (Atwal and Dhaliwal 2015).

Peach leaf curl aphid, *Brachycaudus helichrysi* (Kaltenbach), is the key pest and the predominant aphid species infesting peach throughout the world but is considered an extremely polyphagous pest infesting about 175 plant species belonging to 115 genera of 49 families. The aphid damages floral and vegetative buds of its primary hosts (*Prunus* sp.) by sucking sap due to which the unfolding leaves curl up tightly, remain smaller, get distorted and later turn pale and drop. Heavy infestation can curl almost every leaf on the tree which causes reduction of fruit size as well as production of irregular shoot and twig growth (Fig. 15.1).

Numerous aphids inside the flowers suck the vital sap and render them sickly and dull. Consequently, poor fruit set occurs. There may be premature flower and fruit drop and abnormal development of leftover fruits (Bhalla and Gupta 1993). Newly laid oviparous nymphs are green, but the colour of the adults changes depending on the food, i.e. nymphs feeding on leaves are green while those feeding on bark are brown. The aphid remains on its primary hosts from autumn to beginning of summer (October to April–May) and on secondary hosts during summer (May–September). Depending upon the altitude and temperature, the autumn migrants (gynoparae) fly back on its primary hosts and settle on both sides of leaves preferably on the dorsal side during October–November and deposit a progeny of oviparae. Arrival of the alate males synchronizes with the maturation of oviparous gynoparae, and females are more numerous than males. Apterous oviparae attain sexual maturity and settles down on buds. They are fertilized by alate males and start egg laying singly on scales or young twigs during November–January.

1. Peach leaf curling aphid 2. Green peach aphid 3. Mealy plum aphid

Fig. 15.1 Aphids (Courtesy of Catholic University, Piacenza, Italy)

Green peach aphid, *M. persicae*, is a cosmopolitan and polyphagous pest species infesting numerous vegetable, ornamental and fruit crops and causes more damage as a vector of viral diseases (Blackman and Eastop 1985). It also infests peach, plum, almond, apple, apricot, cherry and pear, but peach is the main host of this species owing to its injury of succulent terminal growth and leaves and feeding on the fruit which can reduce fruit quality (Pascal et al. 2002). However, peach leaf curl aphid is the dominating species on peach which prevents other aphid species from invading the trees (Atwal and Dhaliwal 2015).

Mealy plum aphid has also been reported on peach, apricot and plum. Nymphs and adults assemble on the ventral surface of leaves and suck sap. In case of severe infestations, the affected leaves get twisted, the fruit size is reduced considerably and the fruits drop prematurely. The aphids breed parthenogenetically throughout the year except during winter months when sexual forms appear and eggs are laid on peach twigs around December–January and the aphids overwinter in egg stage. The eggs hatch in the following spring giving rise to apterous viviparous females. During May–June, they fly to the alternate hosts, and during autumn, the alates return to the peach trees and lay the sexuales.

15.2.2 Scale Insects

Scale insects infest a vast variety of plant hosts worldwide and are important pests for all the stone fruits and in fact more or less all orchard trees. Scientists typically have recognized 18 families of scale insects which include several thousand species, but the most common scale insect pests come from three families:

(a) The armoured scales (Diaspididae)
(b) The soft scales (Coccidae)
(c) The mealy bugs (Pseudococcidae)

Armoured scales are generally flat in appearance and are usually cryptic or well camouflaged. They are likely small about 2–3 mm long. They live inside protective covering made of waxes and previously moulted skins except for the adult males and crawlers. This covering protects scales from natural enemies, pesticides and desiccation; e.g. San Jose and white peach scales are armoured scales.

The soft scales are generally convex in shape and are larger (5–10 mm) than armoured scales. Soft scales also secrete a layer of wax that covers their bodies which protects them from natural enemies, e.g. lecanium scales (Hodges 2015).

Two armoured scales are key pests of peaches, viz., the San Jose scale (*Quadraspidiotus perniciosus*) and the white peach scale (*Pseudaulacaspis pentagona*). Other scales that occasionally infest peaches include terrapin scale (*Mesolecanium nigrofasciatum*) (soft scale), European fruit lecanium (*Parthenolecanium corni*) (soft scale), cottony hydrangea scale (*Pulvinaria hydrangeae*) (soft scale), walnut scale (*Quadraspidiotus juglansregiae*) (armoured scale), latania scale (*Hemiberlesia lataniae*) (armoured scale), Forbes scale (*Quadraspidiotus forbesi*) (armoured scale), mining scale (*Howardia biclavis*) (armoured scale), camphor scale (*Pseudaonidia duplex*) (armoured scale), cottony cushion scale (*Icerya purchasi*) (margarodid scale), Comstock mealy bug (*Pseudococcus comstocki*) and taxus mealy bug (*Dysmicoccus wistariae*) (Hodges 2015) (Fig. 15.2).

Scales damage plants directly by feeding injury or indirectly by means of production of honeydew by soft scale and mealy bugs. Scales feed by inserting their piercing/sucking mouthparts and withdrawing nutrients. The feeding damage can cause leaf chlorosis and twig dieback in peaches and even death of trees under heavy infestations. Feeding injury to peach fruit occurs mostly from San Jose scale which produces small, red, measles-like lesions on the skin. But the indirect plant damage such as production of honeydew and subsequently build-up of sooty mould results particularly from soft scales.

San Jose scales overwinter as immature scales, and with the onset of spring season, the tiny-winged males emerge first and mate with the wingless females. Then, about 1 month after the commencement of the male flight, the first crawlers begin to emerge. Each female produces about ten crawlers per day for 2 weeks depending

Fig. 15.2 Scale insects (Courtesy of Catholic University, Piacenza, Italy)

upon the temperature. Eggs are not seen because females give birth directly to live crawlers (ovoviviparous). These tiny yellow insects move around randomly on bark and foliage before settling down permanently. A few days after settling down, crawlers secrete a waxy covering over their body that protects them from most pesticides. The female scales remain stationery throughout their life span, and the males will remain in one location only until maturity. After maturity, the winged males will seek out females for mating and fertilization, and the cycle will begin again (Bessin 2003).

White peach scales overwinter mainly as fertilized females beneath scale coverings and begin laying eggs in February and March. They first lay orange-coloured eggs that will become females and then pinkish eggs, which will become males. Eggs are laid beneath the scale covering. Each female lays 100–150 eggs, usually over a period of 8–15 days. Upon hatching within 2–5 days, crawlers immediately leave the protection of the parent scale and move to new sites to settle. Crawlers often do not move far from their parent, some even anchor under the parent scale. Male white peach scale crawlers are located in clusters on older, lower portions of the tree. Female crawlers are generally more active than male crawlers and may disperse throughout the tree, although they are rarely seen on terminal, green wood or fruit. Crawlers soon anchor at a new site, insert their slender mouthparts and begin to feed. About 7–9 days after hatching from the eggs, anchored crawlers moult and begin forming their own scale covering. The scale covering is cemented firmly to the bark and is relatively impermeable. A third moult gives rise to the adults. Adult females remain under their scale throughout their life. Winged adult males emerge from beneath their scales and seek out females and mate. Males do not feed and die shortly after mating. Mated females soon begin to lay eggs. Females usually die following oviposition. The single generation or life cycle from egg to egg is completed within a period of 50 days at 75 °F. But, developmental time is temperature dependent and varies considerably in the field (Hodges 2015).

Mealy bugs (especially Pseudococcidae) also suck plant juices but generally choose more tender tissues like young shoots and leaf axils as feeding sites. But the main damage due to mealy bugs is not only from the removal of plant product but also from the production of honeydew (liquid drops of excrement rich in simple sugars). Honeydew dripping on fruit can cause fruit russeting in sensitive cultivars or can support the growth of sooty moulds, a superficial but unsightly fungal growth. Both scales and mealy bugs are considered to be induced secondary pests, which would occur only at low levels if their natural enemy complex was not decimated by broad-spectrum pesticides. Moreover, mealy bugs can be extremely persistent once established (usually in large, older trees), and even an intense spray programme can only keep them in check, not eradicate them. Both species will infest the fruit towards the latter part of the season, especially when populations are high. Mealy bugs are recognized by the mealy wax and waxy projections emerging from their bodies. Unlike the armoured and soft scales, which are generally mobile only in the crawler stage, mealy bugs are mobile in all life stages (Hodges 2015).

15.2.3 Wood-Boring Insects

The wood-boring insects in stone fruits include the clearwing moths of several families of Lepidoptera which attack the cambium of the trunk and major scaffold limbs, and prolonged attack can girdle and kill these organs. These clearwing moths (Sesiidae) have several species that attack also different fruit and ornamental trees. The major wood-boring insect pests of stone fruits include peach tree borer, *Synanthedon exitosa* (Say); lesser peach tree borer, *Synanthedon pictipes* (Grote and Robinson); peach twig borer, *Anarsia lineatella* Zeller; flat-headed borer, *Chrysobothris* spp.; and shot-hole/pinhole borer, *Scolytus nitidus and S. rugulosus* (Wilson 2014b).

The adult peach tree borer is a clearwing moth, steel blue with yellow or orange markings. These moths are diurnal and can simply be mistaken for wasps. The females are attracted to trees that have previously been damaged by borers or to which some mechanical injury has occurred. Therefore, it is imperative to avoid damage to the tree trunk in order to reduce borer attack (Fig. 15.3).

Trees with poor growth and vigour because of weed competition or drought stress also seem to be more susceptible to borer attack and damage (Knutson et al. 2018). Peach tree borer and twig borer attack all the stone fruits, but peach and plum are the most susceptible. It is widespread in most fruit-growing areas of the world and is considered the key and most destructive pest of peach, nectarine and apricot. Other hosts for the borer include wild and cultivated cherry, plum, prune and certain ornamental shrubs of the genus *Prunus* (Ames and Maggiani 2013). The larvae of peach tree borer feed in irregular tunnels in the bark and outer wood of the trunk and main root near the ground surface. They chew underneath the bark at the base (crown) of the tree and on larger roots; that is why it is also known as 'peach crown borer'. The injured area produces a great deal of gum which usually contains the casting from the borer. The scratching wounds produced are very extensive which can critically weaken the tree and occasionally cause death of trees. The attack of the borer is identified by the presence of wet spot on the bark or oozing of gummy sap from the affected area (Wilson 2014c). The young larvae upon hatching immediately tunnel into the sapwood of the trees usually through cracks and wounds in the bark and continue to feed until the onset of cold weather and cause extensive damage to trees. Most activity is restricted to the crown and larger roots of the tree. With the onset of warm soil temperature, the larvae resume their activity and cause

Peach Tree Borer damage | Larva of PTB | Adult of PTB

Fig. 15.3 Borer pests (Courtesy of University of Georgia, Bugwood Network)

severe damage to the trees from mid to late spring as the larvae mature. They complete feeding during late May to July and undergo pupation in a cell made of silk, gum and chewed wood fragments located just below the soil surface. The adults emerge within a month which is a type of clearwing moths that resemble wasps. The peach tree borer requires 1 year in completing its life cycle (Cranshaw 2018).

Peach twig borer is a major pest of apricot, peach, plum and prunes. The adult is steel grey moth with white and dark scales. The full-grown larva is about ½ inch long and has alternating bands of light brown and dark brown colour. The head is black, and six legs are clearly visible. The larvae push excrement from their tunnels onto the surface of the fruit where it is readily visible. The larvae of peach twig borer also tunnel the buds and terminal shoots in early summer, but later in the season, the larvae bore into the shoots causing a characteristic flagging or wilting of new growth as the injured areas wilt and die. Flagging of fruit trees diagnoses infestation as it causes severe twig dieback and finally damage to fruits if not controlled. Actually, later generations of larvae infest the stem and may also bore into the peach and plum fruit and feed inside. The pupa is smooth and brown.

The scolytid beetles comprise one of the more serious pests of forest and fruit trees, and several species in the genera *Scolytus* and *Xyleborus* are pests of stone fruits. Attacks are obvious upon careful inspection of tree. They preferentially attack old and/or stressed trees, and the infestation is diagnosed when the vigorous trees have a noticeable clear substance coming out of the small beetle entrance holes, and the less vigorous trees usually have the presence of boring dust (frass) in the bark crevices. They kill branches and trees by feeding in the sapwood, eventually killing the phloem. The larvae form distinctive galleries in the wood, and adults often bore into shoots just below buds, causing weakening and breakage and exposing the tree to other pests and diseases (Agnello 2014a) (Fig. 15.4).

In general, these insects are usually attracted to trees that are already weakened by some other pest or disease, although young trees can suffer damage when they are close to a source of emerging adults. These borers feed on the inner bark of trees, where they may kill the tree by girdling or cause the bark to peel away by stopping the transport of nutrients up and down the tree causing branch or tree death (Fig. 15.5).

15.2.4 Fruit and Seed Feeders

The main pests damaging stone fruits directly include direct pests which damage buds and fruitlets and mature fruit feeders. The direct pests damaging buds and fruitlets include some noctuids like oriental fruit moth, green leaf worms and leaf rollers and some weevils and bugs, while the mature fruit feeders include codling moth, plum curculio and fruit flies. The green leaf worms start egg laying on twigs and developing leaves when the stone fruits reach half-inch green stage. The young larvae feed on new leaves and flower buds and can often be found inside a rolled leaf or bud cluster. Older larvae damage flower clusters during bloom and continue to feed on developing fruits and leaves for 1–2 weeks after petal fall. They then drop

Fig. 15.4 Borer pests (Courtesy of University of Georgia, Bugwood Network)

Fig. 15.5 Borer pests (Courtesy of University of Georgia, Bugwood Network)

to the ground and burrow 2–4 in. beneath the soil surface and pupate. Oriental fruit moth (OFM) is a serious pest of peaches, nectarine and quince worldwide but occasionally attacks other stone fruits like almond, apricot and plum (Strand 1999). OFM overwinters as mature larvae in silken hibernacula on the tree, in dried fruits, on leaves or twigs beneath trees or in other protected areas such as field bins. Pupation takes place in late winter. Adult emergence is frequently initiated a few days before peaches begin to bloom. Egg laying begins a few days after adult emergence. Females lay up to 200 white, flattish eggs, most often on twigs or leaves near the distal ends of shoots (Howitt 1993). OFM larvae feed on succulent terminal growth during the initial spring and subsequently in autumn vegetative growth flushes (Fig. 15.6).

Stem feeding by OFM larvae produces withered or dead shoots, which are referred to as 'flagged'. Vegetative shoot feeding typically destroys the distal or terminal buds of shoots, increasing lateral branching as multiple lateral buds break due to removal of apical dominance. Larvae may continue to burrow in the original shoot or enter other shoots before they reach maturity (Hogmire 1995). Injury from OFM terminal feeding is more important in young trees, because rapid growth of young trees is sought to quickly fill space of each tree, which optimizes early yields. In contrast, the terminal feeding of OFM on mature trees is of little significance.

The importance of OFM as a fruit-feeding pest varies with region; however, it is a key fruit-feeding pest in most of major peach production areas. Depending on fruit phenology, newly hatched OFM larvae burrow into either the tender, succulent tissue at the distal end of shoots or ripening fruits. Newly hatched, neonate OFM

Oriental fruit moth (larva) OFM (Adult) Codling moth (adult)

Fig. 15.6 Oriental fruit moth and codling moth (Courtesy of University of California, USA)

larvae are known for quickly tunnelling beneath the surface and into the fruit. OFM injury to harvested peaches typically falls into two classes (Howitt 1993). 'Old injury' occurs when larvae move into green fruit after abandoning twigs, leaves or tight places where fruit touches. As fruit matures, gum often extrudes from the entrance wounds; this exudate darkens with time and may be seen as a blackish blotch at harvest. Larval entrances for many of these green fruit infestations are through the sides of fruit, which makes them more visible than infestations to more mature fruit. 'New injury' predominates as the season progresses and fruit ripens. Larval infestations in maturing fruit are often difficult to distinguish. Many are associated with the stem cavity, leaving very modest frass accumulation adjacent to the stem. OFM larvae also enter the peach through the stem, leaving no visible entrance wounds and very little indication of infestation until mature larvae exit the fruit (Hogmire 1995).

Leaf rollers feed on the fruit surface rather than the seeds and as such have only an indirect physiological impact on the tree. While the impact on the tree may be negligible, the impact of fruit feeding on grower returns is a direct one. The fruits may be attacked at almost any point during the growing season, from early in the bud stage to harvest. Fruits attacked early in the season are more likely to abscise naturally, or they can be selectively thinned during hand thinning. Fruits attacked during the midseason are more likely to stay on the tree and thus have a higher likelihood of being harvested. Fruits attacked very late may generate sufficient ethylene to abscise prematurely and have a slightly reduced chance of entering the packing or processing plant. Clearly, excessive amounts of fruit drop just before harvest will have a detrimental effect on yield. Overall, pest management programmes are focused most intensely on this group of pests because of their clear and apparent effect on usable yield. Larvae usually web foliage together and many also feed directly on the fruit surface. They first feed on the water sprouts and then move throughout the tree. Those feeding on developing flower buds do so before bloom and continue to consume floral parts during blossom stage. After petal fall, they continue to feed on the developing fruits and show increasing tendency to damaged fruits. Many of these damaged fruits drop prematurely, but a small percentage remains on the trees exhibiting deep corky areas and indentations at harvest. The

Leaf footed bug Brown stink bug Lygus bug Tarnished plant bug

Fig. 15.7 Catfacing insects (Courtesy of Catholic University, Piacenza, Italy)

group of insects feeding directly on fruits attack the fruits leaving either feeding scars or deep entries, potentially serving as an infection site for pathogens (Agnello 2014b).

The plum curculio larvae feed inside the fruit. This pest damages fruit in three ways: by direct feeding or by laying eggs which become larvae and tunnel in the fruit and/or by spreading brown rot. They cause severe damage, and the full-grown grubs lack legs and are dirty white to yellow about half-inch long. The adult is a small beetle ¼ inches long with a short curved snout. It overwinters in wooded areas and flies into the fruit orchards during spring. Female adults chew holes in the fruit during feeding for depositing eggs. Infested fruits fall, or they remain on the trees as deformed and misshapen and are rendered unmarketable. Other catfacing insects include stink bugs, leaf-footed bugs and lygus bugs which penetrate their needle-like mouthparts and suck the juices (Fig. 15.7).

Feeding results in deformed and misshapen fruits, and sometimes, water-soaked areas and gum may exude from the feeding sites on fruit. These bugs feed on many weeds and cultivated crops before flying into fruit orchards. The oriental fruit moth infests late-maturing peach varieties. The larvae feed in growing shoots in the spring, and later generations feed in the fruit. Fruit infestations typically occur once the peaches begin to colour. Because the larvae often enter the peach through the stem, there is no external evidence of damage.

The mature fruit feeders include codling moth, fruit flies and apricot chalcid. The main recorded hosts of codling moth are apple, European pear, nashi (Asian pear), Chinese pear and quince. Walnut and plum are consistently attacked, while peach, nectarine and apricot are also recorded hosts, and damage can be significant in some situations. Differences in the host preference, development, diapause, phenology and population dynamics have been found for strains or races of the moth taken from apple, plum or walnut host plants (Barnes 1991) (Fig. 15.8).

The removal of alternative or abandoned host trees can therefore make an important contribution to control by reducing migration of the pest into smaller orchards. The economic threshold for codling moth is low (1% damaged fruit), even for crops that are not exported. These factors have combined to make this pest one of the greatest scourges for apple growers. Female moths deposit eggs singly on or near developing fruit. On hatching, the young larvae locate apples on the basis of an apple fruit volatile, (*E*,*E*)-α-farnesene (Sutherland and Hutchins 1972), and then

Plum curculio Plum curculio damage Cherry fruitfly Fruitfly Damage

Fig. 15.8 Fruit-feeding insects (Courtesy of University of Georgia, Bugwood Network)

begin to enter the fruit and make their way to the core to feed on the seeds like other members of the genus *Cydia* (Witzgall et al. 1996). The entrance hole is frequently plugged with frass, and the mature larvae emerge from the fruit with a characteristic exit hole. Diapausing fifth-instar larvae overwinter in cocoons in suitably protected locations under the bark of the host tree or on the ground (Beers et al. 2003).

True fruit flies of the family Tephritidae deposit eggs directly into the flesh of developing fruit, particularly fruit approaching readiness for harvest. The tiny puncture made through the skin of fruit during egg laying is difficult to detect without magnification and may remain so even when underlying flesh has decayed substantially during larval feeding (Aluja and Norrbom 2000).

The maggots feed on fruits and destroy much of the flesh, and infested fruits appear normal but when opened show broken brown areas where the maggots are feeding. Commonly infested fruit is detected only after a few days of exposure to room temperature following purchase by an unwary consumer where underlying flesh has decayed substantially. The attack of fruit flies reduces the yield. Adult females lay eggs in small batches of 2–10 inside the ripening fruits by making punctures with their ovipositors. On hatching, the maggots feed on the pulp, and fruit becomes soft, ferments and drops. The attack is more serious on late-maturing varieties. Full-grown maggots come out of the infested fruits and jump to suitable places for pupation (Agnello 2014c).

Another important pest of stone fruits is apricot chalcid (*Eurytoma samsonovi* Vassiliev) in which the grubs bore into the kernels and feed on the inner contents leaving the papery coat intact. As a result, fruit development is arrested, and the fruits fall prematurely with grubs still feeding within the fruits. The insect pest attacks the fruits immediately during fruit set in April–May and deposits the eggs inside the fruits (Fig. 15.9).

On hatching, the grubs feed on the contents of the kernels and complete development during fruit maturation stage, and ultimately, the fruit drops. The fallen fruits show symptoms of flesh shrinkage and are also of reduced size. The larvae are translucent white and feed during April–June and thereafter remain quiescent from July to Feb. Thereafter, they pupate inside kernels, and with the onset of warm weather, the adult chalcid wasps appear after piercing through the stones and oviposit after mating (Rather and Kacho 2011).

Catfacing injury on peaches | Apricot chalcid grubs | Chalcid Adult

Fig. 15.9 Catfacing injury on peaches and chalcid damage in apricots. (Courtesy of University of Georgia, Bugwood Network)

15.2.5 Foliage Feeders

There are multiple groups of arthropods that attack and feed mainly on foliage, with the primary damage being loss of photosynthetic capacity due to loss of chlorophyll and disrupted osmotic balance. From the perspective of plant productivity, specifically yield parameters, the effect of chlorophyll loss is controversial. No clear and uncontested relationships have been established, although it seems clear from the body of literature that there is not a directly proportional relationship between loss of chlorophyll and loss of photosynthetic capacity (Beers et al. 2003). Trees are capable of sustaining a certain degree of foliar damage without any measurable loss in yield. The different types of foliage feeders in stone fruits are as follows.

15.2.5.1 Mesophyll Stylet Feeders

This group feeds on cellular contents (including chlorophyll) by penetrating the leaf surface (often from the underside), killing only one or a group of cells at each feeding site. The damage appears as speckling (leafhoppers) or bronzing (tetranychid and eriophyid mites), depending on the size of the mouthparts and the depth of penetration. Mites feed by inserting their mouthparts into leaf cells to suck out the contents, including the green pigment chlorophyll. The individual spots initially look white, giving the leaves a stippled appearance. As damage progresses, the infested leaves take on a brown hue, commonly called bronzing caused due to combination of chlorophyll loss and/or stomatal closure caused by water loss. Three important species are worldwide pests of stone fruits, including the European red mite (*P. ulmi* (Koch)) and two-spotted spider mite (*T. urticae* Koch) and plum rust mite (*Aculus fockeui*) (Agnello 2014d) (Fig. 15.10).

The different types of mites are considered indirect pests of peach and relative to other deciduous fruits as peaches are more tolerant to mite feeding. Spider mites are tiny mites that feed on the underside of leaves producing a fine webbing and causing the leaves to discolour and eventually dry and drop. Different studies indicate that peach yield and/or fruit quality was not reduced until mite numbers were quite high. Kovach and Gorsuch (1985) reported that more than 48 two-spotted spider mites/leaf were required to reduce the percentage of fruits that reached the more desirable

ERM eggs (OW) European Red Mite Two-spotted spider mite Rust mite

Fig. 15.10 Different mite species on stone fruits (Courtesy of University of Georgia, Bugwood Network)

Japanese beetle Green June beetle Brown tail moth (Silken web)

Fig. 15.11 Bulk leaf feeders (Courtesy of University of Georgia, Bugwood Network)

large-size categories. Bailey (1979) also reported that TSM densities of 40–50 mites/leaf did not reduce peach yield. Moreover, it is generally believed that mites increase on water-stressed plants during hot, dry weather and are more problematic in drier production areas and their outbreaks have been observed in areas where there is indiscriminate use of broad-spectrum insecticides against fruit- and tree-attacking pests (Croft et al. 1987).

15.2.5.2 Bulk Leaf Feeders

This is a varied group, composed mostly of polyphagous Lepidoptera, chaffer beetles and chrysomelid beetles. Many are pests of deciduous forest trees. Examples include hairy caterpillars (*Lymantria obfuscata*), chaffer beetles (*Adoretus* spp. and *Holotrichia* spp.), green June beetles (*Cotinis nitida* L.), Japanese beetle (*Popillia japonica* Newman), flea beetle species (*Altica himensis* Shukla) and brown-tail moth (*Euproctis* spp.). Other gregarious lepidopterans include the tent caterpillars (*Malacosoma americana*, *Malacosoma fragilis* and *Malacosoma disstria*; Lasiocampidae) and the ermine moths (Yponomeutidae, e.g. *Yponomeuta malinellus* (apple ermine moth) and *Yponomeuta rorrella* (willow ermine moth)) (Fig. 15.11).

The hairy caterpillars feed voraciously on leaves, and in case of heavy infestation, complete defoliation occurs. Frequent defoliation may weaken and kill most of the trees. Adult beetles feed on leaves, buds, blossoms and fruitlets. The leaves are perforated. The grubs feed on roots and may cause wilting of the plant. The flea beetle species is also a voracious pest and damages plants by making numerous small holes in the leaves which make them look as if they have been peppered by fine shots or skeletonize the entire leaves. When populations are high, flea beetles can quickly defoliate and kill entire plants. The caterpillars of brown-tail moth also feed gregariously on foliage leaving behind only the midrib and other hard tissues and construct a conspicuous small white silken tent and feed inside.

15.3 Integrated Pest Management of Stone Fruits

Integrated pest management is considered the best practice while dealing with any pest. It includes a set of strategies and tactics which agricultural producers use to manage crop pests. IPM utilizes knowledge of the pest and host plant with different strategies for long-term pest control. It is consistent with sustainable agriculture and uses total farm system approaches to mitigate pest pressure (Damos et al. 2015). The goal of IPM is to increase production efficiency, reduce production costs, reduce worker and consumer exposure to pesticides and protect the environment in order to support sustainable production of marketable products. Developing an IPM programme for stone fruits involves more than putting together a pest spray programme. It involves a proactive approach to growing, beginning with site and cultivar selection, and an understanding of cultural practices that will help to delay, reduce or eliminate potential problems (Wilson et al. 2014). It also involves understanding of the life cycle of the pest (whether it may be insect, mite, disease pathogen, weed, nematode, etc.), meaning which insect vectors and which disease pathogen to enable to identify beneficial and natural enemies; understanding the environmental impact of potential pest problems and plant health management; knowledge of different modes of action of pesticides and their proper timing of application; understanding and utilizing alternative management methods; and understanding economic and injury thresholds (Los and Concklin 2013). The management of insect pests in stone fruits includes the following.

15.3.1 Monitoring of Insect Pests

The first and foremost strategy for management of insect pests in stone fruits includes forecasting and monitoring of insect pests. The growers should monitor their trees carefully for the earliest onset of insect pests. Different threshold limits and monitoring procedures have been set for incidence of different insect pests on stone fruits during the growing season. Monitoring presence of aphid population presence is important before applying a pesticide; e.g. in case of aphids, a few colonies can rapidly infest the entire tree. The threshold of aphids for peaches is two or more colonies per tree between petal fall and shuck-split or post-bloom period and five or more colonies by late May. The threshold for nectarines is one colony per bering tree at any time during the season. Monitoring needs examination of ten fruit clusters from the inner canopy of each ten trees for presence of aphids and damage (curled leaves). Yellow sticky traps are also used for monitoring aphid population and cherry fruit fly when the adult flies begin to fly. But many beneficial organisms, including lacewings, lady beetles, syrphid flies and soldier beetles, are effective predators of the green peach aphid (Los and Concklin 2013).

Monitoring of catfacing insects like adult plant bugs is carried out by using white or pink sticky traps, limb jarring or sweep net sampling of ground cover; however, no thresholds have been established for traps or limb jarring. But in case of sweep net sampling, the groundcover is sampled with a sweep net, taking two sets of 25 sweeps. If a total count exceeds 3–4 combined tarnished plant bugs and native stink bugs, it signals the potential problem of bug infestation. Fruit injury due to bugs and other catfacing insects like plum curculio and oriental fruit moth is also monitored by direct fruit examination. A minimum number of 100–200 fruits per block or ten fruits from ten different trees should be examined for fresh injury. The tentative injury threshold is suggested at 1–2% of new damage. In addition, both old and new feeding should be recorded so that management programmes can be refined or changed if needed (Los and Concklin 2013). Arrival of plum curculio adults in the orchard can also be monitored by direct examination of the fruit for injury (minimum of 200 fruit per block) or by the use of limb jarring for adults. It should also be done from bloom through at least 2 weeks after post bloom. Concentrate monitoring along the edges and border rows of the block. The bulk foliage and fruit damage due to chaffer and June beetles should also be monitored from July to fall (Horton et al. 2008). Monitoring of mites during the season should be carried out by counting number of mites per leaf with a hand lens. About 10–40 leaves should be collected from the canopies of ten trees, and the average number of mites per leaf worked out accordingly. Provisional action thresholds for peaches have been fixed at ten mites per leaf in early season and 20 mites per leaf in late season (Hogmire 1995).

The adults of oriental fruit moth should be monitored by installing pheromone traps at half-inch green stage of peach. Then, monitor terminals for 'flagging' caused by larvae of oriental fruit moth burrowing into new growth. Count the number of flags per tree on a minimum of ten trees. There is currently no threshold established, but monitoring will give an idea of infestation level. Similarly, monitoring of borer infestation is carried out by installing different pheromone traps of at least two traps per block to determine adult flight during the season. However, lesser peach tree borer (LPTB) traps should be installed by petal fall and peach tree borer (PTB) traps by the first week of June. Populations seldom need treatment when trap catches peak at less than 10 moths/trap/week. Traps should always be used in combination with mating disruption (Horton et al. 2008).

The borer infestation levels can also be monitored by observing larvae or pupae infestation per tree. For LPTB, inspect wounded areas on the upper trunk, scaffold limbs and branches for larvae and empty pupal cases protruding from the bark. It is easiest to find pupal cases during peak flight (as indicated by pheromone traps). Control is recommended if 1–2 larvae or empty pupal cases are found per tree. For PTB, inspect the base of the tree for gum containing frass and sawdust. It is best to do this during July through mid-August. Examine the soil at or near base of tree for cocoons and empty pupal cases. Control is recommended for trees up to 3 years old if any evidence of PTB is detected. In older orchards, control is recommended if one or more cocoons or empty pupal cases per tree are found (Los and Concklin 2013).

Pheromone monitoring and use of degree-day developmental models are very important tools for optimizing the timing of IPM strategies in case of damage due to codling moth in peaches. Large-scale monitoring efforts for CM are usually recommended for orchards with a history of CM injury, because these programmes use the CM sex pheromone ((*E*,*E*)-8,10-dodecadien-1-ol) and degree-day accumulations to optimize the timing of IPM response (Epstein et al. 2006). In CM-prone peach orchards, it is recommended that a single insecticide application be applied at 250–300 or 400–500 degree days, depending on whether light or heavy damage is expected (Strand 1999). However, mating disruption of CM is a more challenging and problematic component because the technique is seldom viewed as a reliable stand-alone technology when employed against CM, especially where CM abundance is high. CM control is primarily insecticide-based. Multiple insecticide classes, organophosphates, IGRs and CM granulosis virus, as well as pheromone mating disruption, must sometimes be used in combination to manage resistant CM populations (Knight 2000). Pheromone mating disruption is also a very reliable technique when used for managing OFM in arid peach production areas with relatively narrow complexes of direct pests.

15.3.2 Cultural and Mechanical Management

Cultural practices such as training and pruning, sanitation, variety selection and selecting open, sunny sites for planting are necessary for good pest control. Growing of resistant varieties is the most important strategy for integrated pest management in all stone fruits. Other cultural and mechanical management practices involve scientific training and pruning for rapid drying which aids in reducing disease incidence and keep away shot-hole borers and other pest infestations. It also allows for optimum spray penetration. Pruning should be done at the proper time. Peaches should be best pruned once growth has started in the spring to reduce cold injury and to allow for rapid wound healing. Open wounds are entry sites for disease organisms and borers. Pruning should also be done in dry weather, and small weak shoots in the centre of trees should be removed to avoid the spread of perennial canker and other diseases.

The trees should be kept as free as possible from mechanical wounds, winter injury, crotch separation and cankers, and destroy dead drying fruit trees and branches to avoid borer infestation. Removal and destruction of mummified fruits and cankers are essential to reduce insect outbreak of fruit fly in cherry and other stone fruits. Weed management is also necessary to prevent the attack of flea beetles because few broadleaf weeds in orchard have been observed inhabiting flea beetles. Dormant pruning of infested wood can help to control the overwintering scales, aphids, eggs, mealy bugs and other pest problems during the season (Los and Concklin 2013). The branches inhabiting overwintering webs and tents of bulk foliage feeders like caterpillars of gypsy moth and brown-tail moth should be collected and destroyed by soaking in soapy water or kerosene or by burning. Orchard sanitation by burning fallen leaves during winter also helps in minimizing the pest infestation during the following season. Yellow-coloured sticky traps should be installed for the control of adult cherry fruit fly. Clipping of infested shoots is effective in checking the multiplication of aphids during the growing season. Plastering of open wounds with some protectant paste helps in preventing the entry of disease pathogens and insect pests. Installation of light traps is an important strategy when there is heavy infestation of June beetles and chafers. Treatment against aphids, scales, mealy bugs, etc. is suggested if 30% of the terminals are infested and natural enemies are not present. Maintain healthy trees with appropriate nutrient levels in the soil and trees. Use routine soil and foliar analysis. Avoid late summer and fall nitrogen applications that will encourage trees to grow late into the fall. Maintain optimum potassium levels in the trees as potassium is important in winter hardiness of trees and buds, as well as fruit size. Avoid damaging trees with mowers and other equipment. These wounds are entry sites for borers and canker development. Thinning of fruits should be done for maximum size development, avoiding limb breakage and reducing moisture or relative humidity build-up which is needed for disease development (Hammon and Davidson 2016).

Control broadleaf weeds that inhabit plant bugs, native stinkbugs and flea beetles. Manage weeds also in the tree row as they compete with trees for available water and nutrients.

All stone fruits should be irrigated to avoid stressing trees during periods of drought. Integrated nutrient management practices should be adopted to maintain vigour of the trees which will prevent the attack of insect pests particularly borers and increase the compensation ability of the trees.

15.3.3 Biological Management

Biological control contributes a tremendous potential as a major tactic in regulating pest populations in integrated pest management programmes in fruit orchards. However, it has been recognized that only in few instances biological control reduces a pest from economic to a completely non-economic status. But, in fact, it has been observed that a moderate degree of biological control of certain pests coupled with one or more other tactics of pest population regulation may prove successful (Hoyt and Burts 1974). Natural enemies are most effective in reducing insect pest population in stone fruits. The natural enemies of codling moth include birds, spiders, insects, nematodes, bacteria, fungi, protozoa and viruses (Falcon and Huber 1991). Similarly, aphids have many natural enemies such as lady beetles, lacewings, syrphid flies, predaceous midge larvae and predatory bugs, which can often keep aphid populations under control if they are not disturbed by broad-spectrum insecticide treatments. Natural enemies like *Aphytis proclia* and *Encarsia perniciosi* have been found parasitizing on San Jose scale and should be released at 2000 adults/infested tree at least 15 days after insecticidal or fungicidal spray (Anonymous

2018). Leaf rollers are attacked by a large number of mostly unspecific predators and a wide range of parasitoids comprising specialists and generalists and by different pathogens including bacteria, fungi, protozoa and viruses. Parasitoids include species attacking to all different life stages. Mite predators such as predatory mites, *Stethorus punctum* (a native lady beetle), *Amblyseius fallacis*, *Zetzellia mali* and *Orius insidiosus* (minute pirate bug), can be effective in maintaining mite populations below threshold level (Beers et al. 2003). Shapiro-Ilan et al. (2016) observed entomopathogenic nematodes highly virulent to larvae of many species of Sesiidae including several *Synanthedon* spp. He found entomopathogenic nematode, *Steinernema carpocapsae*, as effective treatment for the curative control of peach tree borers in comparison to chlorpyrifos treatment.

15.3.4 Chemical Management

The first step in chemical management of overwintering scales, aphids, eggs, larvae, mites, etc. is to apply a dormant oil spray or any horticultural mineral oil in the early spring before buds open which kills them by asphyxiation. Oils work by smothering aphid eggs, immatures and adult insects, so complete spray coverage is essential. They also control peach twig borer and oriental fruit moth which spend the winter as larvae in protected spots on the bark. The dormant oils can also be mixed with an organophosphate insecticide to improve their bio-efficacy. They should be followed by foliar insecticide sprays to manage the upsurge of the insect pests during the season; e.g. in case of San Jose scale, foliar insecticide sprays are recommended at the crawler stage of the first generation which is usually in mid-June. If crawlers are still emerging 10 days after the first foliar spray, then an additional spray may be needed. Additional foliar sprays should be directed at crawlers of the second and third generations. Although a foliar spray can be targeted at adult males during their flight and mating period, this is likely to be less effective than a spray directed at the crawlers. In backyard fruit plantings, an all-purpose fruit tree spray can be used for the foliar sprays (Welty 2009).

Under heavy infestation of scales, aphids, mites, eggs, etc. at bud burst stage, a foliar spray is directed with either dimethoate 30 EC at 0.03% or chlorpyrifos 20 EC at 0.02% (Anonymous 2018). A pre-bloom spraying directed against aphids will also provide control against the apricot chalcid and peach twig borers. However, all fallen fruits which may contain the grubs are collected and burnt. If curling of leaves is noticed after petal fall due to aphids, scales or other pests, spray either with dimethoate 30 EC at 0.03% or phosolone 20 EC at 0.02% or quinalphos 25 EC at 0.25% (Anonymous 2018).

Cherry fruit fly management is typically done with a series of contact residual insecticide application aimed at adult flies as they land to lay eggs. This requires spray on a 7–15-day schedule beginning when the flies become active usually as early-maturing cherries begin to show colour. The damage can be controlled by spraying with dimethoate 30 EC at 0.03% or phosolone 20 EC at 0.02% 10 days after petal fall, and repeat the spray with quinalphos 35 EC at 0.35% after sweet varieties begin to develop colour (Anonymous 2018). If cherry fruit fly is left uncontrolled, the fruits will be infested with maggots which have a major impact on quality and storage life. Spraying after petal fall will take care of peach twig borers also (Table 15.2).

Apply carbofuran 3 CG at 32.5 kg/ha in the soil against grubs of chaffer beetles and other soil pests (Anonymous 2018).

Peach twig borer (PTB) and oriental fruit moth (OFM) have similar life histories, and as such, management is directed similar for both the pests. These insects overwinter as larvae in a silken cocoon on protected areas on the trunk of stone fruit trees (Anonymous 2020). The newly hatched first-generation larvae of PTB/OFM bore into newly growing shoots and cause them to wilt and die, referred to as shoot strikes. Larvae of succeeding generations may move to fruit where they chew a small entry hole near the stem end, especially on peaches. Monitor developing fruit for injury from second generation, and apply an insecticide to the entire tree if noticed. If shoot strikes are present in a tree during shoot growth, a foliar insecticide may be necessary to control first-generation larvae.

The different types of borers can also be managed by spraying dormant oils combined with a contact insecticide which are effective in killing the overwintering larvae. Chemical control options for PTB and LPTB include root dips for new plantings and sprays for trunk and scaffold limbs, best applied with a handgun with low pressure and high volume. The aim for the lower trunk at soil level is for PTB and the upper trunk and scaffold limbs for LPTB. Although adult moths are not specifically targeted, insecticides used for other stone fruit pests during the season may also provide some control. Pickel et al. (2006) also observed borer pests on peach as important and easily manageable by adopting commonly used preventive measures prior to fruit set in California. The methods include dormant or delayed dormant application of oil plus an organophosphate insecticide which provided very reliable control. Further, they observed that oil alone did not control the borers but needed two bloom sprays of entomopathogenic bacterium, *Bacillus thuringiensis*, an environmentally safer method.

Mating disruption is a certified organic management option for the borer pests. It works by saturating the orchard with the female moth sex pheromone to delay or prevent mating. A regular monitoring programme is essential when using mating disruption for peach tree borer. Long-term use of mating disruption can reduce pesticide application. Mating disruption can be a highly effective method to lower peach tree borer infestation and reduce insecticide usage. Dispensers are applied either 1 month before predicted biofix (Isomate brand) at a rate of 150 pheromone ties per acre or around June (Checkmate brand). These should be installed at shuck split before LPTB moth flight begins. Use a higher rate (200–250/acre) for outside edges of border rows, areas that haven't been disrupted before and have high populations and in blocks smaller than 5 acres. If a block has more than 30% of trees infested with PTB, regardless of block size, use 200–250/acre for the first year of treatment (Murray 2019). In this situation, a trunk treatment of chlorpyrifos 20 EC would also be advised for the first season to reduce the PTB population. Pheromone traps should also be installed in place for monitoring both PTB and LPTB. If the

Table 15.2 Spray timings for common insect pests of stone fruits

Fruit crop	Phenological stage	Major insect pests	Management action to be taken
Plum and cherry	White bud or popcorn stage (early May)	Aphids, scales, fruit worms, hairy caterpillars, oriental fruit moth	Scouting for aphid or fruit worms and consider contact insecticide application if populations are high
Apricot	First white stage or first bloom (early May)	Aphids, scales, fruit worms, hairy caterpillars, oriental fruit moth, thrips	Scouting for aphid or fruit worms or hairy caterpillars and consider insecticide application if populations are high
Apricot, peach, nectarine, plum and cherry	Bloom (mid-May)	Thrips, aphids, scales, leaf roller or oriental fruit moth	Monitoring of thrips
Apricot, peach, nectarine, plum and cherry	Petal fall (mid-late May)	Thrips, aphids, scales, leaf roller or oriental fruit moth	Scouting for monitoring of insect pests and apply insecticide if needed for aphids, scales, plum curculio, leaf roller or oriental fruit moth
Apricot, peach, nectarine, plum and cherry	Shuck split/post bloom (late May)	Aphids, scales, fruit worms, hairy caterpillars, oriental fruit moth, brown-tail moth, borer pests, apricot chalcid, chaffer beetles	Scouting for monitoring of insect pests and apply any contact insecticide if needed like phosolone 20 EC at 0.02% or quinalphos 25 EC at 0.35%. (Anonymous 2018)
Apricot, peach, nectarine, plum and cherry	Fruit development/fruit maturity		
	First cover (early to mid-June)	Codling moth, fruit flies, plum curculio, borers, bugs and other catfacing insects	Apply any recommended contact or systemic insecticide like dimethoate 30 EC at 0.03% or quinalphos 25 EC at 0.25% (Anonymous 2018)
	Second cover (mid-late June)	–	Repeat the spray of contact insecticide (if needed) for aphids, plum curculio, leaf roller, oriental fruit moth, borers, etc.
	Third cover (early July) or preharvest (mid-July–early August)	–	Repeat the spray of recommended contact insecticide (if needed)

mating disruption is working, no moths would be captured in pheromone traps resulting in trap 'shut-down' (Los and Concklin 2013). Under severe conditions, the damage due to stem borers can be avoided by cleaning and plugging the live holes with cotton or cloth soaked in dichlorovos 76 EC at 3 ml/L and then plastering with mud.

15.4 Conclusions

To grow stone fruits successfully, healthy plants should be grown altogether with management of pest problems. Integrated pest management is the recommended tool in preventing and managing pest problems which include recognition of the most significant pest problem and its close monitoring for worsening symptoms. Accurate identification of the insect pest problem is also necessary to study its life cycle and habits for immediate preventive or curative action. Preventive techniques and control measures must be physical, cultural and chemical. Chemical management is recommended if the level of damage becomes unacceptable. In conclusion, integration of effective monitoring methods and different management tools for insect pests on stone fruits can provide a sustainable and economical control of pests and minimize the damage to the environment. In addition, a large number of problems in stone fruits are cultural and environmental. These abiotic problems include insufficient water and nutrients, lack of space or sunlight, poor soil, low $_{p}$H, temperature extremes and root damage from cultivation. Choosing inappropriate varieties and purchasing poor-quality planting material will also contribute to problems.

References

Agnello, A. (2014a). *Shothole borers. Stone fruit IPM for beginners* (pp. 61–62). Department of Entomology, Michigan State University, North Central IPM Centre. Retrieved from https://bit.ly/stonefruits.ipm.

Agnello, A. (2014b). *Fruit worms and leaf rollers. Stone fruit IPM for beginners* (pp. 50–52). Department of Entomology, Michigan State University, North Central IPM Centre. Retrieved from https://bit.ly/stonefruits.ipm.

Agnello, A. (2014c). *Cherry fruit fly and black cherry fly. Stone fruit IPM for beginners* (pp. 48–49). Department of Entomology, Michigan State University, North Central IPM Centre. Retrieved from https://bit.ly/stonefruits.ipm.

Agnello, A. (2014d). *Mites. Stone fruit IPM for beginners* (pp. 54–56). Department of Entomology, Michigan State University, North Central IPM Centre. Retrieved from https://bit.ly/stonefruits.ipm.

Aluja, M., & Norrbom, A. L. (2000). *Fruit flies: Phylogeny and evolution of behaviour* (p. 999). Boca Raton: CRC Press.

Ames, G. K., & Maggiani, R. (2013). *Plums, apricots and their crosses: Organic and low spray production* (p. 24). National Sustainable Agriculture Information Service. A project of the National Center for Appropriate Technology (NCAT). ATTRA. Retrieved from www.attra.ncat.org.

Anonymous. (2018). *Package of practices* (Temperate fruits) (Vol. 1, p. 199). Srinagar: Directorate of Extension Sher-e- Kashmir University of Agricultural Sciences and Technology of Kashmir Shalimar.

Anonymous. (2020). Insect and mite pests of stone fruits. In *Stone fruits* (Chap. 2). North central regional extension publication 63. Michigan State University Cooperative Extension Service.

Atwal, A. S., & Dhaliwal, G. S. (2015). Pests of temperate fruits. In *Agricultural pests of southeast Asia and their management* (pp. 437–450). Ludhiana: Kalyani Publishers.

Bailey, P. (1979). Effect of late season populations of two spotted mite on yield of peach trees. *Journal of Economic Entomology, 72*, 8–10.

Barnes, M. M. (1991). Codling moth occurrence, host race formation, and damage. In L. P. S. Van der Geest & H. H. Evenhuis (Eds.), *Tortricid pests: Their biology, natural enemies and control* (pp. 313–327). New York: Elsevier.

Beers, E. H., Suckling, D. M., Prokopy, R. J., & Avilla, J. (2003). Ecology and management of apple arthropod pests. In D. C. Ferree & I. J. Warrington (Eds.), *Apples, botany, production and uses* (pp. 489–451). Wallingford: CABI Publishing.

Bessin, R. (2003). *San Jose Scale*. Extension Entomologist. UK Cooperative Extension Service. University of Kentucky, College of Agriculture. ENTFACT-204.

Bhalla, O. P., & Gupta, P. R. (1993). Insect pests of temperate fruits. In K. L. Chadda and O. P. Pareek (Eds.), *Advances in horticulture* (Vol. 3, Part 3, pp. 1576–1589). New Delhi: Malhotra Publishing House.

Blackman, R. L., & Eastop, V. F. (1985). *Aphids on the World's crops*. Chichester: Wiley.

Butani, D. K. (1979). *Insects and fruits* (pp. 316–330). New Delhi: Periodical Expert Book Agency.

Cranshaw, W. S. (2018). *Peach tree borer*. Colorado State University Extension. USDA. FACT SHEET no. 5.566. Insect series. Trees and shrubs.

Croft, B. A., Hoyt, S. C., & Westigard, P. H. (1987). Spider mite management on pome fruits, revisited: Organotin, and acaricide resistance management. *Journal of Economic Entomology, 80*, 304–311.

Damos, P., Colomar, L. A. E., & Loriatti, C. (2015). Integrated fruit production and pest management in Europe: The apple case study and how far we are from the original concept? *Insects, 6*, 626–657.

Epstein, D. L., Stelinski, L. L., Reed, T. P., Miller, J. R., & Gut, L. J. (2006). Higher densities of distributed pheromone sources provide disruption of codling moth (Lepidoptera: Tortricidae) superior to that of lower densities of clumped sources. *Journal of Economic Entomology, 99*, 1327–1333.

Falcon, L. A., & Huber, J. (1991). Biological control of codling moth. In V. D. G. Evenhuis (Ed.), *World crop pest. Tortricid pests: Their biology, natural enemies and control* (Vol. 5, pp. 355–370).

Hammon, R. W., & Davidson, D. (2016). *Backyard orchard. Stone fruits*. Gardening Series fruits and vegetables. Colorado State University Extension. Fact sheet 2.804.

Hodges, G. (2015). *Scale insects*. Gainesville: Florida Department of Agriculture and Consumer Services Division of Plant Industry.

Hogmire, H. W. (1995). *Mid-Atlantic orchard monitoring guide. Northeast regional agricultural engineering service (NRAES-75)*. Ithaca: Cooperative Extension.

Horton, D. L., Fuest, J., & Cravedi, P. (2008). Insects and mites. In D. R. Lyna & D. Bassi (Eds.), *The peach, botany, production and uses* (pp. 467–498). Kings Lynn: CAB International, Biddles.

Howitt, A.H. (1993). Oriental fruit moth. In A. H. Howitt (Ed.) *Common fruit tree pests. North Central Regional Extension Publication No. 63.* (pp. 43–44, 202–207). East Lansing: Michigan State University.

Hoyt, S. C., & Burts, E. C. (1974). Integrated control of fruit pests. *Annual Review of Entomology, 19*, 21–252.

Knight, A. L. (2000). Monitoring codling moth (Lepidoptera: Tortricidae) with passive interception traps in sex pheromone-treated apple orchards. *Journal of Economic Entomology, 93*, 1744–1751.

Knutson, A., Ong, K., & Ree, B. (2018). *Insects and disease pests of peaches, plums and black berries in a small fruit orchard* (pp. 1–15). Texas A& M Agrilife Extension Service. USDA.

Kovach, J., & Gorsuch, C. (1985). Effect of *Tetranychus urticae* populations on peach production in South Carolina. *Journal of Agricultural Entomology, 2*, 46–51.

Looney, N. E., & Jackson, D. I. (1999). Stone fruit part II. In *Temperate and sub-tropical fruit production* (2nd ed., p. 321). New York: CAB International.

Los, L., & Concklin, M. (2013). *IPM guidelines for insects and diseases of stone fruits* (pp. 1–10). UCONN College of Agriculture and Natural Resources. North east USDA Sustainable Agriculture Research and Education (SARE) grant. University of Connecticut IPM Program.

Murray, M. (2019). *Peach tree borer mating disruption*. UTAH Pests Fact Sheet. UTAH State University Extension and UTAH plant pests diagnostics. ENT-172-14PR.

Pascal, T., Pfeiffer, F., Kervella, J., Lacroze, J. P., & Sauge, M. H. (2002). Inheritance of green peach aphid resistance in the peach cultivar 'Rubira'. *Plant Breeding, 121*, 459–461.

Pickel, C., Bentley, W. J., Hasey, J. K., & Day, K. R. (2006). *UC IPM pest management guidelines, Peach*. UC ANR Publication 3454. Davis: University of California.

Rather, B. A., & Kacho, N. F. (2011). Record of new insect pests infesting apricot (*Prunus armeniaca* L.) in Kargil, Ladakh. *Indian Journal of Plant Protection, 39*(2), 155–156.

Shapiro-Ilan, D. I., Cottrell, T. E., Mizell, R. F. I. I. I., & Hortan, D. L. (2016). Curative control of peach tree borer using entomopathogenic nematodes. *Journal of Nematology, 48*(3), 170–176.

Stoven, H. & Bush, M. R. (2020). Plum and prune pests. In John Mellot (Ed.), *Tree fruit crops* (p. 84). PNW insect management handbook.

Strand, L. (1999). *Integrated pest management for stone fruits*. University of California Statewide Integrated Pest Management Program. University of California Agricultural and natural resources, publication no. 3389. Oakland: University of California.

Sutherland, O. R. W., & Hutchins, R. F. N. (1972). α-Farnesene, a natural attractant for codling moth larvae. *Nature, 239*, 170–171.

Thakur, J. R., & Gupta, P. R. (2004). Insect pests of peach, plum and apricot. In L. R. Verma, A. K. Verma, & D. C. Gautam (Eds.), *Pest management in horticulture crops* (pp. 279–294). New Delhi: Asiatech Publishers.

Welty, C. (2009). *San Jose Scale on fruit trees*. Department of Entomology, The Ohio State University extension. HYG-2039-09.

Wilson, J. (2014a). *Aphids. Stone fruit IPM for beginners* (pp. 46–47). Department of Entomology, Michigan State University. North Central IPM Centre. Retrieved from https://bit.ly/stonefruits.ipm.

Wilson, J. (2014b). *Lesser peach tree borer. Stone fruit IPM for beginners* (p. 53). Department of Entomology, Michigan State University. North Central IPM Centre. Retrieved from https://bit.ly/stonefruits.ipm.

Wilson, J. (2014c). *Peach tree borer. Stone fruit IPM for beginners* (p. 59). Department of Entomology, Michigan State University. North Central IPM Centre. Retrieved from https://bit.ly/stonefruits.ipm.

Wilson, J. K., Carroll, J. E. Poucubay, E., Agnello, A and Shane, W (2014). *Stone fruit IPM for beginners. A series of how – To fact sheets for new stone fruit growers and scouts to protect stone fruit orchards from pests* (p. 74). Department of Entomology, Michigan State University.

Witzgall, P., Chambo, J. P., Bengtsson, M., Unelius, R. C., Appelgren, M., Makranczy, G., Muraleedharan, N., Reed, D. W., Hellrigl, K., Buser, H. R., Hallberg, E., Bergström, G., Toth, M., Lofstedt, C., & Lofqvist, J. (1996). Sex pheromones and attractants in the Eucosmini and Grapholitini (Lepidoptera, Tortricidae). *Chemoecology, 7*, 13–23.

Nematodes Associated with Stone Fruits and Their Management Strategies

16

Tarique Hassan Askary, Mudasir Gani, and Abdul Rouf Wani

Abstract

Plant-parasitic nematodes are one of the biological constraints responsible for lowering the production and quality of stone fruit. They cause severe losses to stone fruits all over the world. These nematodes are soilborne microscopic organisms that attack plant roots. They puncture plant roots with the help of their stylet and remove cell contents, and as a result, the capacity of such damaged roots to uptake water and nutrients is drastically reduced. The symptoms exhibited by the affected plants include retarded growth, wilting, and predisposition to infection by other pathogens. The major nematodes of stone fruits are root-knot nematode (*Meloidogyne* spp.), root-lesion nematode (*Pratylenchus* spp.), ring nematode (*Criconemella* spp.), dagger nematodes (*Xiphinema* spp.), spiral nematode (*Helicotylenchus* spp.), and pin nematode (*Paratylenchus* spp.). The growth of peach can be suppressed by 5000 *Criconemella xenoplax*/100 g soil. Similarly, *Pratylenchus vulnus* can suppress the yield of peach by 16%. In cherry and plum, the damage threshold limit for *Pratylenchus penetrans* is 80/100 g soil and 320/100 g soil, respectively. Besides, nematodes such as *Xiphinema* sp. and *Longidorus* sp. are reported to transmit cherry leaf roll virus. Nematodes also aggravate the disease in the presence of other microorganisms. *Meloidogyne javanica* is reported to increase the incidence of crown gall of peach roots caused by bacterium *Agrobacterium tumefaciens*. Therefore, strategic management practices are advised to be adopted to overcome the menace of these tiny microorganisms. These include physical, cultural, biological, and chemical methods. However, the integration of all these methods has been found to yield better results in comparison to any individual practice.

T. H. Askary (✉) · M. Gani · A. R. Wani
Division of Entomology, Faculty of Agriculture, Sher-e-Kashmir University of Agricultural Sciences & Technology of Kashmir, Sopore, Jammu and Kashmir, India

M. M. Mir et al. (eds.), *Production Technology of Stone Fruits*,
https://doi.org/10.1007/978-981-15-8920-1_16

Keywords

Stone fruit · Nematode · Symptom · Disease complex · Life cycle · Management

16.1 Introduction

Stone fruits (*Prunus* spp.) include plum (*P. domestica* L. and *P. salicina* Lindl.), peach (*P. persica* (L.) Batsch), apricot (*P. armeniaca* L.), cherry (*P. avium* L. and *P. cerasus* L.), and almond (*P. amygdalus* Batsch). These are perennial crops generally infested with plant-parasitic nematodes. The younger trees, particularly those that are less than 18 months, are more susceptible that they can die when attacked with root-knot nematodes (Storey 2019). Thus, nematodes are major concern as regards stone fruits particularly from economic point of view.

The three main sources of nematode infestation are soil, plant material, and water. Before planting the stone fruit crops, the previous crops that were present in the field and have been a host range of nematodes that occur on stone fruit may lead to significant crop damage because the population of nematodes will rapidly build up in the presence of stone fruit. The rooted plant material can act as a possible source of nematode infestation, particularly endoparasites, such as root-knot nematode and root-lesion nematode. Besides, ectoparasites, like ring (*Criconema* spp.) and dagger nematodes (*Xiphinema* spp.) possessing long feeding apparatus, cling onto roots and are often transported together with plant roots. Water flowing from a nematode-infested plot to a healthy plot can act as a nematode infestation source. The fluctuation in soil population of nematode depends upon soil temperature and moisture (Askary et al. 2012; Sujata and Sharma 2018). At extremely low temperature, the nematode goes down into the soil, but when favorable temperature comes and soil moisture is sufficient, it becomes active again (Askary et al. 2012).

Use of certified plant materials, resistant root stocks, biological control agents, hot water treatment of planting materials, and soil treatment with chemicals are the strategies adopted to manage these plant-parasitic nematodes.

The present chapter deals with the association of plant-parasitic nematodes with peach, plum, cherry, almond, and apricot. Management strategies for each fruit crop have been also been described.

16.2 Peach (*Prunus persica*)

Plant-parasitic nematodes associated with peach trees are *Aglenchus muktii* (Phukan and Sanwal 1982), *Aglenchus* sp. (Islam et al. 2006), *Criconema serratum* (Khan and Siddiqi 1963), *Criconemella xenoplax* (Raski 1986), *Criconemella* sp. (Chitwood 1949), *Gracilacus peperpotti* (Khan et al. 1989), *Helicotylenchus indicus* (Askary et al. 2014; Maqbool et al. 1988), *H. kashmirensis* (Fotedar and Handoo

1974), *H. pseudorobustus*, *H. platyurus* (Walters et al. 2008), *Helicotylenchus* sp. (Arroyo et al. 2004), *Hemicriconemoides conicaudatus* (Phukan and Sanwal 1982), *Lobocriconema bhowaliensis* (Singh and Khan 1999), *Hoplolaimus* sp. (Walters et al. 2008), *Meloidogyne javanica* (Vlachopoulos 1991; Gomes et al. 2000), *M. incognita* (Rossi and Ferraz 2005), *Meloidogyne* sp. (Neal 1889; Walters et al. 2008), *Mesocriconema xenoplax* (Gomes et al. 2000; Walters et al. 2008), *Paratylenchus prunii* (Sharma and Sharma 1988a), *Paratylenchus* spp. (Askary et al. 2014), *Paratylenchus dianthus* (Walters et al. 2008), *P. brachyurus* (Stokes 1966), *P. hamatus* (Lownsbery et al. 1974), *P. neglectus* (Maqbool et al. 1988), *P. penetrans* (Askary et al. 2014; Walters et al. 2008; Maqbool et al. 1988), *P. projectus* (Walters et al. 2008), *P. vulnus* (Fliegel 1969; Walters et al. 2008), *Scutellonema brachyurus* (Nesmith et al. 1981), *Scutylenchus quettensis* (Maqbool et al. 1988), *Seriepinula truncatum* (Singh and Khan 1998), *Tylenchorhynchus annulatus*, *T. claytoni* (Walters et al. 2008), *Tylenchorhynchus* spp. (Askary et al. 2014), *Tylenchulus palustris* (Eisenback et al. 2007), *T. hamatus* (Walters et al. 2008), *T. prunii* (Gupta and Uma 1981), *Xiphinema americanum* (Wehunt and Good 1975; Walters et al. 2008), and *Xiphinema basiri* (Yadav and Varma 1967; Askary et al. 2014).

However, five plant-parasitic nematodes are of much importance from economic point of view. These are *Circonema*, *Pratylenchus*, *Xiphinema*, *Paratylenchus*, and *Meloidogyne* spp. Yield suppression of 16% due to *P. vulnus* has been reported in peach (Bridge and Starr 2007). The growth of peach can be suppressed by *Paratylenchus neoamblycephalus* at 13/100 g soil (Braun et al. 1975), *Xiphinema* sp. at 100/100 g soil (Bonsi et al. 1984), *P. penetrans* at 5/100 g soil (Barker and Olthof 1976), *Paratylenchus prunii* at 20/100 g soil (Sharma and Sharma 1988a), and >5000 *C. xenoplax*/100 g soil (Raski 1986).

16.2.1 Root-Knot Nematode (*Meloidogyne* sp.)

Root-knot nematodes are considered a potential threat to peach production. They are sedentary endoparasites which cause galling on the roots. The affected plants produce lesser yield with reduced quality. The four species which are associated with peach rootstocks are *M. incognita*, *M. javanica*, *Meloidogyne arenaria*, and *Meloidogyne hapla* (Nyczepir 1991).

Systematic Position

Kingdom: Animalia
Phylum: Nematoda
Class: Secernentea
Order: Tylenchida
Sub-order: Tylenchina
Superfamily: Tylenchoidea
Family: Heteroderidae
Sub-family: Meloidogyninae

Genus: *Meloidogyne* (Goeldi 1892)
Species: *M. incognita* (Kofoid and White 1919)
M. javanica (Treub 1885)
M. arenaria (Neal 1889)
M. hapla (Chitwood 1949)

16.2.1.1 Symptoms

Second-stage juveniles (J_2) are the infective stage that invades plant roots, leading to formation of galls (Bird 1972). Devitalization of root tip takes place, and the affected roots fail to elongate. The affected root system appears dense hairy. Yellowing of leaves, stunted growth, and early senescence of plant are the above ground symptoms that finally result in the lowering of the fruit production and productivity (Jonathan 2010).

16.2.1.2 Life Cycle

The one-celled egg passes through embryogenesis, resulting in a first-stage juvenile within the egg. After first molt, the second-stage juvenile (J_2) hatches out of the egg shell. J_2 is infective stage of the nematode which penetrates host roots, migrates through the root cells, reaches the vascular system, and starts feeding there. The female become sedentary. Due to feeding, several host cells become enlarged, forming specialized feeding cells termed as giant cells (Bridge and Starr 2007). During the feeding stage, females get obese and undergo second and third molts to become third- and fourth-stage juvenile, respectively. The juveniles of the third and fourth stage lack a stylet, but when the nematode undergoes fourth or final molt, the stylet reappears. Mature female lays 200–500 eggs on an average in a gelatinous matrix (secreted by mature female from rectal glands) collectively called an egg mass. Under dry environmental conditions, the gelatinous matrix of the egg sac maintains a high moisture level and prevents the egg from desiccation (Wallace 1968). Reproduction is parthenogenetic. Male juveniles are vermiform. They do not feed or have a role in the reproduction. They leave the root, move in soil, and die after some days. At favorable environmental conditions, hatching of eggs takes place. The second-stage juveniles come out and move in the soil in search of new host. The entire life cycle is completed between 3 and 4 weeks.

16.2.1.3 Disease Complex

Meloidogyne javanica is reported to increase the incidence of crown gall of peach roots caused by bacterium *Agrobacterium tumefaciens* (Nigh 1966). Histopathological examination reveals *A. tumefaciens* inside the cortical cells of such galls (Kaul et al. 1993; Pinochet et al. 2002).

16.2.1.4 Management

(a) Certified planting material: The planting materials selected should free from root-knot nematode infestation.
(b) Hot water treatment: The lethal temperature for nematodes is above 40 °C, but root-knot nematodes are endoparasites and are found inside the plant tissues.

Therefore, a slightly higher temperature is required for killing them into the planting material. Seedling roots treated with hot water (50–52 °C) for 5–10 min can eliminate the root-knot nematode infection (Parvatha Reddy 2008).

(c) Chemicals: The application of carbofuran at 3 kg active ingredient/ha before and after planting of rootstocks in nursery will prove helpful in lowering down nematode infestations (Sharma 2000).

(d) Biological: The combined application of *Glomus mosseae* and *M. javanica* on peach cv. Floridasun has been reported to reduce the population of nematode (Parvatha Reddy 2008)

(e) Host resistance: Several cultivars/rootstocks have been identified resistant to different species of root-knot nematodes. These are Greenpac (Pinochet 2009), Mantianhong (GengRui et al. 2008), Okinawa, R-15-2, Aldrighi (Fachinello et al. 2000), Zhubo 4, Zhubo 5 (Hang et al. 2006), Tsukaba-4, Tsukuba-5 (Wang et al. 2008), GF-31, GxN No. 15, Torinel, AD101, Monopol, Nemaguard, and Cadamum (Pinochet et al. 1996).

16.2.2 Ring Nematode (*Criconemella* sp.)

Ring nematodes are ectoparasitic and possess long stylet that helps them in reaching cortical cells below the root epidermis. These nematodes may be responsible for causing bacterial canker and peach tree short-life syndrome disease. The tree life shortens particularly those which are replanted.

16.2.2.1 Symptoms

Longitudinal cracks are often seen on nematode-infested plant roots that become dark. The serious infection leads to death of finer roots. The affected plants show stunted growth, mineral deficiency syndrome, and susceptibility to water stress. Kamio and Taguchi (2009) reported *Cantharis rustica*-infested peach trees, exhibiting the symptoms of withering and injury of trunks.

16.2.2.2 Life Cycle

Eggs are laid singly by adult female which are deposited in soil. First molting takes place inside the egg, and second-stage juveniles (J_2) hatch out. J_2 is the infective stage of the eggs that feed ectoparasitically on plant roots. After second, third, and fourth molt, the formation of adults takes place. Life cycle completes between 25 and 34 days (Seshadri 1964).

16.2.2.3 Disease Complex

The piercing of stylet into the plant tissue leads to injury which allows the entry of bacterial canker *Pseudomonas syringae* (Lownsbery 1959; Lownsbery et al. 1968, 1973). The nematode is implicated in peach tree short life that may cause death of trees.

16.2.2.4 Management

(a) Cultural: The soil population of *C. xenoplax* can be reduced by manipulating soil with the application of hydrated lime. A change in soil moisture, temperature, and pH will ultimately decrease nematode population (Wehunt et al. 1980; Wehunt and Weaver 1982).
(b) Chemical: Soil fumigation with methyl bromide in sorghum plots (grown with tarp and urea plots) suppressed the population of *C. xenoplax* in soil (Nyczpeir and Rodriguez-Kabana 2007).

16.3 Plum (*Prunus domestica*)

Plant-parasitic nematodes reported from plum are *C. xenoplax* (Majtahedi et al. 1975), *H. indicus* (Maqbool et al. 1988; Askary et al. 2014), *Helicotylenchus dihystera*, *Helicotylenchus thornei* (Sharma et al. 1988), *Longidorus distinctus* (Liskova 2007), *Longidorus* spp. (Askary et al. 2014), *Macroposthonia xenoplax*, *Meloidogyne incognita* (Sharma and Sharma 1990), *M. hapla* (Askary et al. 2014), *Paratylenchus prunii* (Sharma and Sharma 1990), *Paratylenchus* spp. (Askary et al. 2014), *Pratylenchus penetrans* (Askary et al. 2014; Maqbool et al. 1988), *P. neglectus*, *Scutylenchus quettensis* (Maqbool et al. 1988), *Xiphinema diversicaudatum* (Liskova et al. 1993), and *X. basiri* (Askary et al. 2014).

Pratylenchus penetrans at 320/100 g soil is the damage threshold limit on plum (Nyczpeir and Halbrendt 1993; Bridge and Starr 2007).

16.3.1 Disease Complex

It is reported that infestation of plum root by *C. xenoplax* aggravated bacterial canker on the tree caused by *Pseudomonas syringe* (Majtahedi et al. 1975). Transmission of ring-spot nepovirus by the nematode *Xiphinema americanum* has been reported. The virus was associated with brownline disease of plum trees in Chile (Auger 1989).

16.3.2 Management

(a) Chemical: The combined application of methyl bromide and fenamiphos in plum orchard infested with *Xiphinema* sp. resulted in significant reduction in soil population of nematode; however, in case of single application of either nematicide, methyl bromide proved superior over fenamiphos (Halbrendt and Shaffer 1989).
(b) Host resistance: Nematode-resistant cultivars/rootstocks of plum are presented below (Table 16.1).

Table 16.1 Plant-parasitic nematode-resistant cultivars /rootstocks of plum

Resistant cultivars/rootstocks	Nematode	References
P2175	*M. arenaria*	Esmenjaud et al. (1996)
Pixy, San Julian 655-2, Marianna 2624, Damsons PSM 101	*Meloidogyne incognita*	Pinochet et al. (1990)
P2032	*M. javanica, M. hispanica*	Esmenjaud et al. (1996)
Torinel, Red glow	*Pratylenchus vulnus*	Alcaniz et al. (1996)

16.4 Cherry (*Prunus avium/P. cerasus*)

Nematodes reported on cherry plants are *C. xenoplax* (Melakeberhan et al. 1994), *H. indicus* (Askary et al. 2014), *Helicotylenchus* sp. (Waliullah and Kaul 1997), *Longidorus macrosoma* (Buser 1999), *M. hapla* (Waliullah and Kaul 1997; Askary et al. 2014), *P. penetrans* (Hoestra and Oostenbrink 1962; Askary et al. 2014), *Pratylenchus* sp. (Waliullah and Kaul 1997), *Rotylenchus* sp. (Waliullah and Kaul 1997; Askary et al. 2014), *Tylenchorhynchus basiri* (Waliullah and Kaul 1997), *Tylenchorhynchus* spp. (Askary et al. 2014), *X. basiri* (Waliullah and Kaul 1997; Askary et al. 2014), and *X. diversicaudatum* (Liskova et al. 1993).

McElory (1972) reported *P. penetrans* causing injury on roots of cherry plants. *Pratylenchus penetrans* at 80/100 g soil is the damage threshold limit in cherry (Crossa Raynand and Audergon 1987; Bridge and Starr 2007). Brinkman (1977) reported *Paratylenchus hamatus* causing rust brown coloration on cherry roots.

16.4.1 Disease Complex

Xiphinema sp. and *Longidorus* sp. are reported to transmit cherry leaf roll virus (Harrison 1964; McElory 1972). The transmission of virus responsible for cherry rasp leaf disease (Halbrendt 1993) and rosette virus has also been reported (Brown et al. 1994). *Longidorus arthensis* was found to transmit cherry rosette nepovirus, the causal agent of cherry rosette disease in Switzerland (Kunz and Bertschinger 1998).

16.4.2 Management

(a) Hot water treatment: Cherry seedlings treated with hot water (50–52 °C) for 5–10 min can effectively prevent the infection of root-knot nematode (Parvatha Reddy 2008).
(b) Chemical: Walker and Wachtel (1988) observed significant reduction of *Paratrichodorus lobatus* with the application of carbofuran.
(c) Host resistance: Resistant grafts Cob and Colt was found resistant to pfeffinger disease. The causal organism of the disease is raspberry ring-spot nepovirus which is transmitted by nematode (Buser 1999). Cherry replants on mahaleb

rootstock were found resistant to *P. penetrans*; however, cherry rootstocks, viz., English morello, mazzard, montmorency, and Stockton morello, were completely resistant to root-knot nematodes (Parvatha Reddy 2008).

16.5 Almond (*Prunus amygdalus*)

Almond is cultivated in almost all the temperate region of Europe, Asia, and America. Half of the world's almond crop emanates from within the Sacramento and San Joaquin valleys of California (McKenry and Kretsch 1987). It is healthy nourishment that provides the high-quality vegetable proteins, omega-3 fatty acids, phytosterols, tocopherols, fiber, antioxidants, and phenolic compounds. Almond is cholesterol-free and helps in the prevention and treatment of cardiovascular diseases. Besides, it is a source of some vitamins and minerals that is a daily requirement for a person (Thakur and Singh 2013).

Almond is attacked by several species of plant-parasitic nematodes (Table 16.2). But among them, the prominent nematode species are *Pratylenchus vulnus* and *Meloidogyne* sp.

16.5.1 Disease Complex

A little information is available regarding nematode-bacterium disease complex in almond. However, crown gall infection on almond trees has been reported when the soil is heavily populated with nematodes (Sharma and Sharma 1988b).

Table 16.2 Plant-parasitic nematode associated with almond

Nematode	Location	References
Aphelenchus avenae	Turkey	Tan et al. (2018)
Criconema spp.	India	Askary et al. (2014)
Ditylenchus myceliophagus	Turkey	Tan et al. (2018)
Helicotylenchus digonicus	Turkey	Tan et al. (2018)
H. dihystera	India	Khan and Sharma (1992)
H. indicus	India	Askary et al. (2014)
M. floridensis	USA	Westphal et al. (2019)
Meloidogyne hapla	India	Askary et al. (2014)
M. incognita	India	Khan and Sharma (1992)
M. javanica	Pakistan	Khan et al. (2015)
Meloidogyne sp.	Morocco	Abbad et al. (1993)
Paratylenchus spp.	India	Askary et al. (2014)
Pratylenchus neglectus	Spain	Marull et al. (1990)
P. penetrans	Spain	Marull et al. (1990)
	India	Askary et al. (2014)
T. thornei	Spain	Marull et al. (1990)
Scutylenchus rugosus	Pakistan	Khan et al. (2015)
Tylenchorhynchus mashhoodi	India	Khan and Sharma (1992)
Tylenchorhynchus spp.	India	Askary et al. (2014)
Zygotylenchus guevarai	Pakistan	Khan et al. (2015)

16.5.2 Management

(a) Chemical: The application of metam sodium has been found to reduce the soil population of root-knot nematode in almond nurseries (Abbad et al. 1993). Lamberti et al. (2001) reported that soil treatment with methyl bromide 1,3-D and fenamiphos suppressed the population of *P. vulnus* in almond nurseries.
(b) Host resistance: The almond root stocks identified resistant to *M. javanica* are G x N NO.9, d-3-5, and Cachirulo (Marull and Pinochet 1991), whereas root stock resistant to *P. vulnus* is D-3-5.

16.6 Apricot (*Prunus armeniaca*)

Apricot is not very much susceptible to plant-parasitic nematode species. However, some nematode species have been found to be associated with this crop (Table 16.3). Sharma and Kashyap (2009) found problem of *C. xenoplax* and *Pratylenchus* sp. with apricot trees.

16.6.1 Management

(a) Cultural: Apricot trees intercropped with marigold and oat were found helpful in managing the nematode population (Sharma and Kashyap 2009).
(b) Chemical: The application of phorate at 0.03 g active ingredient (a.i.)/m^2 has been reported to reduce the soil population of many plant-parasitic nematodes (Sharma and Kashyap 2009).

16.7 Conclusion

Nematode association with stone fruits requires prior knowledge for understanding nematode-host relationship, and the role of nematodes needs to be assessed for the relative virulence of a particular species, host specificity, and tolerance level in host

Table 16.3 Plant-parasitic nematode associated with apricot

Nematode	Location	References
Criconemella xenoplax	India	Sharma and Kaur (1985)
Criconemella spp.	India	Askary et al. (2014)
Helicotylenchus indicus	India	Askary et al. (2014)
Meloidogyne incognita	Italy	Siniscalco et al. (1976)
M. hapla	India	Askary et al. (2014)
Pratylenchus vulnus	India	Sharma and Kaur (1985)
P. penetrans	India	Sharma and Kaur (1985), Askary et al. (2014)
Rotylenchus spp.	India	Askary et al. (2014)
Tylenchorhynchus sp.	India	Sharma and Kashyap (2009), Askary et al. (2014)
Tylenchulus indicus	Pakistan	Islam et al. (2006)

(Askary et al. 2014). Threshold limit of a particular nematode species needs to be determined precisely for successful nematode management program. However, prior to the commencement of such program, economy and ecology should be taken into consideration. Field efficacy of biocontrol agents should be ascertained for their successful exploitation against nematode pests. Growers should be advised to use resistant rootstocks in the areas having nematode problems (Özarslandan and Tanriver 2018). Emphasis should be laid to identify new sources of plant resistance and their incorporation into crops by traditional breeding or genetic engineering biotechnology (Askary et al. 2011). Finally, different management tactics of nematode management should be integrated, such as deep summer plowing, hot water treatment of planting material, nursery bed treatment with nematicide, the use of potential biocontrol agents, and resistant cultivars/rootstocks.

References

Abbad, F. A., Bajja, M., & Rami, A. (1993). Use of metam-sodium for the control of gall nematodes (*Meloidogyne* spp.) associated with almonds. *Al-Awamica, 80*, 111–122.

Alcaniz, E., Pinochet, J., Fernandez, C., Esmenjaud, D., & Felipe, A. (1996). Evaluation of prunus root stocks for root-lesion nematode resistance. *HortScience, 31*, 1013–1016.

Arroyo, C., Mora, J., Salazar, L., & Quesada, M. (2004). Population dynamics of four peach palm varieties (*Bactris gasipaes* Kunth). *Agronomia Mesoamericana, 15*, 53–59.

Askary, T. H., Banday, S. A., Iqbal, U., Khan, A. A., Mir, M. M., & Waliullah, M. I. S. (2011). Plant parasitic nematode diversity in pome, stone and nut fruits. In E. Lichtfouse (Ed.), *Agroecology and strategies for climate change* (pp. 237–268). Dordrecht: Springer.

Askary, T. H., Waliullah, M. I. S., & Gupta, S. (2012). Population fluctuation of plant parasitic nematodes associated with pome, stone and nut fruit nurseries. *Annals of Plant Protection Sciences, 20*(1), 265–267.

Askary, T. H., Waliullah, M. I. S., & Mir, M. M. (2014). Distribution of phytonematodes associated with stone and nut fruits in Kashmir valley, India. *Journal of Applied Horticulture, 16*(2), 136–140.

Auger, J. (1989). Tomato ring spot virus (TomRsv) associated with brownline disease on prune trees in Chile. *Acta Horticulturae, 235*, 197–204.

Barker, K. R., & Olthof, T. H. A. (1976). Relationship between nematode population densities and crop responses. *Annual Review of Phytopathology, 14*, 327–353.

Bird, A. F. (1972). Quantitative studies on the growth of syncytia induced in plants by root-knot nematodes. *International Journal for Parasitology, 2*, 157–170.

Bonsi, C., Stonffer, R., & Shaffer, R. (1984). Effect of different initial population densities of *Xiphinema americanum* and *Xiphinema rivesi* on growth of apple and peach seedlings. *Phytopathology, 74*, 626.

Braun, A. L., Mojtahedi, H., & Lownsbery, B. F. (1975). Separate and combined effects of *Paratylenchus neoamblycephalus* and *Criconemoides xenoplax* on "Myrobalan" plum. *Phytopathology, 65*, 328–330.

Bridge, J., & Starr, J. L. (2007). *Plant parasitic nematode of agricultural importance: A color handbook* (p. 152). New York: Academic Press/Elsevier.

Brinkman, H. (1977). Nematological observations in 1975 and 1976. *Gewasbischeraning, 8*, 131–136.

Brown, D. J. F., Grunder, J., Hooper, D. J., Klingler, J., & Kunz, P. (1994). *Longidorus arthensis* sp. n. (Nematoda: Longidoridae) a vector of cherry rosette disease caused by a new nepovirus in cherry trees in Switzerland. *Nematologica, 40*, 133–149.

Buser, A. (1999). Pfeffinger disease of cherry trees and its vector nematode *Longidorus macrosoma*. *Obst-und-weinbau, 135*, 42–45.

Chitwood, B. G. (1949). Ring nematodes (Criconematidine) a possible factor in decline and replanting problems of peach orchards. *Proceedings of Heliminthology Society of Washington, 16*, 6–7.

Crossa Raynand, P., & Audergon, J. M. (1987). Apricot rootstocks. In R. C. Rom & R. F. Carlson (Eds.), *Rootstocks of fruit crops* (pp. 295–320). New York: Wiley.

Eisenback, J. D., Reaver, D., & Ashley, J. E., Jr. (2007). First report of the nematode *Tylenchulus palustris*, parasitizing peach in Virginia. *Plant Disease, 91*, 1683.

Esmenjaud, J. D., Minot, J. C., Voisin, R., Bonnet, A., & Salesses, G. (1996). Inheritance of resistance to the root-knot nematode *Meloidogyne arenaria* in Myrobalan plum. *Theoretical and Applied Genetics, 92*, 973–879.

Fachinello, J. C., Silva, C. A. P., Sperandio, C., Rodrigues, A. C., & Strelow, E. Z. (2000). Resistance of rootstock for peach tree and plum to root-knot nematodes (Meloidogyne spp.). *Ciencia–Rural, 30*, 69–72.

Fliegel, P. (1969). Population dynamics and pathogenicity of three species of *Pratylenchus* on peach. *Phytopathology, 59*, 120–124.

Fotedar, D. N., & Handoo, Z. A. (1974). Two new species of *Helicotylenchus* Steiner,1945 (Hoploaimidae :Nematoda) from Kashmir, India. *Journal of Science, 2*, 57–62.

GengRui, Z., Wang, L., & Fanf, W. (2008). Selection and cultivation of a new peach variety Mantianhong both for ornamental and food. *Journal of Fruit Science, 25*, 440–441.

Goeldi, E. A. (1892). Relatorio sobre a molestia do cafeeiro na provincia do Rio de Janeiro. *Archos Mus Nac. Rio de Janeiro, 8*(1887), 1–121.

Gomes, C. B., Campos, A. D., & Almedia, M. R. A. (2000). Occurrence of *Mesocriconema xenoplax* and *Meloidogyne javanica* associated with peach tree short life on plum and reduction of phenol oxidizing enzyme activity. *Nematologia Brasileria, 24*, 249–252.

Gupta, N. K., & Uma. (1981). Description of two new species of Genus *Tylenchorhynchus* Cobb, 1913-family Tylenchorhinchidae (Eliava, 1964), Golden, 1971- from India. *Helminthlogia, 18*, 53–59.

Halbrendt, J. M. (1993). Virus-vector Longidoridae and their associated viruses in the Americas. *Russian Journal of Nematology, 1*, 65–68.

Halbrendt, J. M., & Shaffer, R. L. (1989). Effect of Brom-O-Gas and Nemacur 3 on dagger nematodes and plum replants, 1985-88. *Fungic Nematic Tests, 44*, 143.

Hang, Y., Guojie, L., Lixin, Z., Wenjun, W., & KeGong, J. (2006). 'Zhubo 4' and 'Zhubo 5', 2 new peach rootstocks. *China Fruits, 6*, 63.

Harrison, B. D. (1964). Specific nematode vectors for serologically distinctive forms of raspberry ring spot and tomato black ring virus. *Virology, 22*, 544–550.

Hoestra, H., & Oostenbrink, M. (1962). Nematodes in relation to plant growth. IV. *Pratylenchus penetrans* (Cobb) on orchard trees. *Netherlands Journal of Agricultural Science, 10*, 286–296.

Islam, S., Hamid, F., Khan, A., & Poswal, M. A. (2006). Plant parasitic nematodes associated with different crops of Swat and Malakand Agency, under MRDP area (NWFP), Pakistan. *Proceedings of Parasitology, 41*, 77–83.

Jonathan, E. I. (2010). *Nematology: Fundamentals and applications* (p. 280). Pitam Pura: New India Publishing Agency.

Kamio, S., & Taguchi, Y. (2009). Relationship between withering and injury of trunks in peach and plant parasitic nematode density in rhizosphere soil. *Horticultural Research in Japan, 8*, 137–142.

Kaul, V. K., Chhabra, H. K., Singh, A., Grewal, S. S., & Singh, A. (1993). Response of peach rootstocks to root-knot nematode and bacterial complex. *Plant Disease Research, 8*, 102–109.

Khan, A., Khanzada, K. A., Shaukat, S. S., Khatoon, N., & Sultana, N. (2015). Plant nematodes and their management associated with almond (*Prunus amygdalus* Batsch) seedlings in Balochistan, Pakistan. *Fuuast Journal of Biology, 5*, 243–247.

Khan, E., & Siddiqi, M. R. (1963). *Criconema serratum* n. sp. (Nematoda: Criconematidae)-a parasite of peach trees in Almora, North India. *Current Science, 32*, 414–415.

Khan, M. L., & Sharma, G. C. (1992). Seasonal fluctuations in plant parasitic nematodes in almond, *Prunus amygdalus* rhizosphere. *Indian Journal of Hill Farming, 5*, 10–13.

Khan, M. L., Sharma, G. C., & Sharma, N. K. (1989). *Paratylenchus curvitatus* and *Gracilacus Peperpotti* on apple and plum in Himachal Pradesh. *Nematologia Mediterranea, 17*, 275–276.

Kofoid, C. A., & White, W. A. (1919). A new nematode infection of man. *Journal of the American Medical Association, 72*, 567–569.

Kunz, P., & Bertschinger, L. (1998). Tracing cherry rosette disease using aerial pictures and soil analysis. *Obst-und-Weinbau, 134*, 588–591.

Lamberti, F., Addabbo, T., Sasanelli, N., & Carella, A. (2001). Control of *Pratylenchus vulnus* in stone fruit nurseries. In *Proceedings of the 53rd International Symposium on Crop Protection*, Gent, 8 May 2001. Part II, Mededelingen 66 (pp. 629–632).

Liskova, M. (2007). Morphometrics of females, juveniles and a hermaphrodite of *Longidorus distinctus* Lamberti *et al.*, 1983 (Nematoda: Longidoridae) from Slovakia. *Helminthologia, 44*, 210–213.

Liskova, M., Sabova, M., Valocka, B., & Lamberti, F. (1993). Occurrence of *Xiphinema diversicaudatum* (Nematoda) in the Slovak Republic. *Nematologia Mediterranea, 21*, 107–109.

Lownsbery, B. F. (1959). Studies of the nematode *Criconemoides xenoplax* on peach. *Plant Disease Report, 43*, 913–917.

Lownsbery, B. F., English, H., Moody, E. H., & Schick, F. J. (1973). *Criconemoides xenoplax* experimentally associated with a disease of peach trees. *Phytopathology, 63*, 994–997.

Lownsbery, B. F., Mitchell, J. R., Hart, W. H., Charles, F. M., Gertz, M. H., & Greathead, A. H. (1968). Responses of postplanting and preplanting soil fumigation in California peach, walnut and prune orchards. *Plant Disease Report, 52*, 890–894.

Lownsbery, B. F., Moody, E. H., & Braun, A. J. (1974). Plant parasitic nematodes of California prune Orchards. *Plant Disease Report, 58*, 633–636.

Majtahedi, H., Lownsberg, B. F., & Moody, E. H. (1975). Ring nematodes increase development of bacterial canker in plums. *Phytopathology, 65*, 556–559.

Maqbool, M. A., Qasim, M., Maqbool, M. A., Golden, A. M.,Ghaffar, A., & Krusberg, L. R. (1988) Control of plant parasitic nematodes associated with apple and pistachio trees in Baluchistan. In *Advances in Plant Nematology, Proceedings of the US-Pakistan International Workshop on Plant Nematology*, Karachi (pp. 271–274).

Marull, J., & Pinochet, J. (1991). Host suitability of Prunus to four *Meloidogyne* spp. and *Pratylenchus vulnus* in Spain. *Nematropica, 21*, 185–195.

Marull, J., Pinochet, J., & Verdejo, S. (1990). Response of five almond cultivars to four root-lesion nematodes in Spain. *Nematropica, 20*, 143–151.

McElory, F. D. (1972). Nematodes tree fruits and small fruits. In J. M. Webster (Ed.), *Economic nematology* (pp. 335–373). New York: Academic.

McKenry, M. V., & Kretsch, J. (1987). Survey of nematodes associated with almond production in California. *Plant Diseases, 71*, 71–73.

Melakeberhan, H., Bird, G. W., & Perry, R. (1994). Plant parasitic nematodes associated with cherry rootstocks in Michigan. *Journal of Nematology, 26*, 767–772.

Neal, D. C. (1889). *The root-knot disease of peach, orange and other plants in Florida due to the work of Anguillula*. USDA Division of Entomology, Bulletin No. 2D.

Nesmith, W. C., Zehr, E. I., & Dowler, W. M. (1981). Association of *Macroposthonia xenoplax* and *Scutellonema brachyurus* with the peach tree short life syndrome. *Journal of Nematology, 13*, 220–225.

Nigh, E. L., Jr. (1966). Incidence of crown gall infection in peach as affected by the Javanese root-knot nematode. *Phytopathology, 56*, 150.

Nyczepir, A. P. (1991). Nematode management strategies in stone fruit in the United States. *Journal of Nematology, 23*, 334–341.

Nyczpeir, A. P., & Halbrendt, J. M. (1993). Nematode pests of deciduous fruit and nut trees. In K. Evans, D. L. Trudgill, & J. M. Webster (Eds.), *Plant parasitic nematodes in temperate agriculture* (pp. 381–425). Wallingford: CAB International.

Nyczpeir, A. P., & Rodriguez-Kabana, R. (2007). Preplant biofumigation with sorghum or methyl bromide compared for managing *Criconemoides xenoplax* in a young peach orchard. *Plant Disease, 91*, 1607–1611.

Özarslandan, A., & Tanriver, E. (2018). Evaluation of some stone fruit rootstocks against resistance to root knot nematode (*Meloidogyne incognita*). *International Journal of Agricultural and Natural Science, 1*, 137–141.

Parvatha Reddy, P. (2008). *Diseases of horticultural crops: Nematode problems and their management* (p. 379). Jodhpur: Scientific Publishers.

Phukan, P. N., & Sanwal, K. C. (1982). Two new species of *Hemicriconemoides* Chitwood and Birchfield, 1957 (Nematoda: Criconematidae) from Assam, India. *Indian Journal of Nematology, 12*, 215–220.

Pinochet, J. (2009). A 'Greenpac', a new peach hybrid rootstock adapted to Mediterranean conditions. *HortScience, 44*, 1456–1457.

Pinochet, J., Fernandez, C., Alcaniz, E., & Felipe, A. (1996). Damage by a lesion nematode, *Pratylenchus vulnus* to prunus rootstocks. *Plant Disease, 80*, 754–757.

Pinochet, J., Fernandez, C., Cunill, M., Torrents, J., Felipe, A., Lopez, M. M., Lastra, B., Penyalver, R., Scott, J. R., & Chrisosto, C. H. (2002) Response of new interspecific hybrids for peach to root-knot and lesion nematodes, and crown gall. In *Proceedings of the 5th International Peach Symposium*, 8–11 July 2001, Davis (Vol. 592, pp. 707–716).

Pinochet, J., Marull, J., Verdejo, S., Soler, A., & Felipe, A. (1990). Evaluation of rootstocks of plum, damson, pear and quince on the gall nematode Meloidogyne incognita (Kofoid and white) Chitwod. *Boletin – de Sanidad Vegetal-Phlagas, 16*, 717–722.

Raski, D. J. (1986). *Plant parasitic nematodes of banana, citrus, coffee, grapes and tobacco* (pp. 43–57). North Carolina: Union Carbide Agricultural Products Company Incorporation Publication.

Rossi, C. E., & Ferraz, L. C. C. B. (2005). Plant parasitic nematodes of the superfamily Tylenchus associated with subtropical and temperate fruits in the states of Sao Paulo and Minas Gerais, Brazil. *Nematologia Brasileira, 29*, 171–182.

Seshadri, A. R. (1964). Investigation on the biology and life cycle of *Criconemoides xenoplax* Raski, 1952 (Nematoda: Criconematidae). *Nematologica, 10*, 540–562.

Sharma, G. C., & Kashyap, A. S. (2009). Effect of different intercrops on the nematode populations and yield of apricot var. New Castale. *Indian Journal of Horticulture, 66*, 420–421.

Sharma, G. C., Sharma, N. K., & Khan, M. L. (1988). Plant parasitic nematodes of plum (*Prunus domestica)* in Himachal Pradesh. *Indian Journal of Horticulture, 45*, 355–358.

Sharma, N. K. (2000). Nematode diseases of temperate fruits. In V. K. Gupta & S. K. Sharma (Eds.), *Diseases of fruit crops* (pp. 237–250). New Delhi: Kalyani Publishers.

Sharma, N. K., & Kaur, D. J. (1985). Nematodes associated with temperate crops in Himachal Pradesh. In T. R. Chandha, V. P. Bhutani, & J. L. Kaul (Eds.), *Advances in research on temperate fruits* (pp. 358–362). Solan: UHF-Solan (India) Publication.

Sharma, N. K., & Sharma, G. C. (1988a). Effect of environmental temperature on the population build up of *Pratylenchus prunii* in the peach rhizosphere. *Indian Phytopathology, 41*, 469–470.

Sharma, N. K., & Sharma, G. C. (1988b). Nematode pests of temperate and subtropical fruits and there management. In V. K. Gupta & N. K. Shrma (Eds.), *Tree protection* (pp. 192–210). Solan: ISTS, Publication/UHF-Solan (India) Publication.

Sharma, N. K., & Sharma, G. C. (1990). Distribution of three important nematode species in rhizosphere of plum trees. *Indian Journal of Horticulture, 47*, 207–209.

Singh, M., & Khan, E. (1998). Taxonomic studies on Criconematidae (Nematoda: Tylenchida) associated with fruit crops from north and north-eastern regions of India-I. Description of seven new species. *Indian Journal of Nematology, 28*, 174–191.

Singh, M., & Khan, E. (1999). Taxonomic studies on criconematidae (Nematoda: Tylenchida) associated with fruit crops from north and north-eastern regions of India-II. Description of five new species. *Indian Journal of Nematology, 29*, 48–55.

Siniscalco, A., Lamberti, F., & Inserra, R. (1976). Reaction of peach root stocks to Italian population of root-knot nematodes, *Meloidogyne* spp. *Nematologia Mediterranea, 4*, 79–84.

Stokes, D. E. (1966). Parasitism by *pratylenchus brachyurus* on three peach rootstocks. *Nematologica, 13*, 153.

Storey, S. (2019). *Nematodes in stone fruit: Practical guidelines for the short and long term control of nematodes*. Retrieved from www.nemlab.co.za.

Sujata, & Sharma, G. C. (2018). Status of phyto-parasitic nematodes in major fruits and vegetable crops of the tribal belt Kinnaur in Himachal Pradesh. *Journal of Entomology and Zoological Studies, 6*, 3069–3072.

Tan, A. N., Ocal, A., Ozturk, L., & Elekcioglu, I. H. (2018). Plant parasitic nematodes associated with almond (*Prunus dolcis* Mill.) and walnut (*Juglans regia* L.) orchards in Adiyaman province, Turkey. *International Journal of Molecular Biology, 3*, 295–300.

Thakur, M., & Singh, K. (2013). Walnut (*Juglans regia* L.) a complete health and brain food. *Asian Journal of Bio Science, 8*, 276–288.

Treub, M. (1885). Onderzoekingen over Sereh-Ziek Suikkeriet gedaan in s Lands Plantentium te Buitenzorg. *Mededeelingen uit's Lands Plantentium Batavia, 2*, 1–39.

Vlachopoulos, E. G. (1991). Nemotode species in nurseries of Greece. *Annales de-Institute Phytopathologique Benaki, 16*, 115–122.

Waliullah, M. I. S., & Kaul, V. (1997). Nematodes associated with cherry plants in Kashmir valley. *Indian Journal of Nematology, 27*, 237.

Walker, G. F., & Wachtel, M. F. (1988). Evaluation of nematicides for nematode control in cherries, 1986-87. *Fungic Nematic Tests, 43*, 165.

Wallace, H. R. (1968). The dynamics of nematode movement. *Annual Review of Phytopathology, 6*, 91–114.

Walters, S. A., Bond, J. P., Russell, J. B., Taylor, B. H., & Handoo, Z. A. (2008). Incidence and influence of plant parasitic nematodes in southern Illinois peach orchards. *Nematropica, 38*, 63–74.

Wang, L. Y., Zhu, L. X., & Jia, K. G. (2008). Resistance of several Prunus plants to root-knot nematode Meloidogyne arenaria. *Journal of China Agricultural University, 13*, 24–28.

Wehunt, E. J., & Good, J. M. (1975). Nematodes on peaches. In N. F. Childers (Ed.), *The peach: Varieties, culture, marketing and pest control* (3rd ed., pp. 377–387). New Brunswick: Communications Department, Cook College, Rutgers University.

Wehunt, E. J., Horton, B. D., & Prince, V. E. (1980). Effects of nematicides, lime and herbicide on a peach tree short life site in Georgia. *Journal of Nematology, 12*, 183–189.

Wehunt, E. J., & Weaver, D. J. (1982). Effect of planting site preparation, hydrate lime and DBCP (1,2-dibromo-3-chloropropane) on population of *Macroposthonia xenoplax* and peach tree short life in Georgia. *Journal of Nematology, 14*, 567–571.

Westphal, A., Maung, Z. T. Z., Doll, D. A., Yaghmour, M. A., Chitambar, J. J., & Subbotin, S. A. (2019). First report of the peach root-knot nematode, *Meloidogyne floridensis* infecting almond on root-knot nematode resistant 'Hansen 536' and 'Bright's Hybrid 5' rootstocks in California, USA. *Journal of Nematology, 51*, 1–3.

Yadav, B. S., & Varma, M. K. (1967). New host plants of *Xiphinema basiri* and *X. indicum*. *Nematologica, 13*, 469.

The manufacturer's authorised representative in the EU is Springer Nature Customer Service Centre GmbH, Europaplatz 3, 69115 Heidelberg, Germany. If you have any concerns regarding our products, please contact ProductSafety@springernature.com

Printed and bound by CPI Group (UK) Ltd, Croydon, CR0 4YY

30/09/2026

02225247-0013